ORBITAL THEORIES OF
MOLECULES AND SOLIDS

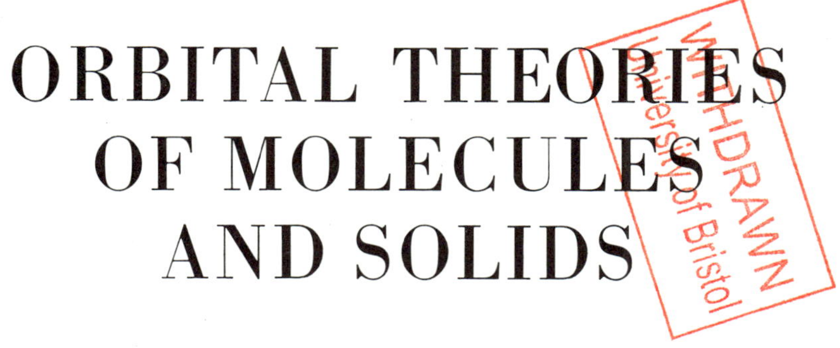

ORBITAL THEORIES OF MOLECULES AND SOLIDS

EDITED BY

N. H. MARCH

DEPARTMENT OF PHYSICS
IMPERIAL COLLEGE OF SCIENCE AND TECHNOLOGY
UNIVERSITY OF LONDON

CLARENDON PRESS · OXFORD

1974

Oxford University Press, Ely House, London W.1

GLASGOW NEW YORK TORONTO MELBOURNE WELLINGTON
CAPE TOWN IBADAN NAIROBI DAR ES SALAAM LUSAKA ADDIS ABABA
DELHI BOMBAY CALCUTTA MADRAS KARACHI LAHORE DACCA
KUALA LUMPUR SINGAPORE HONG KONG TOKYO

ISBN 0 19 855350 1

PRINTED IN GREAT BRITAIN
BY J. W. ARROWSMITH LTD, BRISTOL 3

To
Professor C. A. Coulson

FOREWORD

H. C. LONGUET-HIGGINS

IT IS an honour to be allowed to make a tribute to Charles Coulson in this volume dedicated to him. It was in 1945 that we first met. I had just finished my Part II in Chemistry with R. P. Bell, measuring the rates of some catalytic reactions with no especial competence. Ronnie Bell realized that I would probably be better at theoretical work, and told me that a new I.C.I. Fellow had just joined the Physical Chemistry Laboratory, and suggested that perhaps I might ask to be allowed to do a Ph.D. under him. I had even less idea than Ronnie what theoretical chemistry really was, so that it was with some misgivings that I went to call on Charles Coulson. I need not have worried. This tall, imposing man immediately put me at my ease and made me feel not only that he was prepared to take me on but that he was actually eager to have someone to collaborate with. We started work straight away and the two years that followed were among the most exciting in my life. Above all I was impressed by the simplicity of Coulson's thinking and by his determination to make things so clear that even a novice could grasp the essential ideas. Not that Coulson's mathematics was trivial—far from it. I found myself having to learn all about complex integration and matrix equations, but one could always see where one was going and why the mathematics was relevant. This impression of Charles Coulson's work has never left me, and it is one of his great strengths as a theoretical chemist.

Theoretical chemists were very thin on the ground at that time. Outside Cambridge, where Lennard-Jones's school had begun to make a name, the subject was almost unknown. But in Oxford and other places it was Coulson who really got the subject going. He had already produced a distinguished student in Stanley Rushbrooke, with whom he had discovered the pairing theorem for even alternant hydrocarbons. William Moffitt, though not actually one of Coulson's research students, was obviously much influenced in his thinking by the elegance of Coulson's work in molecular orbital theory. And other contributors to this volume will know how much they themselves have owed to his inspiration.

In 1947 Charles moved to King's College London to become the Wheatstone Professor of Theoretical Physics—an instance of the fact that though a lifelong theoretical chemist he has always had a job in something else. There he founded his first research school, representatives of which are now scattered all over the world. Five years later he moved to Oxford as Rouse Ball Professor of Mathematics, having been already elected a Fellow of the

Royal Society. It was typical of him that his moves hardly interfered at all with his scientific collaborations. One of the earliest of these was with the French theoretical chemists B. Pullman and R. Daudel. There can be little doubt that not only in Britain but also on the continent of Europe it was Coulson who took the lead in introducing ordinary chemists and biochemists to the essentials of the molecular orbital theory.

Coulson's subsequent career, and his monumental scientific output, is too well known to need recording here. But it was in Oxford that his work on solids really developed to maturity, and in Oxford that he started the series of theoretical chemistry summer schools which soon acquired such international renown. Having taken part in some of them I know at first hand how much enthusiasm and appreciation they generated among these—often distinguished scientists—who came in order to acquaint themselves with the latest ideas in the subject. (I still treasure the 'club tie' which Charles himself designed.) But he never allowed the newcomers to the subject to feel out of things, as they might have if the summer schools had been used largely to advertise the work of his own research group. The Oxford summer schools have always been a shining example of how a subject should be taught, as opposed to how scientific empires should be built.

Charles Coulson will undoubtedly be remembered for long as a scientist—for his pioneer calculations on polyatomic molecules, for his concept of bond order in molecular orbital theory, and for his elegant theoretical work on large systems of non-localizable electrons. But he will also be remembered as a great teacher, who put theoretical chemistry on the map almost single-handed, at least in Europe.

Editor's note

Since Professor Longuet-Higgins wrote this, Professor Coulson died, prematurely, on 7 January, 1974. It is only tiny consolation to his friends who have contributed to this volume that he had seen, and approved of, much of the MS of this book. He also gave me, as editor, the benefit of his wisdom with regard to the order of presentation of the material. But, of course, the sole responsibility for this aspect of the book remains mine.

EDITOR'S PREFACE

THIS volume was conceived as one part of plans by Charles Coulson's many friends to honour him on the occasion of his 60th birthday; the other part being a highly successful conference held at Oxford on that occasion. The chapters in this book are all contributed by friends and colleagues who came within his orbit at some stage in their careers. It is to be hoped that they reflect, however imperfectly, at least a little of the inspiration a great many of us derived from his outstanding leadership.

By bringing together material covering areas which cut across the conventional boundaries between quantum chemistry and theoretical solid-state physics, graduate students studying either molecular physics or the physics and chemistry of the solid state should find here a good deal of useful material, much of which is not readily available to them except in the original papers. Additionally, quite a bit of the discussion in this book should be of value to experimentalists researching in boundary areas between chemistry and physics. In particular, if the volume plays some part in increasing the awareness of the importance of synthesizing 'bonds and bands' in descriptions of the solid state, that would be entirely consistent with the viewpoint that Charles Coulson himself adopted throughout his career.

Finally, it is necessary for me to draw attention to the fact that in this work, involving multiple authorship, it did not prove possible to avoid a very wide spectrum of dates on which the original MSS were received. A number of the Chapters (and in particular Chapters 6 and 7) were received as much as two years before the publication date, and therefore cannot include references to work more recent than that.

N. H. M.

London
June 1974

CONTENTS

PART I
CRYSTALLINE SOLIDS

PART II
ISOLATED MOLECULES AND MOLECULAR CRYSTALS

LIST OF AUTHORS

PART 1
CRYSTALLINE SOLIDS

CHAPTER 1: BAND THEORY CALCULATIONS
by L. PINCHERLE, Dept. of Physics, Bedford College, University of London

CHAPTER 2: THE CELLULAR METHOD FOR METALS
by S. L. ALTMANN, Dept. of Metallurgy, University of Oxford

CHAPTER 3: ONE-BODY POTENTIALS IN CRYSTALS
by N. H. MARCH, Dept. of Physics, Imperial College of Science & Technology, University of London

CHAPTER 4: DEFECTS IN CRYSTALLINE SOLIDS
by A. B. LIDIARD, Theoretical Physics Division, Atomic Energy Research Establishment, Harwell, Berks

PART II
ISOLATED MOLECULES AND MOLECULAR CRYSTALS

CHAPTER 5: ON THE ORIGIN OF ELECTRONIC PROPERTIES OF MOLECULES
by R. MCWEENY, Department of Chemistry, The University, Sheffield

CHAPTER 6: ATOMS-IN-MOLECULES
by G. G. BALINT-KURTI* and M. KARPLUS, Dept. of Chemistry, Harvard University, Cambridge, Massachusetts, U.S.A.
*Present address: Department of Theoretical Chemistry, University of Bristol.

CHAPTER 7: INTERMOLECULAR ENERGIES IN THE REGION OF SMALL OVERLAP
by J. N. MURRELL, School of Molecular Sciences, University of Sussex

CHAPTER 8: ELECTRONIC ENERGY LEVELS OF MOLECULAR SOLIDS
by D. P. CRAIG, Research School of Chemistry, Australian National University, Canberra, Australia

PART I
CRYSTALLINE SOLIDS

1

BAND THEORY CALCULATIONS

L. PINCHERLE

1.1. Introduction

THIS chapter examines methods of calculation and modern trends in band theory, aiming at a simple treatment suitable for non-specialists. The problem is that of the energy levels of electrons moving in a perfect lattice of immovable nuclei. A complete many-electron treatment is beyond our present capacity, even though substantial progress has been made in this area. One of the most fruitful approaches to the many-electron problem will probably prove to be the use of the two-electron density matrix. If there are N electrons, the two-electron density matrix is the quantity

$$\int \Psi^*(x_1', x_2', x_3 \ldots x_N)\Psi(x_1, x_2, x_3 \ldots x_N)\,dx_3 \ldots dx_N$$

where Ψ is the many-electron wavefunction and x_i indicate the space and spin coordinates of the ith electron respectively. If we could obtain a good approximation to this matrix, we would be able to perform accurate calculations of the energy, charge density, etc. More precisely, we would be able to calculate the correlation energy, due to the fact that electrons tend to keep away from each other (see Section 1.4 below).

But for the time being the one-electron approach is adopted almost universally, ideally in the self-consistent field scheme. This however is not simply a counsel of despair, because, for a quasi-particle consisting of an electron together with its exchange and correlation holes moving in a perfect crystal, the concepts of **k**-space, Bloch functions, Brillouin zones remain valid, and for the band problem the many-electron corrections are likely to be quite small in many cases. Even in the exact many-body problem it is possible to write down a one-particle-like Schrödinger equation for the energies of quasi-particles, but with a potential energy operator which so far eludes practical computation.

Most calculations aim at obtaining energy levels and wavefunctions for a suitably chosen potential, or pseudopotential, on the assumption that a (usually local) potential can be constructed which, even if not self-consistent, leads to a picture that agrees with experiment. It turns out that the dispersion relation, that is the dependence of the energy on crystal momentum **k** is often sensitive to the form of the potential, particularly in deciding the form of band edges of semiconductors or the details of the Fermi surface of metals.

One also finds that the potential that reproduces these features is not unique. Moreover one hopes that this potential may yield not only the band structure, but also the interpretation of many physical properties. These hopes have been fulfilled to a greater extent than one would have expected (Cohen and Heine 1970). Suitable pseudowavefunctions have provided good interpretations of intensities of optical transitions, phonon spectra, resistivities, superconducting transition temperatures, and some magnetic properties. However we are concerned in this chapter with energy determinations only.

1.2. *a-priori* calculations

First we examine the approaches that are currently used to set up a potential and to solve Schrödinger's equation by assuming only the crystal structures and lattice parameters. Few mathematical details will be given, reference being made instead to review articles containing these details (and the original references). In the review of the methods of calculation of Bloch functions by Ziman (1971), the physical concepts involved are discussed with great clarity and much insight is given into the mathematical techniques. In the present chapter the aim is to give a somewhat easier treatment. Another useful summary, accessible to non-specialists, is that by Fletcher (1971).

Such *a-priori* calculations have had only moderate success and can often be criticized because they have been pushed to an accuracy unwarranted by the basic drastic simplifications. For instance, besides neglect of many-body effects, they treat exchange only approximately, they often neglect relativistic effects, etc. Their main merit is to provide sound theoretical frameworks for semiempirical approaches (e.g. pseudopotentials).

Apart from the cellular method, discussed in full detail in the next chapter, the methods currently in use are based on the nearly free electron (NFE) approximation. They are: orthogonalized plane waves (OPW), augmented plane waves (APW) and variations on the scattered waves theme first put forward by Korringa and then developed by Kohn and Rostoker (KKR method). All have been highly perfected: with the same potential they produce the same $E(\mathbf{k})$ relation to a small fraction of an electronvolt. They are all interrelated, as discussed, for instance, by Ziman (1971). In these methods, in contrast to the cellular method, the periodicity conditions are satisfied automatically, but the wavefunctions are not exact solutions of Schrödinger's equation. The point of departure is Schrödinger's equation in a representation based on the plane waves $\exp\{i(\mathbf{k}+\mathbf{K}_i).\mathbf{r}\}$, where the $\mathbf{K}_i$ are the vectors of the reciprocal lattice; because of periodicity it is in the form of an infinite system of linear homogeneous algebraic equations for the coefficients of the plane waves, involving the Fourier coefficients of the potential $V(\mathbf{K}_n)$ (see for instance Callaway 1964; Pincherle 1971). The compatibility condition for these equations is

$$\mathrm{Det}|(|\mathbf{k}+\mathbf{K}_i|^2-E)\delta_{ij}+V(\mathbf{K}_j-\mathbf{K}_i)| = 0 \tag{1.1}$$

providing, for each value of $\mathbf{k}$, the energies of all the possible bands. At $\mathbf{k}$-points of symmetry, properly symmetrized combinations of plane waves (see Section 5 below) should be used, with the advantage of reducing the dimensions of the secular determinant. Of course the equation is truncated to the size that provides the desired convergence for the levels of interest. For any solution E_n of eqn (1.1) the coefficients of the plane waves can be found. Considered as functions of $\mathbf{k}$ in the extended zone schemes, they are called the momentum eigenfunction $v_n(\mathbf{k})$ of the corresponding band.

Eqn (1.1) neglects spin, i.e. assumes that every eigenvalue is doubly degenerate. This would not be true in magnetic materials, which however we do not consider here. But spin–orbit coupling and other relativistic effects, e.g. variation of mass with velocity, are significant in band structures, particularly of course for heavy elements.

These effects have been taken into account in relativistic versions of the APW and KKR methods (see, for instance, Dimmock 1971). One way is to start from the Dirac equation and, in the region of spherically symmetric potential, expand ψ in terms of 'spherical harmonics with spin' (Dimmock 1971). The wavefunction has two components, one large, which, for $c \to \infty$, reduces to the Schrödinger ψ, and one small, which vanishes for $c \to \infty$. The second can be eliminated and, to order c^{-2}, an equation for the large component is obtained including the various relativistic effects. The same equation can also be obtained by reducing to zero the small component by means of a canonical transformation. However it is difficult to solve, and the tendency has been to neglect those effects that merely shift the energy levels and to consider only spin–orbit coupling which splits degenerate energy levels. This is often introduced through spin–orbit interaction parameters derived from the fine structure of the spectrum lines of the isolated atom. This is legitimate because the integral determining the spin–orbit coupling contains as a factor the gradient of the potential, and this gradient is large near nuclei, where the effect of the binding of the atom in a crystal is negligible. In dealing with spin–orbit interaction, group theory is particularly useful, as it provides symmetry-adapted wavefunctions. In what follows we shall restrict attention to non-relativistic calculations.

1.2.1. *Orthogonalized plane waves (OPW)*

The inner core energies are not generally of interest. Moreover, to obtain the 1s level to, say, 0·1 eV, millions of plane waves would be necessary. In the OPW method (Herring 1940) the core energies are excluded by using, instead of the simple plane waves, combinations of plane waves and inner core functions, $\chi_{\mathbf{k}+\mathbf{K}_i}(\mathbf{r})$—orthogonalized plane waves—these combinations being orthogonalized, by the Schmidt procedure, to the inner core states $b_t(\mathbf{r})$

$$\chi_{\mathbf{k}+\mathbf{K}_i}(\mathbf{r}) = \exp\{\mathrm{i}(\mathbf{k}+\mathbf{K}_i)\,.\,\mathbf{r}\} - \Sigma_t \langle b_{t,\mathbf{k}}|\exp\{\mathrm{i}(\mathbf{k}+\mathbf{K}_i)\,.\,\mathbf{r}\}\rangle b_{t,\mathbf{k}}(\mathbf{r}) \qquad (1.2)$$

the $b_{t,\mathbf{k}}$ being Bloch functions for all the occupied inner states t of energy E_t (only slightly different from the corresponding energy in the free atom)

$$b_{t,\mathbf{k}}(\mathbf{r}) = \Sigma_{\mathbf{R}_s} \exp(\mathrm{i}\mathbf{k}\,.\,\mathbf{R}_s) b_t(\mathbf{r}-\mathbf{R}_s)$$

Let $\langle b_{t,\mathbf{k}}|\exp\{\mathrm{i}(\mathbf{k}+\mathbf{K}_i)\,.\,\mathbf{r}\}\rangle = \beta_{t,i}$. For instance $\beta_{00} = \int \psi^*_{1s}\,\mathrm{d}^3\mathbf{r}$ and, if we take for ψ_{1s} a hydrogen-like function $a^{\frac{3}{2}}\,\mathrm{e}^{-ar}/\sqrt{\pi}$, we have

$$\beta_{00} = a^{\frac{3}{2}}\pi^{-\frac{1}{2}} \int_0^\infty \mathrm{e}^{-ar} 4\pi r^2\,\mathrm{d}r = 8\pi^{\frac{1}{2}}a^{-\frac{3}{2}}.$$

If there are no other terms, $\chi_0 = 1 - 8\,\mathrm{e}^{-ar}$. This is then a (unnormalized) plane wave at $\mathbf{k} = 0$ orthogonalized to a 1s state.

When ψ is expanded in these orthogonalized plane waves,

$$\psi_{n,\mathbf{k}}(\mathbf{r}) = \Sigma_{\mathbf{K}_i} \alpha_{n,\mathbf{k}}(\mathbf{K}_i)\chi_{\mathbf{k}+\mathbf{K}_i}(\mathbf{r}), \qquad (1.3)$$

the matrix elements of the potential between plane waves $V(\mathbf{K}_j - \mathbf{K}_i)$ occurring in (1.1) are replaced by matrix elements between OPWs

$$\chi_i(\mathbf{r}) = \exp\{\mathrm{i}(\mathbf{k}+\mathbf{K}_i)\,.\,\mathbf{r}\} - \Sigma_t\, \beta_{ti} b_{t,\mathbf{k}}(\mathbf{r}) \qquad (1.4)$$

When the Hamiltonian operates on $b_{t,\mathbf{k}}(\mathbf{r})$ the result is $E_t b_{t,\mathbf{k}(\mathbf{r})}$. We have thus for the matrix elements of H between OPWs

$$\begin{aligned}\langle\chi_i|H|\chi_j\rangle = &\int \exp\{-\mathrm{i}(\mathbf{k}+\mathbf{K}_i)\,.\,\mathbf{r}\}(-\nabla^2+V)\exp\{\mathrm{i}(\mathbf{k}+\mathbf{K}_j)\,.\,\mathbf{r}\}\,\mathrm{d}^3\mathbf{r}\\ &-\Sigma_t\Big[\beta^*_{ti}\int b^*_{ti}(\mathbf{r})(-\nabla^2+V)\exp\{\mathrm{i}(\mathbf{k}+\mathbf{K}_j)\,.\,\mathbf{r}\}\,\mathrm{d}^3\mathbf{r}\\ &+\beta_{tj}\int \exp\{-i(\mathbf{k}+\mathbf{K}_i)\,.\,\mathbf{r}\}(-\nabla^2+V)b_{tj}(\mathbf{r})\,\mathrm{d}^3\mathbf{r}\Big]\\ &+\Sigma_{tt'}\beta^*_{ti}\beta_{t'j}\int b^*_{ti}(\mathbf{r})E_{t'}b_{t'j}(\mathbf{r})\,d^3\mathbf{r}.\end{aligned}$$

The first integral is the one occurring for simple plane waves and its value is $|\mathbf{k}+\mathbf{K}_i|^2\delta_{ij} + V(\mathbf{K}_j - \mathbf{K}_i)$. The first sum vanishes because of the orthogonality of the plane waves to the inner core states. The last sum, if the b_t are normalized, reduces to $\Sigma_t \beta^*_{tj}\beta_{ti}E_t$, and thus

$$\langle\chi_i|H|\chi_j\rangle = |\mathbf{k}+\mathbf{K}_i|^2\delta_{ij} + V(\mathbf{K}_j - \mathbf{K}_i) + \Sigma_t E_t \beta^*_{tj}\beta_{ti}.$$

On the other hand

$$\langle\chi_i|\chi_j\rangle = \delta_{ij}+\Sigma_t\beta^*_{tj}\beta_{ti},$$

so that equation (1.1) is replaced by

$$\mathrm{Det}|(|\mathbf{k}+\mathbf{K}_i|^2-E)\delta_{ij}+V(\mathbf{K}_j-\mathbf{K}_i)+\Sigma_t(E_t-E)\beta^*_{tj}\beta_{ti}| = 0. \tag{1.5}$$

This secular equation yields the energy levels with rapid convergence, and the lowest eigenvalue gives the energy of the first non-core level. The β_{ti} are evaluated by expanding the plane wave in spherical harmonics; then, evaluation of the product $\beta^*_{tj}\beta_{ti}$ involves application of the addition theorem. For computational details reference should be made to Woodruff (1957) or Herman, Kortum, Kuglin and Short (1966). As for plane waves, at **k**-points of symmetry the size of the determinant can be reduced by employing symmetrized combinations of OPWs. It should be noted that the method is useless for bands deriving from levels such as 3d or 4f, which are already orthogonal to all states of lower energy.

It is found in practice that the sum in (1.5) cancels out to a large extent the V non-diagonal matrix elements; that is, orthogonalization is equivalent to introducing a repulsive potential. This point is discussed in the next section.

1.2.2. *Pseudopotential*

We may re-interpret (1.5) as the plane wave representation of an equation $(-\nabla^2+V_0+V_{ps}^{OPW})\phi = E\phi$, where V_{ps}^{OPW} is a pseudopotential operator, whose form is given below, such that the matrix elements between plane waves are given by (1.5). Since the convergence of (1.5) is found to be rapid, $V_0+V_{ps}^{OPW}$ must be weak compared to the kinetic energy. To see how this comes about, we can write the wave equation (Cohen and Heine 1970) in terms of a pseudowavefunction defined as

$$\phi_{n,\mathbf{k}}(\mathbf{r}) = \Sigma_{\mathbf{K}_i}\alpha_{n,\mathbf{k}}(\mathbf{K}_i)\exp\{\mathrm{i}(\mathbf{k}+\mathbf{K}_i)\cdot\mathbf{r}\} \tag{1.6}$$

the $\alpha_{n,\mathbf{k}}$ being the coefficients in the expansion (1.3). Then, from (1.2), (1.3), (1.6),

$$\psi_{n,\mathbf{k}}(\mathbf{r}) = \phi_{n,\mathbf{k}}(\mathbf{r})-\Sigma_t\langle b_{t,\mathbf{k}}|\phi_{n,\mathbf{k}}\rangle b_{t,\mathbf{k}}(\mathbf{r}) \tag{1.7}$$

or $\psi = (1-\mathscr{P})\phi$, where $\mathscr{P}$ is the projection operator that projects any state onto the core states, that is that expresses it as a linear combination of core states. If now eqn (1.7) is substituted into Schrödinger's equation with potential V_0, we obtain

$$H\phi+\Sigma_t(E-E_t)\langle b_t|\phi\rangle b_t = E\phi \equiv \{-\nabla^2+V_0+V_{ps}^{OPW}(\mathbf{r})\}\phi.$$

Thus ϕ satisfies a Schrödinger equation with an additional potential operator, V_{ps}^{OPW} meaning formally, when applied to ϕ (but not to other functions)

$$V_{ps}^{OPW}\phi = \Sigma_t(E-E_t)\langle b_t|\phi\rangle b_t = \int \Sigma_t(E-E_t)b_t^*(\mathbf{r}')\phi(\mathbf{r}')b_t(\mathbf{r})\,\mathrm{d}^3\mathbf{r}'. \tag{1.8}$$

Its matrix elements in a plane wave representation are given by the last term in (1.5). The pseudowavefunction ϕ is smoother than ψ, because it includes only plane waves, without the wiggles of the true wavefunction.

Now the Fourier components of the total potential $V_{\text{eff}} = V_0 + V_{\text{ps}}^{\text{OPW}}$ are small, except perhaps for the first few reciprocal lattice vectors, because V_0 is negative, while $V_{\text{ps}}^{\text{OPW}}$, containing $E - E_t$ and the square of the modulus of an atomic orbital, is positive, so that they cancel out to some extent. Thus the process of orthogonalization is equivalent to replacing the rapidly varying true potential by the much weaker and smoother one given by eqn (1.8); so weak that there are no core states, and over a large fraction of the Brillouin zone the energy bands resemble free electron bands. This approach first provided a clear explanation of why the approximation of nearly free electrons is so much better than one would expect, given the strong electric forces in a solid.

The process of constructing the pseudopotential is not unique, because any combination of core functions in (1.8) will do, since the eigenvalues of the non-core states are the same for any pseudopotential of the form

$$V_{\text{ps}}\phi = \Sigma_t \langle F_t | \phi \rangle b_t \tag{1.9}$$

where the F_t are arbitrary functions, and this because, as remarked above, the operator (1.9) projects any state onto the sub-space of the core states. This arbitrariness can then be exploited by choosing the F_t so that the cancellation between V_0 and V_{ps} is improved, though it does not necessarily follow that the weakest pseudopotential is the best for all applications. For a general analysis of the concept of pseudopotential see Harrison (1966) or Heine (1970).† A possible choice, for instance, is $F_t = -V_0 b_t$. Then let

$$V_0\phi = \sum_{s=1}^{n} a_s b_s(\mathbf{r}). \tag{1.10}$$

For an infinite sum eqn (1.10) would be exact; with n terms there is an error depending on how close the b_t are to a complete set. We now have (all sums are over the occupied core states)

$$\begin{aligned} V_{\text{eff}}\phi &= V_0\phi + \Sigma_t \langle -V_0 b_t | \phi \rangle b_t \\ &= V_0\phi - \Sigma_t \langle b_t | \Sigma a_s b_s \rangle b_t = V_0\phi - \sum_{s=1}^{n} a_s b_s \end{aligned}$$

since the b_s are orthogonal. Clearly, if the bound states approximate well to a complete set in the region in which $V_0\phi$ is large, the cancellation is good. Roughly, the more core states, the better the cancellation. Conversely, if there are no core states for a given l, there is no cancellation, e.g. a 3d state in Na is automatically orthogonal to all the core states.

† Pseudopotential theory is discussed also in Chapter 4 by Dr. Lidiard, with particular reference to colour centres.

In practice one treats, at least for a start, the non-local potential in eqn (1.8) as an ordinary potential and the non-local character is introduced as a perturbation only if high accuracy is desired. This is possible because it has been found that the dependence on energy and momentum of eqn (1.8) is weak and has only a small effect on the energy values. For the way in which V_{eff} is generally constructed see Section 1.4 below.

An alternative way of defining pseudopotentials is from the scattering properties of the atoms or ions. When a plane wave of wave vector $\mathbf{k}$ is scattered by a scatterer of radius r_0, the scattered wave can be expressed as a series of Legendre polynomials multiplied by radial functions whose asymptotic behaviour for $r \gg r_0$ is $\sin(kr+\eta_l)$, where $\eta_l = p_l\pi + \delta_l$, $|\delta_l| < \pi/2$, and p_l gives the number of nodes of the radial function of angular momentum l inside the sphere of radius r_0. In our case, p_l is equal to the number of core states of angular momentum l. But the multiple of π does not affect the scattering. The scattering by an angle θ from $\mathbf{k}$ to $\mathbf{k}+\mathbf{q}$ is given by the amplitude

$$f(\theta) = k^{-1}\Sigma(2l+1)\exp i\eta_l(k)\sin\eta_l(k)P_l(\cos\theta).$$

The quantity $(2\pi/\Omega)f(\theta)$, where Ω is the atomic volume, is called the t-matrix. Thus the intensity of any partial wave is determined by $\sin^2\eta_l = \sin^2\delta_l$, so that the true potential can be replaced by a weaker one that produces just the shifts δ_l, or, alternatively, the same logarithmic derivatives of the radial functions at $r = r_0$. To effect this replacement it is convenient to take $V_{ps} = \Sigma_l f_l(r)\mathscr{P}_l$, where $\mathscr{P}_l$ is a projection operator picking out the l angular momentum component of any ψ on which it operates. The f_l-functions are arbitrary. A particularly simple choice has been suggested by Ziman: for each l a δ-function of suitable strength at $r = r_0$. When the true potential has been replaced by a pseudopotential one says that the atom has been replaced by a pseudoatom, without core states, and with scattering properties defined by the shifts δ_l. The pseudopotential so constructed is suitable also for band calculation, particularly when a constant potential outside atomic spheres of radius r_0 is used (muffin-tin potential—MTP). With this potential, all that it is required to know about the radial functions are the logarithmic derivatives $\mathrm{d}(\ln R_l)/\mathrm{d}r = (1/R_l)(\mathrm{d}R_l/\mathrm{d}r)$ at $r = r_0$. These can be expressed through the δ_l and vice versa. All this holds also for a non-periodic arrangement of atoms, provided one can consider atomic spheres outside which the potential is approximately constant.

Finally it should be stressed that extreme caution is needed in identifying plane waves with pseudoplane waves and true potentials with pseudopotentials when one moves outside band calculation, e.g. in dealing with transport properties.

1.2.3. *Augmented plane waves (APW)*

This is a successful alternative to OPW that avoids the difficulties connected with 3d, 4f ... levels. While OPWs consist throughout of a plane wave plus an atomic orbital part, with APWs the plane wave is restricted to the region outside arbitrary 'atomic spheres' drawn around each atom and inside which the solution is of the atomic orbital type. Augmented plane waves (Loucks 1967; Mattheiss, Woods, and Switendick 1968; Dimmock 1971) require, at least for a start, the use of the muffin-tin potential. Slater (1937) first suggested that a simple plane wave $\exp\{i(\mathbf{k}+\mathbf{K}_j)\,.\,\mathbf{r}\}$ outside the atomic spheres of radius r_0 should be made to join to an exact solution, expanded in spherical harmonics, of the Schrödinger equation with a spherically symmetric potential inside each of the spheres. There remains a discontinuity in slope. Using the abbreviation $\mathbf{k}_i = \mathbf{k}+\mathbf{K}_i$, where the components of $\mathbf{k}_i$ in spherical coordinates are k_i, θ_i, ϕ_i and expanding the plane wave $\exp(i\mathbf{k}_i\,.\,\mathbf{r})$ in spherical harmonics $Y_{lm}(\theta, \phi)$, the solution inside r_0 continuous with the plane wave at $r = r_0$ is

$$A_i^{\text{ins}} = 4\pi \sum_{l=0}^{\infty} \sum_{m=-l}^{l} i \frac{j_l(k_i r_0)}{R_l(E_{\text{tr}} r_0)} Y^*_{lm}(\theta_i \phi_i) Y_{lm}(\theta, \phi) R_l(E_{\text{tr}}, r) \tag{1.11}$$

where j_l is a spherical Bessel function, e.g. $j_0(x) = \sin x/x$. The radial functions $R_l(r)$ are generally calculated by numerical integration with the assumed potential and trial values E_{tr} of the energy.

The wavefunction is then expanded in APWs and those energies are determined for which the expectation value of the Hamiltonian is stationary with respect to variation of the coefficients of the APWs. General arguments show that these energies are the best approximation to the true energy levels within the present scheme.

In the above formulation the matrix elements of the Hamiltonian are implicit functions of the energy. To avoid this, Slater (1953) suggested that the energy of *each* APW should be fixed so that the expectation value of the Hamiltonian for this APW is stationary. The average values of the energy is then calculated with a linear combination of APWs belonging to different $\mathbf{k}+\mathbf{K}_i$ and also, for each $\mathbf{k}+\mathbf{K}_i$, to different values of the energy, i.e. to different bands. The average energy is found to converge rapidly when more APWs are included in the linear combination. As a general rule, in each APW it is not necessary to consider spherical harmonics of order larger than about six.

At $\mathbf{k}$-points of high symmetry one can, as for OPWs, take advantage of group-theoretical simplifications for constructing symmetrized linear combinations of APWs (SAPWs) for the irreducible representations of interest, and the size of the secular equation can be drastically reduced. The symmetrization can be done after the initial variation of the single APWs. But experience has shown that the convergence improves if the symmetrization is carried out before the energy variation. Naturally one wishes to construct a linear combination of SAPWs that reduces the discontinuity in slope of the wavefunction at the surface of the atomic spheres. Saffren and Slater (1953) found that, if one proceeds along these lines, the equations obtained become identical to those of the 1937 Slater procedure. A number of computer programmes for this method are available (Mattheiss *et al.* 1968).

The variational procedure is based on the general principle that a solution of

$$(-\nabla^2+V)\psi = E\psi$$

in a volume Ω gives a stationary value to

$$I = \int_\Omega \nabla\psi^* \,.\, \nabla\psi + (V-E)\psi^*\psi \, d^3\mathbf{r},$$

and conversely a function making I stationary is a solution of Schrödinger's equation. When ψ is expanded in any set of basis functions $\psi = \Sigma c_i \phi_i$, I has a stationary value when the infinite system of equations is satisfied

$$\Sigma_j(H_{ij}-ES_{ij})c_j = 0 \qquad i = 1, 2, 3, \ldots$$

where

$$H_{ij} = \int_\Omega [\nabla\chi_i^* \,.\, \nabla\chi_j + V\chi_i^*\chi_j] \, d^3\mathbf{r} \quad \text{and} \quad S_{ij} = \int_\Omega \chi_i^*(\mathbf{r})\chi_j(\mathbf{r}) \, d^3\mathbf{r}.$$

The compatibility condition for these equations is

$$\operatorname{Det}|H_{ij}-ES_{ij}| = 0. \tag{1.12}$$

However, if the ϕ_i are the augmented plane waves A_i, a complication arises because grad A_i is not continuous at $r = r_0$. It may be proved that one must then add to eqn (1.12) an extra term $D_{ij} = \int \chi_i^*\nabla\chi_j \,.\, d\sigma_{\mathbf{r}_0}$ where the integration is over the sphere of radius r_0. For details of the calculations see Mattheiss *et al.* (1968). The final result is that the compatibility equation is again an equation of the type of eqn (1.5), with the last term replaced by

$$\frac{4\pi r_0^2}{\Omega}\left\{-(k_j^2-E)\frac{j_1(|\mathbf{k}_j-\mathbf{k}_i|r_0)}{|\mathbf{k}_j-\mathbf{k}_i|}\right.$$
$$\left.+\Sigma_l(2l+1)P_l(\cos\theta_{ij})j_l(k_ir_0)j_l(k_jr_0)\left\{\frac{R_l'(E,r_0)}{R_l(E,r_0)}-\frac{j_l'(k_jr_0)}{j_l(k_jr_0)}\right\}\right\} \tag{1.13}$$

where Ω is the volume of the unit cell, θ_{ij} is the angle between $\mathbf{k}_i$ and $\mathbf{k}_j$ and R_l is the radial function of angular momentum l; the dash denotes differentiation with respect to the argument. We see that the radial functions enter only through their logarithmic derivatives at $r = r_0$. The matrix elements (1.13) may be called the matrix elements of an APW pseudopotential, analogous to the OPW pseudopotential, showing again that pseudopotentials can be employed within different theoretical frameworks.

If there is more than one atom in the unit cell, extra terms appear in (1.13), but, though the calculation is more complicated, there are no difficulties of principle.

1.2.4. *Muffin-tin potential*

One may wonder how justified it is to achieve great accuracy in the variational calculation when one starts from the MTP whose form has been chosen to make the calculation possible. But in fact, at least in close-packed structures, the potential must be very nearly constant between atomic spheres, and it is also found that, in these structures, its discontinuity at the surface of the spheres is generally less than one electronvolt, so that the approximation is not too rough. However results depend on the arbitrary radius chosen for the atomic spheres. This is generally taken as the largest that will not produce overlapping of spheres. In compounds, the relative size of different atoms must also be taken into consideration.

It would be difficult to think of a better starting point of similar simplicity. However, in loosely packed structures, such as the diamond structure, the potential discontinuity at the spheres may be considerably larger, and errors in the calculated band structure may be expected to be approximately proportional to this discontinuity. A correction must then be introduced, and this is done fairly easily if this correction is assumed to vanish inside the atomic spheres, consisting only of a deviation from constancy of the potential outside. But it is difficult to include it in an iteration procedure designed to achieve self-consistency, and the tendency has been to make self-consistent, by iteration, only the muffin-tin part of the potential, the corrections being introduced at the end in a more or less plausible manner.

As remarked above, the only information required on the radial functions are their logarithmic derivatives, $L_l(r_0)$ at $r = r_0$. This leads to several possibilities, for instance:

(1) The values of the $L_l(r_0)$ can be interpolated from spectroscopic data (quantum defect method). One assumes that the one-electron Hamiltonian in a crystal is, near each atom, the same as for the valence electron in a free atom: the eigenfunctions for $r > r_0$ can then be constructed from the observed term values of the free atom (see for instance Ham 1955).

(2) The $L_l(r_0)$ can be left as disposable parameters, to be determined at the end from appropriate experimental data on the band structure, or from results of more accurate calculations (Section 1.3).

(3) The strong actual potential inside the spheres may be replaced by a weaker pseudopotential, thereby improving the convergence of the secular equation, provided the $L_l(r_0)$ are the same (Section 1.2.2).

1.2.5. *Korringa, Kohn–Rostoker (KKR) method*

The KKR, or Green function, or scattered waves method (see, for example, Ham and Segall 1961; Segall and Ham 1968) starts from the integral equation equivalent to the Schrödinger equation

$$\psi(\mathbf{r}) = \int_{\text{unit cell}} G(\mathbf{r}, \mathbf{r}')V(\mathbf{r}')\psi(\mathbf{r}')\,\mathrm{d}^3\mathbf{r}' \tag{1.14}$$

where the Green function G satisfying the Bloch condition is

$$G(\mathbf{r}, \mathbf{r}') = -\frac{1}{\Omega}\Sigma_n \frac{\exp\{\mathrm{i}(\mathbf{k}+\mathbf{K}_n)\,.\,(\mathbf{r}-\mathbf{r}')\}}{|\mathbf{k}+\mathbf{K}_n|^2 - E(\mathbf{k})}. \tag{1.15}$$

Rather than by iteration, eqn (1.14) is best treated by variational procedures which in fact, for a fixed E, yield a value of $\mathbf{k}$. The calculations however are too complicated to be undertaken, except for a muffin-tin potential. For this potential the equations for the energy values can be derived also without using a variational principle. If the atomic potential outside the atomic spheres is taken equal to zero, as one can always do, the integral in eqn (1.14) extends only to the atomic sphere (we restrict the treatment here to lattices with a single atom per unit cell—the extension to polyatomic lattices is straightforward). Inside the sphere we expand both ψ and G in spherical harmonics. We write

$$\psi(\mathbf{r}) = \Sigma_{l,m} C_{l,m} R_l(r) Y_{lm}(\theta, \phi) \tag{1.16}$$

where the coefficients $C_{l,m}$ are to be determined, while the expansion of G is

$$\begin{aligned} G(\mathbf{r}, \mathbf{r}') &= \Sigma_{l,m}\Sigma_{l',m'}\{\mathrm{i}^{l-l'} B_{ll'}^{mm'} j_l(gr) j_{l'}(gr') + g\delta_{ll'}\delta_{mm'} j_l(gr) n_l(gr')\} \\ &\quad \times Y_{lm}(\mathbf{r}) Y_{l'm'}(\mathbf{r}') \end{aligned} \tag{1.17}$$

where n_l is the Neumann function of order l, $E > 0$, $g = E^{\frac{1}{2}}$, and $r < r' < r_0$. The changes when $E < 0$ are straightforward. The coefficients $B_{ll'}^{mm'}$ can be calculated in a standard way (see Segall and Ham 1968, p. 255). They depend only on the crystal structure.

Now in eqn (1.14) replace $V(\mathbf{r}')\psi(\mathbf{r}')$ by $(\nabla^2 + E)\psi(\mathbf{r}')$ and use Green's formula remembering that $(\nabla^2 + E)G(\mathbf{r}, \mathbf{r}') = \delta(\mathbf{r}-\mathbf{r}')$; we then obtain, for a sphere of radius r_i just inside the sphere of radius r_0,

$$\int_{r_i = r_0 - \varepsilon} \{G(\mathbf{r}, \mathbf{r}')\,\mathrm{grad}\,\psi(\mathbf{r}') - \psi(\mathbf{r}')\,\mathrm{grad}\,'G(\mathbf{r}, \mathbf{r}')\}\,.\,\mathbf{n}\,\mathrm{d}\sigma = 0.$$

Then, using eqns (1.16) and (1.17) we carry out the angular integration remembering the orthogonality properties of the spherical harmonics. The result is multiplied by $Y_{lm}(\mathbf{r})$ and integrated over the sphere of radius r_i.

Finally the limit $\varepsilon \to 0$ is taken, obtaining

$$\sum_{l'=0}^{l_{\max}} \sum_{m'} \mathrm{i}^{l} j_l(gr_0)\{B_{ll'}^{mm'} W[R_{l'}(r_0), j_{l'}(gr_0)] + g\delta_{ll'}\delta_{mm'} W[R_l(r_0), n_l(gr_0)]C_{l'm'} = 0 \tag{1.18}$$

$W[f_1, f_2]$ being the expression $f_1(\mathrm{d}f_2/\mathrm{d}r) - f_2(\mathrm{d}f_1/\mathrm{d}r)$.

The determinant of the coefficients of the C_{lm} must vanish, and this is the secular equation for the energy. Spurious eigenvalues are introduced by the factor $j_l(gr_0)$ and occur also if the energy is near one of the free-electron values $|\mathbf{k}+\mathbf{K}_i|^2$ (cf. eqn 1.15); care is then needed (see Ham and Segall 1961).

The system of eqns (1.18) was first obtained, in a slightly different form, by Korringa (1947) by means of a "scattered wave" method, in which it is assumed that at the surface of each atomic sphere there are both an ingoing, ψ_i, and an outgoing, ψ_0, wave. The ingoing wave is the resultant of the waves coming from all the other atoms, which can be related, through the Bloch condition, to ψ_0. Then ψ_0 is assumed to be related to ψ_i by a scattering matrix, or simply by phase shifts if the potential is spherically symmetric. If ψ_i and ψ_0 are expanded in spherical harmonics, two systems of equations are obtained from the above assumptions for the two sets of coefficients, a_{lm}, b_{lm}. In particular, the scattering condition gives, for a spherically symmetric potential $a_{lm} = \frac{1}{2}(1-\exp(-2i\eta_l))b_{lm}$. These two systems are compatible only for certain values of the energy and this condition of consistency may be proved to be expressed by the equations (1.18) (Kohn and Rostoker 1954). The Korringa approach provides an illuminating picture of the physical problem, but is more difficult to express in a variational form for a general potential.

The eqns (1.18) can be expressed also in terms of the scattering phase shifts δ_l of the muffin-tin potential or pseudopotential. Let

$$R_l(r_0) = A_l\{j_l(gr_0) - \tan\delta_l n_l(gr_0)\}.$$

Then, since

$$W[n_l(gr_0), j_l(gr_0)] = 1/gr_0^2,$$

we obtain, by simple substitution in eqn (1.18) and division by common factors

$$\sum_{l'=0}^{l_{\max}} \sum_{m'} (B_{ll'}^{mm'} \tan\delta_{l'} + g\delta_{ll'}\delta_{mm'})A_l C_{l,m} = 0.$$

Hence the secular equation for a non-zero solution is

$$\mathrm{Det}|B_{ll'}^{mm'} \tan\delta_{l'} + g\delta_{ll'}\delta_{mm'}| = 0. \tag{1.19}$$

Difficulties may arise when there is resonance with a bound state (e.g. a d-band), making the scattering very strong and highly energy-dependent.

As with OPWs and APWs, the KKR method can be reformulated in a plane-wave representation (Ziman 1965), by introducing a KKR pseudopotential $V_{\text{ps}}^{\text{KKR}}$. The matrix elements of $V_{\text{ps}}^{\text{KKR}}$ are

$$(V_{\text{ps}}^{\text{KKR}})_{ij} = \frac{4\pi r_0^2}{\Omega} \sum_l (2l+1) P_l(\cos\theta_{ij}) j_l(k_i r) j_l(k_j r) \left[\frac{R_l'(E, r_0)}{R_l(E, r_0)} - \frac{j_l'(g r_0)}{j_l(g r_0)} \right]. \quad (1.20)$$

Morgan (1966) has noted that the difference between the matrix element (1.13) of the APW pseudopotential and the matrix element (1.20) of the KKR pseudopotential is just the APW matrix element in the limit of a vanishing periodic potential. Eqn (1.20) vanishes for this 'empty lattice' since the KKR wavefunctions in this case are just plane waves.

One advantage of the Green function method is that the quantities $B_{ll'}^{mm'}$ (structure constants) are determined only by the lattice structure, so that, apart from scaling factors, they can be determined once and for all for any structure, for a series of values of **k**, for any given E. Tables are available (see Segall and Ham 1968). The structure constants remain the same when a relativistic treatment is carried out. The other quantities that appear in the secular equation are, as with APW, the logarithmic derivatives of the radial functions at $r = r_0$, which depend on the potential, but not on the lattice structure, and the logarithmic derivatives of the spherical Bessel functions at $r = g r_0$.

The great advantage of the KKR procedure is the rapid convergence of the secular equation, which, it is claimed, rarely requires spherical harmonics with $l > 2$. This is plausible, since the contribution of partial waves with $l > 2$ to the scattering of slow electrons by a spherical atom is expected to be negligible. On the other hand, it contrasts with other methods in which it is found that higher values of l are required for accurate wavefunctions. If $l_{\max}$ is 2, only L_0, L_1, $L_2(r_0)$ are required. The construction of a potential becomes superfluous if L_0, L_1, L_2 are used as disposable parameters, to be fixed from experimental data. Of course, they depend on the energy, but this dependence can be determined fairly accurately with some approximate potential. No complete check however is yet available on how well this procedure accounts, for instance, for the details for the Fermi surface of metals. The KKR method is also convenient for obtaining explicit wavefunctions. As for its restriction to a MTP, more or less similar considerations hold as for APW. Again, a simple perturbation treatment of the corrections appears to be sufficient. Finally, this method provides a point of view which has proved very useful in the treatment of disordered structures.

1.2.6. *Linear combination of atomic orbitals (LCAO) (tight binding) method*

This method is based on Bloch's original suggestion to approximate one-electron wavefunctions in crystals by linear combinations of atomic orbitals (LCAO)†

$$\psi_t(\mathbf{k},\mathbf{r}) = \Sigma_{\mathbf{R}_j} \exp(i\mathbf{k}\,.\,\mathbf{R}_j)\phi_t(\mathbf{r}-\mathbf{R}_j) \tag{1.21}$$

where ϕ_t is an atomic orbital of energy E_t and the $\mathbf{R}_j$ are vectors of the direct lattice. The expression (1.21) assumes that there is only one atom in the unit cell; it describes electrons in a band derived from an atomic level t. Its form was chosen so as to satisfy the Bloch periodicity condition. It is an approximation; it could be made exact by replacing the atomic orbitals by the Wannier functions (Section 1.6). The sum is generally confined only to the nearest, perhaps second nearest, cells to the one under consideration. The method is then called 'tight binding'.

The atomic orbitals have angular factors taken as suitable combinations of spherical harmonics Y_l^m, for instance (unnormalized):

for s-functions	Y_0^0	$=$	1
for p_x-functions	$Y_1^1+Y_1^{-1}$	$=$	$\sin\theta\cos\phi$
for p_y-functions	$Y_1^1-Y_1^{-1}$	$=$	$\sin\theta\sin\phi$
for p_z-functions	Y_1^0	$=$	$\cos\theta$
for d_{xy}-functions	$Y_2^2-Y_2^{-2}$	$=$	$\sin^2\theta\cos\phi\sin\phi$
for d_{yz}-functions	$Y_2^1-Y_2^{-1}$	$=$	$\sin\theta\cos\theta\sin\phi$
for d_{zx}-functions	$Y_2^1+Y_2^{-1}$	$=$	$\sin\theta\cos\theta\cos\phi$
for d_{z^2}-functions	Y_2^0	$=$	$3\cos^2\theta-1$
for $d_{x^2-y^2}$-functions	$Y_2^2+Y_2^{-2}$	$=$	$\sin^2\theta(\cos^2\phi-\sin^2\phi)$, etc.

The radial factors depends on the atomic potential assumed and are generally taken from atomic self-consistent field calculations. The average value of the energy for the wavefunction (1.21) is (see, for instance, Pincherle 1971, p. 134)

$$E_t(\mathbf{k}) = \frac{\Sigma_{R_m} H_t(\mathbf{R}_m)\exp(\mathrm{i}\mathbf{R}_m\,.\,\mathbf{k})}{\Sigma_{R_m} S_t(\mathbf{R}_m)\exp(\mathrm{i}\mathbf{R}_m\,.\,\mathbf{k})}. \tag{1.22}$$

The $S_t(\mathbf{R}_m)$ are overlap integrals

$$\int_{\text{all space}} \phi_t^*(\mathbf{r})\phi_t(\mathbf{r}+\mathbf{R}_m)\,\mathrm{d}^3\mathbf{r}$$

and the $H_t(\mathbf{R}_m)$ are energy matrix elements

$$\int_{\text{all space}} \phi_t^*(\mathbf{r})H\phi(\mathbf{r}+\mathbf{R}_m)\,\mathrm{d}^3\mathbf{r}.$$

† For an account of the LCAO method applied to molecules, see Coulson (1961).

By introducing the difference δV between the potential in the crystal and that of the isolated atom, the $H_t(\mathbf{R}_m)$ can be written

$$H_t(\mathbf{R}_m) = E_t S_t(\mathbf{R}_m) + \int_{\text{all space}} \phi_t^*(\mathbf{r})\, \delta V(\mathbf{r}) \phi_t(\mathbf{r}+\mathbf{R}_m)\, \mathrm{d}^3\mathbf{r}. \qquad (1.23)$$

By restricting the sum to nearest neighbours only, the $E(\mathbf{k})$ relation can be obtained in an analytic form, as a Fourier expansion in reciprocal space. It can then be used as a simple interpolation formula.

The theory can be generalized to the case when the Bravais cell contains several atoms at positions $\mathbf{R}_j + \mathbf{r}_s$ and when it is necessary to consider more than one atomic orbital. Then

$$\psi_{\mathbf{k}}(\mathbf{r}) = \Sigma_{\mathbf{R}_j,t,s} A_{t,s} \phi_t(\mathbf{r}-\mathbf{R}_j-\mathbf{r}_s) \exp\{\mathrm{i}\mathbf{k} \,.\, (\mathbf{R}_j+\mathbf{r}_s)\}$$

and the integrals are more complicated. Minimization of the energy leads to a system of homogeneous equations for the $A_{t,s}$; the resulting secular equation determines the energy levels.

The calculation is simpler at symmetry points and lines in the Brillouin zone, where the secular matrix splits into the direct sum of smaller matrices. Generally only a few of these are of interest and they involve a restricted number of atomic orbitals.

The tight binding method fails completely to describe excited states. When used as an interpolation scheme, the complicated two or three centre integrals are not calculated explicitly, but considered as disposable parameters.† If one should wish to calculate the integrals, reference can be made to Slater (1967) where tables of results of the angular integration are given for nearest neighbours integrals of all s-, p-, and d-functions in cubic and hexagonal lattices. Then, given lattice data, δV, and the radial part of the atomic orbitals, computer programmes are available for the radial integration.

1.3. Empirically-adjusted calculations and model Hamiltonians

From our brief survey of purely first principle calculations, it must be clear that they are frustrating and occasionally even a little illogical. The opposite point of view is to construct an empirical band structure from available experimental information, taking into account general group-theoretical results. This can be achieved by fitting pseudopotentials to empirical data, and a new art has arisen here, discussed for instance by Harrison (1966) and by Cohen and Heine (1970). The $\mathbf{k} \,.\, \mathbf{p}$ formalism can also be employed: this has been used extensively for the study of band edges in semiconductors; recently it has been extended to deal with complete bands (Cardona and Pollak 1966).

† An interesting analysis of the basic approximations involved in the tight-binding and muffin-tin α-resonance models of transition metals has been given by Friedel (1973).

Purely empirical band structures cannot give a complete picture and are often ambiguous. Thus an intermediate approach is now favoured, and will probably be the final solution. This is well exemplified by what Herman calls 'empirically adjusted first principle calculations' (Herman, Kortum, Kuglin, van Dyke, and Skillman 1968). It amounts to adding an empirical step to first principle calculations, carried out, in Herman's case, by the OPW method, and a feedback loop. In practice it means changing by a few per cent a few Fourier coefficients of the potential, and possibly core eigenvalues, so as to obtain exactly about three experimental energy level separations, known accurately from optical studies. It is then assumed that the thereby adjusted calculation constitutes an accurate interpolation. In the above reference impressive evidence is provided of the usefulness and flexibility of the method in providing, for some fifty crystals, band structures of high accuracy and also wavefunctions, charge densities and the distribution of the true potential. These last quantities are open to doubt when a pseudopotential approach has been followed. Another advantage is that, when empirical adjustments are applied to the potential, great accuracy in its initial construction is no longer essential.

A common problem occurs when a reliable band structure has been calculated at symmetry points in the Brillouin zone, where group theory leads to a reduction in the size of the secular equation, and one wishes to interpolate results to all other **k**-points. Interpolation procedures have been considered for a long time, in one form or another. One might argue that simplicity is a very important factor, so that the early suggestion by Slater to use a simplified form of the tight binding method, or, for simple metals, that by Coulson to use a network of conducting paths along chemical bonds, are still useful.

In general, the new $E(\mathbf{k})$ values are obtained as the roots of a 'model' secular equation in which the dependence of the matrix elements on $\mathbf{k}$ is represented by a simple analytical expression in terms of a few disposable parameters. Of course the energy, being a periodic function of $\mathbf{k}$, can always be represented by

$$E(\mathbf{k}) = \Sigma_{\mathbf{R}_i} \exp(\mathrm{i}\mathbf{R}_i \cdot \mathbf{k}) E(\mathbf{R}_i)$$

but, although this formula has been used (Dresselhaus and Dresselhaus 1967), it is desirable to have a smaller number of parameters.

Often the model secular equation is more than an interpolation procedure. The functional relation for the coefficients can bring out the common elements of calculations that have used different potentials, and even of calculations for different crystals. In this case one speaks of a 'model Hamiltonian' (Phillips and Sandrock 1968).

These ideas have often been expressed through pseudopotential form factors (see for instance Cohen and Heine 1970). When the crystal potential is expressed as a sum of atomic potentials, the matrix elements become products of a structure factor depending on the positions of the atoms in the primitive cell, times a form factor

$$V(\mathbf{q}) = \frac{1}{\Omega}\int_{\text{unit cell}} \exp\{-\mathrm{i}(\mathbf{k}+\mathbf{q})\,.\,\mathbf{r}\}\,V(\mathbf{r})\exp(\mathrm{i}\mathbf{k}\,.\,\mathbf{r})\,\mathrm{d}^3\mathbf{r},$$

where Ω is the volume per atom. If $V(\mathbf{r})$ can be taken to be a local potential, rather than an operator, the form factor depends only on $\mathbf{q}$ and is then denoted by $\langle \mathbf{k}+\mathbf{q}|V|\mathbf{k}\rangle$. For metals we are concerned with electrons on the Fermi sphere, so that $|\mathbf{k}| = |\mathbf{k}+\mathbf{q}| = k_F$. Thus, for any $\mathbf{q}$, the angle between $\mathbf{k}$ and $\mathbf{q}$ is fixed and the form factor is a function of q only.

The function $V(q)$ is sketched in Fig. 1.1, omitting the weak oscillating tail which is truncated in band calculations. A simple analytical formula

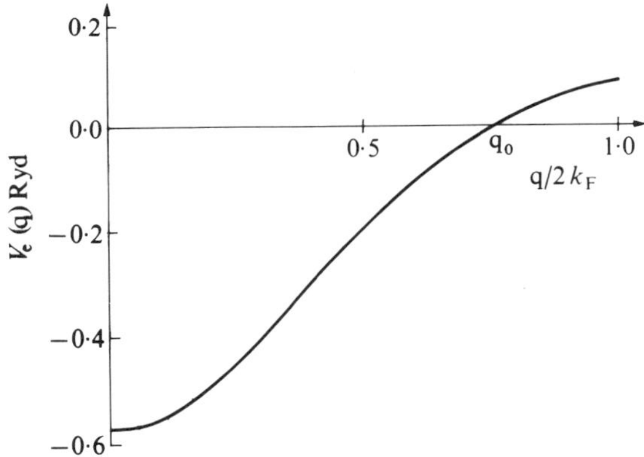

FIG. 1.1. The pseudopotential form factor $V_e(q)$ in the on-Fermisphere approximation with V assumed to be exactly zero inside the atomic core and $-(1/r)$ outside. (After Cohen and Heine 1970.)

(Ashcroft 1966) is $V(q) = \cos qR_0/q^2\varepsilon(q)$ where $\varepsilon(q)$ is the static free electron dielectric function and R_0 is an adjustable radius which is nearly that of the ion core. The limit at $q = 0$ is determined by the model employed. The height of the maximum is 1–2 eV. The Fourier coefficients important in band calculation are near the first zero of $V(q)$, $q_0 \sim \pi/2R_0$, so that the exact position of this zero is very critical. While only these Fourier coefficients are required in band calculations, the whole function is required for the calculation of other physical properties.

The zero at $\pi/2R_0$ appears to be about right for some alkali metals, but one finds that q_0 varies irregularly from element to element. This is a difficulty, since magnitude and sign of the matrix elements of the pseudopotential are so sensitive to the position of q_0. The difficulty is removed by using the $V(\mathbf{K}_n)$ as parameters, to fit values of the energy either deduced from experiment, or calculated by more perfect methods. They can also be fixed on the basis of pseudoatom's (phase shifts). In any case, a simple equation is obtained allowing determination of the $E_n(\mathbf{k})$ at all points in the Brillouin zone. From this, the density of states $N(E)$ can be calculated

$$N(E) = \Sigma_n \frac{1}{4\pi^3} \int_{E_n(\mathbf{k})=E} \frac{\mathrm{d}\sigma}{|\nabla_{\mathbf{k}} E_n(\mathbf{k})|},$$

the integral being over the surface of constant energy E.

Incidentally, the singularities of the integral at points at which $\nabla_{\mathbf{k}} E = 0$ show up as peaks in the absorption spectrum and the fitting of the experimental values of these peaks is another way of determining the disposable parameters in the interpolation scheme.

Interpolation schemes for transition metals have attracted attention (see, for instance, Ehrenreich and Hodges 1968), because here we cannot expect that a nearly free electron or a completely bound electron scheme will be successful. One of these schemes, which has proved useful, describes the 'conduction bands' as a combination of the four plane waves which in the free electron approximation $E_n = |\mathbf{k}+\mathbf{K}_n|^2$ have the lowest energy (in $\frac{1}{48}$ of the Brillouin zone). For a face-centred cubic structure these can be $\exp(\mathrm{i}\mathbf{k}\,.\,\mathbf{r})$; $\exp\{\mathrm{i}y(2\pi/a)\}$; $\exp\{-\mathrm{i}(x+y\pm z)\,.\,\pi/a\}$; they can be orthogonalized to the s- and p-core states and are supposed to interact through a pseudopotential defined by the two coefficients $V(1, 1, 1)$ and $V(2, 0, 0)$.

The d-bands on the other hand are described by a linear combination of the five d-orbitals ϕ_m. Thus the total pseudowavefunction is

$$\phi_{\mathbf{k}}(\mathbf{r}) = \sum_{n=1}^{4} A(\mathbf{k}, \mathbf{K}_n) \exp\{\mathrm{i}(\mathbf{k}+\mathbf{K}_n)\,.\,\mathbf{r}\} f(\sigma)$$
$$+\Sigma_{\mathbf{R}_j} \sum_{m=1}^{5} B_m \exp(\mathrm{i}\mathbf{k}\,.\,\mathbf{R}_j)\phi_m(\mathbf{r}-\mathbf{R}_j) f(\sigma),$$

$f(\sigma)$ being a spin function. The summation over σ has been omitted. Considering spin, there are 18 functions in all. Between functions of the same spin the energy matrix has the form

$$\left(\begin{array}{c|c} \text{OPW–OPW} & \text{OPW–LCAO} \\ \hline (\text{OPW–LCAO})^* & \text{LCAO–LCAO} \end{array}\right)$$

The LCAO–LCAO sub-matrix is evaluated as in the usual tight-binding method. The OPW–LCAO 4×5 matrix describes the interaction between the d-bands and the conduction band and produces hybridization.

In the full 18×18 matrix, spin–orbit interaction can also be introduced. At least 9 parameters are present, which again can be derived from more extensive calculations or from experiment.

Clearly a sharp distinction between first principles and empirical calculations is no longer appropriate. Between the two extreme approaches there is now a continuous range of methods.

1.4. Construction of the effective potential

The ideal aim is to construct a one-electron effective potential or pseudopotential that accounts for as many experimental facts as possible, not only for the band structure. When a pseudopotential is constructed, this means, physically, replacing the true atoms of the crystal by pseudoatoms identified by their scattering shifts or by equivalent parameters. For an excellent discussion of the effectiveness of this approach, reference should be made to Cohen and Heine (1970).

The pseudopotential is generally of the muffin-tin type and is constructed from the atomic potentials of the constituent atoms in the Hartree–Fock scheme. The Coulomb and exchange potentials are treated separately. In each atomic sphere s the Coulomb part includes the spherically averaged atomic potential of the s-atom, and the spherical average of the contribution of the atomic potentials due to the various shells of neighbouring atoms. This contribution flattens the potential near the atomic radius and leads fairly continuously to the constant potential between atomic spheres. There is of course a contribution from the valence charge density in the cell, and to calculate this we ought to know the wavefunctions which are to be calculated. Initially an estimate of this charge density must be introduced, and at the end one must check that the estimate was good, that is, that the potential was self-consistent. If not, one must proceed to iterations. If cells containing different atoms have a net electric charge, the Madelung potential must also be added, also ideally in a self-consistent way.

The exchange potential presents many difficulties, since it depends on $\mathbf{k}$ and on the band index. An average must be used and, following Slater (see also Chapter 3) this is generally the free electron exchange potential given by

$$V_{\text{exch}} = -3(3/8\pi)^{\frac{1}{3}}(n/V)^{\frac{1}{3}} \tag{1.24}$$

for n free electrons in volume V. There have been suggestions, backed by theoretical arguments, that $\frac{2}{3}$ of the Slater exchange term should be used, rather than the full term. In fact the tendency is to take the numerical factor in (1.24) as a disposable parameter. It is remarkable that this apparently rough approximation has led to successful results.

Between the atomic spheres, the Coulomb part of the average potential is obtained immediately. The exchange part requires a three-dimensional numerical integration. The muffin-tin potential thus obtained is generally l-dependent.

Besides the Coulomb and exchange energies, one should consider also the so-called correlation energy due to the mutual electronic repulsions. With the Hartree–Fock scheme, repulsion between electrons with the same spin is taken automatically into account by the Pauli principle, but repulsion between electrons with opposite spin is not considered, though there must be a decrease of the probability of finding the electrons close together; this is sometimes referred to as the 'correlation hole', similar to the 'exchange hole' produced by the exchange potential. One way of dealing with this is to use linear combinations of determinantal functions; the leading Slater determinant is that in which the electrons occupy those orbitals which arise from the self-consistent treatment. In the other determinantal functions one or more orbitals are replaced by different ones (configuration interaction). A better way would be to use techniques of many-body theory, such as the already mentioned density matrix, if we knew how to write it down for the exact wavefunction. It can be written down for a single determinantal function, but this does not include the correlation hole. Asymptotic expressions for the correlation energy can be obtained in the limiting cases of high and low electronic density (see, for instance, Raimes 1961), but there is no general prescription for the intermediate case. In any case 'exchange' and 'correlation' are not separate physical phenomena, but names given to parts of the mathematics. If we could calculate the energy correctly, there would be no unambiguous physical way of determining how much is the 'correlation energy' or how much is the 'exchange energy'. However, the 'correlation energy' is often *defined* as the difference between the exact total energy and the energy calculated by the Hartree–Fock method.

1.5. Application of group theory

Following a general principle of quantum mechanics, the one-electron wavefunctions in crystals must form bases for irreducible representations of the group of operations which commute with the Hamiltonian (see, for instance, Cornwell 1969). This is called a space group and contains: (1) a part consisting of rotations (proper or improper) which is called a point group; (2) a part consisting of translations. If the rotation operation is the identity, the translations are the lattice translations, but for any other rotation R, if the crystal has screw axes (e.g. diamond) one must add a translation $v(R)$ which is not a lattice translation. However we consider only the case $v(R) = 0$ for all R (symmorphic groups). The lattice translations are then an Abelian sub-group of the space group, all the irreducible representations are one-dimensional and, for each translation operator T_i, displacing the lattice by $\mathbf{R}_i$,

$$T_i\psi_{\mathbf{k}}(\mathbf{r}) \equiv \psi_{\mathbf{k}}(\mathbf{r}+\mathbf{R}_i) = \exp(\mathrm{i}\mathbf{k}\,.\,\mathbf{R}_i)\psi_{\mathbf{k}}(\mathbf{r}) \quad \text{(Bloch)}.$$

As for the rotations, we are interested in those in reciprocal space, and for each crystal momentum **k** there is a group of operations, which is a sub-group of the point group of the direct lattice, which sends **k** into itself or into an equivalent vector. For general points in the Brillouin zone this sub-group consists only of the identity operation. For points on symmetry axes or on the surface of the zone it is larger. At the centre of the Brillouin zone it may consist of the whole point group of the direct lattice. At a general point, in order that ψ should be a basis for an irreducible representation of the space group, it is sufficient that it has the Bloch form. Any operation of the point group of the crystal transforms this ψ into a similar ψ characterized by another wave vector of the same 'star'. All these states have the same energy, and hence E has the complete symmetry of the reciprocal lattice, a useful fact in band computations. At points of symmetry ψ must be a basis function for an irreducible representation of the point group of the corresponding **k**. For a list of the irreducible representations of all point groups, and their characters, see, for instance, Koster (1957).

The first application of group theory is then to the classification of states at any **k**-point. There is a conventional notation: points are denoted by Latin or Greek letters, e.g. Γ is the centre of the Brillouin zone, Δ is the [1, 0, 0]-axis in a cubic crystal, etc, and the irreducible representations are denoted by suffices and dashes, e.g. Γ_1, Γ_{15}, Γ'_{12}, etc. Alternatively, symbols like Γ_s are used, the suffix indicating the lowest order of spherical harmonics present in the expansion of ψ in terms of these functions. This may be ambiguous in the case of compounds, where this lowest order may be different for the expansions about different atoms.

Other applications are the determination of the compatibility of states at adjacent points of different symmetry, e.g. between Γ-states and Δ-states, and the splitting of degeneracies at points of lower symmetry. This is achieved by simple inspection of the character Tables of the two groups. Similarly, one can determine the correspondence between the actual states and the limiting cases of states in the free electron approximation or in the tightly bound electron approximation. In the first limit $E_n = |\mathbf{k}+\mathbf{K}_n|^2$, where $\mathbf{K}_n$ is a vector of the reciprocal lattice. In the second limit the states are the ordinary atomic states 1s, 2s, 2p, ... The principle is very simple. Representations of the **k**-group are set up using as basis functions the plane waves or the atomic wavefunctions (spherical harmonics) and the characters are found for all classes in the group. These representations are generally reducible. Inspection of the characters of the known irreducible representations of the **k**-group is sufficient to write down unambiguously the direct sum of irreducible representations into which the given reducible representation splits. Thus one finds, for instance that, at $k = 0$, an atomic d-term splits into a triply-degenerate term whose wavefunctions are of type xy, yz, zx (Γ'_{25}), and a doubly-degenerate term, whose wavefunctions are of type z^2, x^2-y^2 (Γ_{12}).

For the purpose of computation, one of the most important applications is to the determination of symmetry adapted wavefunctions at any **k**-point and for any irreducible representation. The use of these symmetrized functions reduces considerably the number of terms one must include in the expansion of ψ, up to a prefixed order, whatever expansion is used, e.g. in plane waves, in orthogonalized plane waves, in spherical harmonics, etc. The problem consists in constructing linear combinations of these plane waves, spherical harmonics, etc., that transform according to one of the irreducible representations of the group of the wave vector. For instance, suppose, for plane waves, we want one of the combinations, say the s combination for the irreducible representation p at the point **k**

$$\phi_{\mathbf{k}}^{\mathrm{ps}}(\mathbf{r}) = \Sigma_{\mathbf{K}_m} c_{\mathbf{k}}^{\mathrm{ps}}(\mathbf{K}_m) \exp\{\mathrm{i}(\mathbf{k}+\mathbf{K}_m)\cdot\mathbf{r}\}.$$

These symmetrized combinations can be used also in the OPW and APW methods. Group theory shows (Wigner 1959; Heine 1960; Cornwell 1969) that the desired expansion is obtained by operating on $\exp\{\mathrm{i}(\mathbf{k}+\mathbf{K}_m)\cdot\mathbf{r}\}$ with the projection operator

$$\mathscr{P}_{\mathrm{ss}}^{\mathrm{p}} = \frac{l_p}{g(\mathbf{k})}\Sigma[\alpha_{\mathrm{ss}}^{\mathrm{p}}(R)]^* P(R), \tag{1.25}$$

where l_p is the dimension of the representation p, g is the number of operations R in the group of **k** and the sum is over all these operations. $\alpha_{\mathrm{ss}}^{\mathrm{p}}(R)$ is the diagonal element ss of the matrix representing the operation R in the p irreducible representation: if the representation is one-dimensional it is simply the character of the representation for the operation R. $P(R)$ is an operator that, when applied to $\exp\{\mathrm{i}(\mathbf{k}+\mathbf{K}_m)\cdot\mathbf{r}\}$, gives $P(R)\exp\{\mathrm{i}(\mathbf{k}+\mathbf{K}_m)\cdot\mathbf{r}\} = \exp\{\mathrm{i}R(\mathbf{k}+\mathbf{K}_m)\cdot\mathbf{r}\}$. Now $R\mathbf{k}$ is equivalent to **k** because R is an operation of the group of **k** and $R\mathbf{K}_m$ is another vector $\mathbf{K}_n$ of the reciprocal lattice, so that

$$R(\mathbf{k}+\mathbf{K}_m) = \mathbf{k}+\mathbf{K}_n.$$

The length of $\mathbf{k}+\mathbf{K}_n$ is the same as that of $\mathbf{k}+\mathbf{K}_m$.

The above form of the projection operator is quite general: it can be used for instance to find symmetrized combinations of spherical harmonics.

To give an example, consider the centre of the Brillouin zone $k = 0$, and the three-dimensional representation $\Gamma_{d'}(\Gamma'_{25})$ of the body-centred cubic lattice which has basis functions xy, yz, zx. Suppose we want the linear combination of plane waves of type $\exp\{\mathrm{i}(\pm x \pm y)(2\pi/a)\}$ (there are 12 of them) transforming according to the row xy of that representation. First we need the 48 matrices representing the 48 operations of the cubic group when xy, yz, zx are used as basis functions. For instance, a clockwise rotation of $\frac{1}{2}\pi$ about the z-axis changes x into $-y$ and y into x, so that the matrix representing this rotation, which can be denoted by $(-y, x, z)$, is

$$\begin{pmatrix} -1 & 0 & 0 \\ 0 & 0 & 1 \\ 0 & -1 & 0 \end{pmatrix}.$$

Only those operations contribute to the sum for which $\alpha_{11} \neq 0$. They are $(x, y, z), (x, y, -z), (x, -y, z), (x, -y, -z), (-x, y, z), (-x, y, -z), (-x, -y, z), (-x, -y, -z), (y, x, z), (y, x, -z), (y, -x, z), (y, -x, -z), (-y, x, z), (-y, x, -z), (-y, -x, z), (-y, -x, -z)$. The corresponding signs of α_{11} (of modulus one) are: $+, +, -, -, -, -, +, +, +, +, -, -, -, -, +, +$. The application of these operations to the reciprocal lattice vector $2\pi/a(+1, +1, 0)$, which we call $(+1, +1)$ for short, produces the reciprocal lattice vectors $(+1, +1), (+1, +1), (+1, -1), (+1, -1), (-1, +1), (-1, +1), (-1, -1), (-1, -1), (+1, +1), (+1, +1), (+1, -1), (+1, -1), (-1, +1), (-1, +1), (-1, -1), (-1, -1)$. Thus the sum in (1.25) is $4(+1, +1)-4(+1, -1) -4(-1, +1)+4(-1, -1)$, and the required wavefunction, with $l_p = 3$, $g = 48$, is

$$\phi_0^{\Gamma'_{25}(xy)} = \tfrac{1}{2}[\cos\{(2\pi/a)(x+y)\} - \cos\{(2\pi/a)(x-y)\}] = (4\pi^2/a^2)xy + \dots .$$

Thus in the secular equation there is only one row and column labelled with the waves considered, instead of 12.

1.6. Wannier functions

In a book describing one-electron functions in molecules and solids, at least a brief mention must be made of another representation used in solids, in which the 'wavefunction' (Wannier function) $a(\mathbf{r})$ represents the totality of the states of a band. (For a more complete treatment see, for instance, Blount 1962). It is defined as a unitary transformation of the Bloch functions

$$a_n(\mathbf{r}) = \frac{1}{\Omega_B} \int_{BZ} \psi_n(\mathbf{r}, \mathbf{k})\, d^3\mathbf{k} \tag{1.26}$$

the integral being extended to the Brillouin zone of volume Ω_B. One may consider, for each band n, a Wannier function at each lattice point $\mathbf{R}_j$

$$a_n(\mathbf{r} - \mathbf{R}_j) = \frac{1}{\Omega_B} \int_{BZ} \psi_n(\mathbf{r}, \mathbf{k}) \exp(-i\mathbf{R}_j \cdot \mathbf{k})\, d^3\mathbf{k}$$

so that a Wannier function may be considered either as a set of coefficients $a_n(\mathbf{r} - \mathbf{R}_j)$ defined in one cell of the direct lattice, or as a single function $a(\mathbf{r})$ extended to the whole space.

Conversely,

$$\psi_{\mathbf{k},n}(\mathbf{r}) = \Sigma_{\mathbf{R}_j} a_n(\mathbf{r} - \mathbf{R}_j) \exp(i\mathbf{R}_j \cdot \mathbf{k}) \quad \text{(cf. eqn (1.21))}$$

Replacement in (1.26) of $\psi_n(\mathbf{t}, \mathbf{k})$ by its plane-wave expansion shows that for each band the Wannier function $a_n(\mathbf{r})$ is the Fourier transform of the momentum eigenfunction $v_n(\mathbf{k})$. The Wannier functions centred on different $\mathbf{R}_j$ are orthogonal. The matrix element of the one-electron Hamiltonian between two Wannier functions located at $\mathbf{R}_j$, $\mathbf{R}_s$ is equal to the Fourier coefficient $E(\mathbf{R}_s - \mathbf{R}_j)$ of the energy considered as a periodic function of $\mathbf{k}$ with the periodicity of the reciprocal lattice. The $a(\mathbf{r} - \mathbf{R}_j)$ satisfy the system of equations

$$Ha_n(\mathbf{r}) = \Sigma_{\mathbf{R}_j} E_n(\mathbf{R}_j) a_n(\mathbf{r} - \mathbf{R}_j) \tag{1.27}$$

which could, in principle, be used as a basis for band computation. However one can deal with it only by variational methods, and results so far have not been encouraging.

For strongly-bound electrons all the $E_n(\mathbf{R}_j)$ vanish, except $E_n(0)$ which is the atomic eigenvalue, then $Ha_n = E_n(0)a_n$ and the Wannier function coincides with the atomic orbital. The Wannier functions are always localized, thus they provide a good description of inner electrons in a solid and are useful in discussing problems in which the localized nature of these electrons is relevant, e.g. scattering by these electrons.

In (1.26) it is not necessary that all Bloch functions belong to one band, so long as there is one for each $\mathbf{k}$ in the Brillouin zone. The description of N electrons by Bloch functions, one for each $\mathbf{k}$ in the Brillouin zone, is equivalent to their description by the corresponding Wannier functions, one for each $\mathbf{R}_j$, the total charge density being given equivalently by

$$\rho(\mathbf{r}) = \int |\psi(\mathbf{r}, \mathbf{k})|^2 \, \mathrm{d}^3\mathbf{k} \quad \text{or by} \quad \Omega_{\mathrm{B}} \Sigma_{\mathbf{R}_j} |a(\mathbf{r} - \mathbf{R}_j)|^2 \tag{1.28}$$

When we use Bloch functions, we use a 'crystal momentum' representation, where the crystal momentum $\mathbf{P}$ (not the ordinary momentum $\mathbf{p}$) is diagonal $\mathbf{P}\psi_{n,\mathbf{k}}(\mathbf{r}) = \hbar\mathbf{k}\psi_{n,\mathbf{k}}(\mathbf{r})$. An operator $\mathbf{R}$ conjugate to $\mathbf{P}$ can be defined: its eigenstates are the Wannier functions, the eigenvalues are the lattice vectors $\mathbf{R}_j$:

$$\mathbf{R} a_n(\mathbf{r} - \mathbf{R}_j) = \mathbf{R}_j a_n(\mathbf{r} - \mathbf{R}_j).$$

One of the reasons why the Wannier functions have not received much application in band theory is that they are not completely defined by eqn (1.26), because each Bloch function $\psi_{n,\mathbf{k}}(\mathbf{r})$ can be multiplied by an arbitrary phase factor $\exp\{\mathrm{i}\phi_n(\mathbf{k})\}$, so that

$$a_n(\mathbf{r}) = \frac{1}{\Omega_{\mathrm{B}}} \int_{\mathrm{BZ}} \psi_n(\mathbf{r}, \mathbf{k}) \exp\{\mathrm{i}\phi_n(\mathbf{k})\} \, \mathrm{d}^3\mathbf{k} \tag{1.29}$$

providing a different $a_n(\mathbf{r})$ for each different choice of $\phi_n(\mathbf{k})$. In fact $\phi_n(\mathbf{k})$ is not completely arbitrary. It can be proved that the most general admissible $\phi_n(\mathbf{k})$ must be a periodic function of $\mathbf{k}$ with the periodicity of the reciprocal lattice plus the scalar product of $\mathbf{k}$ with some lattice vector (Weinreich 1965).

However, whatever ϕ, eqn (1.28) is always valid. The problem of choosing ϕ_n so that a_n is as localized as possible, for instance so that $\int |a_n(\mathbf{r})|^2 r^2 \, d^3\mathbf{r}$ is minimum, has not received a completely satisfactory solution.

Considering for simplicity the one-dimensional chain of period L, we have, by a known property of Fourier transforms

$$\langle x^{2n} \rangle = \int_{-\infty}^{\infty} |a(x)|^2 x^{2n} \, dx = \int_{-\infty}^{\infty} \left(\frac{d^n v}{dk^n} \right)^2 dk;$$

in particular

$$\langle x^2 \rangle = \int_{-\infty}^{\infty} \left(\frac{dv}{dk} \right)^2 dk, \tag{1.30}$$

provided both $a(x)$ and $v(k)$ are normalized to 1 from $-\infty$ to $+\infty$. Thus, if all derivatives of $v(k)$ are quadratically integrable, the average value of x^{2n} is finite, for all n, and the Wannier function must decay exponentially at infinity. In the case of free electrons the momentum eigenfunctions are discontinuous at the zone boundary; thus the Wannier functions decay only as an inverse power. For instance, for the lowest band in one dimension, $v = 1$ for $-\pi/L \leqslant k \leqslant +\pi/L$ and $v = 0$ otherwise. Then

$$a(x) = L^{-\frac{1}{2}} \frac{\sin(\pi x/L)}{\pi x/L}.$$

In view of the ever-extending use of the nearly-free-electron approximation, it may be interesting to evaluate $\langle x^2 \rangle$, still for the lowest band, when a small periodic potential $V = -2V_0 \cos(2\pi x/L)$ is introduced, where $V_0 \ll \hbar^2/mL^2$. Using perturbation theory in a plane wave representation, we consider, near $k = \pi/L$, the two waves $\exp(ikx)$ and $\exp[i\{k-(2\pi x/L)\}]$. The energy of the lowest band is then given by

$$\frac{2mE}{\hbar^2} = \frac{2\pi^2}{L^2} - \frac{2\pi k_r}{L} + k_r^2 - \frac{2\pi}{L}\left(\frac{\pi}{L} - k_r\right)\left(1 + \frac{m^2 L^2 V_0^2}{\pi^2 \hbar^4 (\pi/L - k_r)^2}\right)^{\frac{1}{2}}$$

where k_r is the reduced wave vector. The momentum eigenfunction is determined by the equations

$$\left\{\frac{\hbar^2}{2m}\left(k + \frac{2\pi n}{L}\right)^2 - E\right\} v\left(k + \frac{2\pi n}{L}\right) + V_0 \left\{ v\left(k + \frac{2\pi(n-1)}{L}\right) + v\left(k + \frac{2\pi(n+1)}{L}\right) \right\} = 0 \tag{1.31}$$

so that approximately, for the lowest band,

$$v\left(k + \frac{2\pi n}{L}\right) = \frac{V_0 v\left(k + \dfrac{2\pi(n-1)}{L}\right)}{\dfrac{\hbar^2}{2m}\left(\dfrac{4\pi n k}{L} + \dfrac{4\pi^2 n^2}{L^2}\right)} \tag{1.32}$$

except near $k = \pi/L$. Here, if we write

$$\gamma = \left| \frac{\pi}{L} - k \; \frac{\pi\hbar^2}{mLV_0} \right| \tag{1.33}$$

we have, for $0 \leqslant k \leqslant \pi/L$,

$$v(\gamma) = \frac{1}{2^{\frac{1}{2}}} \frac{1}{[1+\gamma^2-\gamma(1+\gamma^2)^{\frac{1}{2}}]^{\frac{1}{2}}},$$

and, for $\pi/L \leqslant k \leqslant 2\pi/L$,

$$v(\gamma) = \frac{1}{2^{\frac{1}{2}}} \frac{-\gamma+(1+\gamma^2)^{\frac{1}{2}}}{[1+\gamma^2-\gamma(1+\gamma^2)^{\frac{1}{2}}]^{\frac{1}{2}}} \tag{1.34}$$

while formula (1.32) applies for $k > 2\pi/L$. We want

$$\int_{-\infty}^{\infty} \left(\frac{\mathrm{d}v}{\mathrm{d}k}\right)^2 \mathrm{d}k = \left(\frac{\pi\hbar^2}{mLV_0}\right)^2 2 \int_0^{\infty} \left(\frac{\mathrm{d}v}{\mathrm{d}\gamma}\right)^2 \mathrm{d}\gamma$$

(The coefficient is squared because of the changed normalization)

If V_0 is small, it is sufficient to confine the last integral only to the region where $\mathrm{d}V/\mathrm{d}\gamma$ is appreciable. Then its value is 0·1965. Thus

$$\langle x^2 \rangle = 0{\cdot}393 \frac{\pi^2\hbar^4}{m^2L^2V_0^2}.$$

If we write

$$V_0 = b\frac{\pi\hbar^2}{mL^2},$$

then

$$\langle x^2 \rangle = \frac{L^2}{b^2} 0{\cdot}393, \quad \text{or} \quad \frac{\langle x^2\rangle^{\frac{1}{2}}}{L} = \frac{0{\cdot}63}{b}.$$

For instance, if $L = 4$ atomic units, $V_0 = 0{\cdot}1$ Rydb, then $b = 0{\cdot}25$ and $\langle x^2\rangle^{\frac{1}{2}}$ is $\sim 2{\cdot}5L$. The approximation breaks down for larger values of V_0. It is clear that for very weak potentials the Wannier function spreads out over many lattice cells.

REFERENCES

ASHCROFT, N. W. (1966). *Phys. Letts.* **23**, 48.

BLOUNT, E. I. (1962). *Solid-state physics* vol. 13, p. 306. Academic Press, New York.

CALLAWAY, J. (1964). *Energy-band Theory*. Academic Press, New York.

CARDONA, M. and POLLAK, F. H. (1966). *Phys. Rev.* **142**, 530.

COHEN, M. L. and HEINE, V. (1970). *Solid-state Physics*, vol. 24, p. 38. Academic Press, New York.

CORNWELL, J. F. (1969). *Group Theory and electronic energy bands in solids.* North-Holland, Amsterdam.
COULSON, C. A. (1961). *Valence*, 2nd Edition. Oxford University Press.
DIMMOCK, J. O. (1971). *Solid-state Physics*, vol. 26, p. 104. Academic Press, New York.
DRESSELHAUS, G. and DRESSELHAUS, M. S. (1967). *Phys. Rev.* **160**, 649.
EHRENREICH, H. and HODGES, L. (1968). *Methods in computational physics*, vol. 8, p. 149. Academic Press, New York.
FLETCHER, G. C. (1971). *The electron-band theory of solids.* North-Holland, Amsterdam.
FRIEDEL, J. (1973). *J. Phys.* F**3**, 785.
HAM, F. S. (1955). *Solid-state physics*, vol. 1, p. 127. Academic Press, New York.
—— and SEGALL, B. (1961). *Phys. Rev.* **124**, 1786.
HARRISON, W. A. (1966). *The application of psuedopotentials to metals.* Benjamin, Philadelphia.
HEINE, V. (1960). *Group theory in quantum mechanics.* Pregamon Press, Oxford.
—— (1970). *Solid-state physics*, vol. 24, p. 1. Academic Press, New York.
HERMAN, F., KORTUM, R. L., KUGLIN, C. D., and SHORT, R. A. (1966). In *Quantum theory of atoms, molecules and the solid state*, p. 381. Academic Press, New York.
HERMAN, F., KORTUM, R. L., KUGLIN, C. D., VAN DYKE, J. P., and SKILLMAN, S. (1968). *Methods in computational physics*, vol. 8, p. 193. Academic Press, New York.
HERRING, C. (1940). *Phys. Rev.* **57**, 1169.
KOHN, W. and ROSTOKER, N. (1954). *Phys. Rev.* **94**, 1111.
KORRINGA, J. (1947). *Physica.'s Grav.* **13**, 392.
KOSTER, G. F. (1957). *Solid-state physics*, vol. 5, p. 174. Academic Press, New York.
LOUCKS, T. (1967). *Augmented plane wave method.* Benjamin, Philadelphia.
MATTHEISS, L. F., WOOD, J. H., and SWITENDICK, A. C. (1968). *Methods in computational physics*, vol. 8, p. 64. Academic Press, New York.
MORGAN, G. J. (1966). *Proc. phys. soc. Lond.* **89**, 365.
PHILLIPS, J. C. and SANDROCK, R. (1968). *Methods in computational physics*, vol. 8, p. 21. Academic Press, New York.
PINCHERLE, L. (1971). *Electronic energy bands in solids.* Macdonald, London.
RAIMES, S. (1961). *The wave mechanics of electrons in metals.* North-Holland, Amsterdam.
SAFFREN, M. M. and SLATER, J. C. (1953). *Phys. Rev.* **92**, 1126.
SEGALL, B. and HAM, F. S. (1968). *Methods in computational physics*, vol. 8, p. 251. Academic Press, New York.
SLATER, J. C. (1937). *Phys. Rev.* **51**, 846.
—— (1953). *Phys. Rev.* **92**, 603.
—— (1967). *Quantum theory of molecules and solids*, vol. 2. McGraw-Hill, New York.
WEINREICH, G. (1965). *Solids: Elementary theory for advanced students.* J. Wiley & Sons, New York.
WIGNER, E. P. (1959). *Group Theory and its Application to the Quantum Mechanics of Atomic Spectra* (Transl. by J. J. Griffin). Academic Press, New York.
WOODRUFF, T. O. (1957). *Solid-state physics*, vol. 4, p. 367. Academic Press, New York.
ZIMAN, J. M. (1965). *Proc. phys. Soc. Lond.* **86**, 337.
—— (1971). *Solid-state physics*, vol. 26, p. 1. Academic Press, New York.

2

THE CELLULAR METHOD FOR METALS

S. L. ALTMANN

CHAPTER 1 was concerned with a general review of the methods of calculating one-electron eigenvalues and eigenfunctions for a given crystal potential field. As discussed briefly in Section 1.4, the choice of this field is a difficult matter because adequate account has to be given of such effects as screening, correlation, and exchange. However we shall not be concerned with it in this chapter since it is a problem best treated separately; instead, we shall discuss one of the methods used to solve the eigenvalue problem in a crystal in full detail. As reviewed in Chapter 1, there are at present several such methods, of which the Augmented Plane Wave and the Korringa–Kohn–Rostoker methods are those most often used. On the other hand, the Cellular Method is one of the first techniques used in this field and it will be the subject of this chapter, the plan of which is as follows. The basic properties of crystal wavefunctions are reviewed in Section 2.1. Section 2.2 treats the principles and history of the cellular method and the boundary conditions for the wavefunction which are used are given in Section 2.3. The way in which crystal symmetry is exploited is explained in Section 2.4 and lattices with bases are discussed in Section 2.5. Sections 2.6 and 2.7 give in more detail the methods for fitting the boundary conditions and the form of the boundary conditions matrix. Section 2.8 is one that has general interest in band theory since a very detailed discussion is given in it of the general form of the wavefunction expansion when the potential is centro-symmetric. The convergence of the cellular expansion and a method to deal with non-spherical potentials are discussed in Section 2.9. Sections 2.10 to 2.12 are concerned with practical details of a cellular programme and Section 2.13 discusses tests of the method. Finally, conclusions and a critique are given in Section 2.14.

2.1. The crystal wavefunctions

We shall review briefly in this Section some properties of crystal wavefunctions which are required in the use of the cellular method.

As discussed fully in Chapter 1, the crystal orbitals are labelled by a $\mathbf{k}$ vector, and must have the Bloch form

$$\psi_{\mathbf{k}}(\mathbf{r}) = \exp(i\mathbf{k} \cdot \mathbf{r})u_{\mathbf{k}}(\mathbf{r}), \tag{2.1}$$

where $u_{\mathbf{k}}(\mathbf{r})$ is a function with the full periodicity of the lattice. That is, call

T the vector

$$\mathbf{T} = m\mathbf{a}+n\mathbf{b}+p\mathbf{c}, \tag{2.2}$$

where **a**, **b**, **c** are the unit vectors of the crystal lattice and m, n, p arbitrary integers. Then,

$$u_{\mathbf{k}}(\mathbf{r}+\mathbf{T}) = u_{\mathbf{k}}(\mathbf{r}). \tag{2.3}$$

In the limit of weak potential fields $u_{\mathbf{k}}(\mathbf{r})$ tends to a constant and the crystal orbital $\psi_{\mathbf{k}}(\mathbf{r})$ becomes a plane wave, the energy of which is $E(\mathbf{k}) = \hbar^2k^2/2m$. The problem we must solve is to determine the value of $E(\mathbf{k})$ for any crystal orbital (2.1) in a given potential field.

It will be important for future reference to understand clearly how the Bloch functions change under the translations (cf Section 1.5). Consider an operator $\bar{S}$ that transforms the coordinates $\mathbf{x}$ of a point of configuration space into new coordinates denoted with the symbol $\bar{S}\mathbf{x}$. It is well known (see Wigner 1959) that when $\bar{S}$ acts on the configuration space $\mathbf{x}$ all functions $f(\mathbf{x})$ are transformed into new functions, defined by the symbol $\bar{\bar{S}}f(\mathbf{x})$ and given by the numerical relation

$$\bar{\bar{S}}f(\mathbf{x}) = f(\bar{S}^{-1}\mathbf{x}), \tag{2.4}$$

where $\bar{S}^{-1}$ is the reciprocal of $\bar{S}$. The function space operators $\bar{\bar{S}}$ preserve the multiplication of the configuration space operators. That is, if $\bar{R}\bar{S} = \bar{U}$, then $\bar{\bar{R}}\bar{\bar{S}} = \bar{\bar{U}}$. However, it is important to remember (see Wigner 1959, p. 106) that $\bar{\bar{R}}\bar{\bar{S}}$ cannot be obtained by applying $\bar{\bar{R}}$ on both sides of eqn (2.4). This would be wrong since $\bar{\bar{R}}$ operates on functions whereas both sides of eqn (2.4) are numbers, namely the values of the functions given at the stated points.

Given a vector

$$\mathbf{R} = p\mathbf{a}+q\mathbf{b}+s\mathbf{c}, \tag{2.5}$$

with **a**, **b**, **c** as in eqn (2.2) and p, q, s any real numbers, then the corresponding configuration space operator $\bar{R}$ is defined by

$$\bar{R}\mathbf{r} = \mathbf{r}+\mathbf{R}. \tag{2.6}$$

In eqn (2.6) the so-called *active* picture is used for the configuration space operators. This means that the coordinates of all points of space are directly changed by the operation in question rather than, as in the passive picture, indirectly modified by a change of axes.

The effect of $\bar{\bar{T}}$ on the Bloch function (2.1) is obtained from eqn (2.4):

$$\bar{\bar{T}}\psi_{\mathbf{k}}(\mathbf{r}) = \psi_{\mathbf{k}}(\bar{T}^{-1}\mathbf{r}) \tag{2.7}$$

$$= \psi_{\mathbf{k}}(\mathbf{r}-\mathbf{T}) \tag{2.8}$$

$$= \exp\{i\mathbf{k}\,.\,(\mathbf{r}-\mathbf{T})\}u_{\mathbf{k}}(\mathbf{r}-\mathbf{T}) \tag{2.9}$$

$$= \exp(-i\mathbf{k}\,.\,\mathbf{T})\psi_{k}(\mathbf{r}). \tag{2.10}$$

Here in eqn (2.8), eqn (2.6) is used for $\overline{T}^{-1}$ and in eqn (2.9) eqn (2.1) has been introduced.

Attention should be paid to eqn (2.10), since the sign of the exponent appears reversed quite often; this was the case in the original paper of Bloch, but quite correctly since he used the passive picture. However, many authors have preserved the positive sign when using (as is now common practice) the active convention. This point is discussed by Altmann and Cracknell (1965).

The reader must also beware that some authors use $\bar{S}$ rather than $\bar{S}^{-1}$ on the right hand side of eqn (2.4), on the grounds that this is purely a matter of convention. That this is not so is clear from the discussion given by Wigner (1959, p. 106) and by Altmann and Bradley (1965a).

2.2. Principles and history of the cellular method

Eqn (2.10) shows that the wavefunction is fully determined throughout the crystal, once it is known inside a unit cell. In order to exploit this result in a convenient manner Wigner and Seitz (1933) introduced a special type of unit cell, which is a centred unit cell as opposed to the standard crystallographic one. (See Fig. 2.1). This centred unit cell is called in solid-state theory

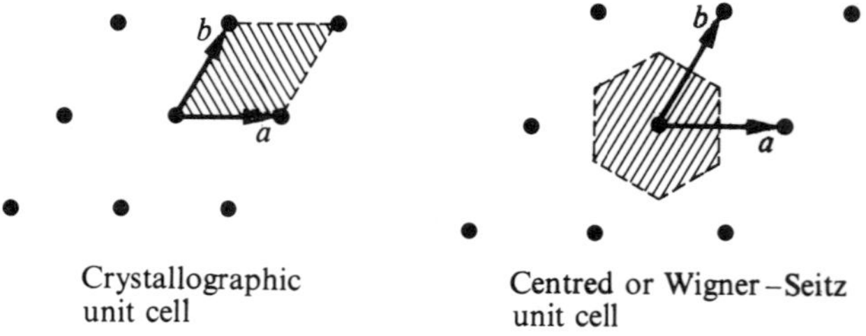

FIG. 2.1. Unit cells.

the Wigner–Seitz cell and it is defined as follows. Take a lattice point as origin and draw the perpendicular bisector planes to all the vectors that join the origin with all neighbours of first, second and higher orders. The smallest polyhedron that is circumscribed by these planes is the Wigner–Seitz cell.

Wigner and Seitz noticed that the centred unit cell does not differ drastically from a sphere for the more common metal structures. As an example, Fig. 2.2 shows the Wigner–Seitz polyhedron for the body-centred cubic lattice, which is a fairly spherical body. Wigner and Seitz were interested in the wavefunction $\psi_{\mathbf{k}}(\mathbf{r})$ for the lowest state in the crystal of sodium metal, which corresponds to $\mathbf{k} = 0$. For this value of $\mathbf{k}$, from eqn (2.10), $\psi_{\mathbf{k}}(\mathbf{r})$ is fully periodic and in order to join smoothly from cell to cell, it must satisfy the boundary condition

$$(\partial\psi/\partial r)_{\mathbf{n}} = 0 \tag{2.11}$$

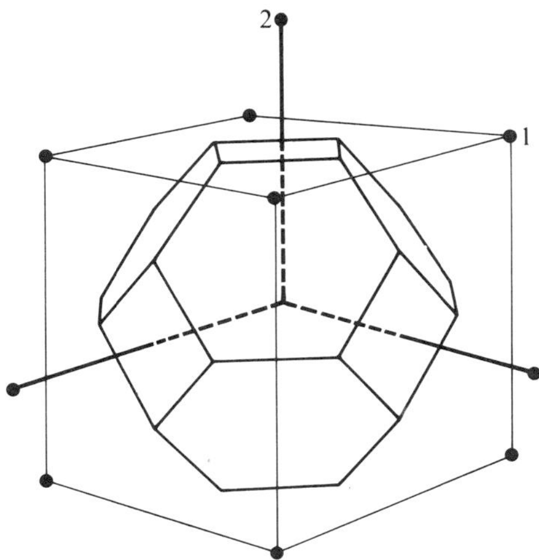

FIG. 2.2. The Wigner–Seitz cell for the body-centred cubic lattice. The hexagonal and square faces are bisectors of the radii to first and second neighbours respectively. One neighbour of each type is labelled with the corresponding number.

at the surface of the Wigner–Seitz cell, where the suffix **n** denotes the normal derivative. As an approximation, the centred unit cell was replaced by a sphere of the same volume, called the equivalent sphere, and eqn (2.11) was most simply verified on the surface of this sphere.

Very soon after the Wigner and Seitz work, Slater (1934) extended it very considerably. He took, again for sodium metal, one s-, three p-, three d- and one f-functions (as opposed to the single, totally symmetrical *s*-function of Wigner and Seitz). Slater then wrote down linear combinations of these eight functions with eight arbitrary coefficients, which he fitted by means of eight boundary conditions at the central points of the eight hexagonal faces of the Wigner–Seitz cell of Fig. 2.2.

The next important result on the cellular method was obtained by Shockley (1937) who argued that if Slater's procedure were applied for a zero potential field, the wavefunctions obtained should coincide with free-electron eigenfunctions and therefore the eigenvalues obtained should agree with the well-known free-electron eigenvalues. His results, alas, were in disagreement with the correct eigenvalues by 30 per cent or so in many cases. This test of Shockley's became known as the empty-lattice test and its results raised serious doubts about the validity of the cellular method.

During World War II a major technical advance was introduced in a calculation that was published a few years later (von der Lage and Bethe 1947). These authors showed how group theory could be used to simplify the cellular eigenfunctions, thereby permitting the use of larger sets of functions and the fitting of the boundary conditions at a greater number of boundary points than before. The main step in their work was the use of combinations of spherical harmonics, which they called kubic harmonics, that belong to given irreducible representations of the group of the **k**-vector and therefore to an irreducible representation of the space group, in accordance with the work of Bouckaert, Smoluchowski, and Wigner (1936). Von der Lage and Bethe obtained these combinations by an *ad hoc* procedure in terms of Cartesian x, y, z coordinates.

The first major advance in the cellular method after World War II came through the work done by Professor H. Jones and his collaborators at Imperial College, London. Owing to the approximate spherical symmetry of the Wigner–Seitz cell the crystal orbital $\psi_{\mathbf{k}}(\mathbf{r})$ can be written as

$$\psi_{\mathbf{k}}(\mathbf{r}) = \sum_{lm} C_{lm} R_l(r;\varepsilon) Y_l^m(\theta, \phi) \tag{2.12}$$

where the functions $R_l(r;\varepsilon)$ are solutions of the radial Schrödinger equation that correspond to the value ε of the energy parameter for the potential field of the metal (see Section 2.12). The functions $Y_l^m(\theta, \phi)$ in eqn (2.12) are spherical harmonics and the C_{lm} are coefficients to be determined through the boundary conditions. Each of the terms in the summation of eqn (2.12) is a solution for a central field potential, so that the expansion can be regarded as a Ritz variational wavefunction and it is the expression used by Slater but written in a more general form. Howarth and Jones (1952) used the kubic harmonics of von der Lage and Bethe but transcribed them into polar coordinates. They considered a small number of boundary points leading to eleven boundary conditions which they used to determine four terms in eqn (2.12) for sodium metal. In each case, they varied ε until 4 out of the 11 boundary conditions were exactly satisfied, thus leading to an energy eigenvalue E. For each eigenvalue, Howarth and Jones took seven or eight combinations of sets of four boundary conditions and they found that in most cases the results coincided within 0·01 Ryd or so, but for some particular sets of boundary conditions discrepancies of more than 0·05 Ryd appeared. To sort this out they computed the error of the empty lattice test for each set of boundary conditions, and disregarded the results given for the metal field by sets of boundary conditions that gave unduly large empty-lattice errors. The results of the other sets were averaged to obtain a final eigenvalue. Although this is a plausible scheme, there is no theoretical reason why a set of boundary conditions that is good for the empty lattice should also be good for a metal field, and the procedure described was later found to be sometimes unreliable by Ham (1954, 1955).

Except for the radial integrations that were performed on the Manchester Ferranti Mark 1 computer, all the work was done by hand and if results were obtained at all, it was due to the skill of the authors. No wonder that some numerical errors crept in, as found later on by Ham (1962). The same method was used once more for copper by Howarth (1953) but the results were discouraging: when computing s and p states the results obtained with different sets of four boundary conditions varied by as much as 100 per cent in some cases. At this time, the Augmented Plane Wave method was being developed by Professor Slater and his collaborators at MIT and Howarth himself joined this group. It was soon found that no reliable agreement could be obtained between the cellular (as then used) and the APW procedures and as a result it came to be generally believed that the cellular method was not an accurate tool.

Not even the fact that Kohn (1952) proved that Slater's scheme for point fitting of the boundary conditions in the cellular method was equivalent to a variational procedure, was enough to lift the despondency about the method and work on it was virtually stopped in the United States, such small efforts as were made continuing only in Europe. For a few years, there was an active group at the Radar Research Establishment at Malvern, U.K., under the leadership of Dr. L. Pincherle. Bell, Hum, Pincherle, Sciama, and Woodward (1953) worked on PbS and adopted a new matching technique, based on minimizing the surface integral

$$\int |\psi(L)-\psi(S)|^2 \, \mathrm{d}\sigma \tag{2.13}$$

on the common face between the lead and sulphur cells for which the wave-functions are $\psi(L)$ and $\psi(S)$ respectively.

Bell (1954) extended considerably the work of von der Lage and Bethe on kubic harmonics. Using essentially their original method and still working in cartesian coordinates, she formed symmetry-adapted combinations of harmonics for several space groups, for which the name of lattice harmonics was later proposed. However, her results go up to $l = 6$ only and the work on the hexagonal close-packed lattice contains some errors.

Jenkins and Pincherle (1954) proposed a very interesting scheme for matching the boundary conditions based on the variational method of Kohn (1952). They devised a practical and fully variational method in which the matching is done through the minimization of a surface integral. This method was applied by Jenkins (1956) to Si.

At the same time, some work continued at Imperial College. Schiff (1955) studied the hexagonal close-packed lattice, which is a rather difficult case, since there are two atoms per unit cell, and some of the expansions that he used were erroneous (Altmann 1956, Schiff 1956). However, Schiff made a very useful contribution to the cellular method. He suggested that, rather than fitting the n coefficients in eqn (2.12) by means of n boundary conditions, a much larger number p of boundary conditions should be used and the coefficients found by a least-squares method.

The next stages of the development of the cellular method took place at the Mathematical Institute, Oxford. From the computational point of view the use of spherical harmonics in spherical coordinates in eqn (2.12) is much to be preferred to that of the kubic harmonics in cartesian form, since standard recurrence relations for them can easily be used. However, the symmetrization procedures of von der Lage and Bethe and of Bell were based on cartesian coordinates and no general procedure was available to obtain lattice harmonics in spherical coordinates. Altmann (1957) gave a general method to symmetrize spherical harmonics in polar coordinates for all point groups and he then showed (Altmann 1958*a*) how this method could be extended to space groups. The method also permitted the use of much higher values of l than so far possible.

More importantly, Altmann (1958*b*) devised a reliable method to fit the boundary conditions, which removed the difficulties that had so far bedevilled the cellular method. Following on Schiff's least-squares procedure, Altmann recognized that the normalization condition for the wavefunction was incorrectly handled and showed that if it was treated as a constraint to be exactly fitted reliable results were consistently obtained, (see Section 2.3), which was confirmed by Pincherle (1958).

Very soon afterwards, Saffren (1959) proved the important result that the expansion (2.12) converges for a class of muffin-tin potentials. This work, although widely discussed, was not published and although it is clear that Saffren's result was well understood by some people, it was ignored by many. A full discussion of it is given in Section 2.9.

During the next decade, Altmann and his collaborators at Oxford developed and applied the method extensively. Group theoretical methods to obtain lattice harmonics were given by Altmann and Bradley (1963*a,b*), Altmann and Cracknell (1965), and Altmann and Bradley (1965*a*). The method was used for Ti by Altmann and Cohan (1958), for Zr by Altmann and Bradley (1964), for Ca by Altmann and Cracknell (1964) and Altmann, Harford, and Blake (1971) and for the hexagonal close-packed metals Sc, Ti, Y, Zr by Altmann and Bradley (1965*b*). Empty lattice and other tests of the method were given by Altmann, Davies, and Harford (1968). (See Section 2.13.)

During this period, some interesting work was carried out in Sweden and the U.S.S.R. on the original Wigner–Seitz form of the cellular method, mainly in order to study the behaviour of metal lattices under pressure. In Stockholm, this work was conducted by Fröman and his collaborators and is reported by Fröman and Berggren (1965) and Berggren and Fröman (1969). Following on this work, Arbman (1968, 1969) developed a scheme to estimate the error introduced by the spherical approximation to the Wigner–Seitz polyhedrons and thus correct the eigenvalues. In the Soviet Union, much work along this line has been done by Gandel'man and his collaborators,

Gandel'man (1963, 1967), Voropinov Gandel'man, and Podval'nyi (1970) as well as by Arkhipov (1966, 1971).

2.3. Boundary conditions

In accordance to the prescription given in Section 2.2, the faces of the Wigner–Seitz cell are perpendicular bisectors of vectors of the lattice **T** (see eqn (2.2)). It is a well-known property of translation lattices that, for any vector **T** of the lattice, the corresponding vector $-\mathbf{T}$ also belongs to the lattice. Therefore, the faces of the Wigner–Seitz cell come in pairs such that the perpendicular distance between two pairs of faces is a vector **T** of the lattice. The Wigner–Seitz cell of Fig. 2.1 is shown in Fig. 2.3 and it can be seen

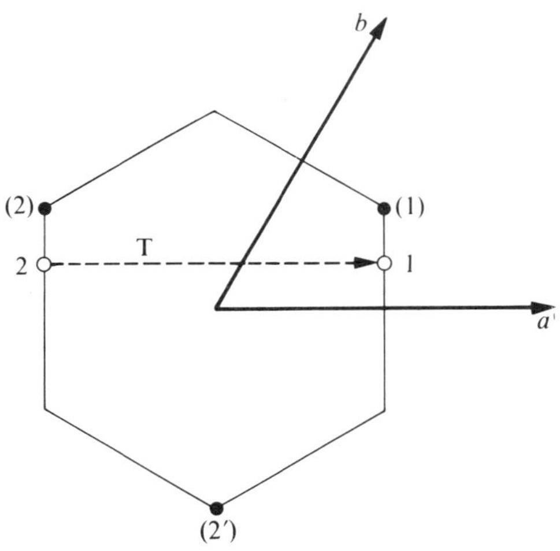

FIG. 2.3. Boundary conditions.

that there are three pairs of faces that differ by the vectors **a**, **b** and $(-\mathbf{a}+\mathbf{b})$ respectively. We now apply eqns (2.8) and (2.10)

$$\psi_{\mathbf{k}}(\mathbf{r}-\mathbf{T}) = \exp(-\mathrm{i}\mathbf{k}\,.\,\mathbf{T})\psi_{\mathbf{k}}(\mathbf{r}). \tag{2.14}$$

As an example, if we identify **r** with the point 1 of Fig. 2.3 and **T** with **a**, then $\mathbf{r}-\mathbf{T}$ denotes the point 2:

$$\psi_{\mathbf{k}}(\mathbf{r}_2) = \exp\{\mathrm{i}\mathbf{k}\,.\,(\mathbf{r}_2-\mathbf{r}_1)\}\psi_{\mathbf{k}}(\mathbf{r}_1). \tag{2.15}$$

Correspondingly, for the outward normal derivative to the face, $\nabla_{\mathbf{n}}$, we have

$$\nabla_{\mathbf{n}}\psi_{\mathbf{k}}(\mathbf{r}_2) = -\exp\{i\mathbf{k}\,.\,(\mathbf{r}_2-\mathbf{r}_1)\}\nabla_{\mathbf{n}}\psi_{\mathbf{k}}(\mathbf{r}_1), \tag{2.16}$$

where the negative sign arises because the direction of the outward normal of one face is opposite to that of the other.

From eqn (2.12) we write

$$\psi_{\mathbf{k}}(\mathbf{r}_1) = \sum_{lm} C_{lm}R_l(r_1\,;\varepsilon)Y_l^m(\theta_1\phi_1), \tag{2.17}$$

and a similar expression for $\psi_{\mathbf{k}}(\mathbf{r}_2)$. Therefore (2.15) takes the form

$$\sum_{lm} C_{lm}[R_l(r_2\,;\varepsilon)Y_l^m(\theta_2\phi_2)-\exp\{i\mathbf{k}\,.\,(\mathbf{r}_2-\mathbf{r}_1)\}R_l(r_1\,;\varepsilon)Y_l^m(\theta_1\phi_1)] = 0, \tag{2.18}$$

and similarly for (2.16). The term in square brackets in eqn (2.18) is a numerical coefficient most easily computed and the C_{lm} are unknowns to be determined. In practice, of course, the infinite expansion (2.17) is truncated and only n terms are retained. To simplify the work that follows we shall write $x_r(r = 1, 2, ..., n)$ for the n unknown coefficients C_{lm} and A_{sr} for the known coefficients in eqn (2.18) and similar, which will take the form

$$\sum_{r=1}^{n} A_{sr}x_r = 0. \tag{2.19}$$

Here s is a label that denotes one particular boundary condition equation. The least-squares procedure will be based on writing p equations of the form (2.19) for a large number of pairs of boundary points, so that $p > n$. These equations will not be fitted exactly but they will rather take the form

$$\sum_{r=1}^{n} A_{sr}x_r = \delta_s, \tag{2.20}$$

where δ_s is a small residual, and the least-squares condition required is the minimization of the residual

$$S^2 = \sum_{s=1}^{p} \delta_s^*\delta_s, \tag{2.21}$$

where the asterisk denotes the complex conjugate (remember that both x_r and A_{sr} are in general complex numbers).

The normalization condition of eqn (2.17) has to be considered, for which an approximation is introduced. The wavefunction has to be normalized within the Wigner–Seitz cell, which is approximated by the equivalent sphere, and the radial wavefunctions $R_l(r\,;\varepsilon)$ as well as the spherical harmonics are normalized within its volume. The normalization condition for eqn (2.17) is therefore

$$\sum_{lm} C_{lm}^*C_{lm} = 1 \tag{2.22}$$

or, in the present notation,

$$\sum_{r=1}^{n} x_r^* x_r = 1. \tag{2.23}$$

This condition has to be verified exactly and it is a constraint subject to which (2.21) has to be minimized.

In order to handle the work that follows matrix notation will be used: $\mathbf{x}$ will be the n-dimensional column vector of the unknown coefficients x_r and A the matrix of the coefficients A_{sr}. The adjoint (conjugate of the transposed) vectors and matrices will be denoted with $\mathbf{x}^\dagger$ and $A^\dagger$ respectively. Then eqn (2.23) takes the form

$$\mathbf{x}^\dagger \mathbf{x} = 1. \tag{2.24}$$

From eqns (2.21) and (2.20),

$$S^2 = \sum_{s=1}^{p} \sum_{r=1}^{n} \sum_{t=1}^{n} A_{sr}^* x_r^* A_{st} x_t \tag{2.25}$$

$$= \sum_{r,t=1}^{n} \sum_{s=1}^{p} A_{rs}^\dagger A_{st} x_r^* x_t \tag{2.26}$$

$$= \sum_{r,t=1}^{n} x_r^* B_{rt} x_t \tag{2.27}$$

$$= \mathbf{x}^\dagger B \mathbf{x}, \tag{2.28}$$

where

$$B = A^\dagger A \tag{2.29}$$

is a square $n \times n$ matrix easily obtained from the rectangular matrix A (n columns by p rows) of the boundary conditions coefficients.

The minimization condition of S^2 is, from eqn (2.27),

$$\partial S^2/\partial x_u = \sum_{r=1}^{n} B_{ru} x_r^* = 0. \tag{2.30}$$

The constraint (2.23) has to be handled by the well-known method of Lagrangian multipliers, that is, eqn (2.23) is multiplied by a constant λ and differentiated with respect to u:

$$\lambda x_u^* = 0 \tag{2.31}$$

which can be written as

$$\sum_{r=1}^{n} \lambda \delta_{ru} x_r^* = 0, \tag{2.32}$$

where δ_{ur} is a standard Kronecker delta. From eqns (2.30) and (2.31),

$$\sum_{r=1}^{n} x_r^*(B_{ru}-\lambda\delta_{ru}) = 0, \tag{2.33}$$

with equations of the same form for $u = 1, \dots, n$. Eqn (2.33) is therefore a homogeneous system of n equations in n unknowns, the compatibility condition of which is

$$\det(B-\lambda\mathbf{1}) = 0, \tag{2.34}$$

where $\mathbf{1}$ is the $n \times n$ unit matrix. This is the well-known condition that gives the n eigenvalues $\lambda_1, \dots, \lambda_n$ of the matrix B. The purpose of the fitting is, of course, to obtain the minimum value of S^2. In order to see how this is achieved, write eqn (2.33) as follows,

$$\mathbf{x}^\dagger B = \lambda\mathbf{x}^\dagger \tag{2.35}$$

and introduce the left-hand side of this equation in (2.28):

$$S^2 = \lambda\mathbf{x}^\dagger\mathbf{x} \tag{2.36}$$

which, on using eqn (2.24), gives

$$S^2 = \lambda. \tag{2.37}$$

Therefore, the minimum value of the residual S^2 will be given by the lowest eigenvalue of the matrix B. The coefficients x_r will be the components of the eigenvector of B that corresponds to its lowest eigenvalue.

In order to complete the work, it has to be remembered that the matrix A and therefore also B depend on the energy parameter ε. The best fitting of the boundary conditions in the least-squares sense, subject to the normalization condition, will therefore be obtained by computing B as a function of ε and determining in each case its lowest eigenvalue $\lambda = S^2$. This will give the minimized S^2 as a function of the energy parameter ε and the successive minima of this function will be the energy eigenvalues corresponding to $\psi_{\mathbf{k}}(\mathbf{r})$ in the successive bands. Since B is positive definite, so are its eigenvalues, whence the residual curve is bounded below by zero. As an example, the curve obtained by Altmann and Bradley (1965*b*) for the irreducible representation A'_{1+} of the point Γ of the Brillouin zone is shown in Fig. 2.4. The three minima correspond successively to the first, second, and third bands.

The method given above is the one introduced by Altmann (1958*b*) and it is useful to comment on the relation between this method and those that preceded it.

Its immediate predecessor is the least-squares procedure of Schiff (1955). In this method the normalization condition (2.22) was not explicitly used. Following Howarth and Jones (1952), Schiff employed a procedure that is often used in performing variational calculations. Condition (2.22) establishes

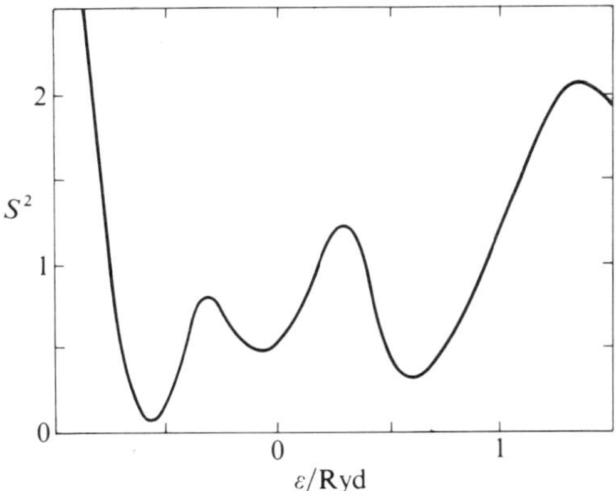

FIG. 2.4. Minima corresponding to the three lower eigenvalues of symmetry $\Gamma A'_{1+}$ for zirconium. (After Altmann and Bradley 1965*b*.)

a relation between the n coefficients so that, in principle, one of them could be eliminated in terms of the others and the minimization carried out. This would give the correct answer in the boundary conditions fitting, although it is rather more laborious than the method described above. On the other hand, since only $n-1$ coefficients are independent, the nth one can be chosen arbitrarily. Thus Schiff took the first coefficient in (2.17) equal to unity, an expedient that works correctly when the coefficients are exactly determined but which fails when they are obtained in the least-squares sense. This follows at once from the theory of the least-squares method with a constraint but it might be useful to illustrate the origin of this difficulty from first principles. When one of the coefficients, say x_j, in eqn (2.19) is taken as unity, we are really dividing both sides of this equation by x_j. However, this equation is not the one that is truly obeyed. Rather, eqn (2.20) is the one that is satisfied and this equation takes the form

$$\sum_r A_{sr} x_r / x_j = \delta_s / x_j, \tag{2.38}$$

which shows that the set of new equations obtained depends on the choice of x_j so that the answer cannot be unique. If one is dealing with n equations for the n unknowns x_r, as Howarth and Jones (1952) did, all the δs in eqn (2.38) vanish and the method could in principle work. However, it must be remembered that Howarth and Jones had p boundary conditions with $p > n$ and fitted them by forming a number of sets of systems of $n \times n$ equations and then averaging. Therefore, the coefficients that correspond to this average cannot satisfy eqn (2.19) but rather eqn (2.20) and there is no guarantee that their normalization is correct.

That the proper normalization procedure of the wavefunction is crucial was demonstrated by Altmann (1958*b*) who obtained the results shown in

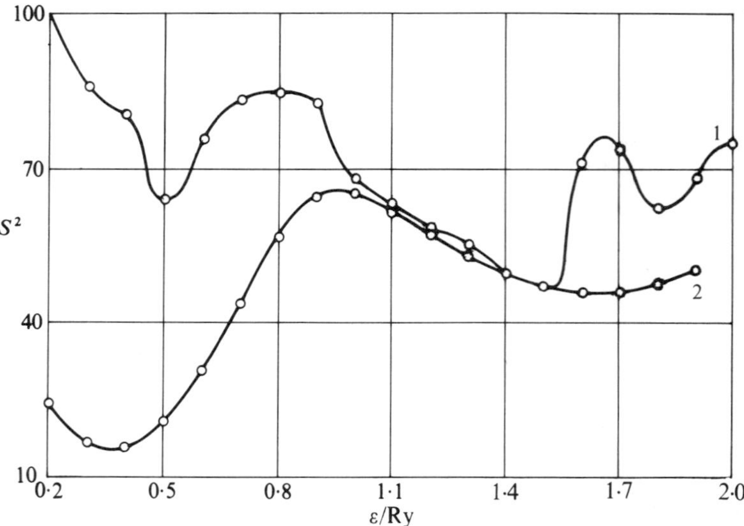

FIG. 2.5. Comparison of methods of fitting the boundary conditions. Curve 1, Schiff's method. Curve 2, Altmann's method. (After Altmann 1958*b*.)

Fig. 2.5, with a potential field for Zr metal. Curves 1 and 2 were obtained respectively by Schiff's and Altmann's methods. Curve 2 shows the minima corresponding to the first and second eigenvalues, whereas the minima in curve 2 are misplaced, difficult to localize and even spurious.

2.4. The use of symmetry

The theory of the space groups given by Seitz (1936) and Bouckaert, Smoluchowski, and Wigner (1936) is well known. Reviews of it have been given by Koster (1957) and Johnston (1960) and a very complete treatment of the theory can be found in the book by Bradley and Cracknell (1972). In this section a brief account will be given of the major results of the theory that directly concern the use of the cellular method.

An irreducible representation of a space group **G** is fully characterized by two labels, which are the **k** vector and the small representation. The latter is defined as follows. Given **k**, the subgroup of operations of the space group is formed that leave **k** invariant or transform it into an equivalent vector. This subgroup is called the group of the **k** vector or little group $\mathbf{G}^{\mathbf{k}}$ and the small representation is one of its irreducible representations. The group of the **k** vector contains the translation group $\mathbf{\Gamma}$ as a subgroup and, when the space group is symmorphic, that is, when it does not contain screw axes or glide planes, all operations of $\mathbf{G}^{\mathbf{k}}$ are products of a translation of $\mathbf{\Gamma}$ times a point group operation $\bar{\alpha}$. The main simplification that is achieved by using group theory is the following one. Since $\psi_{\mathbf{k}}(\mathbf{r})$ in eqn (2.12) is an eigenfunction of the Schrödinger equation, it must belong to an irreducible representation of the space group $\mathbf{G}$ and therefore, in accordance to the results just quoted, to an irreducible representation of $\mathbf{G}^{\mathbf{k}}$. This means that the spherical harmonics $Y_l^m(\theta\phi)$ must form a basis for a small representation or, in other words, that they must be symmetry adapted to $\mathbf{G}^{\mathbf{k}}$.

The technique to form symmetry adapted functions of a group $\mathbf{H}$ of operations h is well known (see Wigner 1959). Call $D^i(h)$, $h \in \mathbf{H}$, the ith irreducible representation of $\mathbf{H}$. The projection operator

$$\sum_{h \in \mathbf{H}} D^i(h)^*_{jj} h \tag{2.39}$$

is such that, when applied on any arbitrary function f, which is called the generator, it will generate zero or a function belonging to the jth column of the ith irreducible representation of $\mathbf{H}$.

For a space group, the label i is double, $\mathbf{k}$, s say, since it depends on the $\mathbf{k}$ vector and the small representation. For a symmorphic space group the expansion (2.39) will take the form

$$\sum_{\Gamma} D^{\mathbf{k}}(\bar{\bar{T}})^* \bar{\bar{T}} \sum_{\mathbf{G}^{\mathbf{k}}} D^{\mathbf{k},s}(\bar{\bar{\alpha}})^*_{jj}\, \bar{\bar{\alpha}} Y_l^m(\theta\phi) = X_{l\nu}^{\mathbf{k}s,j}. \tag{2.40}$$

In eqn (2.40) the effect of the point group operations $\bar{\bar{\alpha}}$ on $Y_l^m(\theta\phi)$ can be obtained by using the representations of the rotation group. These are cumbersome and various short cuts are necessary, which have been fully described by Altmann (1957) and Altmann and Bradley (1963*a*,*b*). The summation over $\mathbf{G}^{\mathbf{k}}$ in eqn (2.40) gives a spherical harmonic of order l, which can be called ${}^{\mathbf{o}}X_{l\nu}^{\mathbf{k}s,j}(\theta\phi)$, adapted to the jth column of the small representation s. The label ν is a serial label that distinguishes the various combinations that belong to the same symmetry type. The left superscript indicates that the polar angles θ, ϕ are measured around the origin of the crystal lattice. Define ${}^{\mathbf{T}}X_{l\nu}^{\mathbf{k}s,j}$ as a function identical with ${}^{\mathbf{o}}X_{ly}^{\mathbf{k}s,j}$, except that it is centred around the lattice point at the end of the vector $\mathbf{T}$ of the lattice. Then $\bar{\bar{T}}\,{}^{\mathbf{o}}X_{l\nu}^{\mathbf{k}s,j}$ has a very simple meaning:

$$\mathbf{T}\,{}^{\mathbf{o}}X_{l\nu}^{\mathbf{k}s,j} = {}^{\mathbf{T}}X_{l\nu}^{\mathbf{k}s,j}. \tag{2.41}$$

Finally, $D^k(\bar{\bar{T}})$ in eqn (2.40) can be obtained from eqn (2.10), and the final form of eqn (2.40) is

$$X_{l\nu}^{\mathbf{k}s,j} = \sum_{\mathbf{T} \in \Gamma} \exp(\mathrm{i}\mathbf{k}\,.\,\mathbf{T})\ {}^{\mathbf{T}}X_{l\nu}^{\mathbf{k}s,j}(\theta\phi). \tag{2.42}$$

This multi-centred expansion is a lattice harmonic and it is clear that it is fully given as long as ${}^{\mathbf{o}}X_{l\nu}^{\mathbf{k}s,j}(\theta\phi)$ is determined. Therefore, for symmorphic groups, it is enough to list the expansions of these functions. They are listed for the cubic groups by Altmann and Cracknell (1965) and for the hexagonal close-packed lattice by Altmann and Bradley (1965*a*). A few misprints in these papers are corrected in the Appendix.

When the lattice harmonics are used for points of symmetry in **k** space (that is points for which $\mathbf{G}^{\mathbf{k}}$ is larger than the identity), eqn (2.12) takes the form

$$\psi^{\mathbf{k}s,j} = \sum_{l\nu} C_{l\nu} R_l(r, \varepsilon)\ {}^{\mathrm{o}}X_{l\nu}^{\mathbf{k}s,j}(\theta\phi). \tag{2.43}$$

An important point to bear in mind is that, whereas the $Y_l^m(\theta\phi)$ are fully orthogonal, the lattice harmonics as defined above need not be so for different ν. This can cause trouble but, fortunately, Altmann and Bradley (1963*b*) have given a method to make these functions fully orthogonal for different ν as well as different l.

The main effect of the use of lattice harmonics is a reduction in the size of the expansion (2.12). As an example, in order to keep all spherical harmonics up to and including $l = 8$, this expansion must contain 81 terms. On the other hand, if the totally symmetrical representation for $\mathbf{k} = 0$ for the cubic lattice is computed on using lattice harmonics, only the following 8 harmonics appear in the expansion

$$Y_0^0,\ Y_4^{0c},\ Y_4^{4c},\ Y_6^{0c},\ Y_6^{4c},\ Y_8^{0c},\ Y_8^{4c},\ Y_8^{8c}, \tag{2.44}$$

where the superscript c denotes the real part of Y_l^m.

2.5. Lattices with bases

The cellular expansion can easily be extended for crystals with more than one atom per unit cell. As an example, the hexagonal close-packed lattice is depicted in Fig. 2.6. In (a) the disposition of the layers is shown, the full

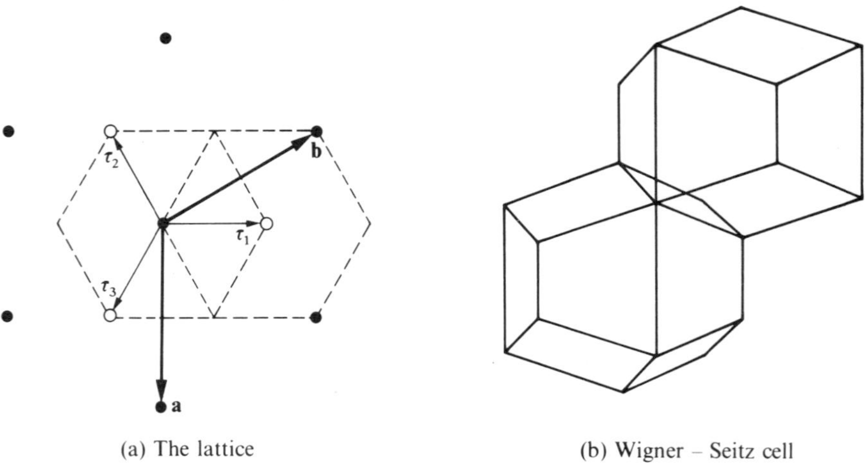

(a) The lattice (b) Wigner – Seitz cell

FIG. 2.6. Hexagonal close packed lattice.

circles denoting the atoms of the first and third layers and the open circles those of the second layer. The crystallographic unit cell is the prism with the unit vectors **a**, **b**, **c** as edges, where **a**, **b** are shown in the figure and **c** is perpendicular to their plane, going from the first to the third layer. The vectors that join the origin of the first layer with the near neighbours of the second one are the $\boldsymbol{\tau}$ vectors shown in the picture. The crystallographic unit cell contains two atoms, one at the origin and one at $\boldsymbol{\tau}_1$. The Wigner–Seitz cell is obtained by a simple extension of the prescription given in Section 2.2, by forming perpendicular bisectors of lattice vectors centred at both atoms of the crystallographic unit cell, as indicated with the broken lines in Fig. 2.6(a), which correspond to the faces shown in full in Fig. 2.6(b). The Wigner–Seitz cell consists of two half-cells centred around each of the atoms of the basis. The half-cells have two types of faces, rhombic and trapezoidal respectively, and they have one rhombic face in common, the centre of which is the mid-point of $\boldsymbol{\tau}_1$, which is a centre of inversion. Thus, one half-cell is obtained from the other by inversion at the centre of inversion.

The cellular expansion (2.12) can easily be extended for a lattice such as the one just described. We must use two centres of coordinates around the first and second atoms of the basis respectively and the corresponding quantities will be labelled with left superscripts 1 and 2 respectively. Since both atoms are related through a centre of inversion, it is convenient to choose the axes **x**, **y**, **z** around the second atom inverted with respect to those of the first atom, in order to avoid a factor of $(-1)^l$ when transforming under inversion spherical harmonics from one to the other atom. To indicate the fact that the coordinate axes are thus inverted, all coordinates around atom 2 will be marked with a tilde (eg. $\tilde{\mathbf{r}}$). With this notation, the cellular expansion takes the form

$$\psi_{\mathbf{k}}(\mathbf{r}) = \sum_{lm} 1C_{lm}\delta(\mathbf{r}, {}^1\mathbf{r})R_l({}^1\mathbf{r}, \varepsilon)Y_l^m({}^1\theta, {}^1\phi)$$

$$\sum_{l'm'} {}^2C_{lm}\delta(\mathbf{r}, {}^2\tilde{\mathbf{r}})R_l({}^2\tilde{\mathbf{r}}, \varepsilon)Y_l^m({}^2\tilde{\theta}, {}^2\tilde{\phi}), \tag{2.45}$$

where $\delta(\mathbf{r}, {}^1\mathbf{r})$ equals unity when **r** belongs to the first half-cell and zero otherwise, and similarly for $\delta(\mathbf{r}, {}^2\tilde{\mathbf{r}})$. In this way, the first and second summations in eqn (2.45) are used for points of the first and second cells respectively.

The boundary conditions are obtained exactly as before, except that there is a continuity condition that must be verified on the rhombic interface:

$$\sum_{lm} {}^1C_{lm}\delta(\mathbf{r}, {}^1\mathbf{r})R_l({}^1\mathbf{r}, \varepsilon)Y_l^m({}^1\theta, {}^1\phi)$$

$$= \sum_{lm} {}^2C_{lm}\delta(\mathbf{r}, {}^2\tilde{\mathbf{r}})R_l({}^2\tilde{\mathbf{r}}, \varepsilon)Y_l^m({}^2\tilde{\theta}, {}^2\tilde{\phi}), \tag{2.46}$$

where the coordinates on both sides refer to the same point in space.

The technique to obtain the lattice harmonics is very similar to the one described in section 2.4 and it is fully described in Altmann and Bradley (1965*a*).

2.6. Some practical details of the boundary conditions fitting

The boundary conditions given in section 2.3 are two-point boundary conditions in the sense that they simultaneously involve the coordinates of a point $\mathbf{r}_1$ on the surface of the Wigner–Seitz cell and of another point $\mathbf{r}_2$ on the same surface, which we call the companion point, chosen to differ from $\mathbf{r}_1$ by a vector of the crystal lattice.

It is possible to apply a point group operation of the crystal lattice to transform $\mathbf{r}_2$ either into $\mathbf{r}_1$ or into a point that has the same polar coordinate θ as $\mathbf{r}_1$. This procedure requires the specification of the irreducible representation matrices in detail, but it reduces the store space required. For this reason, it was used by Altmann (1958*a*) but with modern computers with large stores it is a technique no longer to be recommended since it makes the program more complicated and therefore slower.

Another store-saving device that was used in the early work just quoted is based on the fact that if the basic boundary point is chosen along an edge of the Wigner–Seitz cell, more than one companion point can be associated to it. This can be seen in Fig. 2.3 for the points given in brackets, 2 and 2′ being both adequate companion points of 1. As a result of this situation it is possible to write down two boundary conditions for a single basic point. Again, it is no longer worthwhile to exploit this property, since it is necessary to make a distinction between various types of points on the Wigner–Seitz surface, which complicates the program, whereas the saving of store space effected is no longer of interest.

In summary, it is best with modern computers to stick to the very simple form of the two-point boundary conditions given in Section 2.3 and to choose all the boundary points outside the cell edges and lines or points of symmetry of the faces, a simple precaution which will avoid singularities arising from zeros of the polar coordinates.

The next point to bear in mind is that some of the boundary conditions (2.15) and (2.16) can be trivial for some points of symmetry in $\mathbf{k}$ space, that is points so disposed in the Brillouin zone that their symmetry groups are larger than the identity. This is a point of some importance that will now be discussed.

If the boundary points are chosen in accordance to the prescription given above, the only symmetry operation that can transform a boundary point into its companion point is a plane of symmetry $\boldsymbol{\sigma}$. (Remember that boundary points are always chosen away from lines or points of symmetry.) This is illustrated in Fig. 2.7. The vector $\mathbf{T}$, of course, has to be perpendicular to

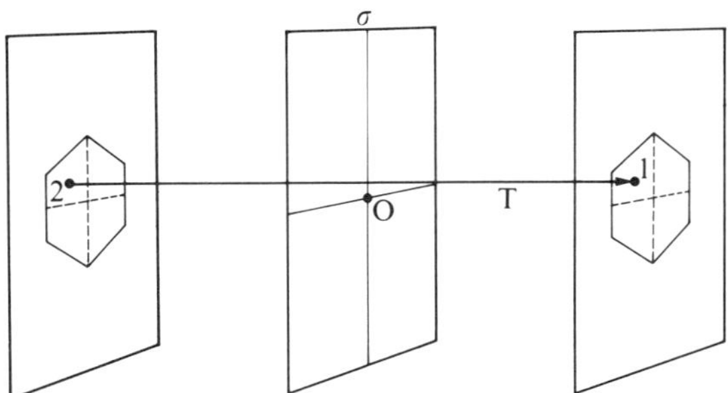

FIG. 2.7. Relation between a point and its companion.

the pair of faces of the Wigner–Seitz cell and its bisector plane can or cannot be a symmetry plane. For the Wigner–Seitz cell of the body-centred cubic lattice, for instance, the first and second alternatives obtain respectively for the square and hexagonal faces shown in Fig. 2.2.

We can now state the necessary (but not sufficient) conditions that must be simultaneously satisfied for a boundary condition to become trivial:

(1) The perpendicular bisector plane to the vector **T** that links a point with its companion is a plane of symmetry $\boldsymbol{\sigma}$ of the lattice.
(2) $\boldsymbol{\sigma}$ belongs to the group of **k**, $\mathbf{G}^{\mathbf{k}}$.
(3) The matrix $D^{\mathbf{k}s}(\boldsymbol{\sigma})$ is a diagonal matrix with ± 1 as diagonal elements.
(4) $\exp(-\mathrm{i}\mathbf{k}\,.\,\mathbf{T}) = \pm 1$.

As an example, take $\mathbf{k} = 0$ (point Γ of the Brillouin zone) for the body-centred cubic lattice, the Wigner–Seitz cell of which is given in Fig. 2.2. For this value of **k**, $\exp(-\mathrm{i}\mathbf{k}\,.\,\mathbf{T}) = 1$ and condition (4) is satisfied. Take also the boundary point on one of the square surfaces, for which (1) is satisfied. Since all point group operations of the lattice belong to $\mathbf{G}^{\mathbf{k}}$ for $\mathbf{k} = 0$, condition (2) is satisfied. Take the totally symmetrical representation of $\mathbf{G}^{\mathbf{k}}$, for which, obviously $D(\boldsymbol{\sigma}) = 1$, and condition (3) is satisfied.

Because the lattice harmonic in eqn (2.43) is adapted to the irreducible representation of $G^{\mathbf{k}}$, the action of $\bar{\bar{\boldsymbol{\sigma}}}$ on $X_{l\nu}^{\mathbf{k}s,j}(\theta\phi)$ must be to multiply it by $+1$. On the other hand, if 1 and 2 are the points so named in Fig. 2.7, $\bar{\bar{\boldsymbol{\sigma}}}$ transforms $X_{l\nu}^{\mathbf{k}s,j}(\theta_1\phi_1)$ into $X_{l\nu}^{\mathbf{k}s,j}(\theta_2\phi_2)$. Therefore,

$$X_{l\nu}^{\mathbf{k}s,j}(\theta_1\phi_1) = X_{l\nu}^{\mathbf{k}s,j}(\theta_2\phi_2). \tag{2.47}$$

Also,

$$R_l(r_1\,;\varepsilon) = R_l(r_2\,;\varepsilon), \tag{2.48}$$

since r_1 and r_2, being measured from 0 in Fig. 2.7, are equal. When eqn (2.43) is used, the lattice harmonics must be substituted for the spherical harmonics in the boundary condition (2.18), which, with the value of the exponential in

question, will take the form

$$\sum_{l\nu} C_{l\nu}\{R_l(r_2\,;\varepsilon)X_{l\nu}^{\mathbf{k}s,j}(\theta_2\phi_2)-R_l(r_1\,;\varepsilon)X_{l\nu}^{\mathbf{k}s,j}(\theta_1\phi_1)\} = 0, \tag{2.49}$$

an equation that becomes a trivial identity when eqns (2.47) and (2.48) are used. However, owing to the change in sign in going from eqn (2.15) to eqn (2.16), the same pair of points yields in this case a non-trivial condition for the derivative of the wavefunction.

Two different strategies can be adopted in dealing with the trivial boundary conditions. One is to ignore the problem entirely, which makes the program very simple but lengthens somewhat the work, owing to the working out by the computer of a fairly large number of trivial, $0 = 0$, equations. However, this will only be the case for **k** values of fairly high symmetry and it does not impair the accuracy of the results. The other strategy is to get the program to rule out all the trivial boundary conditions, which can be a very useful alternative, but only if enough care is taken in programming it so that the search for the triviality condition does not take longer than the writing of the corresponding trivial equation. This can be achieved if enough store space is available so that tags to indicate triviality can be coded and stored for all irreducible representations for each boundary point, thus ensuring that a minimum amount of work has to be done when the boundary condition matrix is formed.

Besides the question of the possible triviality of the boundary conditions, that of their possible redundancy must be considered. This means that two boundary conditions for two different pairs of points can be identical. In early work (Altmann 1958*a*; Altmann and Bradley 1965*b*) redundancy was exploited in order to fit the boundary conditions on a part rather than the whole of the Wigner–Seitz cell, thus saving store space. This is now no longer required, but the problem of redundancy must be borne in mind in order to choose a set of points distributed over the whole surface of the Wigner–Seitz cell such that no two boundary conditions (except possibly for trivial ones) be identical.

Consider a pair of boundary points 1, 2. A symmetry operation $\bar{S}$ will transform them into another pair 1′, 2′, also on the surface of the Wigner–Seitz cell. If in the irreducible representation being computed $D(\bar{S})$ happens to be diagonal, the boundary condition equation for 1′, 2′ will be the same as that for 1, 2.

The highest number of redundancies will occur for the totally symmetrical representation for $\mathbf{k} = 0$, which has the group $\mathbf{G}^0$ of the highest order h. Call the basic domain of the Wigner–Seitz cell a region of its surface equal to $1/h$ of the total surface and such that the whole surface of the polyhedron can be obtained by applying the operations of $\mathbf{G}^0$ on the region. In Fig. 2.8,

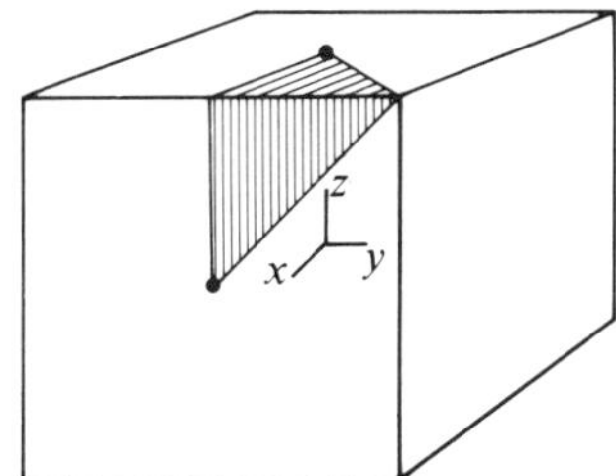

FIG. 2.8. The basic domain of the Wigner–Seitz cell for the simple cubic lattice.

for example the basic domain of the simple cubic lattice is illustrated, the area of which is $\frac{1}{48}$ of that of the total cell.

Given a point 1 in the basic domain and its companion point 2 elsewhere, each of the h operations of $\mathbf{G}^0$ converts this pair into another, 1′, 2′, such that, if the boundary conditions for the totally symmetrical representation of $\mathbf{k} = 0$ are satisfied for the points 1, 2, they are identically satisfied for the points 1′, 2′.

In order to avoid redundancies the following procedure should be adopted, which generates sets of boundary points for which no redundancies exist for the totally symmetric representation for $\mathbf{k} = 0$ and therefore neither for any other representation of any value of $\mathbf{k}$.

Form a suitable grid (see section 2.11) on each face of the basic domain and number the points from 1 to N and the operations of $\mathbf{G}^0$ from 1 to h. Transfer away each successive point of the grid, from 1 to h, by means of the h successive operations of $\mathbf{G}^0$, do the same for the points numbered from $h+1$ to $2h+1$ and go on in the same manner until all the points of the basic domain have been used. The grid of points generated by this prescription covers the whole of the surface of the Wigner–Seitz cell and it is such that no two points of this grid are related by any point group operation, being therefore entirely free of redundancies.

Once the points of the above grid are determined, their companion points must be constructed. It will then be seen that in order to ensure that all point pairs are distinct, the procedure described above to unfold the final grid from that of the basic domain has to be slightly corrected: when dealing with points of a given face of the basic domain any reflection plane parallel to that face must be disregarded.

It should be noticed that, owing to symmetry, the number of pairs of points just defined, N, say, can be equivalent to a much larger number of pairs on the surface. Given a $\mathbf{k}$ vector, if g is the number of operations of $\mathbf{G}^{\mathbf{k}}$ for which their matrix representative is diagonal, the effective number of point pairs on the surface is gN.

One final practical point as regards the fitting of the boundary conditions should be mentioned, that was discussed by Altmann and Bradley (1965*b*).

In accordance to the discussion of Section 2.3 it is in principle sufficient to determine the lowest eigenvalue of the square matrix B, this being the minimum of the residual S^2. However, when two eigenvalues of the same symmetry are very near, it is useful to determine also the second lowest eigenvalue. This is so for the following reason. The eigenvector that corresponds to a given eigenvalue of the matrix varies continuously with the energy parameter ε. It is therefore possible to label an eigenvector as the first eigenvector by continuity from some first eigenvector previously determined. When this is done, a first eigenvector so labelled can in fact be the second. This is illustrated in Fig. 2.9 which gives S^2 as a function of ε. The eigenvalues on the

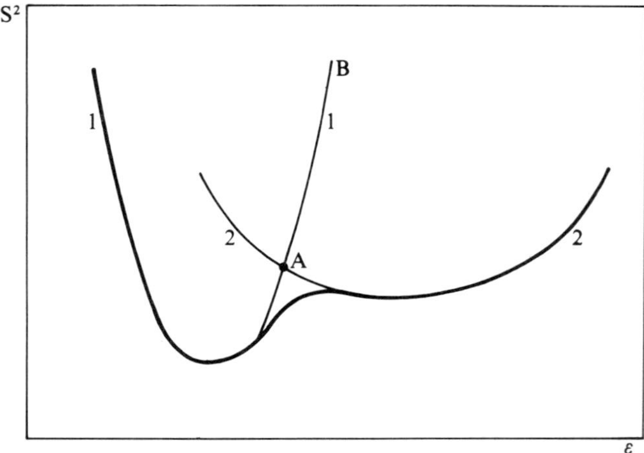

FIG. 2.9. Use of higher eigenvalues of the matrix **B** to determine energy eigenvalues.

curve labelled 1 have been identified with a first eigenvector by continuity. That is, the wavefunction varies smoothly all along this curve. However, from A to B the eigenvalue as determined is the *second* eigenvalue of the square matrix. If only the first eigenvalue is computed, the thick curve of the figure is obtained and, if the two minima were very near, one of them might be almost obliterated. Much greater accuracy is achieved if the minima are instead obtained from the curves 1 and 2 which, of course, requires the use of second eigenvalues.

2.7. The boundary conditions matrix

The handling of the boundary conditions matrix will be described in detail in this section for the case of one atom per unit cell. In this case, from eqns (2.19) and (2.18) the rectangular matrix A of p rows and n columns can be written as

$$A = M + \mathrm{i}N, \tag{2.50}$$

with matrix elements

$$M_{l\nu} = R_l(2)X_{l\nu}(2) - \cos(\mathbf{k}\,.\,\mathbf{T})R_l(1)X_{l\nu}(1), \tag{2.51}$$

$$N_{l\nu} = -\sin(\mathbf{k}\,.\,\mathbf{T})R_l(1)X_{l\nu}(1), \tag{2.52}$$

for the conditions on the wavefunction and

$$M_{l\nu} = \nabla_{\mathbf{n}}\{R_l(2)X_{l\nu}(2)\} + \cos(\mathbf{k}\,.\,\mathbf{T})\nabla_{\mathbf{n}}\{R_l(1)X_{l\nu}(1)\}, \tag{2.53}$$

$$N_{l\nu} = \sin(\mathbf{k}\,.\,\mathbf{T})\nabla_{\mathbf{n}}\{R_l(1)X_{l\nu}(1)\}, \tag{2.54}$$

for the conditions on the normal derivative. In these equations 1 and 2 refer to the polar coordinates with centre at the origin of the boundary point and of its companion point respectively, and $\mathbf{T} = \mathbf{r}_2 - \mathbf{r}_1$. For a general point in $\mathbf{k}$ space, the $X_{l\nu}$ are real and imaginary components of the spherical harmonics in real normalized form, whereas for points of symmetry they are the real lattice harmonics. (See Altmann and Cracknell 1965, where for a general $\mathbf{k}$ vector the $X_{l\nu}$ are labelled as $Y_l^{m,c}$ and $Y_l^{m,s}$.)

The matrix that has to be diagonalized is

$$B = A^{\dagger}A, \tag{2.55}$$

which is clearly hermitean and of dimension $n \times n$. We write

$$B = P + \mathrm{i}Q, \tag{2.56}$$

where P and Q are the $n \times n$ matrices

$$P = M^{\dagger}M + N^{\dagger}N, \tag{2.57}$$

$$Q = M^{\dagger}N - N^{\dagger}M. \tag{2.58}$$

The eigenvalue equation for B is

$$B\mathbf{x} = \lambda\mathbf{x}, \tag{2.59}$$

where the unknown coefficients x are in general complex:

$$\mathbf{x} = \mathbf{u} + \mathrm{i}\mathbf{v}. \tag{2.60}$$

On introducing eqns (2.56) and (2.60) into (2.59), we obtain

$$\begin{cases} P\mathbf{u} - Q\mathbf{v} = \lambda\mathbf{u}, \\ Q\mathbf{u} + P\mathbf{v} = \lambda\mathbf{v}. \end{cases} \tag{2.61}$$

Thus the eigenvalue eqn (2.59) for the $n \times n$ complex B is transformed into the eigenvalue problem of the $2n \times 2n$ real matrix

$$C = \begin{bmatrix} P & -Q \\ Q & P \end{bmatrix}, \tag{2.62}$$

written in supermatrix notation.

It follows from the hermitian property of B that

$$P^{\mathrm{T}} - \mathrm{i}Q^{\mathrm{T}} = P + \mathrm{i}Q, \tag{2.63}$$

where the superscript ^{T} denotes the transpose. From eqn (2.63),

$$P^{\mathrm{T}} = P, \tag{2.64}$$

$$Q^{\mathrm{T}} = -Q, \tag{2.65}$$

so that C is symmetrical. There are a number of standard techniques to find the eigenvalues and eigenvectors of a real symmetric matrix. However, the following property should be noticed. Eqn (2.61) can be written in supermatrix form in two ways,

$$C\begin{bmatrix}\mathbf{u}\\ \mathbf{v}\end{bmatrix} = \lambda\begin{bmatrix}\mathbf{u}\\ \mathbf{v}\end{bmatrix}, \tag{2.66}$$

$$C\begin{bmatrix}\mathbf{v}\\ -\mathbf{u}\end{bmatrix} = \lambda\begin{bmatrix}\mathbf{v}\\ -\mathbf{u}\end{bmatrix}, \tag{2.67}$$

in terms of two different, orthogonal, supervectors. This means that the eigenvalues of eqn (2.61) will appear in degenerate pairs, each corresponding to the two eigenvectors in eqns (2.66) and (2.67), a property which can be incorporated in a standard procedure to diagonalize real symmetric matrices, thereby saving some computing time.

However, it must be noticed that, unless some further simplification can be introduced in the $2n \times 2n$ real matrix C the procedure just described is more lengthy than working directly with the $n \times n$ complex matrix B. (See Wilkinson 1965, p. 342.) The number of multiplications involved in reducing a matrix is of the order of the cube of its dimension. Besides, complex multiplications involve four real operations. Thus the number of multiplications required to reduce the complex matrix B of dimension n is of the order of $4n^3$ and that for the real matrix C of dimension $2n$, $8n^3$. Thus, a saving of time by a factor of the order of 2 is achieved by handling directly the complex matrix.

Nevertheless, it shall be seen in Section 2.8 that if there is a centre of symmetry the real matrix C can be reduced to dimension n in which case it will be about 4 times quicker to use it instead of the complex matrix C. It is in this particular case that the real matrix should be formed and it is for this specific purpose that the work in this section has been carried out.

Another case when the matrix C can be simplified arises when $\mathbf{k} = \frac{1}{2}\mathbf{K}$, the latter being a vector of the reciprocal lattice, for which $\mathbf{K}\,.\,\mathbf{T}$ equals $2\pi m$, where m is an integer or zero. Thus $\mathbf{k}\,.\,\mathbf{T} = m\pi$ whence, from eqns (2.52) and

(2.54) the matrix $N = 0$ and, from eqn (2.58), $Q = 0$. Thus C is super-diagonal and its reduction coincides with that of the single $n \times n$ matrix P. It is also useful to notice that, from eqn (2.61), $\mathbf{u} = \mathbf{v}$ and, since the wave-function and its complex conjugate must be degenerate, it follows that for these values of $\mathbf{k}$ the coefficients of the cellular expansion can be taken to be real. The condition in question, $\mathbf{k} = \frac{1}{2}\mathbf{K}$, is satisfied by the centre of the Brillouin zone and the centres of the Brillouin zone faces.

2.8. Simplifications for a centro-symmetric potential

A centro-symmetric potential $V(\mathbf{r})$ is one that admits of a centre of inversion $\bar{\bar{i}}$:

$$\bar{\bar{i}}V(\mathbf{r}) = V(-\mathbf{r}) = V(\mathbf{r}). \tag{2.68}$$

It will be shown in this section that the cellular expansion can be considerably simplified for such potentials as a consequence of some results which are variously referred to in the literature as well known or easy to prove and which have been extensively used in the cellular as well as in most other methods. However, the present author has failed to find a rigorous proof of these results in published form, or any person who actually knew of such proof. It appears in fact possible that some people have accepted and used the results in question merely on the basis of plausible arguments. Therefore, a complete and rigorous treatment will here be given of the effect of central symmetry on the expansion coefficients. In order to do this some group theoretical results must first be reviewed, which are fully discussed in the references given in Section 2.4.

The basis of an irreducible representation of a space group $\mathbf{G}$ is a row vector, each element of which is a Bloch function. These functions must be labelled in each representation, for which two labels are required.

The first label is the $\mathbf{k}$ vector so that the wavefunction is at first denoted as $\psi_{\mathbf{k}}(\mathbf{r})$, where, from eqn (2.10), the $\mathbf{k}$ vector is determined by the eigenvalue in the equation

$$\bar{\bar{T}}\psi_{\mathbf{k}}(\mathbf{r}) = \exp(-\mathrm{i}\mathbf{k}\,.\,\mathbf{T})\psi_{\mathbf{k}}(\mathbf{r}). \tag{2.69}$$

More than one value of $\mathbf{k}$ appears in each basis, but they can all be obtained from a single value of $\mathbf{k}$ by forming the set $\bar{\alpha}\mathbf{k}$ for all point group operations $\bar{\alpha}$ that appear in the space group, and eliminating from the set all $\mathbf{k}$ values that are equivalent to previously obtained ones. (Two $\mathbf{k}$ vectors are said to be equivalent if they differ by a vector of the reciprocal lattice.) The set of $\mathbf{k}$ vectors thus defined is called a *star*.

The second label of the basis comes from the small representation (see Section 2.4). All the functions $\psi_{\mathbf{k}_i}$ that appear in a basis of $\mathbf{G}$ (all $\mathbf{k}_i$ in a star) must form bases for a given small representation. Thus, if the latter is of dimension $h(h > 1)$, any given $\mathbf{k}_i$ in the basis will be repeated h times and the Bloch functions must be labelled $\psi_{\mathbf{k}_i t_j}$ $(j = 1, 2, \ldots, h)$, the t label corresponding to a given column of the small representation.

For convenience, the row vector of Bloch functions $\psi_{\mathbf{k}_i t_j}$, $\mathbf{k}_i \in$ star, $j = 1, \ldots, h$, will be written as $\langle \psi_{\mathbf{k}t}|$. The matrix representative $\hat{S}$ of a space group operation $\bar{S}$ is defined by

$$\langle \bar{\bar{S}} \psi_{\mathbf{k}t}| \equiv \bar{\bar{S}} \langle \psi_{\mathbf{k}t}| = \langle \psi_{\mathbf{k}t}| \hat{S}. \tag{2.70}$$

(See e.g. Altmann 1962, p. 95.) The matrix $\hat{T}$ that corresponds to a translation of the lattice $\bar{\bar{T}}$ is diagonal, the diagonal elements being the eigenvalues in eqn (2.69), each repeated as many times as the dimension of the small representation.

A sequence of theorems will now be proved of which the last one, Theorem 10, is the crucial one for the simplification of the cellular expansion. These theorems are very much concerned with the properties of two important operators. One is the inversion $\bar{\bar{i}}$, which will always be assumed to belong to the space group, and the other is the conjugation operator or conjugator $\bar{\bar{j}}$. This operator is defined so as to transform every term z in its operand into its complex conjugate z^* and it is therefore not linear. Thus, if m is a constant

$$\bar{\bar{j}} m f(u) = m^* f^*(u^*). \tag{2.71}$$

The importance of the conjugator arises from the fact that if $H\psi = E\psi$ then $\bar{\bar{j}}\psi \equiv \psi^*$ is degenerate to ψ as follows from applying $\bar{\bar{j}}$ to both sides of the Schrödinger equation and using the reality conditions for the hamiltonian H and its eigenvalue E. The conjugator $\bar{\bar{j}}$ is closely related to the time reversal operator, with which it coincides for spinless hamiltonians.

As regards the inversion, it must be remembered that it changes the sign of all vectors $\mathbf{r}$ in configuration space:

$$\bar{\bar{i}}\mathbf{r} = -\mathbf{r}. \tag{2.72}$$

Theorem 1

If $\bar{S}$ is a configuration space operator that can be defined as acting as follows on $\mathbf{r}$,

$$\bar{S}\mathbf{r} = S\mathbf{r}, \tag{2.73}$$

where S is a matrix, then

$$\bar{\bar{i}}\bar{S} = \bar{S}\bar{\bar{i}}. \tag{2.74}$$

Proof. Operate on both sides of eqn (2.73) with $\bar{\bar{i}}$:

$$\bar{\bar{i}}\bar{S}\mathbf{r} = S\bar{\bar{i}}\mathbf{r} = S(-\mathbf{r}) = -S\mathbf{r}. \tag{2.75}$$

On the other hand,

$$\bar{S}\bar{\bar{i}}\mathbf{r} = \bar{S}(-\mathbf{r}) = -\bar{S}\mathbf{r} = -S\mathbf{r}. \tag{2.76}$$

The theorem follows from eqns (2.75) and (2.76). Owing to the conservation of the multiplication rules (see Section 2.1),

$$\bar{\bar{i}}\bar{\bar{S}} = \bar{\bar{S}}\bar{\bar{i}}. \tag{2.77}$$

All point group operations satisfy eqn (2.73) and therefore commute with the inversion. Translations, on the other hand, cannot be expressed as in (2.73) and the following theorem applies to them.

Theorem 2

For all translations $\bar{T}$

$$\bar{i}\bar{T} = \bar{T}^{-1}\bar{i} \tag{2.78}$$

with a corresponding expression for the function space operators.

Proof. By definition,

$$\bar{T}\mathbf{r} = \mathbf{r}+\mathbf{T}. \tag{2.79}$$

From eqn (2.72),

$$\bar{i}\bar{T}\mathbf{r} = \bar{i}(\mathbf{r}+\mathbf{T}) = -\mathbf{r}-\mathbf{T}. \tag{2.80}$$

Also,

$$\bar{T}^{-1}\bar{i}\mathbf{r} = \bar{T}^{-1}(-\mathbf{r}) = -\mathbf{r}-\mathbf{T}. \tag{2.81}$$

Theorem 3

For any symmetry operator $\bar{\bar{S}}$,

$$\bar{\bar{j}}\bar{\bar{S}} = \bar{\bar{S}}\bar{\bar{j}}. \tag{2.82}$$

Proof. Consider eqn (2.4). A function space operator cannot be applied on both sides of this equation because it is an equality between *values* of functions and not between functions themselves. On the other hand, $\bar{\bar{j}}$ can be applied on both sides of eqn (2.4) since it operates on numbers as well as on functions:

$$\bar{\bar{j}}\bar{\bar{S}}f(\mathbf{r}) = \bar{\bar{j}}f(S^{-1}\mathbf{r}) \tag{2.83}$$

$$= f^*\{(S^{-1}\mathbf{r})^*\} = f^*(\bar{S}^{-1}\mathbf{r}). \tag{2.84}$$

(The reality of S and $\mathbf{r}$ is used in the last step.) On the other hand,

$$\bar{\bar{S}}\bar{\bar{j}}f(\mathbf{r}) = \bar{\bar{S}}f^*(\mathbf{r}^*) = f^*(\bar{S}^{-1}\mathbf{r}). \tag{2.85}$$

Theorem 4

For all Bloch functions $\psi_{\mathbf{k}}(\mathbf{r})$,

$$\bar{\bar{i}}\psi_{\mathbf{k}}(\mathbf{r}) = \psi_{-\mathbf{k}}(\mathbf{r}). \tag{2.86}$$

Proof. From Theorem 2,

$$\begin{aligned}\bar{\bar{T}}\bar{i}\psi_{\mathbf{k}}(\mathbf{r}) &= \bar{i}\bar{\bar{T}}^{-1}\psi_{\mathbf{k}}(\mathbf{r})\\ &= \bar{i}\exp(i\mathbf{k}\,.\,\mathbf{T})\psi_{\mathbf{k}}(\mathbf{r})\\ &= \exp(i\mathbf{k}\,.\,\mathbf{T})\bar{\bar{i}}\psi_{\mathbf{k}}(\mathbf{r}),\end{aligned} \tag{2.87}$$

which, on comparison with eqn (2.69), proves that $i\psi_{\mathbf{k}}(\mathbf{r})$ belongs to the eigenvalue that corresponds to $-\mathbf{k}$, thus verifying (2.86), in which, of course irrelevant phase factors are disregarded. We shall now see that the same effect is produced by $\bar{\bar{j}}$.

Theorem 5

For all Bloch functions $\psi_{\mathbf{k}}(\mathbf{r})$,

$$\bar{\bar{j}}\psi_{\mathbf{k}}(\mathbf{r}) = \psi_{-\mathbf{k}}(\mathbf{r}). \tag{2.88}$$

Proof. Introduce $\bar{\bar{j}}\psi_{\mathbf{k}}(\mathbf{r})$ in the left-hand side of eqn (2.69) and use Theorem 3,

$$\begin{aligned}\bar{\bar{T}}\bar{\bar{j}}\psi_{\mathbf{k}}(\mathbf{r}) = \bar{\bar{j}}\bar{\bar{T}}\psi_{\mathbf{k}}(\mathbf{r}) &= \bar{\bar{j}}\{\exp(-\mathrm{i}\mathbf{k}\,.\,\hat{\mathbf{T}})\psi_{\mathbf{k}}(\mathbf{r})\}\\ &= \exp(\mathrm{i}\mathbf{k}\,.\,\hat{\mathbf{T}})\bar{\bar{j}}\psi_{\mathbf{k}}(\mathbf{r}).\end{aligned} \tag{2.89}$$

Comparison of eqns (2.89) and (2.69) proves the theorem.

Theorems 4 and 5 might be understood to imply that $\bar{\bar{i}}\psi_{\mathbf{k}}(\mathbf{r})$ and $\bar{\bar{j}}\psi_{\mathbf{k}}(\mathbf{r})$ are the self-same function. This, however, is not true in general and the following result is the correct one.

Theorem 6

$$\langle\bar{\bar{j}}\psi_{\mathbf{k}t}| = \langle\bar{\bar{i}}\psi_{\mathbf{k}t}|\mathrm{A}, \tag{2.90}$$

where A is a non-singular matrix.

Proof. First, the notation of Theorem 4 and 5 must be refined to include the dimension h of the small representation. What Theorem 4 asserts is that $i\psi_{\mathbf{k}t}(\mathbf{r})$ belongs to the value $-\mathbf{k}$: therefore it must be in general a linear combination of all the functions $\psi_{-\mathbf{k},t_j}$, $j = 1, 2, \ldots, h$, of the basis. (If $-\mathbf{k}$ is not in the star, because $\bar{i}$ belongs to the group of $\mathbf{k}$, then it has to be replaced by a $\mathbf{k}$ vector equivalent to it). Likewise for Theorem 5.

The bases $\bar{\bar{i}}\langle\psi_{\mathbf{k}t}|$ and $\bar{\bar{j}}\langle\psi_{\mathbf{k}t}|$ are both degenerate to $\langle\psi_{\mathbf{k}t}|$, in the case of $\bar{\bar{i}}$ because it is in the space group $\mathbf{G}$ and in the case of $\bar{\bar{j}}$ because of the discussion after eqn (2.71). Also, both bases have the full dimensionality of the degeneracy of the energy eigenvalue, so that they cannot be linearly independent since, if this were so, this dimensionality would be exceeded. Therefore, the two bases must be linearly dependent, as stated by eqn (2.90).

Theorem 7

The matrix A in eqn (2.90) satisfies the condition

$$AA^* = \mathbf{1}. \tag{2.91}$$

Proof. Apply $\bar{\bar{j}}$ on both sides of eqn (2.90). On the left $\bar{\bar{j}}\bar{\bar{j}}$ is the identity and on the right the non-linearity of $\bar{j}$ must be taken into account:

$$\langle\psi_{\mathbf{k}t}| = \langle\bar{\bar{j}}\bar{\bar{i}}\psi_{\mathbf{k}t}|A^*. \tag{2.92}$$

Apply Theorem 3 on the right hand of eqn (2.92):

$$\langle\psi_{\mathbf{k}t}| = \bar{\bar{i}}\langle\bar{\bar{j}}\psi_{\mathbf{k}t}|A^* = \bar{\bar{i}}\langle\bar{\bar{i}}\psi_{\mathbf{k}t}|AA^*, \tag{2.93}$$

(In the last step eqn (2.90) has been used.) In eqn (2.93), $\bar{\bar{i}}\bar{\bar{i}}$ is the identity, whence eqn (2.91) follows.

Theorem 8

Given a symmetry operator $\bar{\bar{S}}$ (assumed real, which is always the case for such operators) that commutes with the inversion,

$$\bar{\bar{S}}\bar{\bar{i}} = \bar{\bar{i}}\bar{\bar{S}}, \tag{2.94}$$

then

$$\hat{S}A = A\hat{S}^*. \tag{2.95}$$

Proof. Apply $\bar{\bar{j}}$ on both sides of eqn (2.70), using on the left the commutation relation of Theorem 3. Then,

$$\bar{\bar{S}}\langle\bar{\bar{j}}\psi_{\mathbf{k}t}| = \langle\bar{\bar{j}}\psi_{\mathbf{k}t}|\hat{S}^* \tag{2.96}$$

Introduce eqn (2.90) in (2.96):

$$\bar{\bar{S}}\langle\bar{\bar{i}}\psi_{\mathbf{k}t}|A = \langle\bar{\bar{i}}\psi_{\mathbf{k}t}|A\hat{S}^*. \tag{2.97}$$

On using first eqn (2.94) and then eqn (2.70) on the left-hand side of (2.97), we obtain

$$\langle\bar{\bar{i}}\psi_{\mathbf{k}t}|\hat{S}A = \langle\bar{\bar{i}}\psi_{\mathbf{k}t}|A\hat{S}^*, \tag{2.98}$$

whence eqn (2.95) follows.

Since translations do not commute with the inversion, Theorem 8 is not applicable for them. However, they satisfy the following similar relation.

Theorem 9

$$\hat{T}A = A\hat{T}. \tag{2.99}$$

Proof. We must first notice that, from the definition of $\hat{T}$ given by eqn (2.70) and from eqn (2.69),

$$\hat{T}^{-1} = \hat{T}^*. \tag{2.100}$$

We now use eqn (2.97) for $\bar{\bar{S}} = \bar{\bar{T}}$,

$$\bar{\bar{T}}\langle\bar{i}\psi_{\mathbf{k}t}|A = \langle\bar{i}\psi_{\mathbf{k}t}|A\hat{T}^*. \tag{2.101}$$

From eqn (2.78),

$$\bar{i}\langle\bar{\bar{T}}^{-1}\psi_{\mathbf{k}t}|A = \langle\bar{i}\psi_{\mathbf{k}t}|A\hat{T}^*. \tag{2.102}$$

Use eqns (2.70) and (2.100) on the left-hand side of eqn (2.102):

$$\langle\bar{i}\psi_{\mathbf{k}t}|\hat{T}^*A = \langle\bar{i}\psi_{\mathbf{k}t}|A\hat{T}^* \tag{2.103}$$

and, on taking complex conjugates it follows that

$$\hat{T}A^* = A^*\hat{T}. \tag{2.104}$$

Multiply both sides of this equation by A on the left and on the right and, on using Theorem 7, eqn (2.99) follows.

We can now prove our fundamental result.

Theorem 10

If all elements of the group of the **k** vector $\mathbf{G^k}$ are either translations $\bar{\bar{T}}$ or operators $\bar{\bar{S}}$ that: (1) commute with the inversion; (2) are represented by real matrices, then

$$\bar{i}\psi_{\mathbf{k}t} = \bar{j}\psi_{\mathbf{k}t}, \tag{2.105}$$

for all **k** in the star of $\mathbf{G^k}$ and each column t of the small representation.

Proof. In order to prove the theorem, it is enough to prove that A is a constant matrix, in which case eqn (2.105) follows from Theorem 6, when account is taken of Theorem 7, from which A must be of unit modulus so that it must introduce at most an irrelevant phase factor in eqn (2.90).

To prove that A is a constant matrix it suffices to observe that, under the conditions stated for the theorem, it commutes with $\hat{S}$ from Theorem 8 and (2). Also, it commutes with $\hat{T}$ from Theorem 9. Therefore, it commutes with all the matrices of the irreducible representation of $\mathbf{G^k}$ and it follows from a general theorem in group theory (see e.g. Altmann 1962, p. 121) that A must be constant.

Assuming that accidental degeneracies are not permitted, the conditions under which Theorem 10 is valid obtain in the following three cases:

(*a*) All general **k** values for all centro-symmetric space groups.

In fact, $\mathbf{G^k}$ in this case contains translations only, for which no restrictions are imposed in Theorem 10. The small representations are one-dimensional in this case, so that eqn (2.105) applies in the form

$$\bar{i}\psi_{\mathbf{k}} = \bar{j}\psi_{\mathbf{k}}. \tag{2.106}$$

(*b*) All *real* small representations of symmetry for all symmorphic centro-symmetric space groups.

For symmorphic groups, the group of the **k** vector contains only point group operations besides translations. From Theorem 1, all point group operations commute with the inversion and hence satisfy condition (1) of Theorem 10. If the small representation is real the second condition is also verified and the theorem is valid.

As an example, all small representations of all cubic groups are real (see e.g. Altmann and Cracknell 1965) and Theorem 10 is valid for them for all points of symmetry as well as, from case A, for general **k** values.

As regards the reality of the representation; it is sufficient of course for the representation to be equivalent to a real one. Such representations are called 'of the first kind' (see Bradley and Cracknell, 1972, p. 20).

(*c*) All one-dimensional representations for any **k** vector of any centrosymmetric space group.

For one-dimensional representations of $\mathbf{G}^{\mathbf{k}}$, each **k** appears only once in the basis of **G**. Therefore $\bar{\bar{i}}\psi_{\mathbf{k}}$ and $\bar{\bar{j}}\psi_{\mathbf{k}}$, corresponding to $\psi_{-\mathbf{k}}$, must appear once and only once in the basis, whence they must be identical, except possibly for a phase factor.

The significance of Theorem 10 follows from its application to the following theorem, which gives an important property of the coefficients of the cellular expansion.

2.8.1. *The coefficient theorem*

Given a general **k** vector of any centrosymmetrical space group with one atom per unit cell, write the cellular expansion with the notation of Section 2.7 as follows,

$$\psi_{\mathbf{k}}(\mathbf{r}) = \sum_{l\nu} C_{l\nu} R_l(r) X_{l\nu}(\theta\phi), \tag{2.107}$$

where the $X_{l\nu}$ are the sin and cos parts of the spherical harmonics, and they are real and normal. Then

$$C^*_{l\nu} = (-1)^l C_{l\nu}. \tag{2.108}$$

Proof. Since the $X_{l\nu}(\theta\phi)$ are linear combinations of the standard spherical harmonics, they transform under the inversion in the well-known manner

$$\bar{\bar{i}} X_{l\nu}(\theta\phi) = (-1)^l X_{l\nu}(\theta\phi). \tag{2.109}$$

Also, because of the central symmetry of the potential,

$$\bar{\bar{i}} R_l(r) \equiv R_l(-r) = R_l(r). \tag{2.110}$$

Apply $\bar{\bar{i}}$ on both sides of eqn (2.107) and use eqns (2.109) and (2.110) on the right-hand side:

$$\bar{\bar{i}}\psi_{\mathbf{k}}(\mathbf{r}) = \sum_{l\nu} (-1)^l C_{l\nu} R_l(r) X_{l\nu}(\theta\phi). \tag{2.111}$$

Also, from (2.107),

$$\bar{\bar{j}}\psi_{\mathbf{k}}(\mathbf{r}) = \sum_{l\nu} C^*_{l\nu} R_l(r) X_{l\nu}(\theta\phi), \tag{2.112}$$

since the $R_l(r)$ and $X_{l\nu}(\theta\phi)$ are real. On applying eqn (2.106) of Theorem 10 to eqns (2.111) and (2.112), eqn (2.108) follows.

As a consequence of the present theorem the $C_{l\nu}$ are real and imaginary respectively for even and odd l. It is therefore possible to write them in the convenient form

$$C_{l\nu} = \mathrm{i}^l \mathscr{C}_{l\nu},\ \mathscr{C}_{l\nu} \text{ real}, \tag{2.113}$$

a result which cuts down the size of expansion (2.107) by one half and which was first stated without proof by Ham and Segall (1961).

2.8.2. *Extensions of the coefficient theorem*

The conditions for the applicability of eqns (2.108) and (2.113) of the coefficient theorem can be somewhat modified, as given in the two cases below.

(1) A **k** vector of symmetry, with a real small representation and real lattice harmonics for a symmorphic centro-symmetric space group with one atom per unit cell.

Proof. $X_{l\nu}(\theta\phi)$ in eqn (2.107 must now be understood as a lattice harmonic belonging to the tth column of the relevant representation and $\psi_{\mathbf{k}}(\mathbf{r})$ in the left of this equation must be replaced by $\psi_{\mathbf{k}t}(\mathbf{r})$. The proof given above is entirely valid, since case b of Theorem 10 applies.

It follows from this result, for instance, that eqn (2.113) is valid for all small representations of all cubic space groups since, as explained in b of Theorem 10, the conditions presently imposed are satisfied.

(2) A **k** vector of symmetry, with a one-dimensional representation and real lattice harmonics for any centrosymmetric space group with one atom per unit cell.

Proof. It follows as in (1), from case c of Theorem 10.

The coefficient theorem, as discussed, deals with the case of one atom per unit cell. We shall now relax this condition.

2.8.3. *Coefficient theorem for two atoms per unit cell*

We shall consider a general **k** vector for any centrosymmetric space group with two atoms per unit cell with the centre of inversion at the midpoint between them. The cellular expansion, given in eqn (2.45) of Section 2.5 will be slightly abbreviated in a self-explanatory notation:

$$\psi_{\mathbf{k}}(\mathbf{r}) = \sum_{lm} {}^1C_{lm}\delta(1)R_l(1)Y_l^m(1) + \sum_{l'm'} {}^2C_{l'm'}\delta(\tilde{2})R_{l'}(\tilde{2})Y_{l'}^{m'}(\tilde{2}).$$

Owing to the choice of axes discussed in Section 2.5, the coefficient $(-1)^l$ does not appear under inversion. Then:

$$\bar{i}\psi_{\mathbf{k}}(\mathbf{r}) = \sum_{lm} {}^1C_{lm}\delta(\tilde{2})R_l(\tilde{2})Y_l^m(\tilde{2}) + \sum_{l'm'} {}^2C_{l'm'}\delta(1)R_{l'}(1)Y_{l'}^{m'}(1). \quad (2.115)$$

Also:

$$\bar{j}\psi_{\mathbf{k}}(\mathbf{r}) = \sum_{lm} {}^1C^*_{lm}\delta(1)R_l(1)Y_l^{-m}(1) + \sum_{l'm'} {}^2C^*_{l'm'}\delta(\tilde{2})R_{l'}(\tilde{2})Y_{l'}^{-m'}(\tilde{2}). \quad (2.116)$$

Since we are dealing with a general $\mathbf{k}$ value, case (a) of Theorem 10 can be applied whence it follows from eqns (2.115) and (2.116) that

$$ {}^2C_{lm} = {}^1C^*_{l,-m}, \quad (2.117)$$

a result which, again, cuts down the size of the cellular expansion by one half.

In the case of one atom per unit cell we saw that the coefficient theorem could be extended from a general $\mathbf{k}$ vector to a $\mathbf{k}$ vector of symmetry, subject to certain conditions. In the present case, such an extension of eqn (2.117) may fail. In the hexagonal close-packed lattice, for instance, there are two atoms per unit cell, with a centre of inversion at the midpoint between them. Eqn (2.117) is valid for general $\mathbf{k}$ but it cannot be extended for points of symmetry in $\mathbf{k}$ space since the space group is not symmorphic and case (b) of Theorem 10 cannot be applied.

2.8.4. *The boundary conditions matrix*

Finally, the form of the boundary conditions matrix will be given for a centrosymmetric field with one atom per unit cell. In Section 2.7 the boundary conditions fitting was shown to require the solution of the eigenvalue equation for the matrix C, of dimension $2n \times 2n$. The eigenvector of the coefficients of the cellular expansion was given in eqn (2.66) in terms of two n-dimensional vectors, $\mathbf{u}$ for the real and $\mathbf{v}$ for the imaginary parts of the coefficients. The rows and columns of P and Q in eqn (2.62), as well as the elements of $\mathbf{u}$ and $\mathbf{v}$ can be labelled by means of the double suffix used in eqn (2.107). It follows from eqn (2.113) that

$$\mathbf{u}_{l\nu} = 0, \quad \text{for } l \text{ odd}, \quad (2.118)$$

$$\mathbf{v}_{l\nu} = 0 \quad \text{for } l \text{ even}. \quad (2.119)$$

When these equations are introduced into eqn (2.66), it can be seen that the $2n \times 2n$ matrix C is reduced into the $n \times n$ matrix

$$D = \begin{bmatrix} P^{\text{ee}} & -Q^{\text{eo}} \\ Q^{\text{oe}} & P^{\text{oo}} \end{bmatrix}, \quad (2.120)$$

where:

$$P^{ee}_{l\nu,l'\nu'} = P_{l\nu,l'\nu'}; \quad l \text{ even}, l' \text{ even}, \tag{2.121}$$

$$Q^{eo}_{l\nu,l'\nu'} = Q_{l\nu,l'\nu'}; \quad l \text{ even}, l' \text{ odd}, \tag{2.122}$$

$$Q^{oe}_{l\nu,l'\nu'} = Q_{l\nu,l'\nu'}; \quad l \text{ odd}, l' \text{ even}, \tag{2.123}$$

$$P^{oo}_{l\nu,l'\nu'} = P_{l\nu,l'\nu'}; \quad l \text{ odd}, l' \text{ odd}, \tag{2.124}$$

and the matrices P and Q are defined by eqns (2.57) and (2.58), with eqns (2.51) to (2.54). The rectangular matrix Q^{oe} is the transposed of Q^{eo} and D is, like C, symmetric.

Besides the conditions for the expansion coefficients discussed in this section, it should be remembered that, as shown in Section 2.7, these coefficients can always be chosen to be real for the centre of the Brillouin zone and the centres of the Brillouin zone faces.

2.9. Convergence of the cellular expansion

The conditions for the validity of the cellular expansion will be examined in this section. Saffren (1959) showed that the standard cellular expansion converges rigorously for a particular type of muffin-tin potential. As well known, a muffin-tin potential is spherically symmetric within a sphere of radius ρ and it is a constant outside it. Saffren's result was independently obtained by Ham in his doctoral thesis in 1954 (see also Ham 1960), where this author pointed out that the cellular expansion may not converge unless the potential is of a muffin-tin type with a specific maximum value of ρ. The same warning was repeated by Ham and Segall (1961). Much confusion, however, has arisen on this point. On one hand, the detailed analyses of Saffren and of Ham were never published. On the other, no serious numerical work was done for many years to ascertain to what extent the alleged convergence difficulties were significant. As a result, statements were often made based on unsupported opinion rather than theoretical analysis and numerical facts. It appears therefore convenient to cover the subject in detail. The treatment that follows is a very close account of the discussion given by Saffren (1959).

We shall need some characteristic radii of the Wigner–Seitz cell, which are defined as follows:

$$R_i = \text{radius of incribed sphere.} \tag{2.125}$$

$$R_c = \text{radius of circumscribed sphere.} \tag{2.126}$$

The solutions of the Schrödinger equation in the crystal are approximated in the cellular method by the expansion (2.12) within the Wigner–Seitz cell. The radial equations of which the functions $R_l(r; \varepsilon)$ are solutions will be given in the abbreviated notation

$$(H_l - \varepsilon)R_l(r\,;\varepsilon) = 0, \qquad (2.127)$$

where H_l is the radial hamiltonian which depends on l and on the potential field. The purpose of this discussion is to study the validity of eqn (2.12) throughout the Wigner–Seitz cell, that is for r up to $r = R_c$.

Eqn (2.12) can be written as an expansion in orthogonal functions

$$\psi(\mathbf{r}) = \sum_{lm} \psi_{lm}(r) Y_l^m(\theta\phi), \qquad (2.128)$$

where we now dispense with the suffix **k**. Because of the completeness of the spherical harmonics, eqn (2.128) is valid for $r < R_i$ and, if $\psi(\mathbf{r})$ is known, it determines uniquely the coefficients $\psi_{lm}(r)$. In practice, however, the $\psi_{lm}(r)$ can be taken as solutions of eqn (2.127),

$$(H_l - \varepsilon)\psi_{lm}(r) = 0, \qquad (2.129)$$

so that

$$\sum_{lm} Y_l^m(\theta\phi)(H_l - \varepsilon)\psi_{lm}(r) = 0. \qquad (2.130)$$

Define

$$\Phi(\mathbf{r}) = \sum_{lm} Y_l^m(\theta\phi)(H_l - \varepsilon)\psi_{lm}(\mathrm{r}). \qquad (2.131)$$

From eqn (2.130),

$$\Phi(\mathbf{r}) = 0, \qquad r < R_i. \qquad (2.132)$$

On the other hand, since the spherical harmonics are a complete set,

$$\Phi(\mathbf{r}) = \sum_{lm} \Phi_{lm}(r) Y_l^m(\theta\phi), \quad \forall r. \qquad (2.133)$$

From eqn (2.132),

$$\Phi_{lm}(r) = 0, \qquad r < R_i, \qquad (2.134)$$

but this is not true for $r > R_i$, since $\Phi(r)$ is not defined over one complete sphere.

From eqns (2.131) and (2.133)

$$(H_l - \varepsilon)\psi_{lm}(\mathrm{r}) = \Phi_{lm}(\mathrm{r}), \qquad (2.135)$$

so that it follows from comparison with eqn (2.129) that, in the region $R_i < r < R_c$ solutions of the radial eqn (2.127) are required that satisfy an inhomogeneous condition. Call $\psi'_{lm}(r)$ such solutions of eqn (2.135). Owing to the need to include these solutions, the correct expansion of the wave-function within the Wigner–Seitz cell, to be denoted with $\Psi(\mathbf{r})$ is given by eqn (2.128) only within R_i,

$$\Psi(\mathbf{r}) = \psi(\mathbf{r}), \qquad 0 < r < R_i, \tag{2.136}$$

and outside this radius an additional term must be added:

$$\Psi(\mathbf{r}) = \psi(\mathbf{r})+\psi'(\mathbf{r}), \qquad R_i < r < R_c. \tag{2.137}$$

In eqns (2.136) and (2.137), $\psi(\mathbf{r})$ and $\psi'(\mathbf{r})$ are given by eqn (2.128) with the solutions $\psi_{lm}(r)$ and $\psi'_{lm}(r)$ respectively.

Eqns (2.136) and (2.137) require the following continuity conditions:

$$\psi'(r) = 0; \qquad r = R_i, \tag{2.138}$$

$$\partial\psi'(r)/\partial r = 0; \qquad r = R_i, \tag{2.139}$$

with similar conditions for the $\psi'_{lm}(r)$. These boundary conditions determine uniquely a solution of the inhomogeneous eqn (2.135) which contains the irregular as well as the regular solution of eqn (2.129).

In order to see how such solutions of the inhomogeneous eqn (2.135) could be found, let us expand the potential field in spherical harmonics,

$$V(\mathbf{r}) = \sum_{\mathrm{LM}} V_{\mathrm{LM}}(r)Y_{\mathrm{L}}^{\mathrm{M}}(\theta\phi), \tag{2.140}$$

Lattice harmonics would in practice be used in eqn (2.140) but this does not affect the discussion below, except in introducing some simplifications of the practical work involved.

For convenience, replace l, m in eqn (2.128) by λ, μ and introduce this equation and eqn (2.140) into the Schrödinger equation

$$\nabla^2\psi(\mathbf{r})+2\{\varepsilon-V(\mathbf{r})\}\psi(\mathbf{r}) = 0, \tag{2.141}$$

from which

$$\sum_{\lambda\mu}\{\nabla^2(\psi_{\lambda\mu}Y_{\lambda}^{\mu})+2(\varepsilon-\sum_{\mathrm{LM}} V_{\mathrm{LM}}Y_{\mathrm{L}}^{\mathrm{M}})\psi_{\lambda\mu}Y_{\lambda}^{\mu}\} = 0. \tag{2.142}$$

Therefore,

$$\sum_{\lambda\mu}\{\mathrm{r}^{-2}\mathrm{d}/\mathrm{d}r(r^2\mathrm{d}\psi_{\lambda\mu}/\mathrm{d}r)Y_{\lambda}^{\mu}-\lambda(\lambda+1)r^{-2}\psi_{\lambda\mu}Y_{\lambda}^{\mu}+2\varepsilon\psi_{\lambda\mu}Y_{\lambda}^{\mu}\}$$

$$= 2\sum_{\lambda\mu}\sum_{\mathrm{LM}} V_{\mathrm{LM}}Y_{\mathrm{L}}^{\mathrm{M}}\psi_{\lambda\mu}Y_{\lambda}^{\mu}. \tag{2.143}$$

Multiply both sides of eqn (2.143) by $(Y_l^m)^*$, integrate over θ and ϕ and use the notation

$$I(l\mathrm{L}\lambda, \mathrm{mM}\mu) = \int (Y_l^m)^* Y_{\mathrm{L}}^{\mathrm{M}} Y_{\lambda}^{\mu} \sin\theta\, \mathrm{d}\theta\, \mathrm{d}\phi, \tag{2.144}$$

The result is

$$r^{-2}\mathrm{d}/\mathrm{d}r(r^2\,\mathrm{d}\psi_{lm}/\mathrm{d}r)-\{l(l+1)r^{-2}-2\varepsilon\}\psi_{lm}$$
$$= 2\sum_{\lambda\mu}\sum_{\mathrm{LM}} V_{\mathrm{LM}} I(l\mathrm{L}\lambda,\, m\mathrm{M}\mu)\psi_{\lambda\mu}. \tag{2.145}$$

The term V_{00} is the spherical part of the potential, which can be identified with the spherical potential in the left-hand side of eqn (2.135). Also,

$$I(l0\lambda,\, \mathrm{m}0\mu) = \delta_{l\lambda}\delta_{m\mu}. \tag{2.146}$$

Therefore, for V_{00}, the summation over λ and μ gives the single term corresponding to l and m. When this is transferred to the left-hand side of eqn (2.145), we obtain

$$r^{-2}\,\mathrm{d}/\mathrm{d}r(r^2\,\mathrm{d}\psi_{lm}/\mathrm{d}r)-\{l(l+1)r^{-2}-2(\varepsilon-V_{00})\}\psi_{lm}$$
$$= 2\sum_{\lambda\mu}\sum_{\mathrm{LM}\neq 0,0} V_{\mathrm{LM}} I(l\mathrm{L}\lambda,\, m\mathrm{M}\mu)\psi_{\lambda\mu}. \tag{2.147}$$

The left-hand side of eqn (2.147) is exactly the left-hand side of eqn (2.135), whence its right-hand side can be identified with the inhomogeneous term in this latter equation. We can thus see that the effect of this term is that of establishing the system of coupled differential eqns (2.147) for the functions ψ_{lm}. These will be uncoupled if and only if

$$V_{\mathrm{LM}} = 0, \qquad \mathrm{L} \neq 0,\, \mathrm{M} \neq 0. \tag{2.148}$$

It is in this case and only in this case that the cellular expansion (2.12) can be used up to $r = R_c$ instead of the more general expression (2.137). Condition (2.148) requires that the potential be spherically symmetrical up to $r = R_c$. This will be the case only for a muffin-tin potential for which the radius ρ of the muffin is

$$\rho < 2R_i - R_c, \tag{2.149}$$

as discussed by Altmann (1970, p. 228) and illustrated in Fig. 2.10. For a muffin-tin potential that satisfies eqn (2.149) and only for such a potential, the cellular expansion (2.12) converges not only from 0 to R_i but also from R_i to R_c, that is over the whole of the Wigner–Seitz cell. This is, in fact, the condition stated by Ham (1960).

It is reasonable to expect that the convergence of the cellular expansion for $r > R_i$ will be maintained for potentials that differ mildly from the correct muffin-tin form that satisfies eqn (2.149). In particular, muffin-tin potentials are often constructed with $\rho = R_i$ (which is the most frequent choice in the APW method) and, if the potential is not too rapidly varying in the region $2R_i - R_c < r < R_i$ the cellular expansion is likely to converge to the correct result.

Of course, even if the potential does not exactly satisfy the stated conditions, there is no reason why one should expect a breakdown of the convergence of

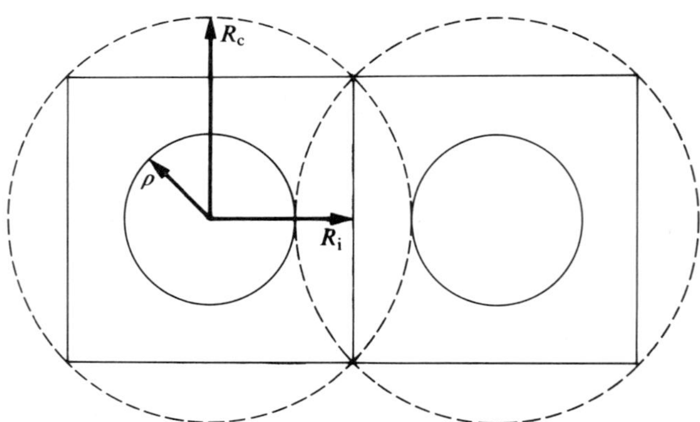

FIG. 2.10. Condition for the muffin-tin radius. $\rho = 2R_i - R_c$.

the cellular expansion, eqn (2.128), immediately after $r = R_i$ has been reached (Anderson and Fricker 1971). The wavefunction ψ satisfies an elliptic differential equation so that it must be analytic wherever the potential is analytic, and therefore eqn (2.128) will continue to represent ψ for $r > R_i$ as long as the series is uniformly convergent.

2.10. The strategy of a cellular program†

A cellular program has to handle a certain number of arrays, such as arrays of spherical harmonics, which can be labelled in two different ways. The value lm of Y_l^m can be fixed as a label and an array $Y_l^m(\mathbf{r}_i)$ formed, where $\mathbf{r}_i$ sweeps over all the boundary points. Or $\mathbf{r}_i$ can be fixed as the label and an array $Y_l^m(\mathbf{r}_i)$ formed where lm sweeps over all values used of these quantum numbers.

The choice between these two schemes must be the first step in deciding the general form that the program will take, and it is far from trivial: it will now be shown that it is very advantageous to label the arrays by the points $\mathbf{r}_i$ rather than by the spherical harmonics lm.

This is so because, if the rectangular matrix A (p rows by n columns) in eqn (2.20) is formed by rows (each row being an equation pertaining to a given pair of boundary points), then the square matrix $B = A^\dagger A$ of eqn (2.29) can be obtained at the same time without having to construct and store the matrix A as a whole. This is achieved by Scheme 2.1.

For general $\mathbf{k}$ vectors the number of columns n (number of coefficients) will be about 100 and the number of rows p (number of equations) about 200. Thus the matrix A will require 20K words of storage space, which in most computer installations will require the transfer of the matrix from the

† Sections 10 to 12 are concerned with programming details and the reader can proceed directly to Section 13 if he so wishes.

SCHEME 2.1

Formation of the square matrix B

Definitions
From eqn (2.29):

$$B_{ij} = \sum_{s=1}^{p} A_{si}^{*} A_{sj} \tag{2.150}$$

p: number of boundary condition equations
n: number of unknowns
$i = 1, 2, \ldots, n$
$j \geqslant i$ (B is symmetric)
Procedure

Set $s = 1$, $\forall B_{ij} = 0$	(2.151)
→Calculate A_{sr}, $r = 1, 2, \ldots, n$	(2.152)
Replace B_{ij} by $B_{ij} + A_{si}^{*} A_{sj}$	(2.153)
(A_{sr} may be discarded)	(2.154)
Replace s by $s+1$	(2.155)
Test current value of s	(2.156)

If $s \leqslant p$ return to eqn (2.152)

If $s > p$ END

primary to a backing store. In the scheme described in Table 2.1 the symmetric matrix B of dimension $n \times n$ is directly formed, which in the example given requires about 5K store space, which can in most cases be held entirely in the primary store without any need for time consuming transfers to the backing store.

An overall description of the program will be given in what follows, leaving numerical details for the next section. The following notation will be used:

ν_T: total number of pairs of points for which arrays are formed and stored. (One array per pair.)

ν: total number of pairs of points actually used in fitting the boundary conditions.

l_{max}: largest value of l included in the cellular expansion.
S^2: residual squared, defined by eqn (2.21).

The overall arrangement of the program is as follows. It commences with two preludes, the purpose of which is to produce and store arrays that relate to the geometry and the field respectively. The program proper is used in two versions. The first version is a preliminary program (Scan) which produces a rapid scan of the band structure in order to provide estimates for the energy eigenvalues. The second version is the main program itself which, starting from the preliminary estimates, produces the final accurate eigenvalues.

2.10.1. *Prelude* 1. *Geometry and harmonics*

This prelude is run once only for each crystal structure. Its results are given in units of the lattice constant and they can be scaled at a later stage for any metal with the given structure. All data produced is stored in magnetic tape for a permanent record.

Data received.

(1) Equations of the faces of the Wigner–Seitz cell, each face being identified by a given label.
(2) The definition of a grid of points on the basic domain of the Wigner–Seitz cell (see Section 2.11).
(3) ν_T.
(4) A set of 3×3 matrices that represent the effect of all the point group operations on the components of vectors defined in a set of orthogonal axes chosen at the centre of the Wigner–Seitz cell.
(5) A table of the coefficients of the lattice harmonics for $\mathbf{k} = \Gamma$.

Program action.

(1) It chooses a total of ν_T points from the grid defined in the basic domain, these points being evenly distributed over the domain in such a way that the number of points on each face of the basic domain is proportional to the area of the face.
(2) On using data (4) and the procedure described in Section 2.6 it distributes the ν_T points over all the faces of the Wigner–Seitz cell and forms a serial list of the corresponding face labels.
(3) It forms coordinates x_i, y_i, z_i and x'_i, y'_i, z'_i $(i = 1, 2, \ldots, \nu_T)$ for each point and companion point respectively.
(4) It forms polar coordinates r_i, θ_i, ϕ_i, r'_i, θ'_i, ϕ'_i $(i = 1, 2, \ldots, \nu_T)$ for each point and companion point respectively.
(5) It forms direction cosines of the outer normal to the face, l_{1i}, l_{2i}, l_{3i}, l'_{1i}, l'_{2i}, l'_{3i}, $(i = 1, 2, \ldots, \nu_T)$ for each point and companion point respectively.

(6) It forms arrays of spherical harmonics $Y_l^m(\theta_i\phi_i)$ and $Y_l^m(\theta_i'\phi_i')$, each array labelled by $\theta_i\phi_i$ and $\theta_i'\phi_i'$ ($i = 1, 2, \dots, \nu_T$) the running index in each array being all the values of l and m for $l = 0, 1, \dots, 12$.

(7) It forms arrays of the angular part of the normal derivatives of the spherical harmonics (see eqn (2.165))

$$\mathrm{r}_i^{-1}\mathrm{l}_{2i}\,\partial Y^m(\theta_i\phi_i)/\partial\theta + (\mathrm{r}_i \sin\theta_i)^{-1} l_{3i}\,\partial Y^m(\theta_i\phi_i)/\partial\phi \tag{2.157}$$

and similarly for the companion points, the arrays being labelled and indexed as in (6).

(8) It forms arrays of lattice harmonics for $\mathbf{k} = \Gamma$, labelled and indexed as in (6).

(9) It forms arrays of derivatives of the form (2.157) for the lattice harmonics for $\mathbf{k} = \Gamma$, labelled and indexed as in (6).

2.10.2 *Prelude 2. Field and radial integrations*

This prelude is run once only for a given metal. All data produced is stored in magnetic tape for a permanent record.

Data received.

(1) Lattice constant a.

(2) Coordinates r_i and r_i' ($i = 1, 2, \dots, \nu_T$) from output (4) of Prelude 1.

(3) Atomic number of metal, N.

(4) Electronic configuration of atom or ion.

(5) Radial functions $R_{ln}(r)$ for the atom or ion computed by a SCF Hartree–Fock program.

(6) Range of energy parameters $\varepsilon_1 < \varepsilon < \varepsilon_2$ for which the radial integrations will be effected.

(7) The step Δ in ε for which the stated range will be covered. (In practice $\Delta = 0{\cdot}01$ Ryd.)

(8) Number of shells of neighbours for which the superposition of the potentials will be effected. (As an example, in the face-centred cubic lattice of Ca it was found sufficient to add up over the first eight shells of neighbours.)

Programme action.

(1) It forms the potential field (see below).

(2) It scales data (2) by means of the lattice constant, datum (1).

(3) From the above result, it decides how far the radial integrations should proceed. (The maximum value of r_i must be found and a suitable excess added to allow for interpolation.)

(4) It forms $R_l(r_i)$ and $R_l(r_i')$ for $i = 1, 2, \dots, \nu_T$, $0 < l < 12$ and all values of ε in the grid defined by data (6) and (7).

(5) It forms $\mathrm{d}R_l(r_i)/\mathrm{d}r$ and $\mathrm{d}R_l(r_i')/\mathrm{d}r$ for $i = 1, 2, \dots, \nu_T$, $0 < l < 12$ and all values of ε in the grid defined by data (6) and (7).

The technique for obtaining the potential field will not be further described in this review, since the method followed is identical with that fully explained by Loucks (1967). There are only two differences. One is that the atomic wavefunctions are obtained from the SCF Hartree–Fock program written by D. F. Mayers. The second is that, in the cellular method, the formation of the muffin-tin potential as described by Loucks is not always essential: if questions of convergence are no worry the spherical potential can be continued right out to R_c.

2.10.3. *Preliminary program (Scan)*

The purpose of this program is to produce preliminary curves of S^2 as a function of ε for a stated set of irreducible representations of a stated set of **k** vectors. Time saving is achieved in three ways. (1) By using a small value for $l_{max}(l = 4$ being suitable for most purposes) and, correspondingly, a small number of boundary point pairs, ν. (2) By using a coarse grid in ε, the step Δ being usually 0·05 Ryd. (3) By dealing simultaneously with a large number of values of **k** and irreducible representations, which permits time saving through forming simultaneously a number of B matrices.

This program is first used for all points of symmetry in **k** space and for a few selected general **k** values, so as to have suitable estimates for the energy eigenvalues for all the bands required. The energy eigenvalues can be estimated from the curves obtained to about 0·01 Ryd in most cases, but in general to better than 0·03 Ryd. The output data is carefully analyzed at this stage to satisfy group-theoretical compatibility rules and to notice possible band crossings that might require some delicate work at a later stage.

Data received.

(1) Output of preludes 1 and 2 in the backing store.
(2) List of **k** vectors and irreducible representations.
(3) l_{max} (normally equal to (4).
(4) ν.
(5) Range in the energy parameter, $\varepsilon_1 < \varepsilon < \varepsilon_2$, to be covered.
(6) Step Δ in ε (normally $\Delta = 0{\cdot}05$ Ryd).
(7) List of reflection planes that belong to the groups of the **k** vectors and of their matrix representatives.
(8) List of coefficients of the lattice harmonics in terms of the lattice harmonics of Γ.

(Data 7 and 8 are prepared beforehand and joined to the output of Prelude 1.)

Program action.

(1) It thins out evenly the ν_T points of the grid into a set of ν points, apportioning a number of points to each face of the basic domain of the Wigner–Seitz cell proportional to its area.

(2) It forms arrays of lattice harmonics and derivatives for each point and companion point, from 1 to v, each array containing all the values of l and m that correspond to $0 < l < l_{max}$.
(3) It follows the procedure described in Section 2.6 to form a code that indicates for each pair of points whether the boundary condition for the wavefunctions or its normal derivative is trivial, this code to be used to step over the corresponding equation when forming B.
(4) It forms, for each value of ε in the range stated, all the B matrices for all the **k** vectors and irreducible representations stated and computes the corresponding values of S^2.

2.10.4. *Main program (determination of eigenvalues)*

The purpose of this program is to find accurate minima of S^2 as a function of ε for a given **k** and a given irreducible representation. The program starts from an estimate ε_0 of the eigenvalue, provided by the preliminary scan. (This value must belong to the grid for ε, in steps of 0·01 Ryd, used by Prelude 2.) By adding $\Delta = 0{\cdot}01$ Ryd to ε_0 a new residual is formed and it is possible to decide in which direction to change ε in order to move towards the minimum. Sequences of three residuals are formed for three successive values of ε until a non-monotonic set is obtained, and a parabola is then fitted to this set to determine the minimum.

Data received.

(1) Output of preludes 1 and 2 stored in backing store.
(2) **k** vector.
(3) Irreducible representation.
(4) ε_0 from Scan and Δ (usually 0·01 Ryd).
(5) l_{max}.
(6) v.
(7), (8) Same as (7), (8) respectively of Preliminary Program.

Program action.

(1), (2), (3) Same as actions so numbered of Preliminary Program. The remaining actions are described in the flow diagram given in Scheme 2.2.

If the wavefunctions are required in complete detail an appendix must be added to the program shown in Scheme 2.2. For the value $\varepsilon = E$ the radial functions are computed and stored for $0 < r < R_c$, B is formed, and a new ρ and eigenvector computed.

Details of the geometry and of the radial integrations will be discussed in Sections 2.11 and 2.12.

2.10.5. *Numerical accuracy*

The numerical accuracy of the program will now be briefly discussed. The matrix elements of B are obtained by forming scalar products of vectors of, at most, 300 components. If δ is the relative error of a vector component,

SCHEME 2.2

Main program flow diagram

Definitions

$\varepsilon_3 > \varepsilon_2 > \varepsilon_1$; $\varepsilon_3 - \varepsilon_2 = \varepsilon_2 - \varepsilon_1 = \Delta$; $S^2 = \rho$.

$\varepsilon_3, \varepsilon_2, \varepsilon_1$ and their residuals ρ_3, ρ_2, ρ_1 are stored in fixed locations.

$a \rightarrow b$ means: replace contents of location b by contents of location a.

Procedure

Set and print $\varepsilon_1 = \varepsilon_0$
Form B
Obtain and print ρ_1
Form and print $\varepsilon_2 = \varepsilon_1 + \Delta$
Form B
Obtain and print ρ_2
Compare ρ_1 and ρ_2

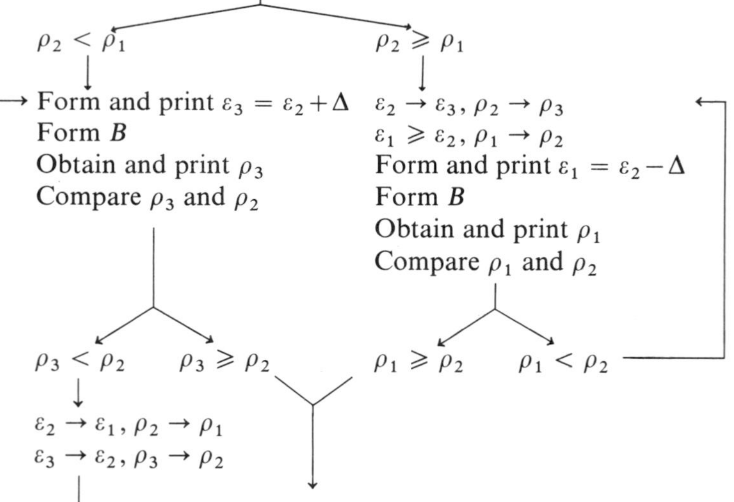

Fit parabola to points $(\rho_1\varepsilon_1)$, $(\rho_2\varepsilon_2)$, $(\rho_3\varepsilon_3)$
Find coordinates (ρ, ε) of apex
Print eigenvalue $E = \varepsilon$
Print eigenvector corresponding to the value ε_1 or ε_2 or ε_3 which is nearest to ε
END

the relative error of a matrix element is $300 \times 2\delta$, say $10^3\delta$. On using a Householder routine to get the eigenvalue, the relative error of the eigenvalue will be of the order of $10^5\delta$. Since we want this error to be of the order of 10^{-3} the relative error of one term of the cellular expansion should be about 10^{-8}. These terms, therefore, must be computed with 8 significant figures. However, since experience shows that the contribution of the terms over $l = 4$, although significant, is about two orders of magnitude smaller than that of the terms for $l < 4$, it is to be expected that a slightly reduced accuracy for the terms of higher l will be acceptable. We aim therefore at keeping no less than 8 significant figures up to $l = 4$ and about 7 for higher l values. Tests of the method show that this gives an accuracy of the eigenvalues far in excess of the stated limit of 10^{-3}. (See Section 2.13.)

2.11. Geometry and harmonics

The first question to be tackled is the choice of a grid of points in the basic domain of the Wigner–Seitz cell surface (see Fig. 2.8). For this purpose, each type of face must be separately considered and a grid formed on it. If the face is square an orthogonal grid is used but in other cases it is most convenient to use non-orthogonal axes parallel to the sides of the face. In Fig. 2.11 the grid for the top face of Fig. 2.8 is shown. A convenient point is

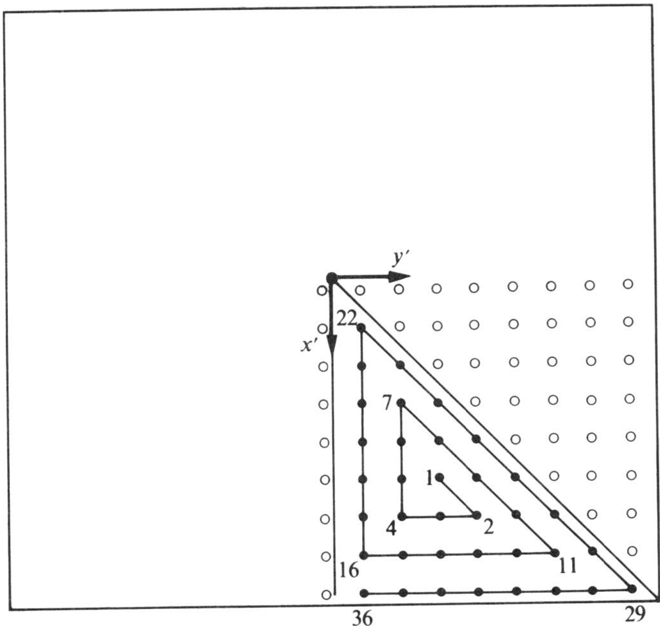

FIG. 2.11. The basic grid on a Wigner–Seitz face.

taken as origin of the count near the centre of the basic area and the points of the grid are labelled spiral-wise from it: this is the label that counts the points serially in the various arrays. The advantage of this spiral counting is that when the points are either spread out to cover the whole cell, or thinned out, and this is done serially in the order defined, then no unwanted periodicities in the positioning of the points are introduced. That is, an element of randomization is incorporated in the final set of points.

It should be noticed that points on the edges of the basic area are not included in the initial array, so as to avoid singularities that might appear in some of the polar coordinates.

As the initial array is formed, the programme records the coordinates of the points in the local axes x', y' of the face. This is done for the various types of faces and the next step is to reduce the number of points in the grid to the desired number. It has been found in practice that 256 points are ample for all purposes. All points on the basic area on each type of face are counted and the number assigned to each face is made proportional to its area. Once this is done, the initial array is thinned out to reduce the total number of points to 256 or to whatever number ν_T that might be chosen. Of course, the spiral count of the points is preserved and for each point the local coordinates as well as a label denoting the face are recorded.

The next step is to obtain the cartesian coordinates x, y, z of each point in the central axes of the Wigner–Seitz cell (see Fig. 2.8). This is easily done by obtaining rotation matrices that map the given face onto the x, y plane.

Once this is done the ν_T points are spread out over the whole Wigner–Seitz cell surface, avoiding possible redundancies as explained in Section 2.6. In order to do this, the 3×3 matrices of the point group operations are used and they are serially applied to the points as listed in the array (always on the spiral count). The result is now a non-redundant set of ν_T points spread out over the whole of the Wigner–Seitz cell.

The next step is to form the parallel array of the ν_T companion points. In order to do this the program must be provided with the normal vector to each Wigner–Seitz face and the distance from the origin of the cell to the centre of the face, which are data given by the equations of the face. If $\mathbf{r}$ is the vector corresponding to a point, the face label must be inspected and from it a vector $\mathbf{t}$ constructed parallel to the outer normal to the face and of modulus equal to twice the distance from the origin to the face centre. The vector $\mathbf{r}'$ corresponding to the companion point is given by $\mathbf{r}' = \mathbf{r} - \mathbf{t}$.

For all points and companion points spherical coordinates

$$r = (x^2 + y^2 + z^2)^{\frac{1}{2}} \tag{2.158}$$

$$\theta = \tan^{-1}\{(x^2 + y^2)^{\frac{1}{2}} z^{-1}\} \tag{2.159}$$

$$\phi = \tan^{-1}(yx^{-1}) \tag{2.160}$$

are formed.

Possible singularities for $x = z = 0$ are avoided by ensuring, as described above, that the grid contains no lines of symmetry.

As explained in Section 2.10 the number of points ν_T of the basic grid is in practice reduced to a smaller number ν before the boundary conditions are fitted. This is done by thinning out the grid serially and keeping to the same reduction factor face by face, to maintain the proportionality between the number of points in a face and its area. It should be noticed that, if $\mathbf{G^k}$ is larger than the identity the effective number of pairs of points on which the boundary conditions are fitted is ν times the number of point group operations in $\mathbf{G^k}$.

In order to form the outer normal derivatives of the wavefunction, the operator $\mathbf{n} \cdot \nabla$ must be formed, where $\mathbf{n}$ is the unit vector in the direction of the outer normal to the face. In polar coordinates

$$\nabla = \mathbf{u}_r \,\partial/\partial r + r^{-1}\mathbf{u}_\theta \,\partial/\partial\phi + (r \sin\theta)^{-1}\mathbf{u}_\phi \,\partial/\partial\phi, \tag{2.161}$$

where $\mathbf{u}_r$, $\mathbf{u}_\theta$ and $\mathbf{u}_\phi$, the spherical polar unit vectors, have the following cartesian coordinates (x, y, z), respectively in the order given below,

$$\mathbf{u}_r = (\sin\theta\cos\phi, \sin\theta\sin\phi, \cos\theta), \tag{2.162}$$

$$\mathbf{u}_\theta = (\cos\theta\cos\phi, \cos\theta\sin\phi, -\sin\theta), \tag{2.163}$$

$$\mathbf{u}_\phi = (-\sin\phi, \cos\phi, 0). \tag{2.164}$$

Therefore,

$$\mathbf{n} \cdot \nabla = l_1 \,\partial/\partial r + r^{-1} l_2 \,\partial/\partial\theta + (r\sin\theta)^{-1} l_3 \,\partial/\partial\phi, \tag{2.165}$$

where the direction cosines l_1, l_2, l_3 are

$$l_1 = \mathbf{n} \cdot \mathbf{u}_r, \qquad l_2 = \mathbf{n} \cdot \mathbf{u}_\theta, \qquad l_3 = \mathbf{n} \cdot \mathbf{u}_\phi. \tag{2.166}$$

For each point the vectors (2.162), (2.163) and (2.164) are worked out and from them the direction cosines (2.166) are obtained and stored.

2.11.1. *The harmonics*

We shall now briefly discuss the spherical and lattice harmonics. For convenience of calculation we define modified associated Legendre polynomials as follows,

$$\mathscr{P}_l^m(x) = \left\{\frac{2l+1}{4\pi}\frac{(l-m)!}{(l+m)!}\right\}^{\frac{1}{2}} P_l^m(x); \qquad m \geqslant 0, \tag{2.167}$$

so that the *normalized* spherical harmonics are

$$Y_l^m(\theta\phi) = \mathscr{P}_l^{|m|}(\cos\theta)\exp(im\phi). \tag{2.168}$$

The polynomials $\mathscr{P}_l^m(\cos\theta)$ for $l = 0, 1, 2, \ldots, 12$, and $m = 0, 1, \ldots, l$, are obtained from the recurrence relation

$$\mathscr{P}_{l+1}^m(\cos\theta) = \{(2l+1)(2l+3)/(l-m+1)(l+m+1)\}^{\frac{1}{2}} \cos\theta\,\mathscr{P}_l^m(\cos\theta)$$
$$-\{(l-m)(l+m)(2l+3)/(2l+1)(l-m+1)(l+m+1)\}^{\frac{1}{2}}\mathscr{P}_{l-1}^m(\cos\theta). \tag{2.169}$$

Starting values are provided from

$$\mathscr{P}_0^0 = (2\sqrt{\pi})^{-1} \tag{2.170}$$

by the recurrence relations

$$\mathscr{P}_{l+1}^{l+1} = \{(2l+3)/(2l+2)\}^{\frac{1}{2}} \sin\theta\mathscr{P}_l^l, \tag{2.171}$$

$$\mathscr{P}_{l+1}^l = (2l+3)^{\frac{1}{2}} \cos\theta\mathscr{P}_l^l. \tag{2.172}$$

All the values $\mathscr{P}_l^m$ for $l = 12$ are computed directly and compared with the results of the recurrence relation; if the difference is larger than a unit in the ninth significant place an appropriate output is printed so that the run can be revised.

The derivatives of the Legendre polynomials are computed by the relation

$$\mathrm{d}\mathscr{P}_{l+1}^m(\cos\theta)/\mathrm{d}\theta = (l+1)\cot\theta\mathscr{P}_{l+1}^m(\cos\theta)$$
$$-(\sin\theta)^{-1}\{2l+3)(l-m+1)(l+m+1)/(2l+1)\}^{\frac{1}{2}}\mathscr{P}_l^m(\cos\theta). \tag{2.173}$$

Starting values are provided from

$$\mathrm{d}\mathscr{P}_0^0/\mathrm{d}\theta = 0 \tag{2.174}$$

by the recurrence relations

$$\mathrm{d}\mathscr{P}_{l+1}^{l+1}/\mathrm{d}\theta = (l+1)\cot\theta\mathscr{P}_{l+1}^{l+1}, \tag{2.175}$$

$$\mathrm{d}\mathscr{P}_{l+1}^l/\mathrm{d}\theta = (l+1)\cot\theta\mathscr{P}_{l+1}^l - (2l+3)^{\frac{1}{2}}(\sin\theta)^{-1}\mathscr{P}_l^l. \tag{2.176}$$

As a result of this work an array is formed for each point with all $Y_l^m(\theta\phi)$ up to $l = 12$ and another one for the angular derivatives in the form given by eqn (2.157).

The lattice harmonics $X_{l\nu}(\theta\phi)$ for Γ are formed from the following information stored as part of the program data:

(1) A list of l values that appear in all $X_{l\nu}(0 < l < 12)$.
(2) A list of m values that appear in each $X_{l\nu}$.
(3) The ϕ-dependence of each representation.
(4) The coefficients of the linear combinations of $Y_l^m(\theta\phi)$ that appear in each $X_{l\nu}$, as given by Altmann and Cracknell (1965) and recomputed by Bradley and Cracknell (1972).

In all the work concerning the geometry the lattice constant is taken to be $a = 2\pi$, a convenient value for empty-lattice calculations (see Section 2.13).

All the arrays are output onto magnetic tape and when they are input from this tape to backing store by the master programme for a given metal they are simultaneously scaled for the corresponding lattice constant.

In principle, one could also work out and store all the lattice harmonics for all representations. If a very large backing store is available they could be worked out from those of Γ as the geometry tape is read onto the backing store by the master programme. This should normally be quicker and more convenient than keeping all the lattice harmonics in magnetic tape.

2.12. The radial integrations

The radial equation to be solved takes the form

$$\{d^2/dr^2 + 2rZ_p(r) + \varepsilon - l(l+1)r^{-2}\}rR_l(r) = 0, \tag{2.177}$$

where the potential $V(r)$ is given in terms of the effective nuclear charge

$$Z_p(r) = r^{-1}V(r) \tag{2.178}$$

and the radii r are measured in atomic units and the energy parameter ε in Rydberg units. There are now available a number of accurate procedures and programs to solve this equation but it might nevertheless be worthwhile to give the details of a scheme used by the author and collaborators, since the functions $R_l(r)$ must be determined to eight significant figures in accordance to the discussion at the end of Section 2.10. Since the potential field is only known to about three or four significant figures it might appear inconsistent to integrate eqn (2.177) to the accuracy stated. However, the higher order figures in $V(r)$, whether zeros or not are kept constant throughout a band calculation and they will merely introduce a systematic error, whereas it is the random error in the matrix elements that must be kept within the stated limits. As discussed in Section 2.10 the relative error of a term in the cellular expansion must be 10^{-8}. Since the harmonics are computed to 10^{-9} it is enough to obtain $R_l(r)$ with eight significant figures, although, as explained in Section 2.10 about seven for $l > 4$ should suffice.

In order to avoid having to use a variable interval in the integration of eqn (2.177) it is convenient to introduce a logarithmic transformation in the variable (Altmann 1955),

$$\rho = \ln r \tag{2.179}$$

so that eqn (2.177) takes the form

$$d^2G_l(\rho)/d\rho^2 + \{2Z_p(\rho)\exp\rho + \varepsilon\exp(2\rho) - (l+\tfrac{1}{2})^2\}G_l(\rho) = 0, \tag{2.180}$$

where

$$G_l(\rho) = R_l(r)\exp(\tfrac{1}{2}\rho). \tag{2.181}$$

Eqn (2.180) contains no first derivative and can be integrated by Numerov's method (Hartree 1957). However, this method is not very accurate near the nucleus where the potential varies very rapidly. Admittedly, the integration is very stable with respect to small errors in the starting region but, in order to obtain the high accuracy required, it is worthwhile to start the integration by the slower but more accurate Runge–Kutta method. (It must be remembered that the computation of a complete band structure requires about 200 values of ε and 12 of l, that is about 2000 radial integrations only, so that the total computational time involved is not very long in any case.) We use the transformation

$$H_l(\rho) = G_l(\rho)\exp\{-(l+\tfrac{1}{2})\rho\}, \tag{2.182}$$

which yields the pair of simultaneous linear differential equations

$$\mathrm{d}H_l(\rho)/\mathrm{d}\rho = Q_l(\rho). \tag{2.183}$$

$$\mathrm{d}Q_l(\rho)/\mathrm{d}\rho = -(2l+1)Q_l(\rho)-\{2Z_p(\rho)\exp\rho+\varepsilon\exp 2\rho\}H_l(\rho). \tag{2.184}$$

These equations are integrable by the Runge–Kutta method.

The integrations are started at $\rho = -6.125$ and continued until half a dozen steps beyond the value of ρ that corresponds to R_c. The step in the integration is taken as $\Delta = \frac{1}{64}$ but for the first hundred steps computed by the Runge–Kutta procedure 2Δ is used, the intermediate values of the various tabulated quantities being in any case required in the computations by this method. At the end of this range Everett interpolation is used to obtain values of $G_l(\rho)$ at two adjacent grid points to serve as starting values for the Numerov routine. The radial functions computed are stored for each value of ε, for six steps before and beyond respectively R_i and R_c, in all for about 30 values in a typical case. From these values, $R_l(r)$ for the values of r corresponding to all boundary points is computed by Everett interpolation and stored. Also, the values $\mathrm{d}R_l(r)/\mathrm{d}r$ are computed by a six-point central difference formula (Hartree 1958).

We must return to the first part of the integration. One of the advantages of the Runge–Kutta method is that it is easier to start it than the Numerov procedure. Hartree (1957) has given a method to obtain the initial values which is based in a linear approximation to the potential near the origin. For the reasons stated above a more accurate method was devised which in any case has a negligible effect on the computing time.

The effective nuclear charge is expanded as follows:

$$Z_p(r) = N+v_1r+v_2r^2+v_3r^3+\mathrm{O}(r^4), \tag{2.185}$$

where N is the atomic number and the coefficients v_1, v_2, v_3 are obtained by a least-squares fit to the numerical values of the potential very near the origin. Corresponding to eqn (2.185) a similar expansion is written for $H_l(\rho)$:

$$H_l(\rho) = 1 - a \exp \rho + b \exp 2\rho - c \exp 3\rho + d \exp 4\rho + \mathrm{O}(\exp 5\rho), \quad (2.186)$$

so that, from eqn (2.183),

$$Q_l(\rho) = -a \exp \rho + 2b \exp 2\rho - 3c \exp 3\rho + 4d \exp 4\rho + \mathrm{O}(\exp 5\rho). \quad (2.187)$$

On substituting the above expansions into eqn (2.177), we obtain

$$a = N(l+1)^{-1}, \quad (2.188)$$

$$b = \{2Na - (2v_1 + \varepsilon)\}\{2(2l+3)\}^{-1}, \quad (2.189)$$

$$c = \{2Nb - (2v_1 + \varepsilon)a + 2v_2\}\{6(l+2)\}^{-1}, \quad (2.190)$$

$$d = \{2Nc - (2v_1 + \varepsilon)b + 2v_2 a - 2v_3\}\{4(2l+5)\}^{-1}. \quad (2.191)$$

The coefficients a, b, c, d are computed from eqns (2.188)–(2.191) so that eqns (2.186) and (2.187) permit the values of H_l and Q_l to be computed for the first value of ρ of the grid, ρ_0. The Runge–Kutta procedure can now be started.

The normalization integral for $R_l(r)$,

$$\mathcal{N} = \int_{-\infty}^{\rho_e} G_l(\rho) \exp 2\rho \, \mathrm{d}\rho, \quad (2.192)$$

where ρ_e corresponds to the radius of the equivalent sphere, is obtained in three parts. From $-\infty$ to ρ_0 the power series expansion is used and the integral computed analytically. From ρ_0 to ρ_f, the nearest grid value in ρ to ρ_e, the integral is given by a finite differences formula (Hartree 1958, p. 102) extended to sixth differences. From ρ_f to ρ_e the contribution (which can be positive or negative, depending on whether ρ_f is larger or smaller than ρ_e) is estimated by a trapezium rule.

The accuracy of the integration was checked by using the program for $V(r) = 0$, in which case the solutions are spherical Bessel functions, which are tabulated. In order to avoid any effect of the normalization integrals, ratios of the Bessel functions were obtained at two values of r, and compared with the corresponding ratios for the tabulated values. The results (Harford 1972) are shown in Table 2.1.

From these and similar results it follows that the integrations are correct to about 10^{-7} in relative errors, which is within the order of magnitude desired for this high value of l. The normalization integrals were similarly checked against Bessel functions and the relative errors found to be 10^{-8} for $l = 0$ going up to 10^{-7} for $l = 8$ (Harford 1972).

TABLE 2.1
$R_l(r_1)/R_l(r_2)$ *for* $l = 8$ *and* $r_1 = 2{\cdot}2186$ *a.u.*, $r_2 = 3{\cdot}0325$ *a.u.*

Eigenvalue Ryd	$R_l(r_1)/R_l(r_2)$ Program	Tables	Difference
0	0·08208486	0·08208500	14×10^{-8}
1	0·09204499	0·09204510	11×10^{-8}
2	0·10367067	0·10367074	7×10^{-8}
3	0·11734201	0·11734204	3×10^{-8}
4	0·13355525	0·13355524	1×10^{-8}

2.13. Tests of the method

2.13.1. *The empty lattice test*

The first test that will be discussed in this section is the so-called empty lattice test first used by Shockley (1937). An account of the method is given by Jones (1960) and a rather more general treatment of it was suggested by Altmann and Bradley (1965*b*).

The energy $E_{\mathbf{k}}$ of a free electron in a state given by a plane wave $\psi(\mathbf{r}) = \exp(i\mathbf{k}\,.\,\mathbf{r})$ is

$$E_{\mathbf{k}} = \mathbf{k}^2 \tag{2.193}$$

in atomic units with the Rydberg as unit of energy. If in a cellular program we take the potential field $V(r) = 0$ and specify a value of $\mathbf{k}$ and an irreducible representation and obtain the eigenvalue that corresponds to a given band, the result obtained should be a good approximation to eqn (2.193). From the point of view of the numerical calculations entailed in the cellular method, a vanishing potential field introduces no simplification whatever. On the contrary, the accuracy of the radial integrations is considerably strained by the very rapid variation of the spherical Bessel functions near their nodes. So, the empty-lattice test is a very powerful check of the method.

In order to use the empty-lattice test we must be able to identify the irreducible representation or representations that correspond to the free-electron eigenvalue for a given $\mathbf{k}$ vector and a given band. Let us write

$$\mathbf{k} = \boldsymbol{\kappa} + \mathbf{K} \tag{2.194}$$

where $\boldsymbol{\kappa}$ is a vector of the first Brillouin zone and $\mathbf{K}$ a vector of the reciprocal lattice. In the extended zone scheme, if $\mathbf{K} = 0$, $E_{\mathbf{k}}$ is an eigenvalue in the first band and the eigenvalues of the successive bands are obtained by taking $\mathbf{K}$ in such a way that $\mathbf{k}$ is in the second and higher Brillouin zones successively.

The method used will be more easily understood by considering a specific example. For the simple cubic lattice, with lattice constant a, the unit vectors of the reciprocal lattice are, as shown in Fig. 2.12,

$$\mathbf{x}^* = 2\pi a^{-1}(1, 0, 0), \tag{2.195}$$

$$\mathbf{y}^* = 2\pi a^{-1}(0, 1, 0), \tag{2.196}$$

$$\mathbf{z}^* = 2\pi a^{-1}(0, 0, 1). \tag{2.197}$$

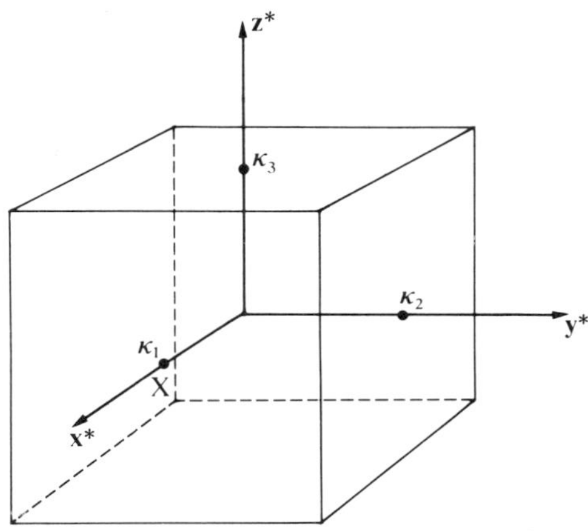

FIG. 2.12. First Brillouin zone for the simple cubic lattice. The points X and κ coincide.

Since the lattice constant is immaterial for the empty lattice test, we take

$$a = 2\pi \text{ a.u.} \tag{2.198}$$

A vector **k** will be given as

$$\mathbf{k} = k_x\mathbf{x}^* + k_y\mathbf{y}^* + k_z\mathbf{z}^*. \tag{2.199}$$

The energy $E_{\mathbf{k}}$ is given by $\mathbf{k}^2$ in Rydberg units (with lengths measured in reciprocal atomic units). Therefore from eqns (2.195) to (2.197) and eqn (2.199),

$$E_{\mathbf{k}} = k_x^2 + k_y^2 + k_z^2. \tag{2.200}$$

Thus, for **k** given by eqn (2.194),

$$\begin{aligned} E_{\boldsymbol{\kappa}+\mathbf{k}} &= (\kappa_x + K_x)^2 + (\kappa_y + K_y)^2 + (\kappa_z + K_z)^2 \\ &= E_{\boldsymbol{\kappa}} + E_{\mathbf{k}} + 2(\kappa_x K_x + \kappa_y K_y + \kappa_z K_z). \end{aligned} \tag{2.201}$$

Consider the point X of the first Brillouin zone shown in Fig. 2.12, for which

$$\boldsymbol{\kappa}_1 = (\tfrac{1}{2}, 0, 0). \tag{2.202}$$

If we take in eqn (2.201)

$$\mathbf{K} = (0, \pm 1, 0) \text{ or } (0, 0, \pm 1) \tag{2.203}$$

it follows at once that the four vectors $\mathbf{k} = \boldsymbol{\kappa}_1 + \mathbf{K}$ so defined correspond to the same energy $E = 1{\cdot}25$ Ryd. It is also easy to identify them as the vectors that correspond to X in the third band, the first and second bands having the same energy at this point.

Likewise, we take the other two vectors

$$\boldsymbol{\kappa}_2 = (0, \tfrac{1}{2}, 0), \tag{2.204}$$

$$\boldsymbol{\kappa}_3 = (0, 0, \tfrac{1}{2}), \tag{2.205}$$

in the star of X and we find for each of them four values of $\mathbf{K}$ that lead to the same degenerate eigenvalue $E = 1{\cdot}25$ Ryd. There are now twelve vectors $\boldsymbol{\kappa} + \mathbf{K}$ that correspond to this eigenvalue and they must span a reducible representation of the group of the $\mathbf{k}$ vector X. This representation must be reduced and the irreducible representations that appear after reduction are those that correspond to the energy eigenvalue $E = 1{\cdot}25$ Ryd in the third band.

The only problem that remains is a standard one, namely that of reducing the representation spanned by the twelve $\mathbf{k}$ vectors just described. The wavefunctions that correspond to these vectors are of the form $\exp\{\mathrm{i}(\boldsymbol{\kappa}+\mathbf{K})\,.\,\mathbf{r}\}$ and it is straightforward to find the characters of the reducible representation for all the operations of the group of the $\mathbf{k}$ vector X. Therefore, on using tables of irraducible representations it is possible in a manner which is standard in group theory to predict the representations that appear in the reducible one. Alternatively, group theoretical projection operators can be applied on the plane waves of the basis and the representations that appear on projection are the required ones. For low-lying eigenvalues the assignment of the irreducible representations can be considerably simplified by using group theoretical compatibility conditions.

The practical importance of the empty lattice test depends on the following three points.

(1) It checks the theoretical validity of all aspects of the method, except those that concern the spherical approximation of the potential.

(2) It provides a necessary and sufficient condition for the correctness of the numerical analysis, in so far as it verifies the number of significant figures that are kept by the computation.

(3) It provides a necessary and sufficient condition for a cellular program to be free from programming errors, except in the section concerned with the derivation of the potential field.

Points (2) and (3) are of extraordinary importance in the process of debugging a cellular program and of ensuring sufficient numerical accuracy. For this purpose, the test is applied for all small representations of all $\mathbf{k}$ vectors. As an example, Table 2.2 gives an analysis of the results obtained for the body-centred cubic lattice by Altmann, Davies, and Harford (1968) and given in full by Harford (1972).

TABLE 2.2
Empty lattice test for all small representations of all **k** *vectors of symmetry of the b.c.c. lattice.*

Range of eigenvalue	0–1	1–2	2–3	3–4	4–5	5–6	Totals
Error							
10^{-6}–10^{-5}	8	23	10	2			43
10^{-5}–10^{-4}	1	2	2	4	1	2	12
10^{-4}–10^{-3}	1					3	4
10^{-3}–2×10^{-3}						2	2

The numbers that appear in this Table are the number of eigenvalues computed with the stated error for each range of energy shown in the heading. The eigenvalues and their errors are in Rydbergs.

For the **k** vectors of symmetry given in Table 2.2 the expansions for l_{max} = 12 are quite small in most cases and have been used in general. However, for the points of lower symmetry l_{max} has been taken as either 8 or 9. The number of boundary points is adjusted to the number of unknowns, so as to make the number of equations roughly 50 per cent larger than it. For most points of symmetry 64 pairs of points more than satisfy this requirement.

Table 2.2 shows that in the large majority of cases five significant figures in the eigenvalues are kept. The two maximum errors found, between 0·001 and 0·002 Ryd appear for very high eigenvalues.

Table 2.3 shows the results of Altmann, Davies, and Harford (1968) and Altmann, Harford, and Blake (1971) for a general **k** vector of the face-centred cubic lattice. It can be seen that, even for the higher bands, errors of 0·0002 or 0·0003 Ryd are obtained if l_{max} = 8 is used.

TABLE 2.3
Empty lattice test for a general **k** *vector of the f.c.c. lattice.*

Band	Exact eigenvalue	Calculated eigenvalue	
		$l_{max}=6$	$l_{max}=8$
1	0·35	0·349997	
2	1·55	1·550566	
3	1·95	1·95311	
4	2·35	2·355	
5	2·75	2·762861	2·749796
6	3·55	3·569161	3·550307

k = (0·2, 0·3, 0·4). The lattice constant is 2π a.u. Eigenvalues in Rydbergs.

2.13.2. *The effect of the boundary points*

Changes in the boundary points used introduced drastic errors in the older cellular methods and on the light of such obsolete evidence it has been alleged until not long ago that this was a serious difficulty. Altmann and Bradley (1965) carried out a very drastic test indeed for the hexagonal close-packed lattice. They computed the eigenvalue MB' for the empty lattice, the theoretical value of which is 1 Ryd, for two sets of 96 boundary points grotesquely chosen. In each case the points covered only 30 per cent of the boundary area, in one set near the centre and in the other near the edges. The eigenvalues obtained for $l_{max} = 6$ were 1·012 Ryd and 1·024 Ryd respectively.

The choice of boundary points should be more significant for a metal potential field. Accordingly, Altmann, Davies, and Harford (1968) computed two eigenvalues for the f.c.c. lattice with a copper potential with six different sets of boundary points (4 of 64 pairs and 2 of 128 pairs). The results were constant for X_2 in all cases to better than 10^{-4} Ryd and for L_1 the root-mean-square deviation of the results was a little over 10^{-4} Ryd. In fact, it follows from the results of Altmann, Davies, and Harford (1968) that the eigenvalues stay constant within 10^{-4} Ryd when increasing the number of pairs of boundary points from 64 to 128 or 256.

As a result of general experience of this type it is recommended to use just enough boundary point pairs so as to have 50 per cent more equations than unknowns. (Although excellent results have often been obtained for much smaller margins.)

2.13.3. *Convergence of the expansion*

From the discussion of Section 2.9 the cellular expansion is fully convergent for a constant potential so that its convergence should be tested with a metal potential. Altmann, Davies, and Harford (1968) used a Cu potential constructed by Burdick (1963) and described later on in this section. They computed the eigenvalue L_3 for $l_{max} = 10$ and $l_{max} = 12$ and they obtained a difference of 0·0001 Ryd, a result which is typical of a large number of similar checks. For general **k** it is found that $l_{max} = 8$ ensures in most cases convergence within 0·001 Ryd.

2.13.4. *Accuracy of the fitting*

It is useful to assess the goodness of the fitting as far as it concerns the wavefunctions for which truncation of the cellular expansion must lead to some discontinuities at the boundary of the Wigner–Seitz cell. Their magnitude can be assessed in two ways. Firstly, an estimate of them can be gauged from the value of the residual of the fitting or, better still, the quantity $\sigma = S/\sqrt{p}$, where p is the number of boundary condition equations. Altmann and

Bradley (1965) give some values of σ when fitting eigenvalues in the empty h.c.p. lattice. For expansions with l_{max} equal to 8 or 10, σ is of the order of 10^{-6} and experience with other lattices shows typical values of σ around 10^{-5}.

A second procedure to assess the fitting results from direct inspection of the coefficients of the cellular expansion which, in its turn, can be done in two ways. For example, the theoretical form of the expansion for MB' for the empty hexagonal close-packed lattice shows that all coefficients for even l must vanish. Altmann and Bradley (1965) found that the largest such computed coefficient was 8×10^{-3}, although only $l_{max} = 6$ was used. Another way is to compute, for any field, **k** vectors of symmetry as if they were general. Differences of the order of 0·001 Ryd are found and the coefficients that should vanish by symmetry take values of 10^{-4} to 10^{-3}.

It follows from this evidence that the fitting is accurate enough to prevent the appearance of unphysical discontinuities in the wavefunction at the Wigner–Seitz cell boundary.

2.13.5. *Comparison with other methods*

The final test of the method must come from a comparison of the eigenvalues obtained for exactly the same potential by the cellular as well as other methods. This has been done by Altmann, Davies, and Harford (1968) and their results are presented in Table 2.4 in the form given by Connolly (1971).

TABLE 2.4

Copper eigenvalues for the Chodorow potential (Ryd).

Method	X_5–X_1	X'_4–Γ_1	Γ'_{25}–Γ_1	Secular equation size
APW[a]	0·249	0·804	0·399	27
APW[b]	0·2516	0·8071	0·4027	59
MOPW[c]	0·254	0·808	0·398	24
Cellular[d]	0·2511	0·8084	0·4031	7–21

[a] Burdick (1963); [b] Arlinghaus (1965); [c] Butler, Bloom, and Brown (1969); [d] Altmann, Davies, and Harford (1968).

Connolly chose to compare the various methods with reference to the differences of eigenvalues given in the table, because they represent respectively the d-band width, the sp-band width and the d–s separation. The size of the secular determinant quoted by Connolly for the cellular method, 98, is in error and has been corrected: the numbers given are the minimum size of the determinant (for Γ_1) and its maximum (for X'_4). The computations of Butler *et al.* use a modified form of the OPW method.

All computations in the table were conducted with the identical potential used by Burdick. The computations of Altmann *et al.* were conducted in order to compare with this author and without knowledge of the unpublished calculations of Arlinghaus, who used a much larger APW expansion than Burdick. The fact that the results of Arlinghaus reduce substantially the already small difference between the APW and cellular results shows up most clearly the convergence of both methods and, since the calculations were performed entirely independently, this evidence is unimpeachable. Clearly, both methods give the same results to better than 0·001 Ryd.

Altmann *et al.* also show similar agreement, to within 0·002 Ryd, between the cellular and Green's-function method.

Burdick's potential is a muffin tin, for which the muffin radius was chosen as

$$\rho = R_i, \tag{2.206}$$

which is 2·415 a.u., larger, of course, than the maximum muffin radius of eqn (2.149) required for the full convergence of the cellular expansion in the spherical approximation, which is 1·450 a.u.

What do we learn, then, from the remarkable agreement between the cellular and APW results for this potential? We know that the cellular expansion is fully convergent for a muffin-tin radius of 1·450 Å but, for $\rho = 2{\cdot}415$ Å there are non-spherical parts of the potential within the radius R_c of the circumscribed sphere which ought to be taken into account. These are fairly small lunettes bounded by the circumscribed sphere centred on one atom and the inscribed sphere centred on the next one (*cf.* Fig. 2.10). These lunettes should be expanded in spherical harmonics whereby coupled differential equations appear as shown in Section 2.9. The agreement between the APW and cellular methods shows that the effect of this coupling is negligible to better than a 10^{-3} Ryd, for the copper potential in question. This is not a specially flat potential (it varies from about $-2{\cdot}6$ at the correct $\rho = 1{\cdot}450$ a.u. to -1 at $\rho = R_i = 2{\cdot}415$ a.u. It follows, therefore, that it would be unreasonable to expect the simple form of the cellular expansion to show up convergence errors larger than 0·001 Ryd for any metal whose potential does not vary more rapidly than Burdick's, even when the muffin-tin radius is chosen to coincide with R_i.

2.14. Conclusions and critique

We shall discuss in this section advantages and disadvantages, real and alleged, of the cellular method.

Let us say straight away that the cellular method has no remarkable merit of its own that makes it superior to any other method. On the contrary, it suffers from a drawback, as compared with the APW or Green function methods, namely that it does not easily lend itself to theoretical interpretation and extension. Ziman (1971) has very rightly written as a comment on the cellular method: 'Mathematical techniques in theoretical physics are not to be judged simply on the accuracy of the answers they give when correctly programmed into a very powerful computer: it is important that we should understand clearly what is going on, and have the means for quick approximate calculations in simplified cases.'

If the reader is in the market for such goods he should address himself to other sellers: the cellular method is not for him. On the other hand, it appears to the reviewer that it is an established fact that people exist who go to computing laboratories in order to purchase band structures, Fermi surfaces, and such things and that there are people who write and operate programs for such customers. And it also appears to the reviewer that, when considering this particular activity and none other, it is legitimate to ask the question whether customer and seller alike might find it more advantageous, in a given case or for a given computer configuration, to use method A rather than method B in order to produce the goods that are transacted.

When dealing with this specific problem details must be taken into consideration that have no place whatever in Ziman's remark: speed, storage space, numerical accuracy, facility of programming and debugging, etc.

In the rest of this section we shall discuss methods in band structure purely and entirely from the point of view of the activity just described.

Pro

We here summarize such features of the cellular method as might appear advantageous.

(1) The theory of the method is extremely simple. Thus all steps in the computation are quite straightforward, which makes it easy to write a cellular program.

(2) The storage space required both as main and as backing store is quite small, 10–20 K of the first and under 100 K of the second. Successful calculations have in fact been carried out with a fraction of these facilities.

(3) The program can be completely checked both for bugs and for defects in the numerical analysis by the empty lattice test, which is a very distinctive advantage of the method. In the APW method, for instance, this test provides only a sufficient but *not necessary* condition for the correctness of the programme (Mattheiss, Wood, and Switendick 1968).

Anyone who has ever debugged a program will appreciate the importance of this fact. Consumers, on the other hand, will grasp the advantage of objective credentials as compared with consensus.

It is true, of course, that 'going up to spherical harmonics of order 12 and ensuring continuity at 256 boundary points, we arrive extremely close to the trivial result $E = k^2$' (Ziman 1971.) But what a blessing this is! Although one might reflect that not too long ago the cellular method was castigated for not being able to do precisely what Ziman says. Incidentally, $l_{max} = 12$ is also unblushingly used by the APW method and, as explained in Section 2.13, 256 boundary points are grossly in excess of the number normally used.

(4) It is extremely easy to extract the complete wavefunction from a cellular calculation. The coefficients of the expansion, of course, are obtained without any extra work and they give useful information about the wavefunction.

(5) An SCF procedure can be carried out by the method without getting involved in producing a discontinuous potential from a continuous wavefunction, as in other methods.

(6) In the last few years refinements to the APW method have been introduced to include deviations from the constancy of the potential outside the muffin. Such approximations are naturally incorporated in the cellular method and further refinements as regards the lack of sphericity of the potential can be obtained by using the method of Section 2.9.

Contra

We shall now discuss objections raised against the cellular method.

(1) Since the days of Shockley (1937) it has been alleged that the cellular eigenvalues depend significantly on the choice of boundary points. This is of course entirely obsolete as demonstrated by the evidence given in Section 2.13.

(2) Somewhat vague objections have been voiced about the trouble taken in fitting boundary conditions at a large number of points. As a practical point this is of no significance; when a computer is used it is exactly the same to fit the boundary conditions at two points or at two hundred.

(3) It has been alleged that the convergence of the cellular method is too slow (Herman 1958). Two points have to be noticed. First, it has been shown in Section 2.13 that satisfactory convergence is achieved from $l_{max} = 8$ to $l_{max} = 12$, which compares very well with the APW method in which the latter value is used.

Secondly, to discuss in isolation the value of l_{max} used might be misleading. The computer time depends crucially on three factors: size of the matrix, complexity of the matrix elements, and number of matrices that have to be computed to get one eigenvalue and it is only the first factor that depends on l_{max}. As regards this single factor the sizes of the matrices quoted in Table 2.6 appear to favour in fact the cellular method, but it should be noted that all **k** vectors therein given admit of a fair amount of symmetry, which simplifies considerably the cellular expansions. For general points the numbers would be comparable or would probably favour the APW method.

As regards the third factor above mentioned, it should be remembered that in general 3 to 4 matrices suffice to get one eigenvalue in the cellular method, a figure that appears to be exceeded in other methods.

(4) It has been pointed out (Ziman 1964, p. 85) that on fitting the boundary conditions at the surface of the Wigner–Seitz cell discontinuities are introduced in the very region where the potential is flattest. Two points must be made.

First, as shown in Section 2.13 the fitting of the boundary conditions is good enough to ensure that any remaining discontinuities in the wavefunctions are not significant.

Secondly, when in the APW the muffin-tin radius is taken as that of the inscribed sphere, large discontinuities in the wavefunction are unavoidably left at the centre of the Wigner–Seitz face, the very regions under discussions.

The argument, nevertheless is of some importance and it has recently been re-stated by the same author (Ziman 1971, p. 24) roughly to the effect that the Wigner–Seitz polyhedron is an artificial mathematical construct and therefore inadequate to fit physical boundary conditions. This might well be so, but if it were so, something pretty horrible ought to happen when attempting to do such a fitting. This is far from being the case, as has been shown in Section 2.13.

It appears to the present writer that, on the contrary, the success of the Wigner–Seitz surface in fitting boundary conditions can easily be understood *a posteriori* (*a priori* predictions of doom or success being, of course, of little use). First, a good deal of the surface is quite near the muffin-tin sphere when, as usual $\rho = R_i$. Secondly, the surface closely reflects the crystal symmetry which is so influential in this context. Thirdly, although the surface may be in a somewhat featureless region, fittings in such regions have successfully been made for a long time, ever since Hartree showed how to join the outgoing and incoming solutions of the radial Schrödinger equation well outside the 'featured' region in which the wavefunction is oscillatory and the potential rapidly varying.

(5) We now come to the use of the spherical approximation and the related problems of convergence, discussed in Section 2.9. From the references therein given it is evident that the problem has been very clearly understood, at least by some people, for many years. Unfortunately much confusion continues to arise owing to incomplete statements. As an example: 'The cellular wavefunctions will not converge to the true wavefunction solutions of the periodic potential problem except within a sphere inscribed within the Wigner–Seitz cell. Outside that sphere some other form of the expansion must be used.' (Dimmock 1971.)

What is not mentioned here is that the statement is valid *only* for potentials that do *not* obey the conditions discussed in Section 2.9.

It appears in fact to the present writer that Saffren's proof of the convergence of the cellular expansion for a special muffin-tin potential is not as widely known as it deserves to be. As an example, another recent quotation: 'More recently Altmann *et al.* using a muffin-tin potential have obtained a band structure for copper which agrees to 0·002 Ryd with the results of APW and KKR calculations for the same potential. While this comparison undoubtedly justifies the method by hindsight *in this particular case* [their italics] the authors remark that "the alleged lack of convergence of the cellular expansion outside the inscribed sphere entirely disappears if a muffin-tin potential is used" is misleading because even for a periodic muffin-tin potential one has in general no *a priori* knowledge that (2) converges outside the inscribed sphere.' (Anderson and Fricker 1971.)

The present writer wishes to make two comments. First, the phrase quoted from Altmann *et al.* was not used, as the quotation appears to imply, to validate the agreement found. It was given in the introduction to the paper merely to stress the point that convergence 'can be no more than a property of the potential used and not of the cellular method itself', words that immediately follow the quotation given. Obviously, to validate this desired conclusion no further details of the muffin-tin were necessary.

Secondly, it is clear from the discussion in Section 2.13 that the agreement found is to be expected for a large family of potentials and not just 'in this particular case', a statement which appears to ignore the fact that convergence is completely rigorous for a whole class of muffin-tin potentials and most reasonably to be expected therefore for potentials that differ little from those in this class. (See Section 2.9).

In any case, the whole question of convergence of the cellular method has to be seen in the right perspective. It is now well known that substantial changes appear in APW eigenvalues if allowances are made for the non-constancy of the potential outside the muffin. Such errors are in fact similar to the convergence errors in the cellular method: just as an allowance for warping of the constant potential is introduced in the APW method, so the non sphericity of the potential outside R_i can be treated by coupling differential equations in the cellular method, as shown in Section 2.9.

(6) Finally, it appears to be believed that the cellular 'method depends upon the fullest use of group theory ... this means that the method is mainly useful for the calculation of energy eigenvalues in symmetry directions of the Brillouin zone' (Ziman 1971.)

The present writer wishes to deny this emphatically. The impression may have been created because the cellular method was used for years on computers that were much smaller than those employed for other methods. As a result, there was a tendency both to refine simplifications provided by group theory and to deal with the simpler **k** vectors of symmetry. Evidence has been presented in this review and elsewhere to show that general **k** eigenvalues are now easily computed by the cellular method. Also, the amount of group theory used in the cellular method is practically identical to that employed in a good APW program (see Mattheiss, Wood, and Switendick 1968).

Admittedly, some band programs have treated all **k** vectors as general ones, thus dispensing with group theory. It is the writer's opinion that this is unwise. It is not so much that time is saved and accuracy gained by using group theory for points of symmetry, but that the states can be unequivocally identified thus ensuring that compatibilities are correctly preserved. This is most important in getting a clear picture of the nature of band crossings which, however accurate a method might be, could otherwise be easily overlooked or misidentified.

2.15. Acknowledgements

I should like to acknowledge the invaluable help given me by past students and collaborators, from whose works I have freely drawn to prepare this review: Drs C. J. Bradley, J. E. Jeacocke, A. P. Cracknell, J. S. Rousseau, B. L. Davies, R. C. Harford, A. R. Harford and Messrs R. G. Blake and J. B. Robinson. I am particularly grateful to Dr Bradley for much discussion and correspondence.

Part of this work was done at the Institute of Theoretical Physics of the University of Stockholm and I am particularly grateful to Professor Inga Fischer–Hjalmars and Dr S. Flodmark for their kind hospitality. This work was continued and the whole of the paper written at the Institute di Fisica Guglielmo Marconi, University of Rome, where I enjoyed the hospitality of Professor F. G. Bassani, which it is my pleasure to acknowledge.

This work was made possible by a grant of the Swedish Science Research Council and a Nato Visiting Professorship at Rome University, which are gratefully acknowledged.

Finally, and most importantly, the work which is described in this review was started in 1953 at the Mathematical Institute, University of Oxford, under the sponsorship of Professor C. A. Coulson, without whose help and encouragement it would never have been done. It is therefore with great pleasure and gratitude that I contribute this review to a volume dedicated to him.

Appendix: Corrigenda to papers on lattice harmonics

All lines in the Tables are counted disregarding headings.

Altmann and Bradley (1963*b*)

p. 210, Table 8, line 4 from bottom of Table:
61689600(2) − 3240(6) − 1(10) *should read*
61689600(2) + 3240(6) + 1(10).

p. 211, Table 10, line 14 from top of Table:
30880(2) − 90(6) *should read*
308880(2) − 90(6).

p. 211, Table 10, line 17 from top of Table:
231517440(0) *should read*
2315174400(0).

p. 211, Table 10, line 5 from bottom of Table:
$-\sqrt{3}/6$ *should read* $+\sqrt{3}/6$.

p. 211, Table 10, line 3 from bottom of Table:
17244057600(2) − 950400(6) − 144(10) *should read*
17244057600(2) + 950400(6) + 144(10).

p. 213, Table 11, line 7 from top of Table:
5080320(1) + 60480(3) + 864(5) − 18(7) + 1(9) *should read*
5080320(1) − 60480(3) + 864(5) − 18(7) + 1(9).

Altmann and Cracknell (1965)

Owing to the above misprints the following tables contain some errors: Table I(*a*), Table I(*c*), Table I(*d*).

These errors are corrected in the tables of Bradley and Cracknell (1972), where greater numerical accuracy is also given.

p. 30, Table IV, Notes, right hand column, line 8 from bottom;
Y_1^0 *should read* Y_0^0.

Altmann and Bradley (1965*a*)

p. 42, Table IV, Notes, left hand column, line 13 from bottom.
After: 'in the table' *insert*:
, except for *K*, *H* and *P* where *l* is given mod(2).

REFERENCES

Altmann, S. L. (1955). *Proc. Phys. Soc. Lond.* **68**, 987.
—— (1956). *Proc. Phys. Soc. Lond.* **69**, 184.
—— (1957). *Proc. Cambridge phil. Soc.* **53**, 343.
—— (1958*a*). *Proc. R. Soc. A* **244**, 141.
—— (1958*b*). *Proc. R. Soc. A* **244**, 153.
—— (1962). In *Quantum theory* (ed. D. R. Bates), vol. II. Academic Press, New York.
—— (1970). *Band theory of metals*. Pergamon Press, Oxford.
—— and Bradley, C. J. (1963*a*). *Phil. Trans. R. Soc. A* **255**, 193.
—— —— (1963*b*). *Phil. Trans. R. Soc. A* **255**, 199.
—— —— (1964). *Phys. Rev.* **135**, *A* 1253.

—— —— (1965*a*). *Rev. Mod. Phys.* **37**, 33.
—— —— (1965*b*). *Proc. Phys. Soc.* **86**, 915.
—— —— (1967). *Proc. Phys. Soc.* **92**, 764.
—— and COHAN, N. V. (1958). *Proc. Phys. Soc.* **71**, 383.
—— and CRACKNELL, A. P. (1964), *Proc. Phys. Soc.* **84**, 761.
—— —— (1965). *Rev. mod. Phys.* **37**, 19.
——, DAVIES, B. L., and HARFORD, A. R. (1968). *J. Phys. C.* **1**, 1633.
——, HARFORD, A. R., and BLAKE, R. G. (1971). *J. Phys. F.* **1**, 791.
ANDERSON, P. W. and FRICKER, H. S. (1971). *J. chem. Phys.* **55**, 5028.
ARBMAN, G. (1968). *Lattice symmetry corrections to energy bands from a spherical model.* I. *Theory*. FOA 4 report, C 4352–23. Stockholm.
—— (1969). *Lattice symmetry corrections to energy bands from a spherical model.* II. *Applications to electron gas and some simple metals.* FOA 4 report. C 4380–23. Stockholm.
ARKHIPOV, R. G. (1966). *Sov. Phys. JETP.* **22**, 1095.
—— (1971). *Sov. Phys. JETP.* **32**, 931.
ARLINGHAUS, F. J. (1965). *Ph.D. Thesis, MIT.* Unpublished.
BELL, D. G. (1954). *Rev. mod. Phys.* **26**, 311.
——, HUM, D. M., PINCHERLE, L., SCIAMA, D. W., and WOODWARD, P. M. (1953). *Proc. R. Soc. A* **217**, 71.
BERGGREN, K. -F. and FRÖMAN, A. (1969). *Ark. Fys.* **24**, 355.
BOUCKAERT, L. P., SMOLUCHOWSKI, R., and WIGNER, E. (1936). *Phys. Rev.* **50**, 58.
BRADLEY, C. J. and CRACKNELL, A. P. (1972). *The mathematical theory of symmetry in solids.* Clarendon Press, Oxford.
BURDICK, C. A. (1963). *Phys. Rev.* **129**, 138.
BUTLER, F. A., BLOOM, F. K., and BROWN, E. (1969). *Phys. Rev.* **180**, 744.
CONNOLLY, J. W. D. (1971). In *Computational methods in band theory* (eds. P. M. Marcus, J. F. Janak, and A. R. Williams) p. 3. Plenum Press, New York.
DIMMOCK, J. O. (1971). *Solid St. Phys.* **26**, 103.
FRÖMAN, A. and BERGGREN, K. -F. (1965). *Compressed atoms.* FOA 4 report. A 4444. Stockholm.
GANDEL'MAN, G. M. (1963). *Sov. Phys. JETP.* **16**, 94.
—— (1967). *Sov. Phys. JETP.* **24**, 99.
HAM, F. S. (1954). *Ph.D. Thesis, Harvard University.* Unpublished.
—— (1955). *Solid St. Phys.* **1**, 127.
—— (1960). In *The Fermi surface* (eds. W. A. Harrison and M. B. Webb), p. 9. John Wiley and Sons, New York.
—— (1962). *Phys. Rev.* **128**, 82.
—— and SEGALL, B. (1961). *Phys. Rev.* **124**, 1786.
HARFORD, A. R. (1972). *D. Phil. Thesis, Oxford University.* Unpublished.
HARTREE, D. R. (1957). *The calculation of atomic structures.* Wiley, New York.
—— (1958). *Numerical analysis.* Clarendon Press, Oxford.
HERMAN, F. (1958). *Rev. mod. Phys.* **30**, 102.
HOWARTH, D. J. (1953). *Proc. R. Soc. A* **220**, 513.
—— and JONES, H. (1952). *Proc. Phys. Soc. Lond. A* **65**, 355.
JENKINS, D. P. (1956). *Proc. Phys. Soc. Lond. A* **69**, 548.
—— and PINCHERLE, L. (1954). *Phil. Mag.* **45**, 93.

JOHNSTON, D. F. (1960). *Rep. Prog. Phys.* **23**, 66.

JONES, H. (1960). *The theory of Brillouin zones and electronic states in crystals.* North Holland, Amsterdam.

KOHN, W. (1952). *Phys. Rev.* **87**, 472.

KOSTER, G. F. (1957). *Solid. St. Phys.* **5**, 173.

LOUCKS, Ṫ. (1967). *Augmented plane wave method.* Benjamin, New York.

MATTHEISS, L. F., WOOD, J. H., and SWITENDICK, A. C. (1968). In *Methods in computational physics, vol.* 8, *Energy bands of solids.* (eds. B. Alder, S. Fernbach, and M. Rotenberg) p. 63. Academic Press, New York.

PINCHERLE, L. (1958). *Proc. Phys. Soc. Lond.* **72**, 281.

SAFFREN, M. M. (1959). *Ph.D. Thesis. MIT.* Unpublished.

SCHIFF, B. (1955). *Proc. Phys. Soc. Lond. A* **68**, 686.

—— (1956). *Proc. Phys. Soc. Lond. A* **69**, 185.

SEITZ, F. (1936). *Ann. Math.* **37**, 17.

SHOCKLEY, W. (1937). *Phys. Rev.* **52**, 866.

SLATER, J. C. (1934). *Phys. Rev.* **45**, 794.

VON DER LAGE, F. C. and BETHE, H. A. (1947). *Phys. Rev.* **71**, 612.

VOROPINOV, A. I., GANDEL'MAN, G. M., and PODVAL'NYI, V. G. (1970). *Sov. Phys: Usp.* **13**, 56.

WIGNER, E. P. (1959). *Group theory.* Academic Press, New York.

WIGNER, E. and SEITZ, F. (1933). *Phys. Rev.* **43**, 804.

WILKINSON, J. H. (1965). *The algebraic eigenvalue problem.* Clarendon Press, Oxford.

ZIMAN, J. M. (1964). *Principles of the theory of solids.* University Press, Cambridge.

—— (1971). *Solid State Physics,* **26**, 1.

3

ONE-BODY POTENTIALS IN CRYSTALS

N. H. MARCH

3.1. Introduction

IT is now well established that for a wide class of crystalline materials, energy band theory, based on Bloch's theorem and Brillouin zones, and discussed at length in Chapters 1 and 2, is very successful in enabling a variety of experimental observations to be understood.

If we leave aside, for the present, non-equilibrium properties, and we do not consider disorder, introduced either by impurities or phonons, then the central problem of energy band theory which remains is the construction of a suitable periodic potential for use in the one-electron Schrödinger equation.

While it is recognized that it may well be necessary to refine such a theory, it has been customary to work with a common potential $V(\mathbf{r})$ in which all the electrons are assumed to move, rather than with an energy or wave vector dependent potential. Thus, the philosophy most usually adopted is nearer to a Hartree point of view than the Hartree–Fock. Reasons for this will become clearer later.

In order to see the motivation for, and the (partial) justification of, such an approach, we shall start from the problem of a uniform electron gas, in a slowly-varying potential field. This will allow us to expose some of the essential ideas, while keeping the mathematical detail to a minimum.

3.2. Thomas–Fermi theory

Such a problem of a uniform electron gas in a slowly-varying potential field can be treated by the Thomas–Fermi theory. If we recall the result that we can divide the phase space (essentially a product of the volume of the crystal $\mathscr{V}$ and the volume of momentum space) into cells of volume h^3, into each of which we can put, in principle, two electrons having opposed spins, then at absolute zero, all momentum space will be occupied out to a radius p_f, the Fermi momentum. Thus, if we have N electrons in the volume $\mathscr{V}$ of the crystal we can write in a non-magnetic situation

$$N = 2 \, . \frac{\frac{4}{3}\pi p_f^3 \mathscr{V}}{h^3} \tag{3.1}$$

or the electron density $\rho_0 = N/\mathscr{V}$ is related to the maximum momentum p_f

by

$$\rho_0 = \frac{8\pi}{3h^3} p_f^3. \tag{3.2}$$

Now we 'switch on' to this electron gas a slowly varying potential $V(\mathbf{r})$. By slowly varying, we imply that $V(\mathbf{r})$ changes by only a small fraction of itself over a distance of the order of the de Broglie wavelength h/p_f for an electron at the Fermi energy. Then, it is clear that the density ρ_0 will become inhomogeneous and we denote the modified density by $\rho(\mathbf{r})$. If the spatial variation is slow, we can then take over eqn (3.2) into the inhomogeneous situation and write:

$$\rho(\mathbf{r}) = \frac{8\pi}{3h^3} p_f^3(\mathbf{r}), \tag{3.3}$$

the maximum momentum evidently now becoming a function of position. At this stage we employ the classical energy equation for the fastest electron, namely

$$E_f = \frac{p_f^2(\mathbf{r})}{2m} + V(\mathbf{r}). \tag{3.4}$$

It will be noted that, while the terms on the right-hand side depend individually on position $\mathbf{r}$, the left-hand side does not. For if the maximum or Fermi energy E_f varied from point to point, then electrons could spill over from one region of space with a high energy into another with a lower energy to reduce the total energy. Thus E_f in the equilibrium situation is independent of $\mathbf{r}$.

Combining eqns (3.3) and (3.4) we can thereby relate the density $\rho(\mathbf{r})$ to the potential $V(\mathbf{r})$ by

$$\rho(\mathbf{r}) = \frac{8\pi}{3h^3}(2m)^{\frac{3}{2}}[E_f - V(\mathbf{r})]^{\frac{3}{2}} \tag{3.5}$$

which is the usual Thomas–Fermi relation between density ρ and potential $V(\mathbf{r})$.

3.2.1. *Definition of potential for interacting electrons*

Let us suppose now that we want to treat a many-body system of interacting electrons. If again we switch a slowly varying perturbation on to a uniform gas, then the density $\rho(\mathbf{r})$ will vary only slowly in space and we could use eqn (3.5) operationally to define a one-body potential $V(\mathbf{r})$. Though $\rho(\mathbf{r})$ is a many-body density, we could, in principle, measure it in a crystal by finding the intensity of X-ray scattering at the Bragg reflections. Hence, from such a many-body density, we could extract a one-body potential. Of course, if this one-body potential would only reproduce the measured electron density, we would not have made useful progress. But we shall argue below that such a one-body potential will enable us to do a variety of calculations of physical interest in crystalline solids, both when perfect and when perturbed by impurities or a phonon.

We stress again that the above method of extracting a one-body potential is *not* adequate in real crystals, because the density $\rho(\mathbf{r})$ varies too rapidly through the unit cell, from a high value at the nucleus to a much lower value at the unit cell boundary. In the presence of such substantial gradients of electron density, the above theory is not valid and we must of course transcend the density–potential relation (3.5).

The way to do this will occupy quite a bit of this chapter, but before going on to consider that question let us return to the original one-body treatment of Thomas and Fermi and set up the total energy in that approximation.

3.2.2. *Total energy in one-body Thomas–Fermi theory*

Suppose we simply regard the electrons as moving without interaction in the common one-body potential $V(\mathbf{r})$. Then the total potential energy is evidently

$$U = \int \rho(\mathbf{r})V(\mathbf{r})\,\mathrm{d}\mathbf{r} \tag{3.6}$$

and we can write the total electronic energy as

$$E = T + U, \tag{3.7}$$

where T is the kinetic term. In the spirit of the theory sketched above, we can use free electron relations locally, and if we calculate the mean kinetic energy of an electron in a non-interacting uniform gas we find that the answer is simply $\frac{3}{5}E_{\mathrm{f}}$. Thus the kinetic energy per unit volume, or kinetic-energy density, is evidently

$$\frac{3}{5}E_{\mathrm{f}}\frac{N}{\mathscr{V}} = \frac{3}{5}\frac{p_{\mathrm{f}}^2}{2m}\rho_0 \tag{3.8}$$

and using eqn (3.2) this is readily seen to yield for the total kinetic energy T the approximate result

$$T = c_k \int \rho^{\frac{5}{3}}\,\mathrm{d}\mathbf{r}; \qquad c_k = \frac{3}{10m}\left(\frac{3h^3}{8\pi}\right)^{\frac{2}{3}} \tag{3.9}$$

where we have written eqn (3.8) as $c_k\rho_0^{\frac{5}{3}}$ and have then gone over to the kinetic energy density of the inhomogeneous gas. It should not need emphasizing that eqn (3.9) is only correct when the conditions of applicability of the Thomas–Fermi theory are satisfied.

Now we write, therefore

$$E = T + U = c_k \int \rho^{\frac{5}{3}}\,\mathrm{d}\mathbf{r} + \int \rho V\,\mathrm{d}\mathbf{r} \tag{3.10}$$

and we show next that the Thomas–Fermi density–potential relation (3.5) is regained if we vary E with respect to the charge-density ρ, subject only to the condition that the total number of electrons is a given constant N, related to $\rho(\mathbf{r})$ evidently by

$$N = \int \rho \, d\mathbf{r}. \tag{3.11}$$

Thus, we write

$$\delta(E - \mu N) = 0, \tag{3.12}$$

μ being the usual Lagrange multiplier which takes care of the subsidiary condition (3.11) in the density variation. Evidently, we find from eqns (3.10)–(3.12) that

$$\tfrac{5}{3}c_k \int \rho^{\frac{2}{3}} \, \delta\rho \, d\mathbf{r} + \int \delta\rho V \, d\mathbf{r} - \mu \int \delta\rho \, d\mathbf{r} = 0 \tag{3.13}$$

and if this is to be true for arbitrary small variations $\delta\rho$ then we must have

$$\tfrac{5}{3}c_k \rho^{\frac{2}{3}} = (\mu - V). \tag{3.14}$$

This is readily shown to be the same as eqn (3.5), provided the Lagrange multiplier μ is identified with the Fermi energy E_f. That this identification is indeed consistent is readily seen, since from eqn (3.12) it follows that $\mu = \delta E/\delta N$. In other words, μ is the chemical potential.

This argument, spelt out above for this very elementary case of non-interacting particles moving in a slowly varying potential, is, in fact, the prototype of a very basic formulation of the many-body problem of the electron gas in a crystalline solid, in terms of the electron density.

3.2.3. *Introduction of classical Coulomb interaction between electrons*

Of course, for electrons, it is essential to incorporate at least some average account of electron–electron interactions. Let us begin in the spirit of the Hartree self-consistent method by including in the total energy E the classical Coulomb interaction energy

$$U_{ee} = \frac{e^2}{2} \int \frac{\rho(\mathbf{r})\rho(\mathbf{r}')}{|\mathbf{r} - \mathbf{r}'|} \, d\mathbf{r} \, d\mathbf{r}', \tag{3.15}$$

the factor of $\frac{1}{2}$ being introduced as usual to avoid counting electron–electron interactions twice. The potential energy of interaction between the electrons and the nuclear potential V_N, say, has the form (3.6) and hence for the total electronic energy E we now have

$$E = c_k \int \rho^{\frac{5}{3}} \, d\mathbf{r} + \int \rho(\mathbf{r}) V_N(\mathbf{r}) \, d\mathbf{r} + \frac{e^2}{2} \int \frac{\rho(\mathbf{r})\rho(\mathbf{r}')}{|\mathbf{r} - \mathbf{r}'|} \, d\mathbf{r} \, d\mathbf{r}'. \tag{3.16}$$

If we again carry through the variation with respect to ρ, we regain the Euler equation (3.14), where V is now the sum of the nuclear potential energy V_N and the potential energy V_e due to the electron density $\rho(\mathbf{r})$, that is

$$V = V_N + V_e \tag{3.17}$$

with

$$V_e(\mathbf{r}) = e^2 \int \frac{\rho(\mathbf{r}')\,\mathrm{d}\mathbf{r}'}{|\mathbf{r}-\mathbf{r}'|}. \tag{3.18}$$

The field corresponding to eqn (3.17) is simply the Hartree field in the limit as $N \to \infty$ (since the correction in this theory for the fact that an electron does not act on itself is order $1/N$ for non-localized wavefunctions).

Clearly, the above theory allows the electron density to be calculated for a given nuclear framework, that is a given V_N.†

3.3. Improved approximations for kinetic energy

As we stressed earlier, the above theory is only true for slowly-varying potentials. It might seem natural to attempt to extend this theory to deal with density gradients and such an approach was advocated by von Weizsäcker (1935) who proposed to write for the kinetic-energy density $t_{\mathbf{r}}[\rho]$

$$\underset{\text{Weizsäcker}}{t_{\mathbf{r}}\ [\rho]} = c_k \rho^{\frac{5}{3}} + \frac{\lambda}{8} \frac{(\nabla\rho)^2}{\rho} \frac{\hbar^2}{m} \tag{3.19}$$

Of course, higher order terms in grad ρ really should be included but von Weizsäcker's argument was that if the Thomas–Fermi term $\alpha\rho^{\frac{5}{3}}$ was a good starting point, then it makes sense to introduce simply the first gradient correction (a term linear in grad ρ does not contribute to the total kinetic energy and is therefore not included).

The original von Weizsäcker theory gave $\lambda = 1$ whereas Kirznits (1957) subsequently obtained the proper value $\lambda = \frac{1}{9}$ for such a gradient expansion.

The status of these two choices of λ has been clarified recently by Jones and Young (1971). They use the expression (3.19) to find the density change in a uniform electron gas resulting from a change in potential δV. We can then write, as usual, in linear response theory

$$\delta\rho(\mathbf{r}) = \int F_W(\mathbf{r}-\mathbf{r}')\,\delta V(\mathbf{r}')\,\mathrm{d}\mathbf{r}' \tag{3.20}$$

† For example, before modern computational facilities were available, such a calculation gave a useful first-approximation to electronic clouds in both solids and molecules (e.g., for benzene; March 1952).

the von Weizsäcker response function F_W then following simply from eqn (3.19) for a general λ. The result, in Fourier transform, is readily shown to be

$$F_W(k) = -\frac{k_f}{\pi^2}\left[\frac{1}{1+3\lambda\eta^2}\right]; \qquad \eta = \frac{k}{2k_f} \tag{3.21}$$

where $p_f = \hbar k_f$, k_f then being the Fermi wave number of the electron gas.

It can be shown (see, for example, Kittel 1963) that the *exact* response function in the non-interacting case is, in fact

$$F(k) = -\frac{k_f}{2\pi^2}\left\{1+\frac{1-\eta^2}{2\eta}\ln\left|\frac{1+\eta}{1-\eta}\right|\right\}. \tag{3.22}$$

If we compare eqns (3.21) and (3.22) for small k and for large k, we find that F_W and F agree at small k, that is for long wavelength perturbations, when $\lambda = \frac{1}{9}$. This is evidently the limit of slowly-varying potential and is therefore the proper value in the spirit of gradient expansions. On the other hand, for large k, $F_W \to F$ with $\lambda = 1$.

The situation is shown in Fig. 3.1, and it is seen that, in a crystal, since the first reciprocal lattice vector is often near to $2k_f$, it may well be that the von Weizsäcker theory could give better numerical results than the Kirznits value. This is simply saying that gradient expansions are not good in a crystal, because of the periodic nature of the potential and density, at least for realistic metallic densities.

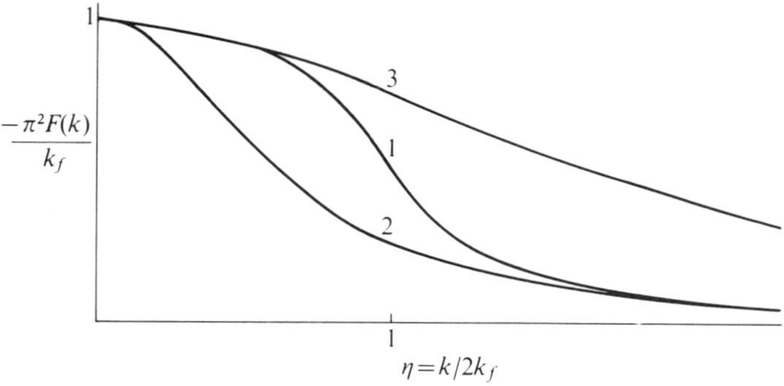

FIG. 3.1. Schematic form of response function $F(k)$ for uniform electron gas: (1) exact result; (2) von Weizsäcker result ($\lambda = 1$); (3) von Weizsäcker result ($\lambda = \frac{1}{9}$).

Though we have given the above argument in some detail, we want to stress that methods are now available for treating (usually numerically it is true) the single-particle kinetic-energy density $t_r[\rho]$ essentially exactly, and so this does not constitute a real problem in practice.

The major problem, of course, is to treat the electron–electron interactions beyond the Hartree self-consistent field approximation and it is to this problem that much of the rest of this chapter is devoted.

3.4. Introduction of exchange into Thomas–Fermi theory

As Dirac (1930) first recognized, it is of interest to study the effect of exchange interactions in the Thomas–Fermi theory. This then transcends, for slowly-varying potentials and densities, the Hartree self-consistent field theory, and provides us with at least a first approximation to Hartree–Fock theory.

There is little doubt, that, in atoms and molecules, Hartree–Fock is a significant improvement over Hartree theory. However, it is necessary to warn the reader at this point that in metals it is essential to screen the electron–electron interactions in dealing with exchange terms. In fact, straightforward application of Hartree–Fock theory to a uniform electron gas leads to certain objectionable properties such as zero density of states at the Fermi level, a logarithmic variation of electronic specific heat with temperature rather than the linear variation found experimentally, and a similar anomalous behaviour of the diamagnetic susceptibility. This is in contrast to the predictions of Hartree theory, which are at least qualitatively correct.

Nevertheless, in spite of such difficulties, there is now no doubt that a contribution to the total energy density from the exchange energy is of considerable importance in making band-structure calculations.

We shall begin the discussion by studying the analogue of the $\rho^{\frac{5}{3}}$ Thomas–Fermi term in the kinetic energy, for the exchange energy.

This we can approach just as before by noting that the exchange energy per particle in a uniform electron gas is in fact

$$-\frac{0{\cdot}916}{r_s}\ \text{Ryd},$$

with r_s in atomic units (that is in units of the first Bohr radius a_0 for hydrogen). Here r_s is the mean interelectronic spacing in the electron gas, and is related to the density ρ_0 through

$$\rho_0 = 3/4\pi r_s^3. \tag{3.23}$$

The above result for free electrons is derived, for example, by Seitz (1940). It leads directly, for slowly-varying densities, to the exchange energy density $\varepsilon_x(\rho)$ as

$$\varepsilon_x(\rho) = -c_e\rho^{\frac{4}{3}}: \qquad c_e = \frac{3}{4}e^2\left(\frac{3}{\pi}\right)^{\frac{1}{3}}. \tag{3.24}$$

This result, apart from the constant, can be got by dimensional analysis.

Whereas, from the wave equation, the kinetic energy must involve h^2/m, the exchange potential energy comes from an appropriate average of e^2/r_{ij}, the electrostatic interaction between electrons separated by r_{ij}. Thus the exchange energy is proportional to e^2 as in eqn (3.24) and then the density raised to the power $\frac{4}{3}$ follows.

The same argument based on dimensional analysis shows that the analogue of the von Weizsäcker term $(\nabla\rho)^2/\rho$ in correcting the exchange energy density $\propto \rho^{\frac{4}{3}}$ is in fact $(\nabla\rho)^2/\rho^{\frac{4}{3}}$, as appears to have been first pointed out by Herman, van Dyke, and Ortenburger (1969).

3.4.1. *Variational derivation of Thomas–Fermi—Dirac relation*

To obtain the generalizations due to Dirac of the Thomas–Fermi theory, we now write the total electronic energy E as (omitting nuclear–nuclear terms)

$$E = c_k \int \rho^{\frac{5}{3}}\,\mathbf{dr} + \int \rho V_{\mathrm{N}}\,\mathbf{dr} + \frac{e^2}{2}\int \frac{\rho(\mathbf{r})\rho(\mathbf{r}')}{|\mathbf{r}-\mathbf{r}'|}\,\mathbf{dr}\,\mathbf{dr}' - c_{\mathrm{e}}\int \rho^{\frac{4}{3}}\,\mathbf{dr}, \qquad (3.25)$$

leaving out gradient corrections. Using the variational principle (3.12) once again, we obtain

$$\tfrac{5}{3}c_k\rho^{\frac{2}{3}} = \mu - V + \tfrac{4}{3}c_{\mathrm{e}}\rho^{\frac{1}{3}} \qquad (3.26)$$

where we have written the Euler equation in this form to show that we can interpret the effect of exchange as modifying the potential energy V by a term (cf Chapter 1, eqn 1.24)

$$-\tfrac{4}{3}c_{\mathrm{e}}\rho^{\frac{1}{3}}. \qquad (3.27)$$

This term, first given by Dirac (1930), was rediscovered by Slater (1951) with a slightly different coefficient. Dirac's coefficient was later obtained by Gáspar (1954) and by Kohn and Sham (1965). We shall refer to a term $\propto \rho^{\frac{1}{3}}$ in the one-body potential as a Dirac–Slater exchange potential.

Such a modification of the 'Hartree' potential $V(\mathbf{r})$ proves to be essential if one is to make successful energy band calculations. We want to stress that such calculations using the Dirac–Slater exchange potential do *not* make Thomas–Fermi approximations for the kinetic energy.

In view of the success of this '$\rho^{\frac{1}{3}}$' exchange potential, it is obviously a sensible further step to enquire whether such an exchange potential can be systematically corrected. To do this, we will have to consider a generalization of the density $\rho(\mathbf{r})$, which was first effected by Dirac (1930).

3.5. Dirac density matrix and gradient expansions

In terms of the eigenfunctions of a one-body Schrödinger equation, say $\psi_i(\mathbf{r})$, we can write the electron density $\rho(\mathbf{r})$ as

$$\rho(\mathbf{r}) = \sum_i^{E_{\mathrm{f}}} \psi_i^*(\mathbf{r})\psi_i(\mathbf{r}). \qquad (3.28)$$

Dirac observed that it was not possible directly from the one-electron Schrödinger equation for $\psi_i(\mathbf{r})$ to obtain a useful equation from which $\rho(\mathbf{r})$ can be calculated. Thus he proposed to work with the generalized density

$$\rho(\mathbf{r}\mathbf{r}_0) = \sum_i^{E_f} \psi_i^*(\mathbf{r})\psi_i(\mathbf{r}_0) \tag{3.29}$$

which evidently reduces to $\rho(\mathbf{r})$ in eqn (3.28) when we put $\mathbf{r}_0 = \mathbf{r}$. Eqn (3.29) defines the Dirac *density matrix*,† and we notice that if the ψ_is, as we are assuming, form a complete orthonormal set, then

$$\int \rho(\mathbf{r}\mathbf{r}')\rho(\mathbf{r}'\mathbf{r}_0)\,d\mathbf{r}' = \rho(\mathbf{r}\mathbf{r}_0), \tag{3.30}$$

which in matrix language is the condition of idempotency:

$$\rho^2 = \rho. \tag{3.31}$$

It proves possible now to generate a perturbation expansion of $\rho(\mathbf{r}\mathbf{r}_0)$ in powers of the one-body potential $V(\mathbf{r})$, based on plane waves $\exp(i\mathbf{k}\,.\,\mathbf{r})$. Unfortunately, it has not so far proved possible to sum this series exactly except in certain special cases. However, if we put $\mathbf{r} = \mathbf{r}_0$, and assume $V(\mathbf{r})$ is slowly varying, then it proves possible to sum the series, when we regain the Thomas–Fermi ρ–V relation, as of course we must.

It is therefore of obvious interest to inquire whether this perturbation expansion (see March, Young and Sampanthar 1967 for details) can be used to transcend the Thomas–Fermi theory. The answer is that it can, and we want to consider how it can be used, in particular, to generate the exchange energy to a higher order than the Dirac–Slater $\rho^{\frac{4}{3}}$ term in the exchange-energy density.

The result obtained for the total exchange energy in terms of the density matrix is (Dirac 1930)

$$E_x[\rho] = -\frac{e^2}{4}\int\int \frac{[\rho(\mathbf{r},\mathbf{r}_0)]^2}{|\mathbf{r}-\mathbf{r}_0|}\,d\mathbf{r}\,d\mathbf{r}_0. \tag{3.32}$$

We evidently want to obtain from this the Dirac–Slater exchange energy $-c_e\int \rho^{\frac{4}{3}}\,d\mathbf{r}$ and to learn how to systematically correct this. To see how to begin such a calculation, let us denote the 'displaced charge' $\rho(\mathbf{r})-\rho_0$ due to the introduction of the one-body potential $V(\mathbf{r})$ by $\Delta(\mathbf{r})$ and then we can write formally that

$$\begin{aligned}\rho(\mathbf{r},\mathbf{r}_0) &= \rho_0(\mathbf{r},\mathbf{r}_0)+\tfrac{1}{2}\int A_1(\mathbf{r},\mathbf{r}_0,\mathbf{s})\Delta(\mathbf{s})\,d\mathbf{s}\\ &\quad+\tfrac{1}{4}\int\int A_2(\mathbf{r},\mathbf{r}_0,\mathbf{s}_1,\mathbf{s}_2)\Delta(\mathbf{s}_1)\Delta(\mathbf{s}_2)\,d\mathbf{s}_1\,d\mathbf{s}_2+0(\Delta^3)\end{aligned} \tag{3.33}$$

† For further discussion of the density matrix, reference may be made to Chapter 5.

where A_1 and A_2 are known, but highly complicated, functions, which are given explicitly by Stoddart, Beattie, and March (1971). The important thing to note is that, in such a one-body potential theory, $\rho(\mathbf{r}\mathbf{r}_0)$ can be expressed as a functional of the displaced charge Δ. Unfortunately, of course, we are not really interested in such a perturbation expansion, but rather in a gradient expansion. Nevertheless, it is clear that in the limited sense of an expansion in the displaced charge we can insert the result (3.33) into eqn (3.32) to find an expansion for the exchange-energy density $\varepsilon_x[\rho]$ in powers of Δ. The result we obtain can be formally written as (with $e = 1$)

$$\varepsilon_x[\rho] = -\frac{3}{4}\left(\frac{3}{\pi}\right)^{\frac{1}{3}} \rho_0^{\frac{4}{3}} + \int \mathrm{d}\mathbf{s}\, B(\mathbf{r}-\mathbf{s})\Delta(\mathbf{r})\Delta(\mathbf{s}) + 0(\Delta^3) \tag{3.34}$$

where B is again a very complicated, but known, function, given explicitly by Stoddart, Beattie, and March (1971).

Again we stress that such a perturbative result is not directly useful except in the linear response regime. We therefore turn to enquire how this result (3.34) can be used to generalize Dirac–Slater exchange theory.

3.6. Partial summations of gradient series

Hohenberg and Kohn (1964; see also Kohn and Sham, 1965) pointed out that the gradient series for the kinetic energy density could be partially summed into the form (with $\hbar = m = 1$)

$$t_{\mathbf{r}}[\rho] = \frac{3(3\pi^2)^{\frac{2}{3}}}{10}\{\rho(\mathbf{r})\}^{\frac{5}{3}} + \tfrac{1}{8}\int \mathrm{d}\mathbf{r}'\, K(\mathbf{r}', \rho(\mathbf{r}))[\rho(\mathbf{r}+\tfrac{1}{2}\mathbf{r}') - \rho(\mathbf{r}-\tfrac{1}{2}\mathbf{r}')]^2 \tag{3.35}$$

where K is again a known function recorded in their papers. Kohn and Sham wrote down a formal generalization of eqn (3.35) for the exchange energy, but did not obtain the corresponding kernel. This can be done using the expression (3.34), as shown by Stoddart, Beattie, and March (1971), and the result takes the form (with $e = 1$)

$$\varepsilon_x[\rho] = -\frac{3}{4}\left(\frac{3}{\pi}\right)^{\frac{1}{3}}\{\rho(\mathbf{r})\}^{\frac{4}{3}} - \tfrac{1}{2}\int \mathrm{d}\mathbf{r}'\, B(\mathbf{r}', \rho(\mathbf{r}))[\rho(\mathbf{r}+\tfrac{1}{2}\mathbf{r}') - \rho(\mathbf{r}-\tfrac{1}{2}\mathbf{r}')]^2 \tag{3.36}$$

where $B(\mathbf{r}', \rho(\mathbf{r}))$ is obtained from B in eqn (3.34). The corresponding exchange potential is obtained by forming $\delta\varepsilon_x[\rho]/\delta\rho$ when one finds

$$V_x(\mathbf{r}) = -\left(\frac{3}{\pi}\right)^{\frac{1}{3}}\{\rho(\mathbf{r})\}^{\frac{1}{3}}$$
$$-\frac{1}{2}\int d\mathbf{r}' \frac{\partial B(\mathbf{r}', \rho(\mathbf{r}))}{\partial \rho(\mathbf{r})}\left[\rho\left(\mathbf{r}+\frac{1}{2}\mathbf{r}'\right)-\rho\left(\mathbf{r}-\frac{1}{2}\mathbf{r}'\right)\right]^2$$
$$-2\int d\mathbf{r}'\, B\left(\mathbf{r}-\mathbf{r}', \rho\left(\frac{\mathbf{r}+\mathbf{r}'}{2}\right)\right)[\rho(\mathbf{r})-\rho(\mathbf{r}')]. \tag{3.37}$$

We want to stress that the 'correction' terms exhibited to the Dirac–Slater exchange potential in eqn (3.37) are non-local in the sense that the potential $V(\mathbf{r})$ is not solely determined now by the density at $\mathbf{r}$, which is surely a consequence of the Schrödinger equation.

We emphasize that the above argument has resulted from the use of a Dirac density matrix obtained from a one-body potential $V(\mathbf{r})$ which is designed to yield the exact density $\rho(\mathbf{r})$. An alternative method of approach is to construct the density matrix $\rho(\mathbf{r}, \mathbf{r}_0)$ by solution of the Hartree–Fock equations. This problem has been considered by Beattie, Stoddart, and March (1971), who show that the form (3.37) is regained, though of course with a different kernel B. It is obviously going to be of considerable interest to examine how sensitive the 'correction' to the Dirac–Slater exchange potential is to the choice of the kernel B.

One point which follows from both methods of calculating B is that $B(\mathbf{r}, k_f)$ appears to fall off very slowly with distance, in fact like r^{-5}. This stems from the fact that the Fourier transform $\tilde{B}(q, k_f)$ appears to have a small q expansion of the form

$$\tilde{B}(q, k_f) = \tilde{K}^2(q, k_f)[\alpha_0+\alpha_1 q^2+\alpha_2 q^2 \ln q + \ldots] \tag{3.38}$$

where part of $\tilde{B}$ has been approximated by a power series. It is the $q^2 \ln q$ singularity at $q = 0$ which leads to the long range of $B(r, k_f)$ in $\mathbf{r}$ space. We shall refer to arguments below which suggest that, when correlations are included, B most probably drops off exponentially. This means that a local theory is a better approximation in a theory where correlations are properly incorporated than in the case when only exchange interactions are considered (see Hedin and Lundqvist 1969, for a basic many-body discussion). We shall consider below how correlations can be introduced into the argument.

3.7. Density functional theory of many-body problem

So far the theory has really been based on a one-electron framework, that is a single Slater determinant for our approximation to the many-body wave function.

At this stage, it is right to ask the question 'Can correlation effects be properly treated, within such a description in terms of the electron density?' The answer is in the affirmative as shown by Hohenberg and Kohn (1964). Various workers, especially Gombás and his school, had anticipated this result and had already generalized the Thomas–Fermi–Dirac theory discussed above to include correlation effects. These attempts to include genuine correlation in fact used uniform electron gas theory in lowest order, the Wigner (1938) interpolation formula (see also eqn 3.50), namely

$$\varepsilon_c(\rho) = \frac{-0{\cdot}056\rho^{\frac{4}{3}}}{0{\cdot}079+\rho^{\frac{1}{3}}} \tag{3.39}$$

being a favourite starting point. It turns out that such a philosophy is well based, though because of basic limitations in the Thomas–Fermi treatment of kinetic energy it was not apparent from the early work that the approach could be successful quantitatively.

This basic justification of an approach to the many-body problem is sufficiently important for us to digress from practical matters of one-body potentials to deal with the description of a genuinely many-electron system in terms of the charge density.

3.7.1. *Proof that a unique charge density exists for each external potential*

Before proceeding to the (brief!) discussion of the way electron correlation can be approximately incorporated in such a description, we shall give here a proof that a unique charge density exists for each external potential.

Suppose that corresponding to two external potentials V and V_1 we have many-body wavefunctions Ψ and Ψ_1. Then if the corresponding forms of the Hamiltonian are H and H_1, we can evidently write for the ground-state energy E of H, (we assume a non-degenerate ground state)

$$E = \int \Psi^* H \Psi \, d\mathbf{r} < \int \Psi_1^* H \Psi_1 \, d\mathbf{r}. \tag{3.40}$$

Now let us suppose that the charge densities associated with wavefunctions Ψ_1 and Ψ are the same. Evidently then we can write

$$\int \Psi^*[V-V_1]\Psi \, d\mathbf{r} = \int \Psi_1^*[V-V_1]\Psi_1 \, d\mathbf{r} = \int [V-V_1]\rho(\mathbf{r}) \, d\mathbf{r} \tag{3.41}$$

in this case. Also

$$H = H_1 + (V - V_1)$$

and hence we find

$$E < \int \Psi_1^*[H_1]\Psi_1\,d\mathbf{r} + \int [V - V_1]\rho(\mathbf{r})\,d\mathbf{r}.$$

Using the variational principle for H_1, with energy E_1,

$$E_1 < \int \Psi^* H \Psi\,d\mathbf{r} + \int [V_1 - V]\rho(\mathbf{r})\,d\mathbf{r}. \tag{3.42}$$

By addition these yield $E + E_1 < E + E_1$ and we conclude that two different external potentials cannot generate the same charge density. But E is uniquely determined by the external potential and hence we deduce that the ground-state energy is a unique functional of the electron density $\rho(\mathbf{r})$.

3.7.2. *General Euler equation, including exchange and correlation*

To formally complete the theory therefore, we can evidently write, from the above theorem, the total electronic energy E in the form

$$E = \int t_{\mathbf{r}}[\rho]\,d\mathbf{r} + \int \rho V_{\mathrm{N}}\,d\mathbf{r} + \frac{e^2}{2}\int \frac{\rho(\mathbf{r})\rho(\mathbf{r}')}{|\mathbf{r}-\mathbf{r}'|}\,d\mathbf{r}\,d\mathbf{r}' + \int \varepsilon_{\mathrm{xc}}[\rho]\,d\mathbf{r}. \tag{3.43}$$

We want to stress that $t_{\mathbf{r}}[\rho]$ is a single-particle kinetic energy associated with density ρ and therefore that the exchange and correlation energy density $\varepsilon_{\mathrm{xc}}[\rho]$ includes a contribution from correlation kinetic energy. We have seen already that we can write down an Euler equation for the density by varying E with respect to ρ when we neglect correlation. Provided only that we assume the existence of the functional derivative $\delta\varepsilon_{\mathrm{xc}}[\rho]/\delta\rho$ we can clearly generalize the argument to obtain

$$\frac{\delta t_{\mathbf{r}}[\rho]}{\delta\rho} + V_{\mathrm{Hartree}} + \frac{\delta\varepsilon_{\mathrm{xc}}[\rho]}{\delta\rho} = \mu. \tag{3.44}$$

It is clear that if we put $\varepsilon_{\mathrm{xc}}[\rho] = -c_{\mathrm{e}}\rho^{\frac{4}{3}}$ then we regain the Dirac–Slater exchange potential added to the Hartree term. In general, however, we shall want to transcend this approximation to include an account of correlation.

3.7.3. *Correlation energy with small density gradients*

It seems clear that, to make progress, we must again try to build the theory up from the knowledge we have of correlations in a uniform electron gas. Here, as Gell–Mann and Brueckner (1957) have shown, the correlation energy per particle in the high-density limit takes the form

$$\varepsilon_{\mathrm{c}} = A \ln r_{\mathrm{s}} + c + 0(r_{\mathrm{s}}). \tag{3.45}$$

The result eqn (3.45) is exact to the order shown and, for slowly-varying densities which are sufficiently high so that $r_{\mathrm{s}} \ll 1$ Bohr radius, we can use eqn (3.45) by making the replacement

$$\rho_0 = \frac{3}{4\pi r_{\mathrm{s}}^3} \rightarrow \rho(\mathbf{r}) \tag{3.46}$$

to obtain $E_c[\rho]$. This would evidently add to the Dirac–Slater term $-c_e\{\rho(\mathbf{r})\}^{\frac{4}{3}}$.

However, for lower density, eqn (3.45) is not valid and no exact calculations of the correlation energy are available to use in the range of intermediate densities (for real metals $2 < r_s < 6$).

3.7.4. *Low-density electron gas and Wigner lattice*

The Hartree–Fock energy per particle is, as we have seen, in a uniform gas

$$\frac{E_{HF}}{N} = \frac{2{\cdot}21}{r_s^2} - \frac{0{\cdot}916}{r_s}. \tag{3.47}$$

From this result, it is perfectly clear that the Fermi energy, the first term on the right-hand side, dominates in the high-density limit, the exchange term αr_s^{-1} making only a small contribution at high densities. On the other hand, in the low-density electron gas, the potential energy becomes dominant and it turns out, as Wigner (1934; 1938) was the first to emphasize, that the exchange energy given in eqn (3.47) is a poor approximation. This is simply because the calculation of the exchange was based on plane waves, whereas in the low-density or strong correlation limit we should evaluate the potential energy using localized orbitals.

Physically, since the kinetic energy becomes unimportant at low densities, we must seek a way in which electrons can avoid one another as far as possible in the extreme low-density limit. The most effective way for this to come about is for an electron lattice to form: the so-called Wigner lattice. In the limit of extreme strong coupling, that is as $r_s \to \infty$, the energy of such an electron lattice is purely electrostatic. Of the lattice structures so far investigated, the lowest energy is found for a body-centred cubic lattice (see, for instance, Fuchs 1935). The energy per particle in Ryd is given by

$$\frac{E}{N} = \frac{-1{\cdot}792}{r_s} \tag{3.48}$$

which is substantially lower than the exchange energy $-0{\cdot}916/r_s$. Adopting the conventional definition of correlation energy as the difference between the exact energy and the Hartree–Fock value, we see that as $r_s \to \infty$

$$\frac{E_{corr}}{N} \to \frac{-1{\cdot}792}{r_s}\frac{+0{\cdot}916}{r_s} \doteqdot \frac{-0{\cdot}88}{r_s}. \tag{3.49}$$

As recorded previously in eqn (3.39) Wigner proposed an interpolation formula to extend the range of r_s and came up with the result

$$\frac{E_{corr}}{N} = \frac{-0{\cdot}88}{r_s+7{\cdot}8}\,\text{Ryd}. \tag{3.50}$$

By comparison with the Gell–Mann and Brueckner high-density limit given above, we see that this form is actually not correct as $r_s \to 0$ but it gives very useful results over a substantial range of r_s.

Gombás and others have extended the Thomas–Fermi theory by replacing r_s in favour of the local density in the above formula, and, after adding this to the total Thomas–Fermi–Dirac energy, they have again carried through the minimization procedure to obtain a new Euler equation. This is the most elementary example of the way we can include a partial account of correlation in the density functional theory.

We wish to stress again that we are advocating here the use of local density approximations, plus correction terms arising from partial summations of gradient series, for the exchange plus correlation energy density only; not for the single-particle kinetic energy. The motivation for this is the knowledge that the Dirac–Slater exchange potential has had notable success in energy band theory, and it therefore appears logical to try to correct it systematically:

(*a*) in lowest order for correlation effects using a local density approximation;

(*b*) in higher order, by adding selected sums of gradient series.

The outcome of all this is to suggest that we can write the exchange and correlation energy density in the form (cf eqn 3.36)

$$\varepsilon_{xc}[\rho] \doteqdot \varepsilon^0_{xc}[\rho] - \tfrac{1}{2}\int d\mathbf{r}'\, B_{xc}(\mathbf{r}', \rho(\mathbf{r}))[\rho(\mathbf{r}+\tfrac{1}{2}\mathbf{r}') - \rho(\mathbf{r}-\tfrac{1}{2}\mathbf{r}')]^2. \qquad (3.51)$$

Here, $\varepsilon^0_{xc}[\rho]$ is the local density form obtained from the uniform electron gas. It should be noted, of course, that in the correction term in eqn (3.51), the kernel B_{xc} now includes correlation. We earlier considered approximate forms of B for exchange alone, and it seems that the introduction of correlation has the effect of annulling the slow fall-off of B at large r as r^{-5}, and leads to an exponential decay of this kernel with r. Relevant work here is that of Overhauser (1971) and Ma and Brueckner (1968). Because of the importance of attempting to approximate to the correlation effects in this way, we shall discuss this work in general terms below. Actually, Overhauser's work is very closely connected with the above development, whereas the investigation of Ma and Brueckner is more in the spirit of the von Weizsäcker and Herman *et al.* approximations discussed earlier.

3.8. Gradient expansion of Ma and Brueckner

The approach of Ma and Brueckner essentially combines the density functional philosophy expounded in detail above with the use of diagrammatic techniques in many-body theory. It is not our purpose here to give the detailed diagrammatic arguments (see for example Jones and March 1973, for a summary of this work†) but rather to give an outline of the approach and to summarize the results. As before one writes

$$E[\rho] = E_{\mathrm{HF}}[\rho]+E_{\mathrm{c}}[\rho], \tag{3.52}$$

where evidently $E_{\mathrm{c}}[\rho]$ is the correlation energy as a functional of ρ.

Within the spirit of the von Weizsäcker approach, one now writes

$$E_{\mathrm{c}} = \int \mathrm{d}\mathbf{r}[\varepsilon_{\mathrm{c}}^{0}[\rho(\mathbf{r})]+B\{\rho(\mathbf{r})\}|\nabla\rho(\mathbf{r})|^{2}], \tag{3.53}$$

where the first term, as discussed above, is from the uniform gas theory, and our object must be to calculate $B\{\rho(\mathbf{r})\}$ as a functional of the density.

The main steps in the argument are then as follows. We consider an electron gas, and for convenience we take the volume of the gas to be unity. The electron gas is then perturbed by an external static potential, to yield

$$\phi(\mathbf{r}) = \sum_{\mathbf{k}} \tilde{\phi}_{\mathbf{k}} \exp(\mathrm{i}\mathbf{k}\,.\,\mathbf{r}). \tag{3.54}$$

The perturbing Hamiltonian H' then takes the form

$$H' = \sum_{\mathbf{k}\neq 0} \tilde{\phi}_{\mathbf{k}}\rho_{-\mathbf{k}} \tag{3.55}$$

where $\rho_{\mathbf{k}}$ is the density operator. To second order in this perturbing potential the energy is given by

$$E = E_0(\rho_0)+\sum_{\mathbf{k}\neq 0}\tilde{\phi}_{\mathbf{k}}\tilde{\phi}_{-\mathbf{k}}\sum_{m}\frac{\langle 0|\rho_{\mathbf{k}}|m\rangle\langle m|\rho_{-\mathbf{k}}|0\rangle}{E_0-E_m} \tag{3.56}$$

where $|0\rangle, E_0, |m\rangle$ and E_m are the exact ground- and excited-state wave-functions and energies of the uniform gas. This eqn (3.56) can be written in terms of the density response function $\mathscr{F}(\mathbf{k})$ say (compare eqn 3.20), as

$$E = E^{0}(\rho_0)+\tfrac{1}{2}\sum_{\mathbf{k}\neq 0}\tilde{\phi}_{\mathbf{k}}\tilde{\phi}_{-\mathbf{k}}\mathscr{F}(\mathbf{k}). \tag{3.57}$$

In terms of this response function, the density may be written, to first order in ϕ as (cf eqn 3.20)

$$\rho_{\mathbf{k}} = \tilde{\phi}_{\mathbf{k}}\mathscr{F}(\mathbf{k}). \tag{3.58}$$

The Hartree–Fock energy can now be separated from the total energy, following Ma and Brueckner. First, they noted that the results in eqns (3.57) and (3.58) could be obtained by minimizing the functional

$$E[\rho] = E^{0}[\rho_0]+\sum_{\mathbf{k}\neq 0}\tilde{\phi}_{-\mathbf{k}}\rho_{\mathbf{k}}-\tfrac{1}{2}\sum_{\mathbf{k}\neq 0}\rho_{\mathbf{k}}\rho_{-\mathbf{k}}\mathscr{F}^{-1}(\mathbf{k}). \tag{3.59}$$

† *Note added in proof.* A valuable account of many-body methods is that by J. Linderberg and Y. Öhrn (Propagators in Quantum Chemistry: Academic Press, New York, 1973).

Similarly, by minimizing the functional

$$E_{\mathrm{HF}}[\rho] = E_{\mathrm{HF}}[\rho_0]+\sum_{\mathbf{k}\neq 0}\tilde{\phi}_{-\mathbf{k}}\rho_{\mathbf{k}}-\tfrac{1}{2}\sum_{\mathbf{k}\neq 0}\rho_{\mathbf{k}}\rho_{-\mathbf{k}}\mathscr{F}_{\mathrm{HF}}^{-1}(\mathbf{k}) \tag{3.60}$$

we obtain the density and energy in the Hartree–Fock approximation, $\mathscr{F}_{\mathrm{HF}}$ being the density response function in this theory. Hence we find, by subtraction, from eqns (3.59) and (3.60)

$$\begin{aligned}E[\rho]-E_{\mathrm{HF}}[\rho] &= E_{\mathrm{c}}[\rho]\\ &= E_{\mathrm{c}}^{0}[\rho_0]-\tfrac{1}{2}\sum_{\mathbf{k}\neq 0}\rho_{\mathbf{k}}\rho_{-\mathbf{k}}[\mathscr{F}^{-1}(\mathbf{k})-\mathscr{F}_{\mathrm{HF}}^{-1}(\mathbf{k})].\end{aligned} \tag{3.61}$$

Expanding eqn (3.53) we can write

$$E_{\mathrm{c}}[\rho] = E_{\mathrm{c}}^{0}[\rho_0]+\frac{\mathrm{d}^2}{\mathrm{d}\rho_0^2}E_{\mathrm{c}}[\rho_0]\frac{1}{2}\sum_{\mathbf{k}\neq 0}\rho_{\mathbf{k}}\rho_{-\mathbf{k}}+B(\rho_0)\sum_{\mathbf{k}\neq 0}k^2\rho_{\mathbf{k}}\rho_{-\mathbf{k}}. \tag{3.62}$$

Diagrammatic analysis of the response functions in eqn (3.61) now allows eqns (3.61) and (3.62) to be related, and after lengthy calculations Ma and Brueckner find the result

$$B_{\mathrm{corr}}(\rho) = [8{\cdot}47\times 10^{-3}+O(\rho^{-\frac{1}{3}}\ln\rho)]\rho^{-\frac{4}{3}}\ \mathrm{Ryd} \tag{3.63}$$

where lengths are in atomic units. The leading term in eqn (3.53) is thus of the form of an energy density $(\nabla\rho)^2/\rho^{\frac{4}{3}}$ as proposed for exchange by Herman *et al.* It would seem that any partial summation of gradient series, such as embodied in eqn (3.51), should contain within itself the result (3.63) for the correlation contribution.

3.9. Applications of density functional theory

Having set up this density functional scheme, we are left in pure crystals with a periodic potential, obtained from eqn (3.51), through

$$V(\mathbf{r}) = V_{\mathrm{Hartree}}(\mathbf{r})+\frac{\delta\varepsilon_{\mathrm{xc}}[\rho]}{\delta\rho} \tag{3.64}$$

which, at least in principle, we can use to calculate energy bands. It ought really to be clear from the outset that this potential enters the expression for the chemical potential or Fermi energy. We have given no basic justification for using such a potential away from the Fermi energy. Thus, without further discussion, we had no right to suppose that we could use $V(\mathbf{r})$ to calculate the local density $\rho(\mathbf{r}E)$ of electrons lying below energy E, except of course when $E = E_{\mathrm{f}} = \mu$.

Actually therefore, even if we could get $\varepsilon_{\mathrm{xc}}[\rho]$ in eqn (3.64) exactly, which we manifestly cannot at present, we ought really only to calculate $\rho(\mathbf{r}E_{\mathrm{f}})$ which anyway we can measure by X-ray scattering experiments. Nevertheless, the first principles work of S. Lundqvist and Hedin on the one hand, and the success of Dirac–Slater exchange in band structure calculations on the other, show that eqn (3.64) is very useful in many cases away from the Fermi level.

However, we now want to indicate some other applications of such an approach, based on the density, in the physics of crystals. These applications, however, have primarily to do with perturbing the system, say by a phonon, or by an impurity.

3.9.1. *Phonon theory and electron density*

We want here to point out that the above approach using one-body potential theory which includes in the term $\varepsilon_{xc}[\rho]$ the effect of exchange and correlation forces can be used to formulate the lattice dynamical problem, within the Born–Oppenheimer and the harmonic approximations.

The essential quantity one needs, within such an approximational framework, is the change in the electron density due to small displacements of the ions from their perfectly periodic lattice sites.

As is well known in lattice dynamics (see Born and Huang, 1954; also Vosko, Taylor, and Keech, 1965) the phonon frequencies ω are determined by the eigenvalue equation

$$\sum_{\beta} D_{\alpha\beta}(\mathbf{k})\varepsilon_{\mathbf{k}}^{\beta} = \omega^2 \varepsilon_{\mathbf{k}}^{\alpha}, \tag{3.65}$$

where $\boldsymbol{\varepsilon}_{\mathbf{k}}$ is a polarization vector and D is the dynamical matrix given in the harmonic approximation by

$$D_{\alpha\beta}(\mathbf{k}) = \frac{1}{M}\sum_{l} \Phi_{\alpha\beta}(\mathbf{l})[\exp(-i\mathbf{k}\,.\,\mathbf{l})-1], \tag{3.66}$$

where M is the ionic mass, and

$$\Phi_{\alpha\beta}(\mathbf{l}'-\mathbf{l}) = \left(\frac{\partial^2\Phi}{\partial u_{\mathbf{l}'}^{\alpha}\partial u_{\mathbf{l}}^{\beta}}\right)_0, \tag{3.67}$$

Φ being the total potential energy governing the motion of the ions. The vector $\mathbf{l}$ denotes the equilibrium lattice sites and the derivative in eqn (3.67) is evaluated at these equilibrium positions.

The electronic contribution to $\Phi_{\alpha\beta}$ can be calculated from Feynman's theorem relating the Hamiltonian H to the energy E when a parameter λ appears in the Hamiltonian H by

$$\frac{\partial E}{\partial \lambda} = \left\langle \Psi_{\lambda}\left|\frac{\partial H}{\partial \lambda}\right|\Psi_{\lambda}\right\rangle \tag{3.68}$$

in an obvious notation.

If $V(\mathbf{r})$ is the potential due to an ion, this equation yields

$$\frac{\partial E}{\partial u_{\mathbf{l}'}^{\beta}} = \int \frac{\rho(\mathbf{r})\partial V(\mathbf{r}-\mathbf{l}'-\mathbf{u}_{\mathbf{l}'})}{\partial u_{l}^{\beta}}\,\mathrm{d}\mathbf{r} \tag{3.69}$$

and therefore

$$\frac{\partial^2 E}{\partial u_{\mathbf{l}}^{\alpha}\partial u_{\mathbf{l}'}^{\beta}} = \int \frac{\partial\rho(\mathbf{r})}{\partial u_{\mathbf{l}}^{\alpha}}\frac{\partial V(\mathbf{r}-\mathbf{l}'-\mathbf{u}_{l'})}{\partial u_{\mathbf{l}'}^{\beta}}\,\mathrm{d}\mathbf{r} + \delta_{\mathbf{l}'\mathbf{l}}\int \rho(\mathbf{r})\frac{\partial^2 V(\mathbf{r}-\mathbf{l}-\mathbf{u}_{l})}{\partial u_{\mathbf{l}}^{\alpha}\partial u_{\mathbf{l}}^{\beta}}\,\mathrm{d}\mathbf{r}. \tag{3.70}$$

The second term on the right-hand side, involving the second derivative of the ionic potential (and representing in fact intrinsic two-phonon processes, which we need not go into further here) is not required for the determination of the phonon frequencies, as eqn (3.66) reveals.

The essential point which now allows us to make further progress in this lattice dynamical theory is that the density involved in eqn (3.70) can be calculated exactly from the one-body potential (3.64), at least in principle.

3.9.2. *Generalized rigid-ion model*

In early work on lattice dynamics, it was almost always assumed that an electron cloud could be assigned to each ion in a crystal and that this was such that, at least for the small displacements we are concerned with here, this charge distribution moved rigidly with the nucleus when it was displaced. Thus, one wrote the periodic density $\rho(\mathbf{r})$ as a sum of 'rigid ion' densities $\sigma(\mathbf{r})$, centred on the lattice sites $\mathbf{l}$, namely

$$\rho(\mathbf{r}) = \sum_{\mathbf{l}} \sigma(\mathbf{r}-\mathbf{l}). \tag{3.71}$$

Such a decomposition into localized distributions does not involve approximation, but unfortunately $\sigma(\mathbf{r})$ is by no means uniquely determined. For it is clear that $\rho(\mathbf{r})$ is completely characterized by its Fourier components $\rho_{\mathbf{K}_n}$ at the reciprocal lattice vectors $\mathbf{K}_n$, through the usual Fourier series

$$\rho(\mathbf{r}) = \sum_{\mathbf{K}_n} \rho_{\mathbf{K}_n} \exp(\mathrm{i}\mathbf{K}_n\,.\,\mathbf{r}) \tag{3.72}$$

However, knowledge of the $\rho_{\mathbf{K}_n}$s only allows the Fourier components $\sigma(\mathbf{k})$ (now continuous) of the localized distribution $\sigma(\mathbf{r})$ to be determined at $\mathbf{k} = \mathbf{K}_n$, and many different Fourier transforms can be chosen which all have the property $\sigma(\mathbf{K}_n) = \rho_{\mathbf{K}_n}$, and therefore all reproduce the periodic density.

However, as indicated above, the early workers in lattice dynamics were essentially assuming that there is a choice of σ which not only allows $\rho(\mathbf{r})$ in the perfectly periodic crystal to be built up, but which also allows, via the rigid-ion model, the density to be built up when the ions are displaced to new sites $\mathbf{l}+\mathbf{u}_l$. It is already physically clear that if such a rigid-ion model is valid, then one can define pairwise forces between the ions. We shall see from the one-body potential formulation that in general this is not possible, but that a suitable exact generalization of the rigid ion model in fact exists.

Let us now proceed to calculate the first-order density change $\rho_1(\mathbf{r})$ when the ions are moved through small displacements $\mathbf{u}_l$ from the sites $\mathbf{l}$ following Jones and March (1970). It is then clear that the first-order density change

$$\rho_1(\mathbf{r}) = \sum_{\mathbf{l}} \mathbf{u}_l \cdot \frac{\partial \rho(\mathbf{r})}{\partial \mathbf{u}_l} \tag{3.73}$$

is given by linear response theory as (compare eqn (3.20))

$$\rho_1(\mathbf{r}E) = \int \Delta V^{(1)}(\mathbf{r}')F(\mathbf{r}\mathbf{r}'E)\,\mathrm{d}\mathbf{r}' \tag{3.74}$$

where (see, for example, Stoddart, March, and Stott 1969), F can be expressed through the result

$$\frac{\partial F}{\partial E} = 2\,\mathrm{Re}\left[G_0(\mathbf{r}\mathbf{r}_1E_+)\frac{\partial \rho_0(\mathbf{r}_1\mathbf{r}E)}{\partial E}\right] \tag{3.75}$$

where G_0 is the perfect lattice Green function and $\rho_0(\mathbf{r}\mathbf{r}'E)$ is a Dirac density matrix whose diagonal element $\rho_0(\mathbf{r})$ is the *exact* crystal density. We want to emphasize that the response function F in eqn (3.74) is determined solely by one-body band theory, and *not* by many-body theory. The local one-body potential which generates the exact density completely determines F. Of course, the price we must pay is that, in eqn (3.74), the potential $\Delta V^{(1)}(\mathbf{r})$ is the change due to the ionic displacements in the one-body potential (3.64) incorporating the electron–electron interactions. Since we can always write for a given ionic configuration, that the one-body potential has the form (eqn 3.64)

$$V(\mathbf{r}) = V_{\text{electrostatic}} + V_{\text{exchange}} + V_{\text{correlation}} \tag{3.76}$$

where these are solely functionals of the electron density, we can express $\Delta V^{(1)}$ to first order in ρ_1 as

$$\Delta V^{(1)} = \Delta V^{(1)}_{\text{electrostatic}} + \int U(\mathbf{r}\mathbf{r}')\rho_1(\mathbf{r}')\,\mathrm{d}\mathbf{r}' \tag{3.77}$$

where

$$\Delta V^{(1)}_{\text{electrostatic}} = \int \frac{\rho_1(\mathbf{r}')}{|\mathbf{r}-\mathbf{r}'|}\,\mathrm{d}\mathbf{r}' + \sum_{\mathbf{l}}\left[\frac{Ze}{|\mathbf{r}-\mathbf{l}-\mathbf{u}_l|} - \frac{Ze}{|\mathbf{r}-\mathbf{l}|}\right] \tag{3.78}$$

and Ze represents the charge on each ion. More generally, we could take a potential describing the nucleus plus core electrons, instead of Ze/r. Eqns (3.73) and (3.78) are now to be solved simultaneously and to do so we write (cf. Johnson 1969)

$$\rho_1(\mathbf{r}) = \sum_l \mathbf{u}_l \cdot \frac{\partial \rho(\mathbf{r})}{\partial \mathbf{u}_l} = \sum_l \mathbf{u}_l \cdot \mathbf{R}_l(\mathbf{r}). \tag{3.79}$$

Now if $\mathbf{u}_0 \cdot \mathbf{R}(\mathbf{r})$ represents the density change resulting from a displacement $\mathbf{u}_0$ of the ion at the origin say, the physical equivalence of every site implies that the displacement of the ion at $\mathbf{l}$ will result in the same density change referred to $\mathbf{l}$ as a new origin. Thus it follows that

$$\mathbf{R}_l(\mathbf{r}) = \mathbf{R}(\mathbf{r}-\mathbf{l}). \tag{3.80}$$

Using this eqn (3.80) in (3.79), we can write from eqns (3.77) and (3.78)

$$\begin{aligned} \Delta V^{(1)}(\mathbf{r}) = & \sum_{\mathbf{l}} \int \frac{\mathbf{u}_l \cdot \mathbf{R}(\mathbf{r}'-\mathbf{l})}{|\mathbf{r}'-\mathbf{r}|}\, \mathrm{d}\mathbf{r}' \\ & + \sum_{\mathbf{l}} \left[\frac{Ze}{|\mathbf{r}-\mathbf{l}-\mathbf{u}_l|} - \frac{Ze}{|\mathbf{r}-\mathbf{l}|} \right] \\ & + \sum_{\mathbf{l}} \mathbf{u}_l \int U(\mathbf{r}\mathbf{r}')\mathbf{R}(\mathbf{r}'-\mathbf{l})\, \mathrm{d}\mathbf{r}'. \end{aligned} \tag{3.81}$$

We wish now to demonstrate, by expanding ionic terms to order $\mathbf{u}_l$, that this can be written in a form analogous to eqn (3.79), namely

$$\Delta V^{(1)}(\mathbf{r}) = \sum_l \mathbf{u}_l \cdot \mathbf{P}(\mathbf{r}-\mathbf{l}). \tag{3.82}$$

The first two terms on the right-hand side of eqn (3.81) present no difficulty. Furthermore, it is straightforward in the exchange and correlation term to demonstrate that

$$U(\mathbf{r}+\mathbf{l}, \mathbf{r}'+\mathbf{l}) = U(\mathbf{r}\mathbf{r}') \tag{3.83}$$

and hence

$$\begin{aligned} \int U(\mathbf{r}\mathbf{r}')\mathbf{R}(\mathbf{r}'-\mathbf{l})\, \mathrm{d}\mathbf{r}' &= \int U(\mathbf{r}, \mathbf{r}'+\mathbf{l})\mathbf{R}(\mathbf{r}')\, \mathrm{d}\mathbf{r}' \\ &= \int U(\mathbf{r}-\mathbf{l}, \mathbf{r}')\mathbf{R}(\mathbf{r}')\, \mathrm{d}\mathbf{r}'. \end{aligned} \tag{3.84}$$

Thus eqn (3.82) results, with $\mathbf{P}(\mathbf{r})$ given by

$$\mathbf{P}(\mathbf{r}) = \int \frac{\mathbf{R}(\mathbf{r}')\, \mathrm{d}\mathbf{r}'}{|\mathbf{r}-\mathbf{r}'|} - \frac{Ze\mathbf{r}}{r^2} + \int U(\mathbf{r}\mathbf{r}')\mathbf{R}(\mathbf{r}')\, \mathrm{d}\mathbf{r}'. \tag{3.85}$$

As we remarked above, we can generalize this eqn (3.85) by replacing $Ze\mathbf{r}/r^2$ by the gradient of an appropriate ionic potential V_b (for example, a Herman–Skillman (1965) potential).

But we can also readily prove that

$$F(\mathbf{r}+\mathbf{l}, \mathbf{r}'+\mathbf{l}) = F(\mathbf{r}\mathbf{r}'), \tag{3.86}$$

and hence we can write, after a little manipulation

$$\rho_1(\mathbf{r}) = \sum \mathbf{u}_l \cdot \int \mathbf{P}(\mathbf{r}')F(\mathbf{r}-\mathbf{l}, \mathbf{r}')\,\mathrm{d}\mathbf{r}'. \tag{3.87}$$

Comparing eqn (3.87) with eqn (3.79) one obtains the basic integral equation of the theory:

$$\mathbf{R}(\mathbf{r}) = \int \mathbf{P}(\mathbf{r}')F(\mathbf{r}\mathbf{r}')\,\mathrm{d}\mathbf{r}'. \tag{3.88}$$

If we knew the exchange and correlation function $U(\mathbf{r}\mathbf{r}')$ then eqns (3.85) and (3.88) could be solved iteratively to yield $\mathbf{R}(\mathbf{r})$.

3.9.3. *Approximations leading to rigid-ion model*

This is now the appropriate point to establish contact with the rigid-ion model. If we return to eqns (3.79) and (3.80), and put all the $\mathbf{u}_l$s equal, then we have made a uniform translation of the lattice, and the density change is directly related to $\nabla\rho_0(\mathbf{r})$, through:

$$\nabla\rho_0(\mathbf{r}) = \sum_{\mathbf{l}} \mathbf{R}(\mathbf{r}-\mathbf{l}). \tag{3.89}$$

Comparing this with eqn (3.71) of the 'rigid-ion' model, we see that, from that equation

$$\nabla\rho_0(\mathbf{r}) \doteqdot \sum_{l} \nabla\sigma(\mathbf{r}-\mathbf{l}). \tag{3.90}$$

If we could write, by comparing eqns (3.89) and (3.90), that

$$\mathbf{R} = \nabla\sigma \tag{3.91}$$

then obviously the rigid-ion model is regained, with the corresponding simplification of a pairwise force field. It is clear from eqn (3.89) that

$$\operatorname{curl} \nabla\rho_0(\mathbf{r}) = \operatorname{curl} \sum_{\mathbf{l}} \mathbf{R}(\mathbf{r}-\mathbf{l}) = 0, \tag{3.92}$$

but the rigid-ion model (3.91) insists on the stronger condition

$$\operatorname{curl} \mathbf{R} = 0. \tag{3.93}$$

Eqn (3.92) is exactly valid, whereas (3.93) is only one (albeit the simplest) way of satisfying eqn (3.92) and it is not generally true.

However, it is straightforward to show from eqns (3.89) and (3.72) that if we Fourier transform $\mathbf{R}(\mathbf{r})$ then the Fourier components $\mathbf{R}(\mathbf{k})$ have the property that

$$\mathbf{R}(\mathbf{K}_n) = \mathrm{i}\mathbf{K}_n\rho_{\mathbf{K}_n} \tag{3.94}$$

at the reciprocal lattice vectors $\mathbf{K}_n$. This is an exact result, and, as we have remarked earlier, the Fourier coefficients of the periodic density are, in principle, known from X-ray scattering experiments. Clearly, the result (3.94) must be contained in eqns (3.85) and (3.88) and is to be used to check choices of the exchange and correlation function U.

Whereas the rigid-ion model gives $\mathbf{R}(\mathbf{k})$ in the direction of $\mathbf{k}$ for all $\mathbf{k}$, this is not generally true except when $\mathbf{k}$ is a reciprocal lattice vector $\mathbf{K}_n$. Deviations of $\mathbf{R}(\mathbf{k})$ from the direction of $\mathbf{k}$ reflect the existence of many-body forces in the ion–ion interaction and the theory presented here affords a systematic basis for their inclusion in lattice dynamics. No division into two-body, three-body, etc., forces is necessary if we work with the generalized rigid-ion model, based on the vector $\mathbf{R}(\mathbf{r})$. In practical calculations at present, however, combining this formally exact theory with pseudopotentials, one will in fact then classify corrections to the pairwise force model in terms of 3, 4, etc., body forces.

3.10. Elementary theory of pairwise interactions

As an elementary example of the above theory, let us use the response function F as calculated for a uniform electron gas, and derive the corresponding pair forces (Jones and March 1970). To do so, we write for the pair potential $\phi(\mathbf{X})$ the expression

$$\phi(\mathbf{X}) = \int \sigma(\mathbf{r})V_{\mathrm{b}}(\mathbf{r}-\mathbf{X}) \tag{3.95}$$

and then we find

$$\frac{\partial^2\phi(\mathbf{X})}{\partial X_\alpha \partial X_\beta} = \int \frac{\partial\sigma(\mathbf{r})}{\partial x_\alpha}\frac{\partial V_{\mathrm{b}}(\mathbf{r}-\mathbf{X})}{\partial X_\beta}\,\mathrm{d}\mathbf{r} \tag{3.96}$$

and it is clear that

$$R_\alpha(\mathbf{r}) = \frac{\partial\sigma}{\partial x_\alpha} \tag{3.97}$$

which is equivalent to the assumption of rigid ions. For a uniform electron gas we have

$$F(\mathbf{r},\mathbf{r}') = F(\mathbf{r}-\mathbf{r}') \tag{3.98}$$

and

$$U(\mathbf{r}, \mathbf{r}') = U(\mathbf{r}-\mathbf{r}'). \tag{3.99}$$

Hence

$$V(\mathbf{r}) = \int U(\mathbf{r}-\mathbf{r}')\Delta\rho(\mathbf{r}')\,\mathrm{d}\mathbf{r}' + V_{\mathrm{b}}(\mathbf{r}) + \int \frac{\Delta\rho(\mathbf{r}')\,\mathrm{d}\mathbf{r}'}{|\mathbf{r}-\mathbf{r}'|} \tag{3.100}$$

or in Fourier transform

$$\begin{aligned} V(\mathbf{k}) &= U(\mathbf{k})\Delta\sigma_{\mathbf{k}} + V_{\mathrm{b}}(\mathbf{k}) + \Delta\sigma_{\mathbf{k}}k^{-2} \\ &= U'(\mathbf{k})\Delta\sigma_{\mathbf{k}} + V_{\mathrm{b}}(\mathbf{k}) \end{aligned} \tag{3.101}$$

where we have denoted $U(\mathbf{k})+k^{-2}$ by $U'(\mathbf{k})$. Now we have further that

$$\Delta\sigma_{\mathbf{k}} = F(\mathbf{k})V(\mathbf{k}) \tag{3.102}$$

which can be rewritten using eqn (3.101) as

$$\Delta\sigma_{\mathbf{k}} = F(\mathbf{k})U'(\mathbf{k})\Delta\sigma_{\mathbf{k}} + F(\mathbf{k})V_{\mathrm{b}}(\mathbf{k}). \tag{3.103}$$

Representing the relation between $\Delta\sigma$ and V_{b} in the conventional way by a dielectric function $\varepsilon(\mathbf{k})$, we have

$$\Delta\sigma_{\mathbf{k}} = k^2 V_{\mathrm{b}}(\mathbf{k})[\varepsilon^{-1}(\mathbf{k}) - 1] \tag{3.104}$$

and hence, by comparing eqns (3.103) and (3.104) we find

$$\varepsilon^{-1}(k) = \frac{1 - F(k)U(k)}{1 - F(k)U'(k)}. \tag{3.105}$$

Explicitly for a uniform electron gas we have

$$F(\mathbf{r}, \mathbf{r}', E) = -\frac{k^2}{2\pi^3} j_1 \frac{(2k|\mathbf{r}-\mathbf{r}'|)}{|\mathbf{r}-\mathbf{r}'|^2}: \qquad E = \frac{k^2}{2}, \tag{3.106}$$

where $j_1(x) = (\sin x - x\cos x)/x^2$ and Fourier transforming this we regain the Lindhard theory (see eqn 3.22) for a uniform electron gas when $U'(k)$ is replaced by k^{-2}.

From eqn (3.106) we may write, adding the ion–ion interaction ϕ_{ii}:

$$\begin{aligned} \phi_{\mathrm{total}}(\mathbf{X}) &= \iint \Delta\sigma_{\mathbf{k}} \exp(\mathrm{i}\mathbf{k}\,.\,\mathbf{r})\,\mathrm{d}\mathbf{k}\, V_{\mathrm{b}}(\mathbf{r}-\mathbf{X})\,\mathrm{d}\mathbf{r} + \phi_{\mathrm{ii}} \\ &= \int \Delta\sigma_{\mathbf{k}} V_{\mathrm{b}}(\mathbf{k}) \exp(\mathrm{i}\mathbf{k}\,.\,\mathbf{X})\,\mathrm{d}\mathbf{k} + \phi_{\mathrm{ii}} \\ &= \int \frac{k^2 V_{\mathrm{b}}^2(\mathbf{k})}{\varepsilon(\mathbf{k})} \exp(\mathrm{i}\mathbf{k}\,.\,\mathbf{X})\,\mathrm{d}\mathbf{k}. \end{aligned} \tag{3.107}$$

This reduces to the $\mathbf{r}$-space result of Corless and March (1961) for point ions, whereas for ions with structure the result is the same as that given, for example, by Ziman (1964), with $V_b(\mathbf{k})$ appropriately interpreted as a pseudopotential,† though in the present case the dielectric function includes exchange and correlation effects through eqn (3.105). A very convenient representation of the interaction $U(k)$ is afforded by the work of Singwi, Tosi, and Sjolander (1968, 1970), using an effective-field approximation. We must reluctantly refer the reader to these papers for further details (see also, Fig. 4.4 for the kind of potential which would be obtained from eqn 3.107 for Na metal).

3.11. Defects in crystals

Dr. Lidiard, in the following chapter, describes in detail the effects of impurities and defects in crystals. We shall therefore restrict ourselves here to some brief comments on the usefulness of one-body potentials in this area. First of all, electron theory classifies a defect rather basically by the charge it displaces. Thus, if ρ_0 is the periodic charge density, we want to calculate the new charge density $\rho(\mathbf{r})$ in the defect lattice, the displaced charge $\Delta(\mathbf{r})$ being defined by $\Delta(\mathbf{r}) = \rho(\mathbf{r}) - \rho_0(\mathbf{r})$. If the defect represented a genuine perturbation on the perfect lattice, then we could express $\Delta(\mathbf{r})$ immediately as in eqns (3.74) and (3.75). Thus a knowledge of the one-body potential (3.64) would enable us to calculate the one-body response function F, and if we could set up a model for ΔV in eqn (3.74) we could map out the charge displaced round an impurity. Some prescriptions have been developed for estimating ΔV. For example, in a free-electron matrix, the defect potential must satisfy the Friedel sum rule (cf Kittel, 1963)

$$Z = \frac{2}{\pi} \sum_l (2l+1)\eta_l(k_f) \tag{3.108}$$

where Z is the excess charge on the impurity centre, while $\eta_l(k_f)$ is the phase shift of the lth partial wave due to the presence of the defect potential, evaluated at the Fermi momentum k_f. Obviously, if the response function F were known from a Bloch wave calculation, the angularity of the charge density round a defect could be mapped out. However, we must stress that whereas for a 'defect' which is a phonon, the perturbation is small because the displacements $\mathbf{u}_l$ are small, for impurities in crystals the perturbations are generally too strong to be treated by linear response theory. Nevertheless, we also want to stress that, for properties which can be described in terms of the displaced charge, one-body potential theory is again appropriate.

3.12. Fermi surface and one-body potential theory

Let us first summarize what the potential (3.64) can do for us. It can give the exact charge density $\rho(\mathbf{r})$ in an N-electron system from the sum of the squares of N one-body wavefunctions. Secondly, we can use the knowledge of

† See Chapter 1 and Chapter 4 for details.

$\rho(\mathbf{r})$ to estimate the ground-state energy approximately, using the exchange and correlation energy functionals.

Thirdly, as we have seen, we can, in principle, from a knowledge of exchange and correlation, calculate the phonon dispersion relations. We now want to emphasize that one of the methods which can be used to map out the Fermi surface in a metal is from the so-called Kohn anomaly in the lattice dispersion relations. This arises from the 'kink' in the Lindhard dielectric constant at $k = 2k_f$, which, in turn, is a consequence of the sharp Fermi surface. And since, as we have seen, the phonon dispersion curves can be got from the density, in essence, it seems clear that this Fermi surface must be correctly generated by the one-body potential (3.64) (cf. Jones and March 1973). We have not seen how to prove that the Fermi surfaces which are measured by different techniques, such as the de Haas–van Alphen effect or cyclotron resonance must be precisely the same as that determined from the Kohn anomaly (Kohn 1959), but the experimental evidence available tells us that any differences are small and presumably relatively uninteresting at this stage in the development of the theory.

There is a further point we would like to make here. If the electron density is high, the Landau quasi-particle picture† is valid and a Fermi surface is well defined. However, if we could lower the density of a metal sufficiently, we would at some stage pass through a transition to an insulating phase, after which the concept of a Fermi surface could no longer be useful. It is not presently clear to us how one-body potential theory could herald this, even though it might be possible to characterize the new phase by a charge density with broken symmetry. This aspect of one-body potential theory needs further investigation.

3.13. k-Dependent potentials and energy bands

It is often argued in metals that the concept of one-electron states is only useful relatively near to the Fermi surface, at which the quasi-particle states have infinite lifetimes. It might appear therefore that it is not meaningful or useful to calculate energy bands over a wide range of energies away from the Fermi surface, because of the short lifetimes of quasi-particle states well away from the Fermi surface. This does not, however, appear to be the case, for band structure information over wide energy bands appears to be very helpful, for example, in interpreting experiments on optical properties of crystals.

If we adopt this pragmatic point of view, then it is, of course, natural to ask whether the potential eqn (3.64), which is appropriate at the Fermi energy, can be used to calculate energy bands over a wide range of energy.

† See, for example, the account of Jones and March (1973).

We certainly see no basic reason why this must be so in general, though it is reasonable to expect that properties near the Fermi surface might be usefully described in this way.

One reason for fearing that very different potentials might have to be used as we move away from the Fermi surface is that Hartree–Fock theory leads to a potential which varies strongly with E or $\mathbf{k}$. However, in metals, as we have already remarked, such variation leads to qualitative disagreement with experiment and, starting from the work of Bohm and Pines (1953), many authors have argued that the effect of correlations is to weaken the energy dependence.

Nevertheless, a variety of prescriptions exist for setting up $\mathbf{k}$ dependent potentials. We can only mention again, in conclusion, the fundamental work of Hedin and S. Lundqvist, which, very fortunately, they have summarized themselves in an excellent review article (Hedin and Lundqvist, 1969; see also Hedin and Lundqvist B. I., 1971 and the Lundqvists and Hedin, 1971) together with the work on non-local effects in Fermi surface studies of Li by Vosko and Rasolt (1974).

REFERENCES

BEATTIE, A. M., STODDART, J. C., and MARCH, N. H. (1971). *Proc. R. Soc. A* **326**, 97.
BOHM, D. and PINES, D. (1953). *Phys. Rev.* **92**, 609.
BORN, M. and HUANG, K. (1954). *Dynamical theory of crystal lattices*, Clarendon Press, Oxford.
CORLESS, G. K. and MARCH, N. H. (1961). *Phil. Mag.* **6**, 1285.
DIRAC, P. A. M. (1930). *Proc. Cambridge Phil. Soc.* **26**, 376.
FUCHS, K. (1935). *Proc. R. Soc. A* **151**, 585.
GÁSPÁR, R. (1954). *Acta Phys. Hung.* **3**, 263.
GELL-MANN, M. and BRUECKNER, K. A. (1957). *Phys. Rev.* **106**, 364.
HEDIN, L. and LUNDQVIST, B. I. (1971). *J. Phys. C.*, **4**, 2064.
HEDIN, L. and LUNDQVIST, S. (1969). *Solid state physics*. (ed. F. Seitz, D. Turnbull, and H. Ehrenreich), vol. 23, p. 1. Academic Press, New York.
HERMAN, F. and SKILLMAN, S. (1965). *Atomic structure calculations*. Prentice Hall, New Jersey.
——, VAN DYKE, J. P., and ORTENBURGER, I. B. (1969). *Phys. Rev. Letts* **22**, 807.
HOHENBERG, P. and KOHN, W. (1964). *Phys. Rev.* **136**, B864.
JOHNSON, F. A. (1969). *Proc. R. Soc. A* **310**, 101.
JONES, W. and MARCH, N. H. 1970, *Proc. R. Soc. A* **317**, 359.
—— —— (1973). *Theoretical solid-state physics*, Wiley-Interscience, London, Volume 1, p. 202.
—— and YOUNG, W. H. (1971). *J. Phys. C.* **4**, 1322.
KIRZNITS, D. A. (1957). *Sov. Phys. JETP.* **5**, 64.
KITTEL, C. (1963). *Quantum theory of solids.* John Wiley, New York.
KOHN, W. (1959). *Phys. Rev. Letts.* **2**, 393.
—— and SHAM, L. J. (1965). *Phys. Rev.* **140***A*, 1133.

LUNDQVIST, S., HEDIN, L., and LUNDQVIST, B. I. (1971). *Solid St. Communs.* **9**, 537.
MA, S. and BRUECKNER, K. A. (1968). *Phys. Rev.* **165**, 18.
MARCH, N. H. (1952). *Acta Crystallogr.* **5**, 187; *see also* 1957, *Adv. Phys.* **6**, 1.
——, YOUNG, W. H., and SAMPANTHAR, S. (1967). *The many-body problem in quantum mechanics*. University Press, Cambridge.
OVERHAUSER, A. W. (1971). *Phys. Rev. B* **3**, 1888.
SEITZ, F. (1940). *Modern theory of solids*. McGraw Hill, New York.
SINGWI, K. S., SJOLÄNDER, A., TOSI, M. P., and LAND, R. H. (1970). *Phys. Rev. B* **1**, 1044.
——, TOSI, M. P., LAND, R. H., and SJOLÄNDER, A. (1968). *Phys. Rev.* **176**, 589.
SLATER, J. C. (1951). *Phys. Rev.* **81**, 385.
STODDART, J. C., BEATTIE, A. M., and MARCH, N. H. (1971). *Int. J. Quantum Chem.* **4**, 35.
——, MARCH, N. H., and STOTT, M. J. (1969). *Phys. Rev.* **186**, 683.
VON WEIZSÄCKER, C. F. (1935). *Z. Phys.* **96**, 431.
VOSKO, S. H. and RASOTT, M. (1974), *Phys. Rev. Letts.*, **32**, 297.
VOSKO, S. H., TAYLOR, R., and KEECH, G. H. (1965). *Can. J. Phys.* **43**, 1187.
WIGNER, E. P. (1934). *Phys. Rev.* **46**, 1002.
—— (1938). *Trans. Faraday Soc.* **34**, 678.
ZIMAN, J. M. (1964). *Adv. Phys.* **13**, 89.

4

DEFECTS IN CRYSTALLINE SOLIDS

A. B. LIDIARD

4.1. Introduction

CHAPTERS 1 and 2 have concerned themselves with the electronic states of perfect crystalline solids, i.e. with non-localized one-electron Bloch states and their energies. Here we are concerned with some of the consequences of having simple defects in such solids, particularly vacant lattice sites, atoms in interstitial sites and impurity, or foreign, atoms.† Such defects may be introduced thermally or by irradiation with energetic particles or in other ways, and may be studied by a wide variety of physical measurements, e.g. by studies of atomic transport, by spectroscopic and resonance methods, etc. From a theoretical point of view they constitute an interesting bridge between solid-state theory and molecular theory since these defects can often be regarded as molecule-like systems embedded in a crystal lattice. For example, the F-centre, an electron trapped in the field of an anion vacancy (Fig. 4.1), is the analogue of a free hydrogen atom and as such occupies

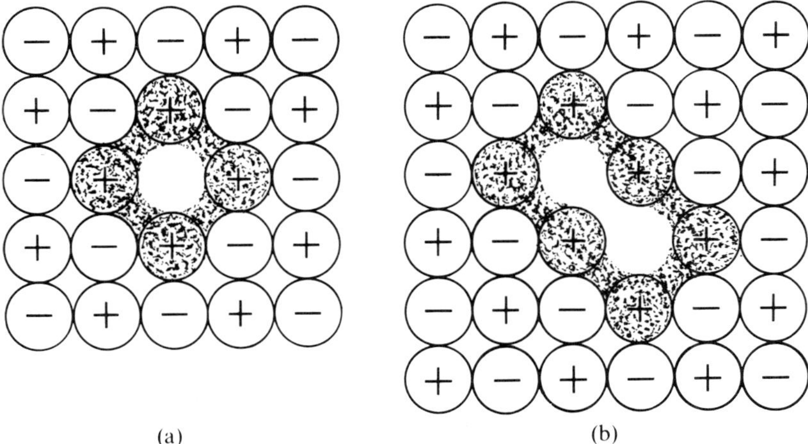

FIG. 4.1. Schematic diagram of (a) the F-centre, an anion vacancy which has trapped an electron; and (b) the M- or F_2-centre, a pair of anion vacancies which have trapped two electrons. (It should be noted that several notations for the description of defects are in use; the above terminology is however universal for the alkali and alkaline earth halides even though variations are in use for equivalent centres in other materials. For example in MgO the centre (a) would be referred to as an F^+ centre since it carries a net charge of $+e$.

† Collectively called point defects, a term taken from continuum elasticity theory and distinguishing them from linear defects, i.e. dislocations.

a position in defect studies comparable to that of hydrogen in chemical theory. This analogy to molecular systems has often been very fruitful; for example the ideas and methods of molecular theory have provided substantial guidance to the field of colour centres, i.e. to situations where we are particularly concerned with the electronic levels and other properties of spectroscopic interest.

Yet the fact that these defects are in a crystal lattice is of fundamental importance. The coupling of the defect electrons to the lattice is often strong as shown by the large Stokes shifts between absorption and emission bands which are frequently seen (e.g. from 2·3 eV to 1·0 eV for the F-centre in KCl). Sometimes it is so strong that even the symmetry of the local atomic configuration cannot be predicted by simple arguments. It is not generally the case that the lattice simply gives rise to a crystal field which splits otherwise predetermined atomic or molecular levels although, of course, there are such examples (e.g. transition metal and rare earth solute ions). The stability and constraints imposed by the lattice structure can lead to unusual molecular analogues; for example among the aggregates of F-centres one sees not only the analogues of molecular hydrogen (so-called $M \equiv F_2$-centres) but also higher aggregates ($R \equiv F_3$- and $N \equiv F_4$-centres). Also it is true that the balance of forces can change qualitatively with the nature of the matrix with the result that the 'same' defect can have very different properties in two different substances; for example the principal ($1s \rightarrow 2p$) transition of the F-centre is at 2·3 eV in KCl but at nearly 5·0 eV in MgO. Conversely different defects in the same substance can have similar properties. Thus a great variety of possibilities arise in practice. For these reasons the task of the solid-state physicist often contrasts with that of the molecular physicist. He must not only determine the detailed properties of defects of specified 'composition' and structure. He must also determine that composition and structure in the first place. For example although the nature of the F-centre has been known with reasonable certainty for 30 years, only in the last dozen years or so have the M-, R-, and N-centres been shown to be simple F-aggregates (F_2, F_3, and F_4). The methods which the solid-state physicist uses are thus often diverse and the theoretical arguments detailed.

It is not possible in the space available here to go into this analytical aspect of the subject except by way of illustration, but we refer the reader to the recent book by Flynn (1972) for a wide-ranging and comprehensive review. For a more elementary treatment of many of the important ideas basic to our discussion we refer to the book by Kittel (1971). Instead we shall take for granted the interest in defect energy levels, activation and formation energies and other related quantities and assume the importance and interest in having reliable theoretical predictions as an aid to the identification and understanding of defects and processes such as diffusion and radiation damage which are determined by them.

This chapter is divided into two main parts. The first deals with the response of the lattice to defect forces. Often these forces can be specified from the models adopted for the solid; as for example with simple vacancies and interstitial atoms in the solid rare gases or with 'closed-shell' ion defects in strongly ionic solids. In these cases we can calculate directly defect formation energies, activation energies for motion and other thermodynamic and kinetic properties of interest.† When, however, we are interested in 'open-shell' defects and in spectroscopic properties we must calculate the electronic states of the defect; this constitutes the second part of the chapter. Such calculations should ideally also allow us to calculate the forces which the defect exerts upon the lattice. The Franck–Condon principle, of course, allows us to equate transition energies calculated in a static lattice with the energy of the maximum in simple absorption bands (e.g. the F-band) but for many properties one needs also to know the coupling of the defect to the lattice and the nature and extent of the lattice relaxation.

Both areas of work described in this chapter are still being actively developed.

4.2. Response of the lattice

It will be apparent from what we have already said that the coupling between the defect states and the crystal lattice in which the defects are embedded is important. The defect exerts forces on the rest of the crystal which depend on its electronic state and which are absent in the perfect crystal. The distortion and polarization of the crystal under the influence of these forces results in general in large, additional energy terms. In this section we shall review the calculation of the response of crystal lattices to such forces. We shall take the Born–Oppenheimer approximation as our basis. Furthermore we shall limit our explicit calculations to static-lattice models. One of the reasons for doing so is that many experiments on colour centres are carried out at very low temperatures simply in order to freeze out lattice vibrations. However the limitation is not so severe as might at first appear. For example it is not difficult to show that if one may make a quasi-harmonic approximation to the lattice vibrations then at high temperatures the (internal) energies of defect formation and activation are simply the corresponding potential energies in an equivalent static lattice.‡ Thus, properly interpreted, calculations for a static lattice cover many of the situations of direct interest.

† Although these same models and methods can be applied for defects such as dislocations we shall not include them in our discussion mainly because their physical interest lies in a quite different field. The interactions of point defects with dislocations are, of course, often important, e.g. in mechanical behaviour and in some aspects of radiation damage.

‡ Though it should be noted in such an application that the equivalent static lattice is not in mechanical equilibrium at the actual lattice parameter.

There is, of course, a considerable interest in the dynamic response of a lattice to time-varying forces and in changes in lattice-vibration frequencies caused by defects (see e.g. Flynn 1972). These effects are also directly relevant to the properties of colour centres, e.g. for vibrational side-bands. However, we shall not pursue the formalism needed to describe these effects, limiting ourselves instead to the basic problem of the distortion of a static lattice.

Several approaches to this question have been evolved. The strategy of all of them can, however, be described very simply as follows: divide the lattice into two regions, (*i*) an inner region (I) containing the defect and as many of its immediate neighbours as desired (this choice will depend, as we shall see, on the method used), (*ii*) the rest of the crystal (region II) which is assumed to be far enough from the defect that a harmonic approximation is valid (Fig. 4.2). In this strategy, we can evidently write the total energy of the defect lattice as,

$$E = E_{\mathrm{I}}(\boldsymbol{\lambda}\,;\mathbf{x})+E_{\mathrm{I,II}}(\boldsymbol{\lambda}\,;\mathbf{x},\boldsymbol{\xi})+E_{\mathrm{II}}(\boldsymbol{\xi}) \tag{4.1}$$

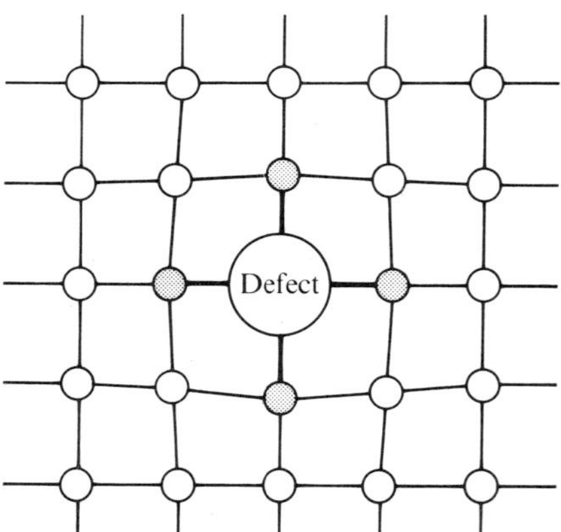

FIG. 4.2. Schematic diagram showing the division of the lattice into two regions (I and II) for the purposes of calculating lattice distortion, energy of relaxation etc. In this Figure region I is shown as the defect and its immediate neighbours while region II is the rest of the lattice. In any given calculation the choice of region I is guided by the method used (Section 4.2.2) but should be large enough that the results are insensitive to the exact boundary between I and II.

where we use $\mathbf{x}$ and $\boldsymbol{\xi}$ respectively to denote the sets of region I and region II displacements,† while $\boldsymbol{\lambda}$ stands for any variational parameters used in determining the defect electronic wavefunction. We can evidently rearrange terms in $\boldsymbol{\xi}$ between $E_{\mathrm{I,II}}$ and E_{II} so that the harmonic approximation to E_{II} takes the purely quadratic form

$$E_{\mathrm{II}} = \tfrac{1}{2}\boldsymbol{\xi}^{T}\mathbf{A}\boldsymbol{\xi} \tag{4.2}$$

† In general these will include both nuclear displacements and displacements of the electronic distribution relative to the atomic nuclei (polarization moments).

corresponding to the energy of a distorted region II filled with a perfect and undistorted inner region I. The term $E_{\mathrm{I,II}}$ then represents the change in interaction between I and II resulting from the presence of the defect and the distortions in I.

The mathematical problem is to minimize E with respect to $\boldsymbol{\lambda}$, $\mathbf{x}$, and ξ. For the moment we deal only with the third of these. The equilibrium condition for region II is then

$$\mathbf{F} \equiv -\frac{\partial E_{\mathrm{I,II}}}{\partial \xi} = \mathbf{A}\xi \tag{4.3}$$

which can be solved formally as long as it is adequate to expand the left-hand side only as far as the linear terms in ξ. Thus with

$$-\frac{\partial E_{\mathrm{I,II}}}{\partial \xi} \equiv \mathbf{F}^{(0)}(\boldsymbol{\lambda}, \mathbf{x}) + \mathbf{F}^{(1)}(\boldsymbol{\lambda}, \mathbf{x})\xi, \tag{4.4}$$

we obtain

$$\mathbf{A} - \mathbf{F}^{(1)}(\boldsymbol{\lambda}, \mathbf{x})\xi = \mathbf{F}^{(0)}(\boldsymbol{\lambda}, \mathbf{x}) \tag{4.5}$$

or

$$\xi = \mathbf{G}\mathbf{F}^{(0)}(\boldsymbol{\lambda}, \mathbf{x}), \tag{4.6}$$

where

$$\mathbf{G} = \{\mathbf{A} - \mathbf{F}^{(1)}(\boldsymbol{\lambda}, \mathbf{x})\}^{-1} \tag{4.7}$$

is the perturbed static Green's function. We observe that, while the unperturbed Green's function

$$\mathbf{G}^{(0)} = \mathbf{A}^{-1} \tag{4.8}$$

requires only a harmonic analysis of the perfect crystal dynamics and can even be obtained empirically (e.g. from neutron scattering experiments; Tewary and Bullough 1971), the perturbed Green's function requires a model from which to calculate the perturbation $\mathbf{F}^{(1)}$ of the force constant matrix. Obviously too $E_{\mathrm{I}}(\boldsymbol{\lambda}, \mathbf{x})$ requires the specification of the interactions among the atoms in the anharmonically distorted inner region. This shows that to proceed it is necessary to consider explicitly the interatomic forces in these crystals as well as calculating quantum mechanically the energy levels of the defect itself. We therefore next discuss the information obtainable about these interatomic interactions and then return to the methods of solving the equilibrium equations.

4.2.1. *Physical models*

We discuss briefly the four different solid types (1) molecular crystals as exemplified by the rare-gas solids, (2) ionic crystals, (3) valence crystals, and

(4) metals. In view of our later applications, we give most attention to ionic and valence crystals but we include some remarks on the others for completeness. It should be noted that the models used are empirical to a greater or lesser extent; while the form of the equations used is often determined by fundamental considerations, empirical data is used to determine characteristic parameters. Space does not permit us to do more than touch on these fundamentals and we therefore refer the reader to general sources such as Coulson (1961), Pauling (1960), and Slater (1939) for the necessary general background. Up-to-date references to more particular discussions are given where appropriate.

4.2.1.1. *Rare-gas Solids.*† Central pairwise potentials of the Lennard-Jones type i.e.

$$\phi(r) = \frac{A}{r^n} - \frac{B}{r^m}, \tag{4.9}$$

have long been used to represent the interactions of rare-gas atoms, generally with $m = 6$. Despite the lack of any fundamental justification for this form—except that when $m = 6$ the attractive term $-B/r^6$ may be said to represent van der Waals forces—it appears reasonably successful in allowing predictions of the lattice vibrations and of thermodynamic quantities. From a fundamental point of view the closed-shell repulsive term would be better represented over the separations of relevance here by an exponential Born–Mayer function, thus

$$\phi(r) = A\,e^{-r/\rho} - \frac{B}{r^6}. \tag{4.10}$$

Thus several quantum mechanical calculations of greater or lesser sophication agree in the prediction of repulsion energies which closely follow the above Born–Mayer form, typically from $r \sim 1$ Å to normal interatomic separations in the crystal (see e.g. Abrahamson 1964, 1969; Wedepohl 1967; Gilbert and Wahl 1967). In these regions ρ is typically of the order of 0·3 Å. At closer separations the repulsion increases more rapidly than the Born–Mayer form predicts while the r^{-6} form for van der Waals interactions is also inapplicable. However this region is only of interest in high-energy events (e.g. radiation damage cascades) and does not concern us in dealing with defects undergoing optical transitions or in systems in thermodynamic equilibrium at normal temperatures. The potential (4.10) appears to describe gaseous properties well (Dymond 1968). Both potentials (4.9) and (4.10) are, of course, central pairwise potentials and there is substantial evidence that higher-order interactions, especially three-body terms, are significant for the cohesion and equilibrium lattice structure of the solid rare gases. For example,

† For a review of potentials of interaction of rare gas atoms in relation to their physical properties generally see e.g. Pollack (1964), and in relation to the solid rare gases see Horton (1968).

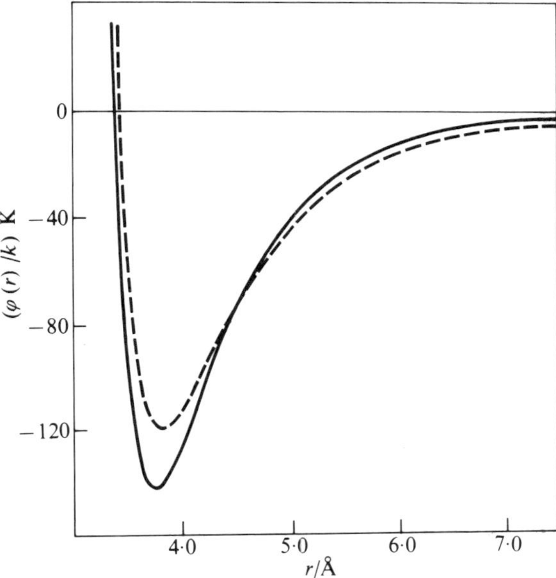

FIG. 4.3. Two empirically determined two-body interatomic potentials for Ar: --- a Lennard-Jones 6–12 potential and —— a more complex form fitted to a wider variety of empirical data assuming the presence of three-body forces described in the Axilrod–Teller approximation. (After Barker, Fisher, and Watts 1971.)

potentials of the type (4.9) and (4.10) even with empirically determined constants predict the stable phase to be a hexagonal close-packed structure instead of face-centred cubic as observed. Three-body interactions are necessary to stabilize the f.c.c. structure and may contribute about 10–30 per cent of the cohesive energy. Their presence is also confirmed by the fact that the elastic constants of Ar do not satisfy the Cauchy relation $c_{12} = c_{44}$, at 0 K, c_{12} exceeding c_{44} by about 10 per cent (Keeler and Batchelder 1970). These terms are, however, difficult to estimate accurately and thus are often ignored. Any purely pairwise potential model of the f.c.c. solids is thus potentially unstable with respect to transformation to the hexagonal phase. However, it is unlikely that the point defects and colour centres we are concerned with here will give rise to a transformation to the h.c.p. structure—in general, as has been shown by many calculations, the distortions of the rare-gas lattices and the energies of relaxation which they give rise to, are small compared to other materials. Nevertheless, the difficulty of estimating three-body forces introduces some uncertainty into the theory of defects in the solid rare-gases at the present time. Essentially two different sources of such terms have been proposed: (i) multiple dipole interactions resulting from long-range mutually induced polarization, i.e. high-order van der Waals interactions (Axilrod and Teller 1943, Axilrod 1951); (ii) short-range three- (or more) atom overlap and exchange interactions (Jansen 1964, Lombardi and Jansen 1968; but see also Swenberg 1967). The relative importance of these different terms remains uncertain, though the triple-dipole term is the one most extensively included so far. Fig. 4.3 shows two empirically determined two-body potentials for Ar, one obtained by fitting to (4.9) ignoring many-body effects, the other obtained by allowing for triple-dipole interactions (Barker *et al.* 1971). Most other empirically determined two-body potentials lie close to the curves shown in Fig. 4.3.

4.2.1.2. *Ionic crystals.* Although there exist orbital theories of the cohesion of ionic crystals (Lowdin, 1956; Lundqvist 1957) which give important insight and guidance into the nature of ionic bonding in solids, in calculations of lattice relaxation we use semi-empirical models. These are based on the fact that, for many of their physical properties in both solid and liquid states, ionic substances can be regarded as assemblies of ions whose closed-shell electronic structure ensures that their non-electrical interactions are repulsive at short distances and of van der Waals type farther away. Even when the cations have open-shell structures, e.g. transition-metal ions, the general pattern is often remarkably close to this classical ionic model—exchange energies and covalency effects, though vital for properties such as magnetism, are small on the scale of energies we may be concerned with. The existence of electrical charges on the ions means, however, that we are often concerned with defects which themselves carry net or effective electrical charges, with the result that the electrical polarization of the lattice is important. The correct representation of the dielectric behaviour of these crystals is thus vital to the accurate prediction of defect energies in these cases. In general, reasonably satisfactory models of these substances can be constructed by the analysis of four main physical properties: (1) cohesion; (2) (some) elastic constants; (3) dielectric constants; and (4) lattice vibrational frequencies, particularly those at long wavelength, e.g. the Reststrahl frequency.

The classical approach to this analysis is, of course, that of Born who many years ago showed how one could accurately describe cohesion of ionic crystals in terms mainly of electrostatic interactions among the ions (see e.g. Born and Huang 1954). In this approach the lattice spacing and the compressibility are analyzed to derive the non-Coulombic terms in the energy. The most important of these are the closed-shell overlap repulsion terms. As with the rare gases, modern quantum mechanical calculations of the energy of interaction of pairs of ions demonstrate that the overlap repulsion depends exponentially on the internuclear separation r over an appreciable range (see e.g. Catlow and Hayns 1972; Keeton and Wilson 1973). But of course this exponential or Born–Mayer form has been used as an empirical expression of the overlap repulsion ever since the paper by Born and Mayer (1932). In fact, this paper not only set down the exponential form but also joined it to the concept of ionic radii by proposing that the overlap potential between a pair of ions i, j could be represented by

$$\phi_R^{ij} = bc_{ij}\exp(r_i + r_j - r)/\rho \tag{4.11}$$

in which r_i and r_j are the ionic radii of the two ions. The factor c_{ij} originally introduced by Pauling in 1928 was taken to be

$$c_{ij} = \left(1 + \frac{Z_i}{n_i} + \frac{Z_j}{n_j}\right) \tag{4.12}$$

in which the Z_i are the ionic charges and the n_i are the numbers of electrons in the outermost filled shells of the ions, i.e. those whose overlap is responsible for the repulsion ϕ^{ij}. The usefulness of a form such as (4.11) lies in the constancy of b and ρ throughout a family of compounds, specific properties then being determined principally through the radii parameters (and to a lesser extent through the Pauling factors).

The notion of ionic radii, of course, plays an important part in the chemistry and crystallography of ionic compounds (see e.g. Pauling 1960) and through (4.11) may also be a useful way of estimating interionic interactions in conditions when more direct ways are impracticable. It should nevertheless be borne in mind that different sets of radii are in use (see, e.g. Tosi 1964). Also the ionic radius does not seem to be as independent of crystal environment as would be convenient. Thus the anion wavefunction in the crystal is less spread out than it is for the free ion. Also modern work which aims to describe a variety of crystal properties including lattice vibrations shows that (4.11) has been over–revered in the past. In particular it may be misleading for second neighbours in the crystal; their net interactions appear often to be attractive in sign and cannot be accounted for on the basis of (4.11) plus normal van der Waals interactions.† This may be a consequence of configuration interaction between the perfectly ionic state and excited 'charge-transfer' states formed by the transfer of one electron from an anion to a cation leading to attractive terms in the energy from sharing of the electron hole between pairs of neighbouring anions and of the electron between pairs of neighbouring cations.

In this classic Born approach the compressibility (i.e. $(c_{11}+2c_{12})/3$) is the only elastic constant generally employed in the analysis. But since the models used are central force models with the stability condition imposed, the failure of many of the cubic crystals to satisfy the Cauchy relation $c_{12} = c_{44}$ is a sign that such models can never be perfectly satisfactory. In

† One could quote many references in this connection, but the point is well illustrated by the following: for the analysis of lattice vibrations see e.g. Sangster, Peckham, and Saunderson (1970) for MgO, Saunderson and Peckham (1971) for CaO, Elcombe and Pryor (1970) for CaF_2; for the analysis of 'long-wave' data only see, e.g. Faux (1971a) for alkali halides with the NaCl structure, Catlow and Norgett (1973) for the alkaline earth fluorides.

practice it seems that for many defect problems the uncertainties thereby introduced are not very important. Much more important—at any rate for charged defects—is the consistency of the model with the actual dielectric properties. There are three models (or classes of models) which have been used and which aim to describe dielectric as well as elastic properties. They are (*i*) polarizable point-ion models, (*ii*) deformation dipole models, and (*iii*) shell models, of which the last are the most satisfactory when sufficient empirical data is available. The starting point of polarizable point-ion models is generally the analysis of the high-frequency dielectric constants of a class of compounds to give individual ion polarizabilities α_i independent of the other ions present in the structure. These analyses generally give good evidence for the applicability of the Clausius–Mosotti relation

$$\frac{(\varepsilon_\infty - 1)}{(\varepsilon_\infty + 2)} = \frac{4\pi}{3v_m}\alpha_m$$

where α_m is the sum of the individual ion polarizabilities in the unit cell (see e.g. Tessman, Kahn, and Shockley 1953; and Pirenne and Kartheuser 1964). These ionic polarizabilities are used to describe the electronic dipoles induced on the ions by the internal electric fields generated by the defects or other sources. However, although this model is satisfactory for the analysis of ε_∞, when one comes to low-frequency or static phenomena it has some severe drawbacks. Firstly it has been known for a long time that it cannot generally describe the phonon spectrum nor the low-frequency dielectric constant ε_0 correctly. For example, if one evaluates the displacement polarizability of the ions using potential functions derived from elastic constants, etc., and then calculates ε_0 from the Clausius–Mosotti formula, ε_0 is too large—in some cases dramatically so. In addition, the model is liable to self-polarization catastrophes (see Faux 1971*b*). Early attempts to reduce the ionic polarizability by consideration of ionic overlap were described by Mott and Gurney (1940) but a more successful model, the deformation dipole model, was introduced by Szigeti (1950). In this it was supposed that the relative motion of two ions against one another led to a deformation of their electron-charge distributions, additionally to any polarization caused directly by electric fields; the most important term in this deformation was presumed to be the dipole term. The model has been quite successfully developed for the prediction of the phonon spectra in alkali halides (see particularly Hardy 1962; and Karo and Hardy 1963, 1968). However, the formulation of the deformation dipole model is such that it does not introduce any dependence of the overlap repulsion upon the polarization of the ions. It is therefore liable to the same polarization catastrophes as the polarizable point-ion model. Furthermore, questions of the location and representation of the deformation moments in situations of non-uniform distortion and imperfect co-ordination are not clearly answered.

The shell model does not suffer from these ambiguities, and in addition it is now widely used to represent experimentally obtained phonon data. It assumes each ion to be composed of a core (charge Xe) and a shell (of charge Ye) coupled together harmonically. Non-coulombic interactions between adjacent ions are principally shell–shell interactions. Although core–shell and core–core interactions are also sometimes assumed, the simplest shell models with only shell–shell interactions are often quite successful. Indeed when, as in defect calculations, one wishes to go beyond the harmonic approximation, to use the simplest shell model has some evident advantages—thus the Born–Mayer form $b \exp(-r/\rho)$ can be used for the shell–shell interaction directly rather than partitioned into several different terms. Since the shell model is a model of the ions and their mutual interactions there are evidently no ambiguities when a defect structure is being considered. At present the balance of advantage for both defect and lattice dynamical calculations lies with the shell model, even though this is still being elaborated (see, e.g. Bilz 1971; Sangster 1973).

4.2.1.3. *Valence crystals.* Such crystals are distinguished by having strongly directional covalent bonding between nearest-neighbour atoms with only much weaker interaction between more distant atoms. The traditional representation of the energy of a covalent bond as a function of interatomic distance is the Morse potential

$$\phi(r) = D[\exp\{-2\alpha(r-r_0)\} - 2\exp\{-\alpha(r-r_0)\}], \tag{4.13}$$

and such potentials have been used as models of the lattices of diamond, silicon, and germanium, among others, in calculations of the distortion of imperfect crystals. In keeping with the idea of saturated forces, (4.13) is assumed to act only between nearest neighbours. Each bond in the perfect lattice is in equilibrium so that r_0 in (4.13) is in fact the nearest-neighbour distance. Likewise D then becomes the dissociation energy per bond, i.e. $\frac{1}{2}$ the sublimation energy per atom. However, according to the usual theory of hybridized sp^3 covalent bonds this is not the usual thermodynamic energy of sublimation required to separate atoms in their ground states, since in the solid the atoms are presumed to be in sp^3 hybrid states. D in (4.13) is thus the energy required to separate the atoms while keeping them in these sp^3 states. To the thermodynamic sublimation energy should therefore be added the 'promotion energy' required to prepare them in their correct valence states (Table 4.1). However, some papers using this model do not include this term. The coefficient α is generally determined by fitting the compressibility. A simple generalization of (4.13) obtained by taking different unrelated constants α and β in the two exponents, has been used by Seeger and Swanson (1968); in this case the second constant was found from the pressure derivative of the compressibility.

TABLE 4.1

Energies of promotion (in eV*) from the spectroscopic ground states to the hybridized* sp^3 *valence state (a) as obtained from Herman–Skillman self-consistent Hartree–Fock calculations and (b) as estimated from spectroscopic data by Öpik.*

Element	Promotion energy	
	Hartree–Fock	Öpik
C	7·46	7·55
Si	6·06	—
Ge	6·64	—

This Morse function model is thus simple, intuitive and in a convenient form for numerical calculation. Unfortunately it does not tell us how to predict forces coming from a change in the angles between bonds and has thus in practice been taken as a central force potential—which is clearly inconsistent with the idea of saturated covalent bonds, and in any case is inconsistent with the observed lattice dynamics. However, within the range of harmonic distortions there is another more complete expression of the valence force model. This is due to McMurry, Solbrig, Boyter, and Noble (1967) and to Solbrig (1971). The potential energy of a crystal is written as a sum of terms representing the stretching of bonds and the changing of the angles between bonds.

$$\phi = \tfrac{1}{4}F_r \sum_{i=1}^{n}\sum_{j=1}^{4}(\Delta r_{ij})^2 + \sum_{i=1}^{n}\sum_{j=1}^{3}\sum_{k=j+1}^{4}\{\tfrac{1}{2}r_0^2F_\theta(\Delta\theta_{jik})^2 + f_{rr}\Delta r_{ij}\Delta r_{ik} + r_0 f_{r\theta}(\Delta r_{ij}\Delta\theta_{jik} + \Delta r_{ik}\Delta\theta_{jik})\} \tag{4.14}$$

plus much smaller terms of second order in $\Delta\theta$. Here the change in bond length between nearest neighbours i and j from the equilibrium distance, r_0, is Δr_{ij} while the change in angle between bonds i–j and i–k is $\Delta\theta_{jik}$. This potential function gives a good representation of the lattice dynamics of diamond and of Si. It implies that each bond of the lattice is in equilibrium, i.e. that more distant interactions can be neglected. This is supported by the observation that the nearest-neighbour distances in diamond and Si are almost exactly equal to the mean single bond length found in the molecules of saturated hydrocarbons and similar compounds (1·544 Å in the diamond lattice compared to 1·537 Å and 2·35 Å in the Si lattice compared to 2·32 Å). Also the bond stretching force constant F_r in diamond is only about 10 per cent less than in the hydrocarbons, although the angle bending constant F_θ is changed rather more.

These valence force models, however, do not have anything to say about the dielectric properties of the solids; in that sense they are 'rigid' bond models analogous to rigid ion models of alkali halides. The model will not, therefore, be adequate for charged defects. The shell model has however been shown to give a good fit to the lattice dynamics of these lattices as well as allowing the atoms to be polarizable. As long as one remains within the harmonic approximation it is possibly the most satisfactory empirical model to use, although the physical significance of the force constants is not yet clear. For details of this model see, e.g. Cochran (1971).

4.2.1.4. *Metals*. Since in this article we are principally concerned with localized states it is not appropriate to refer to metals except rather briefly for the sake of comparison and contrast. Yet in the pseudopotential model of metals we have an important approach which allows, if not the prediction, at least the determination of interatomic potentials (cf. Chapter 3). Furthermore, to a first approximation it allows us to represent the energy of a metal as a sum of a configuration-dependent term and a term dependent purely on the volume of the metal. The configuration-dependent term, to a first approximation, can be written as a sum of two-body central potentials, which can either be calculated from a knowledge of the electronic states or else determined by analysis of experimental data (at constant volume) on solid or liquid metals. Examples of such a potential for Na are shown in Fig. 4.4.

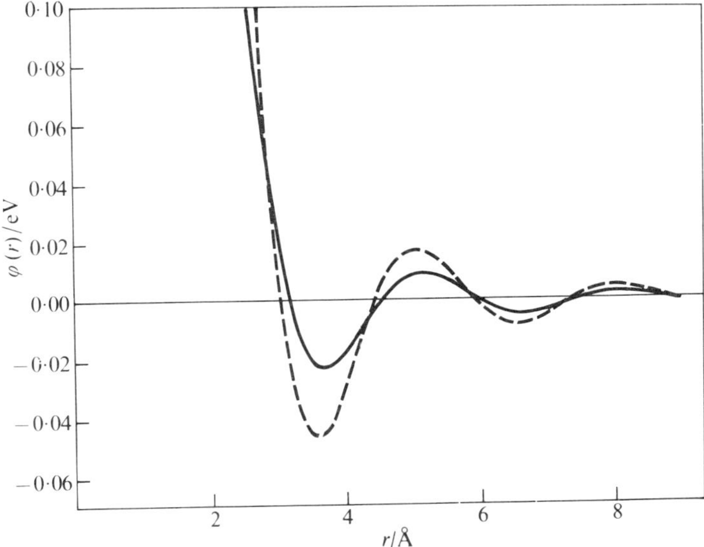

FIG. 4.4. Two interatomic two-body potentials for Na obtained by analysis of (constant volume) empirical data on liquid sodium. The potential represented by the solid line is probably the better one. (After Paskin and Rahman 1966.)

Unfortunately, the pseudopotential method does not apply in the same simple form to noble metals or to the transition metals which are often of interest. The only models in use for such metals are then entirely empirical and are obtained by fitting as much relevant experimental data as possible and assuming that the metallic electron contributions are subsumed into these forms. An example of such a potential for Cu is shown in Fig. 4.5.

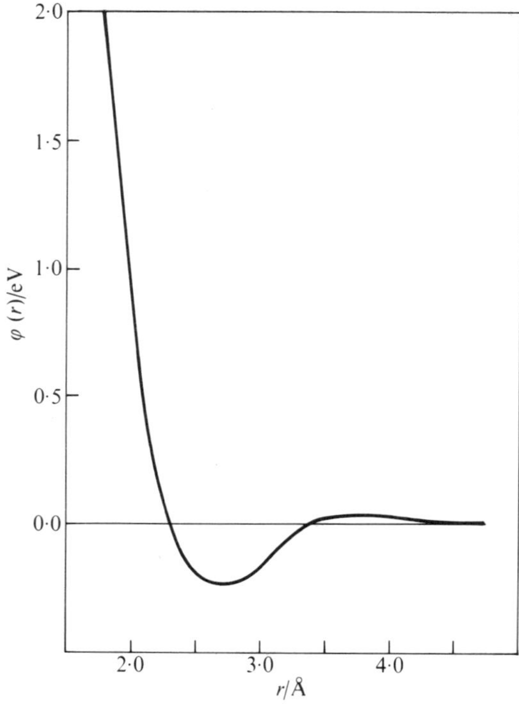

FIG. 4.5. An empirical interatomic two-body potential for Cu obtained by fitting a variety of empirical solid state data including some known defect energies such as vacancy formation and stacking fault energies. Lattice stability under this potential alone is not however assumed, i.e. additional forces dependent only on volume are also present. (After Englert, Tompa, and Bullough 1970.)

4.2.2. *Mathematical methods*

We have already set out the general strategy of the methods of calculating the relaxation of the lattice in the presence of a defect. There are two principal methods in use in practice, one deriving from Mott and Littleton (1938) and one from Kanzaki (1957). Both have been applied quite widely since the original papers. The idea behind the Mott–Littleton method is to make region I large enough that it is sufficient to use only the limiting continuum solution of (4.3) in region II. This approach requires the facilities and techniques for finding the minimum of a function of many variables. The

Kanzaki method on the other hand attempts to solve (4.5) by making a Fourier analysis; this reduces the force constant matrix A to the same block-diagonal form as arises in the dynamics of the perfect crystal and this is clearly a convenience. One aim of the method is to make region I as small as possible and indeed it is often taken as no more than the defect itself. A third method in the same spirit as Kanzaki has recently been proposed by Tewary (1973) who showed how to simplify the evaluation of the perturbed Green's function G once the unperturbed Green's function $G^{(0)} \simeq A^{-1}$ was known. We comment on these methods in turn.

4.2.2.1. *Generalized Mott–Littleton method.* This requires first the specification of the distortion and polarization far from the defect and then the minimization of E with respect to the x-variables.

We shall first discuss region II. In materials such as cubic crystals having no non-zero piezoelectric or electrostrictive constants one can separate the limiting displacement ξ into elastic and electric terms

$$\xi = \xi_{\text{elas}} + \xi_{\text{elec}}. \tag{4.15}$$

In addition there will be electronic polarization terms. For defects bearing a net or effective electric charge these electrical terms are important. The Mott–Littleton method in effect obtains these by dividing up the macroscopic polarization per unit cell in proportion to the appropriate polarizabilities. Thus for a single defect such as an empty vacancy in KCl or an F^+-centre in MgO bearing a net charge q the polarization at a distance $\mathbf{r}$ away is

$$\mathbf{P} = \frac{1}{4\pi}\left(1 - \frac{1}{\varepsilon_0}\right)\frac{q\mathbf{r}}{r^2}, \tag{4.16}$$

and the total polarization $\mathbf{P}v_{\text{m}}$ per unit cell can be divided into its displacement and electronic polarization components. Of course, to do so requires expressions for the relevant polarizabilities and these will change with the physical model adopted, e.g. whether we are using a polarizable point-ion model or a shell model. In any case they can be found by analyzing the response of the model to long wavelength longitudinal optical forces. Study of particular cases, has, however, shown that for charged defects it is most imimportant that the model used should correctly describe the dielectric properties. This is simply a result of the long range of electrical interactions. When this requirement is satisfied, region I need not be very large; in favourable cases one or two shells of neighbours may be quite adequate. Indeed in really desperate situations where it is difficult to arrive at a satisfactory microscopic model for region I, for example through lack of adequate data, it may be sufficient to use the limiting continuum results right up to the defect if only a rough estimate is required.

To complete the description of region II we need in principle to consider ξ_{elas}. However we shall not discuss this term in detail partly because such a discussion is of necessity rather complicated and partly because for many calculations one may omit this term without causing much error. In fact the error in the energy E arising from the neglect of the elastic *distortion* energy in region II falls off as $O(R^{-3})$ where R is the 'radius' of the inner region I. By contrast the energy of polarization falls off very slowly, i.e. only as $O(R^{-1})$ and may not be neglected. For uncharged defects ($q = 0$) this term is absent of course and thus, provided R is large enough, region II may then be treated as undistorted; this is, in fact, often done. For discussions of the elastic term for cases where it is required (e.g. in calculations of volume change caused by defects) we refer the reader to Tewary (1969) and to Faux and Lidiard (1971).

We have so far only discussed the description of region II. It remains now to evaluate the displacements x in region I. These can be found by minimizing E with respect to x numerically. In this field of minimization there is a choice of several well-tried methods and for an up-to-date introduction to the extensive literature on this topic we refer the reader to a recent review by Fletcher (1972). These methods divide into three main groups (*i*) those which evaluate only the function itself (search procedures) (*ii*) those which also evaluate the first derivatives of the function (e.g. steepest descents) and (*iii*) those which evaluate second derivatives as well (variable metric or modified Newton–Raphson methods). We shall comment briefly on the use of these methods in defects calculations.

In the simplest search procedure one seeks the minimum of the energy function by a cycle of linear searches in the variables of the problem, $x_1 \ldots x_n$. This is equivalent to the successive relaxation of equivalent shells of atoms or ions as applied to point defect studies by a number of authors. However the convergence of this method is known to be bad, although computation times can be reduced ($\sim n/2$ times) by not evaluating the total energy function at each step but only that part involving the shells of ions being relaxed. Various improvements of this direct search procedure have therefore been developed, of which the most widely used is probably that of Powell (1964). The general idea here is that of a linear transformation to new variables at the end of each cycle, which, by incorporating the results of that cycle, defines new, less strongly-coupled variables. This 'conjugate direction' method retains the advantage of not requiring derivatives, but converges satisfactorily. Even so the required number of evaluations of the energy increases as the square of the number of variables n and for long-range forces a reasonable practical limit is then reached with ca. 10–20 variables. These search procedures are convenient for small computers as they do not require large storage.

Methods which utilize derivatives of the energy function, i.e. the forces on the ions, in general converge faster than search methods, the number of iterations being reduced by a factor $\sim n$. The calculation of the function derivatives at any point specifies a local optimum direction of search. The choice of a suitable step length in this direction may be made according to several criteria. Thus a linear search may be made to minimize the energy and the various shells of ions may be relaxed independently or collectively—when the method is simply the application of the method of steepest descent. It is known, however, that the convergence of the steepest descent method is poor; and it is also well established in general that it is possible to gain substantially improved convergences within the limitations of this class of methods by using the method of Fletcher and Reeves (1964). This method has been used in a study of defect structure by Norgett, Perrin, and Savino (1972) and shown to be very much faster than search methods. It should undoubtedly be used more widely, particularly when a computer with a large store is not available.

Thirdly, we turn to methods derived from the Newton–Raphson method. These incorporate information on the first *and* second derivatives at each iteration and coverge particularly rapidly. The basic equation relating the ith iteration to the $(i+1)$th is

$$\mathbf{x}^{(i+1)} = \mathbf{x}^{(i)} - \mathbf{H}^{(i)}\mathbf{g}^{(i)} \tag{4.17}$$

where $\mathbf{g}$ is the vector of derivatives of the energy and $\mathbf{H}$, the 'Hessian' matrix, is the inverse of the matrix of second derivatives $\mathbf{W}$. A straightforward application of (4.17) is however ruled out since the calculation and inversion of the matrix of second derivatives at each iteration is far too time-consuming. Fortunately this is not necessary. General methods designed to refine $\mathbf{H}$ during iteration from any well-conditioned approximation to the Hessian have been developed. These even avoid the initial calculation of the second derivatives (Fletcher and Powell 1963). However in a crystal lattice with pairwise interactions the initial evaluation of $\mathbf{H}$ is not difficult since each second derivative concerns only one pair of atoms or ions. A convenient adaption of the method specially suited to defects calculations has been devised by Norgett and Fletcher (1970). The method has the important advantage that the number of iterations required to converge is only slowly dependent on the number of variables although, of course, since forces as well as energy must be calculated, each iteration takes somewhat longer (typically about two times) than in a search method which evaluates the energy alone. It then compares very favourably with search methods, e.g. for 12 variables it is roughly 10–20 times faster while the time of calculation increases only slowly with increasing number of variables. The penalty is that considerable storage for the large Hessian matrix is required, so that the method is not suitable for small computers. Given that this storage requirement can be met, these variable metric methods are speedy and especially suitable for large numbers of variables and for models involving long-range forces, i.e. for ionic crystals.

Previously, with the slower search methods the large amount of computing necessary has meant that it has been possible to study only a limited range of simple symmetric defects and furthermore that only a rather small region I could be used. The application of these faster methods means that it is now feasible to construct programs ('packages') that will calculate characteristic defect energies and configurations for a wide range of cases. The first general program based on these fast methods (although not the first general defect program) is one due to Norgett called HADES.† This is a comprehensive modular program for ionic crystals which will evaluate the relaxation about an arbitrary defect configuration made up of vacancies, interstitials and substitutional atoms or ions. The program allows a choice of polarizable point ion or shell models and is designed to require the specification of only essential crystal data (lattice and defect type, physical constants, size of region I, etc.). This HADES program shows clearly what is now possible in this field, namely the accurate calculation of the energies and configurations of defects in specified models; any uncertainties about the validity of such calculated results then repose clearly in the physical model employed.

4.2.2.2. *Kanzaki method.* The aim of the method introduced originally by Kanzaki (1957) is to take advantage of the full harmonic approximation for region II and not simply to rely on the continuum limit as in the Mott–Littleton method.‡ This enables us to use a much smaller region I. Tewary (1969) observed that the original method could with a change of emphasis be brought into close contact with recent developments in the calculation of lattice response functions. The method has been further developed especially for applications to colour centres by Stoneham and Bartram (1970). In all three formulations one considers separately the terms $\mathbf{F}^{(0)}$ and $\mathbf{F}^{(1)}$ in eqns (4.3)–(4.7). The equation of equilibrium (in region II) under the action of the forces $\mathbf{F}^{(0)}$ is

$$\mathbf{F}^{(0)} = \mathbf{A}\xi. \tag{4.18}$$

In the original Kanzaki method we take the Fourier transform of (4.18) to obtain

$$-\mathscr{F}(\mathbf{k})+\mathscr{A}(\mathbf{k})\mathbf{Q}(\mathbf{k}) = 0, \tag{4.19}$$

† HArwell Defect Evaluation System.

‡ At the time it also represented the first attempt at a rigorous and mathematically accurate method, the Mott–Littleton method being then still rather intuitive and of uncertain accuracy.

where

$$\mathbf{Q}(\mathbf{k}) = \sum_{l} \xi(l) \exp(-\mathrm{i}\mathbf{k} \cdot \mathbf{R}_l), \tag{4.20}$$

and

$$\mathscr{F}(\mathbf{k}) = \sum_{l} \mathbf{F}^{(0)}(l) \exp(-\mathrm{i}\mathbf{k} \cdot \mathbf{R}_l). \tag{4.21}$$

Conversely

$$\xi(l) = \frac{1}{N} \sum_{\mathbf{k}} \mathbf{Q}(\mathbf{k}) \exp(i\mathbf{k} \cdot \mathbf{R}_l), \tag{4.22}$$

and

$$\mathbf{F}^{(0)}(l) = \frac{1}{N} \sum_{k} \mathscr{F}(\mathbf{k}) \exp(\mathrm{i}\mathbf{k} \cdot \mathbf{R}_l). \tag{4.23}$$

Since $\mathbf{A}$ is defined as the matrix of harmonic force constants for the perfect crystal, $\mathscr{A}(\mathbf{k})$ is simply the dynamical matrix of lattice dynamics i.e.

$$\mathscr{A}(\mathbf{k}) = \sum_{l-l'} \exp\{\mathrm{i}\mathbf{k} \cdot (\mathbf{R}_l - \mathbf{R}_{l}')\} \mathbf{A}_{ll'}. \tag{4.24}$$

Inversion of (4.19) yields the solution for $\mathbf{Q}(\mathbf{k})$ as

$$\mathbf{Q}(\mathbf{k}) = \mathscr{A}^{-1}(\mathbf{k}) \mathscr{F}(\mathbf{k}). \tag{4.25}$$

Since the matrix $\mathscr{A}(\mathbf{k})$ is only of order $3n \times 3n$, where n is the number of particles (atom cores plus shells) in the lattice cells, this inversion is quite readily accomplished. This solution for the transform of the displacement field is directly useful if one is interested in diffuse X-ray scattering or long-wavelength neutron scattering from defects since with these wave-scattering techniques one studies directly the Fourier amplitudes of the displacement field (see Kanzaki 1957; MacDonald 1972). In other cases it remains to apply the inverse transformation (4.22) to obtain the actual displacements $\xi^{(0)}$. In general of course, this must be done numerically, by sampling $\mathbf{k}$-points in the Brillouin zone. An examination of the effect of varying the fineness of the sample has been made by Flocken and Hardy (1970) but in general, at least as far as energies and displacements near the defect are concerned, one will obtain satisfactory accuracy with say 1000 points in the Brillouin zone. Since periodic boundary conditions are used, this is actually equivalent to solving for a superlattice of defects with one defect per 1000 cells. To obtain limiting results, e.g. to give the continuum limits for use in the generalized Mott–Littleton method, one may, however, proceed largely analytically; for example, the inverse transformation (4.21) can be replaced by an infinite integral over all **k**-space since only the region around $\mathbf{k} \to 0$ contributes significantly.

The above procedure is equivalent to the formal solution of (4.18)

$$\xi = \mathbf{G}^{(0)}\mathbf{F}^{(0)} \tag{4.26}$$

where the perfect lattice Green's function

$$\mathbf{G}^{(0)} \equiv \mathbf{A}^{-1} \tag{4.27}$$

is expressed as

$$\mathbf{G}_{ll'}^{(0)} = \frac{1}{N}\sum_{\mathbf{k}} \mathscr{A}^{-1}(\mathbf{k}) \exp\{i\mathbf{k} \,.\, (\mathbf{R}_l - \mathbf{R}_{l'})\} \tag{4.28}$$

as follows from insertion of (4.25) and (4.21) into (4.22). This Green's function may thus be calculated if the dynamical matrix is known (as from a force constant model) or obtained directly if the eigenvalues and eigenvectors have been determined experimentally as for example by neutron scattering experiments. When $\mathbf{G}^{(0)}$ is so determined then ξ is readily evaluated from (4.26) particularly when only short-ranged forces $\mathbf{F}^{(0)}$ are involved.

It remains to consider the additional terms in the force, $\mathbf{F}^{(1)}, \xi$. By (4.5) the addition of these terms modifies the solution $\xi^{(0)}$ we have obtained to

$$\xi^{(1)} = (1 - \mathbf{A}^{-1}\mathbf{F}^{(1)})^{-1}\xi^{(0)} \tag{4.29}$$

or equivalently the determination of the static Green's function for the imperfect lattice (4.7). This step from $\xi^{(0)}$ to $\xi^{(1)}$ is often relatively simple in practice. Thus for localized forces, $\mathbf{F}^{(1)}$ will only be significant for immediate neighbours of the defect and is thus of low rank. The use of symmetry-adapted coordinates will reduce this further. For example, in a problem such as the ground state of the F-centre which only involves totally symmetric distortions (A_{1g}), $\mathbf{F}^{(1)}$ will only be of rank 1×1 if only nearest neighbours are significant or 2×2 if second neighbours also must be considered. The inversion of the factor $(1 - \mathbf{A}^{-1}\mathbf{F}^{(1)})^{-1}$ thus presents no problems under these conditions.

The Kanzaki method has been applied successfully to problems of simple defects such as vacancies, foreign ions and F-centres in both polar and non-polar lattice models. In all these applications region I has been assumed to be just the defect itself, e.g. the vacancy or impurity (which is sometimes going beyond the validity of the harmonic approximation for region II; see Tewary 1970). Also the forces $\mathbf{F}^{(0)}$ and $\mathbf{F}^{(1)}$ have been short-ranged in most cases. For charged defects and thus long-range forces special methods must be used to evaluate the transforms of $\mathbf{F}^{(0)}$: one can adopt the Ewald method of lattice summation for this purpose (Hardy and Lidiard 1967; Karo and Hardy 1971). The method is not so convenient when complex and unsymmetrical defects must be studied. For instance, it is not likely to be suited to cases where further region I minimizations (in λ, x) remain to be carried out. Probably for this reason no general program packages capable of dealing with defects of arbitrary composition and symmetry have been developed.

In conclusion, we can say that both the generalized Mott–Littleton and the Kanzaki method will yield accurate predictions of the energies of relaxation about defects in polar and non-polar crystals. The Kanzaki method has also proved a convenient formalism in which to derive analytic results, e.g. expressions for defect interactions and various limiting results such as the expressions for the elastic strengths of defects. These have removed the more intuitive features of the original Mott–Littleton formulation for example. At the present time it seems possible to produce fast general purpose programs in the Mott–Littleton scheme more conveniently, and that this technique is the most suitable for the study of complex and unsymmetrical defects.

4.2.3. *Results for some simple defects*

The simplest defects are those involving only closed-shell atoms or ions e.g. vacancies and interstitials in the solid rare gases or in the alkali halides. These defects are simple because they do not involve any separate calculation of electronic states and allow us to take over directly the models we have described in Section 4.2.1. They thus provide independent checks upon the validity of these physical models, especially as the methods of calculation of the properties of defects in these models are now quite accurate. They have therefore naturally engaged a fair amount of attention. Here we shall comment briefly on some of the results.

4.2.3.1. *Solid rare gases.* In these solids the calculations have mostly been directed to the evaluation of the energies of vacancy formation and motion (u_f and u_m) and thus to the activation energy for self-diffusion, $Q = u_m + u_f$. The energy of formation of vacancies, u_f, has been determined directly experimentally for Ne and Kr by the Simmons–Balluffi technique of simultaneously measuring changes in X-ray lattice parameters and macroscopic dimensions of single crystals as they are warmed up. The measured value of u_f for Ne is slightly greater than H_s, the heat of sublimation (at 0 K), while that for Kr is only about two-thirds of H_s (Table 4.2). In the past, it has often been assumed that one could estimate the values of u_f for Ar and Xe from that for Kr by assuming proportionality to the melting temperature or other corresponding states arguments. Although the incompatibility of Ne with Kr in this respect throws doubt on such estimates we nevertheless include them in Table 4.2 since they are not inconsistent with estimates of u_f for Ar obtained less directly from other thermodynamic properties. The activation energies of self-diffusion Q have been measured by various techniques and are also given in Table 4.2. These values are all about twice the

TABLE 4.2

Experimental values of the energies of formation u_f of vacancies in rare-gas solids together with independently-determined activation energies for self-diffusion, Q. The vacancy migration activation energies u_m are estimated from the relation $Q = u_f + u_m$. Values of u_f in parentheses are estimated from the value for Kr *by assuming proportionality to the melting point. (All values in cals/mole.)*

Rare gas	u_f	u_m	Q
Ne	475[a]	472	947[b]
Ar	(1270)	2330–2595	3600[c]–3865[d]
Kr	1770[e]	3030	4800[f]
Xe	(2470)	4880	7350[g]

[a] Schoknecht (1971); [b] Henry and Norberg (1972); [c] Berne, Boato, and de Paz (1966); [d] Parker, Glyde, and Smith (1968); [e] Losee and Simmons (1968); [f] Chadwick and Morrison (1970); [g] Yen and Norberg (1963).

corresponding heats of sublimation. We make the following remarks about the calculations which have been made.

Firstly, they have generally assumed Lennard-Jones two-body potentials (eqn 4.9) with $n = 12$ or 13 and $m = 6$ and have been carried out in the static lattice approximation (Kanzaki 1957; Burton and Jura 1966). Attempts to avoid the static lattice approximation were made by Nardelli and Repanai–Chiarotti (1960) and by Glyde and Venables (1968), while Burton (1969), Druger (1971) and Sinclair, Niebel and Venables (1973) have included certain three-body and higher order interaction terms in the potential energy function. In general all these calculations give values of u_f which are too high—mostly being only a few per cent less than the sublimation energy, H_s. There is clearly too little relaxation in these models compared to the real solids. In fact the relaxation of the nearest neighbours to the vacant site are found in these calculations to be only ~ 1 per cent of the nearest-neighbour distance (inwards), i.e. smaller than in all other substances.

It is rather surprising therefore that the same calculations of $Q = u_f + u_m$ are generally correct i.e. the values obtained are $\sim 2H_s$. It thus seems that these models can calculate the absolute energy of the activated configuration better than they can represent the ground configuration of the vacancy. It is not clear why. The superficial indications are that the three-body (and higher) terms are not the reason though further work is necessary to be sure of this. Anharmonic effects, which one knows to be important in the thermodynamic behaviour of these solids at the temperatures where diffusion and vacancy formation can be studied experimentally, may offer an explanation. Molecular dynamical computer simulations rather than the semi-analytic methods we have described here may enable that to be tested.

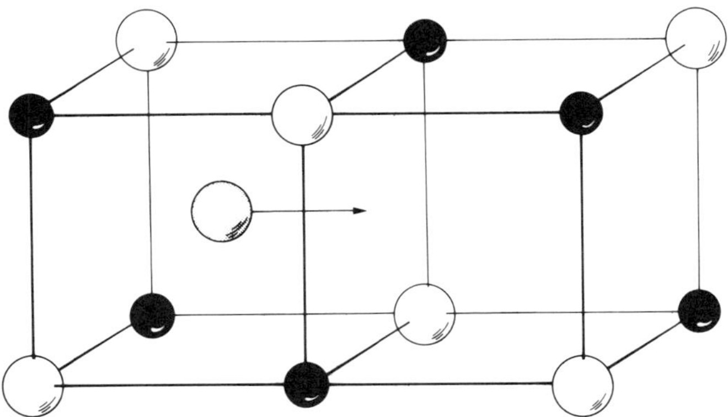

FIG. 4.6. Schematic diagram showing the mode of migration of rare gas interstitials in an ionic crystal with the NaCl structure. The interstitial atom is shown shaded at its normal interstitial position. The arrow indicates the path to the saddle point.

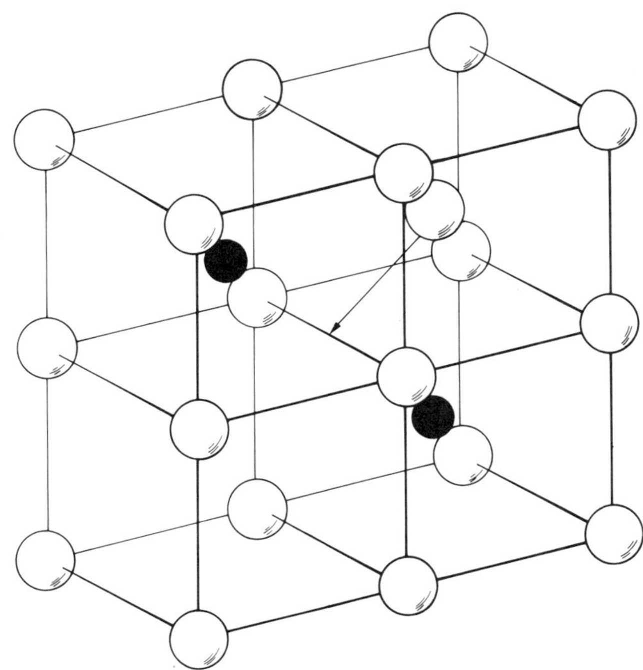

FIG. 4.7. Schematic diagram showing the mode of migration of rare gas interstitials in an ionic crystal with the CaF_2 structure. The more restricted path compared to that shown in Fig. 4.6 leads to higher activation energies.

4.2.3.2. *Rare gas impurities in ionic crystals.*† There is increasing interest in the behaviour of rare-gas atoms in ionic crystals and calculations by methods such as we have described are playing a significant part in the interpretation of experiments. We list several of the more important results.

(1) The calculations show that rare-gas interstitial atoms diffuse through perfect alkali halide lattices (Figure 4.6) of the rocksalt structure very rapidly, the activation energies for this diffusion being only $\sim 0{\cdot}1$–$0{\cdot}3$ eV (Norgett and Lidiard 1968). This also appears to be true in the CsCl structure (Müller and Norgett 1973) but in the more tightly bound fluorite structure (Fig. 4.7), the energies of activation of gas interstitials are high, being ~ 2–3 eV (Norgett 1971*a,b*).

(2) However one cannot observe gas atoms in a *perfect* lattice since thermally produced intrinsic defects, i.e. Schottky or Frenkel defects, and other defects are necessarily present in all real crystals. Furthermore, since in most experiments the rare-gas atoms are generated by nuclear transmutations (e.g. (n, p) reactions on the alkalis) under conditions which also lead to radiation damage, additional defects will be present —at least up to temperatures where these have sufficient mobility to anneal out. It is therefore important to consider the interaction of gas atoms with other defects. The calculations show that trapping of gas atoms into vacancies and vacancy aggregates is particularly important since much or all of the distortion energy associated with the gas interstitial is thereby released. If the gas atom is smaller than the vacancy hole then effectively all the distortion energy is released, otherwise only a part is given up. For example, the binding energy of Ar into K^+ vacancies in KF is calculated to be *ca.* 1·8 eV whereas into the much smaller F^- vacancies it is only *ca.* 0·5 eV. Additional trapping energy into vacancy aggregates may result from the polarization of the gas atom in the internal electric (Madelung) field; thus the binding energy of Ar into a vacancy pair in KF is *ca.* 2·3 eV. The consequence of this trapping for the observed diffusion of gas atoms via the interstitial mechanism is to increase the effective activation energy. A simple application of statistical mechanics then allows one to account for a wide variety of observed behaviour in alkali halides of the NaCl structure and in the alkaline earth fluorides on the basis of these calculated energies. The underlying mechanism in all cases is the migration of the gas atoms as interstitials between periods when they are trapped into vacancies or vacancy aggregates and effectively immobilized.

† Many of the calculations by Norgett *et al.* referred to in this section have been carried out by means of the HADES program applied to the models described in Section 4.2.1.1.

(3) This picture however is not correct for the Cs-halides with the CsCl structure (Müller and Norgett 1972, 1973). It is striking, for instance, that the experimentally determined diffusion coefficients of Ar, Kr, and Xe are markedly different in the high-temperature β-phase of CsCl (which has the rocksalt structure) while in the low-temperature α-phase they are practically equal to one another with a common activation energy of *ca.* 1·2 eV (Fig. 4.8). In fact, although the activation energies for *free* gas interstitial motion are calculated to be small in the CsCl structure, because this structure is more closely packed the trapping energies are high ($\sim$2–3 eV). The result is that the effective activation energies in a crystal containing a thermal equilibrium concentration of Schottky defects would range from 1·5 eV (Ar) to 2·6 eV (Xe), compared to the (single) experimental value of 1·2 eV. The explanation lies in the mobility of vacancy pairs *which have trapped a gas atom.* Thus the calculations show that the activation energies for the migration of such complex defects in CsCl vary only from 1·2 eV (Ar) to 1·3 eV (Xe). The large size of the cavity which the vacancy pair represents relative to the gas atom within it means that the energy

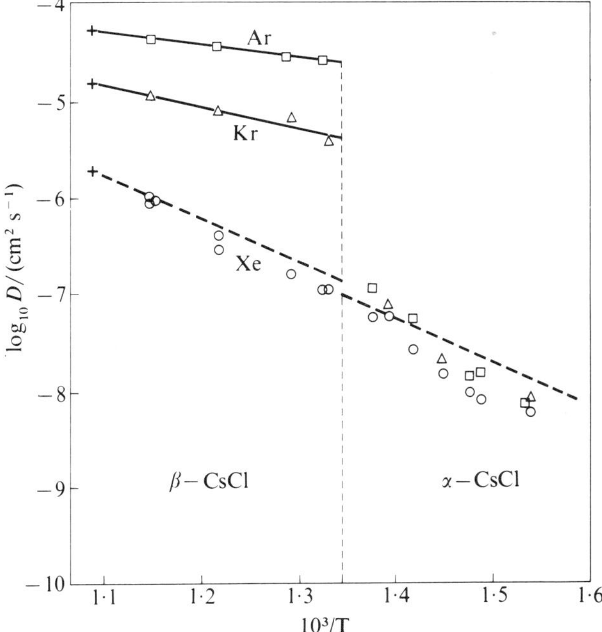

FIG. 4.8. An Arrhenius plot of experimental results giving the diffusion coefficients of Ar (□), Kr (△), and Xe (○) in both phases of CsCl as a function of temperature. We draw attention to the almost identical behaviour of the three gases in the CsCl structure compared with the very different diffusion coefficients in the phase with the NaCl structure. (After Felix and Meier 1969.)

required to activate an ion to jump into the vacancy pair is only slightly dependent on the size of the gas atom.

In summary then, we see that these calculations of the properties of rare-gas atoms in ionic crystals provide additional insight into their diffusion behaviour and agree quantitatively with observed energies remarkably well. Somewhat similar results have been obtained also for rare gases in metals (Wilson and Johnson 1972).

4.2.3.3. *Schottky and Frenkel defects in ionic crystals.* Ever since the original paper of Mott and Littleton (1938) these defects have been continuously studied. In view of the large amount of experimental information available they provide important tests of the models and methods described in Section 4.2. Table 4.3 shows formation energies of Schottky and Frenkel defects in

TABLE 4.3

Energies of formation of Frenkel and Schottky defects (*in* eV) *in four alkali halides as calculated by Karo and Hardy* (1971) *and by Schulze and Hardy* (1972*a*, *b*) *using deformation dipole models. For the Schottky defects different versions of this model were used and this is responsible for the range of values quoted. These same variations were not tested for the Frenkel defects. In all cases the Schottky energies are the lowest.*

Substance	Cation Frenkel	Anion Frenkel	Schottky
NaCl	2·88	4·60	1·79–2·34
NaBr	2·56	4·84	1·66–2·20
KCl	3·46	3·73	1·90–2·20
KBr	3·16	4·17	1·81–2·13

four alkali halides calculated by means of the Kanzaki method (Section 4.2.2.2) applied to deformation dipole models (Section 4.2.1.2). In these four substances, as in all the other alkali halides with the NaCl structure, the formation energies of both anion and cation Frenkel defects are found to be higher than the corresponding Schottky defect energies. This is consistent with the experimental findings of Schottky defects as the intrinsic defects. For these same four substances, a comparison of calculated and experimentally determined energies of formation is given in Table 4.4. In addition to the results of Hardy *et al.* already given in Table 4.3 this table also includes some results obtained for a shell-model by using a real-space method of the Mott–Littleton type as described in Section 4.2.2.1. It will be seen that this model agrees somewhat better with experiment. Both methods of calculation give essentially exact descriptions of the underlying models. The relaxations around ion vacancies found in these calculations are generally several

TABLE 4.4

Calculated and experimentally-determined Schottky formation energies (in eV) in four alkali halides. The calculated values have been obtained by using deformation dipole models (Karo and Hardy 1971; *Schulze and Hardy* 1972*a, b) and simple shell models (Faux and Lidiard* 1971). *For sources of experimental data see the reviews by Barr and Lidiard* (1970) *and by Corish and Jacobs* (1973).

Substance	Calculated: Deformation dipole	Calculated: Shell model	Experimental
NaCl	1·79–2·34	2·23	2·18–2·50
NaBr	1·66–2·20	2·00	1·72
KCl	1·90–2·20	2·34	2·26–2·59
KBr	1·81–2·13	2·25	2·30–2·53

TABLE 4.5

Values of the calculated relaxation displacements of the nearest neighbours of cation and anion vacancies in four alkali halides expressed as percentages of the nearest neighbour anion–cation lattice separation. In all cases the relaxations are outward *from the vacancy. The ranges of values given result from the use of different models and other approximations but the deformation dipole models have invariably led to the highest values. In general there is less relaxation of the anions neighbouring a cation vacancy than of the cations neighbouring an anion vacancy.*

Substance	Cation vacancy	Anion vacancy
NaCl	4·8–7·6	8·4–12·5
NaBr	3·8–6·6	8·4–12·6
KCl	4·6–8·4	7·1–11·1
KBr	4·0–7·5	7·1–11·1

per cent of the nearest neighbour separation *outwards* (as a result of the dominance of electrostatic forces): see Table 4.5.

Table 4.6 gives some similar results for CaF_2 taken from the work of Catlow and Norgett (1973) who used the HADES program to evaluate shell models of the alkaline earth fluorides. It will be seen that the intrinsic defects are correctly predicted to be anion Frenkel defects—Schottky defects and

TABLE 4.6
Calculated energies of formation (in eV) of Frenkel and Schottky defects in three different shell models of CaF_2 *(Catlow and Norgett 1973). Anion Frenkel defects have the lowest formation energy.*

Defect	Formation energy
Anion Frenkel	2·6–2·7
Schottky	7·0–8·6
Cation Frenkel	8·5–9·2

cation Frenkel defects having higher formation energies. Table 4.7 shows the comparison of the calculated and the experimentally determined energies of formation of anion Frenkel defects; again there is good numerical agreement.

TABLE 4.7
Comparison of calculated and experimentally-determined energies of formation (in eV) of anion Frenkel defects in alkaline earth fluorides (Catlow and Norgett 1973).

Substance	Calculated	Experiment
CaF_2	2·6–2·7	2·2–2·8
SrF_2	2·2–2·4	1·7–2·3
BaF_2	1·6–1·9	1·9

4.3. Electronic states

In the last section we considered the general problem of calculating the relaxation of a lattice containing a defect and some particular defects where the physical model itself told us the form of the energy terms E_{I} and $E_{\mathrm{I,II}}$ and thus of the interactions of the defect with the lattice. We must now consider other cases where we must calculate the electronic states of the defect in a given lattice configuration from first principles in order to be able to specify these terms. We shall be concerned with localized states and thus with states which necessarily lie in the forbidden energy gaps between the bands of Bloch states. The non-localized states in the perturbed lattice may have an enhanced or a diminished amplitude in the vicinity of the defect but there are no states with amplitude in the vicinity of the defect which do not also have normal Bloch-wave like behaviour at large distances away if the energy is within an allowed energy band. Quasi-localized or resonant states which have a very large amplitude in the vicinity of the defect but which are still Bloch-like far away are, however, often important (for example d-resonances associated with transition metal impurities in other metals). Here we shall be concerned primarily with strictly localized states, described by wavefunctions falling off exponentially as we go away from the defect. We can make a physical distinction at the outset between shallow states, i.e. those which lie close to a band edge, and deep states, i.e. those which lie around the middle of a forbidden band. Shallow states correspond to a weak perturbation and have wavefunctions which spread out over many atoms round the defect. Well-known examples are provided by the shallow donors and acceptors in the semiconductors Si and Ge. Deep states on the contrary correspond to strong perturbations and have wavefunctions which are concentrated in the immediate region of the defect. Well-known examples are the transition metal ions in ionic and covalent matrices and colour centres such as the F-centre in ionic crystals. Other important examples are provided by various vacancy centres in silicon and diamond. In the one-electron approximation it is possible to develop the formal theory of these cases together; see, for example, Koster and Slater (1954) and Kohn and Luttinger (1955). For the shallow states all such formulations lead to the same 'effective mass' equations but for the deep states these unified approaches are not especially convenient. Indeed, many of the useful results on deep states owe more to molecular theory than to solid-state theory. In this section therefore we first review some general features of quantum calculations familiar from molecular theory. We then describe the effective mass theory for shallow states before reviewing a few of the special methods developed for deep states. Our aim is to bring out a number of the analogies to molecular systems rather than to produce a unified formal theory or to be comprehensive.

4.3.1. *Calculation of electronic states, general*

We are concerned here with the calculation of the new states and changes of energy which occur when we introduce a specified defect into an otherwise perfect crystalline solid. We are also interested in the configurations of these defects, the extent and symmetry of lattice distortions, the relative energies of different configurations, etc. We thus require a theory which gives us both energies and states of these defects as well as the means to calculate symmetric and symmetry-lowering (Jahn–Teller) distortions resulting from the forces exerted by the defect on the perfect lattice.

As usual any quantum theory of these systems must in principle start from the Schrödinger equation

$$\mathscr{H}\Psi \equiv (T+V)\Psi = E\Psi \tag{4.30}$$

with the requirement that the wavefunction Ψ be antisymmetric in the co-ordinates of all the electrons (space and spin). In practice, of course, one can only discuss solutions of this equation on the basis of one-electron functions and even then, except for small systems, one will solve a reduced problem in which only a part of the total electron system is explicitly considered and the rest is treated as providing (mean field) potential terms in V. The wavefunction is then only antisymmetrized in the coordinates of those electrons explicitly considered. Many examples of this could be cited from molecular theory (e.g. $\pi-\sigma$ separation in unsaturated and aromatic molecules; Coulson 1961). In the present connection it means we treat the localized defect electrons as moving in a potential field provided by the rest of the lattice. Exchange forces may be included in this mean field.

The object now is to construct the necessary many-particle wavefunction Ψ for the chosen reduced problem from one-electron wavefunctions and to determine the best approximation within this class. This is done by standard methods extensively applied in atomic, molecular, and solid-state theory (Slater determinants and configuration interaction). Each assignment of electrons to a given set of one-electron orbitals and spin states, is a *configuration* and a correct wavefunction for a particular configuration in general is a sum of determinants of these one-electron functions, which belongs to a particular irreducible representation ${}^{2S+1}\Gamma_i$ of the symmetry group of the defect. There will usually be several such states for each configuration, as a consequence of spin (and orbital) degeneracy of the one-electron levels, but for a filled shell no spin or orbital degeneracy remains and the wavefunction is then a single determinant. These single-configuration states, however, still do not allow for the correlations among the motion of the individual electrons which results from their Coulomb repulsions. A necessary further step to allow for these Coulomb correlations is to take linear combinations of single-configuration states of the same spin and spatial symmetry ${}^{2S+1}\Gamma_i$ to give 'configuration interaction' wavefunctions, the extent of the configuration mixing and the energy levels being determined by solving the secular problem on the (reduced) Hamiltonian.

The determination of the one-electron orbitals is the starting point in any such program and, of course, how good they are may considerably influence the rate of convergence of the configuration interaction sequence. A wide variety of approaches to the determination of these one-electron functions exists at present in molecular theory. The optimum set for a single determinantal wavefunction are Hartree–Fock orbitals satisfying the equation

$$H_{\mathrm{HF}}\phi_i \equiv \left(-\frac{\hbar^2}{2m}\nabla_i^2 + V_{\mathrm{F},i} + V_{\mathrm{C},i} + V_{\mathrm{Exch},i}\right)\phi_i = E_i\phi_i \tag{4.31}$$

where $V_{\mathrm{F},i}$, $V_{\mathrm{C},i}$, and $V_{\mathrm{Exch},i}$ are the nuclear plus core (or framework) potential, the Coulomb potential with other electrons and the exchange potential with other electrons respectively. Since the other electron wavefunctions appear in $V_{C,i}$ and $V_{Exch,i}$ eqn (4.31) can only be solved iteratively to obtain self-consistency. In practice this is only feasible for relatively few electrons unless some constraints on the ϕ_i are assumed. Most schemes, in fact, aim to determine ϕ_i as linear combinations of suitable atomic, i.e. localized, orbitals χ_ν

$$\phi_i = \sum_\nu c_{\nu i}\chi_\nu. \tag{4.32}$$

Eqn. (4.31) then becomes an equation for the coefficients c

$$\mathbf{Fc} = \mathbf{ESc} \tag{4.33}$$

where $\mathbf{F}$ and $\mathbf{S}$ are the matrix of the Fock-operator, H_{HF}, and the overlap matrix on the atomic orbitals, χ respectively. The secular equation determining the energy levels E_i is then formally

$$\det \|\mathbf{F}-\mathbf{ES}\| = 0 \tag{4.34}$$

but of course since $\mathbf{F}$ is of second order in the coefficients c, (4.34) can only be solved iteratively. A number of methods of solving (4.34) approximately have been devised and extensively tested and evaluated on molecules in recent years. They range in sophistication from the IBMOL and ATMOL programs based on large sets of Gaussian orbitals χ_ν to simple parameterizations of (4.34) such as the extended-Hückel method.

From these generalities we turn now to the application to some particular defects beginning with one-electron centres for which the complications resulting from exchange and correlation effects among the defect electrons are absent. Instead we are concerned only with the lattice potential although we should remember that in general this will include exchange of the defect electron with the lattice electrons and may thus be non-local (as it is in the Hartree–Fock approximation though it is local in other approximate schemes). We begin with shallow states, i.e. weakly bound, diffuse states, where the effects of the lattice are rather conveniently parameterized.

4.3.2. *Effective mass equation*

The simplest form of the effective mass equation, that appropriate to a defect electron trapped by a weak field into states lying just below an isotropic non-degenerate band minimum (Fig. 4.9), is fairly obvious physically. We have then a simple Schrödinger-like equation

$$\left(\frac{-h^2}{2m^*}\nabla^2 + V_p\right)\chi(r) = E\chi(r) \tag{4.35}$$

where m^* is the effective mass of the Bloch states at the band minimum, V_p is the perturbing potential and E is the energy of the state $\chi(r)$ *relative to the band minimum*. The function $\chi(r)$ is not the total wavefunction; this is the

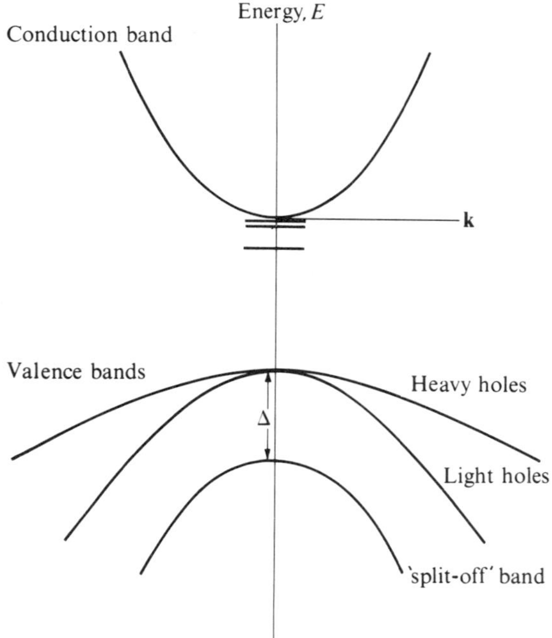

FIG. 4.9. A schematic diagram showing the location of shallow donor states near a band extremum; E represents the one-electron energy and **k** the wave-vector of the non-localized Bloch states.

product of the envelope $\chi(r)$ with the Bloch function $u_0(r)$ at the band minimum. A similar equation holds for a hole trapped into a localized state just above an isotropic non-degenerate band maximum. In both these cases the perturbing potential will be accurately Coulombic, except in the immediate vicinity of the donor or acceptor atom, i.e.

$$V_p(r) = -e^2/\varepsilon r. \tag{4.36}$$

where ε is the dielectric constant. The eigensolutions of (4.35) will then be the usual hydrogenic functions with the free electron mass m replaced by m^* and e^2 by e^2/ε. For example, the energy levels become

$$E_n = \frac{-\text{Ryd}}{n^2}\left(\frac{m^*}{m\varepsilon^2}\right) \qquad (n = 1, 2, 3, \dots) \tag{4.37}$$

where Ryd is the Rydberg of energy ($\frac{1}{2}$ a.u.). Evidently for the large dielectric constants and small effective masses arising in many semiconductors the level spacing is very small compared to free atom spacings (often only a few meV). Correspondingly the localized electron wavefunction $\chi(r)$ is very spread out, the equivalent Bohr radius being $\varepsilon m/m^*$ times the actual Bohr radius. In this situation distortion of the lattice in the immediate vicinity of the impurity is unlikely to affect the pattern of energy levels very much, except possibly for the lowest 1s state.

These qualitative features persist even when the band extremum is not isotropic (as can happen in cubic solids when it lies away from the zone-centre, i.e. at wave vector $\mathbf{k} \neq 0$) or when it is orbitally degenerate (as is often true of the valence band maximum), even though in these situations the equations are more complicated than (4.35). A more formal analysis is necessary to derive these generalizations but as the methods and the results for shallow donors and acceptors have been well described in the literature it would be superfluous to repeat them at length here (but see e.g. Bassani 1971). The nature of the agreement between this generalized effective mass theory and the experimental energy levels is illustrated for donor impurities in Si and Ge in Table 4.8. It will be seen that it is very good. It will also be

TABLE 4.8

Comparison of the experimental and the theoretical (effective mass) energy levels of shallow (Group V) donors in Si *and* Ge *(in units of* 10^{-3} *eV below the conduction band minima). In the calculations the perturbing potential has been taken to be given by eqn* (4.36). *The experimental ground levels are split as a result of the mixing of states from different but equivalent conduction band minima. This effect lies outside the effective-mass approximation and is unobservably small for the excited states. (After Bassani* 1971.)

System	State							
	1S	$2P_0$	2S	$2P_1^{\pm}$	$3P_0$	3S	$3P_1^{\pm}$	$4P_0$
Si (theory)	31·27	11·51	8·83	6·40	5·48	4·75	3·12	3·33
Si (P)	45·5, 33·9, 32·6	11·45	—	6·39	5·46	—	3·12	3·38
Si (As)	53·7, 32·6, 31·2	11·49	—	6·37	5·51	—	3·12	—
Si (Sb)	42·7, 32·9, 30·6	11·52	—	6·46	5·51	—	3·12	—
Ge (theory)	9·81	4·74	3·52	1·73	2·56	2·01	1·03	1·67
Ge (P)	12·89, 9·88	4·75	—	1·73	2·56	—	1·05	—
Ge (As)	14·17, 9·96	4·75	—	1·73	2·56	—	1·04	—
Ge (Sb)	10·32, 10·01	4·74	—	1·73	2·57	—	1·04	—

observed that the threefold degeneracy of the usual hydrogenic *n*p levels is removed in these cases, because the surfaces of constant energy around the conduction band minima (which are *not* at $\mathbf{k} = 0$) are ellipsoidal and not spherical.

In a covalent crystal there is, of course, only the one dielectric constant but in polar materials where we have to consider both static and optical dielectric constants it is a matter for analysis which should be used in the expression for V_p. In simple terms the question is, does the defect electron move in its orbit rapidly or slowly compared with the speed with which the ions of the lattice respond to a changing electric field. The usual adiabatic assumption would lead us at once to suppose that the electron motion is much faster and that the ionic or displacement polarization is determined by the average field of the centre plus trapped electron. We then conclude that the perturbing potential should include no lattice response and set $V_p = -e^2/\varepsilon_\infty r$. Indeed such an assumption sometimes allows one to represent even the F-centre fairly satisfactorily in the effective mass approximation† (Herman, Wallis, and Wallis 1956). A little consideration shows us however that this is too simple a representation. The lattice polarization must be determined self-consistently with the defect electron wavefunction; inside the defect electron orbit the field of the centre is not screened and the ions will be displaced whereas outside the orbit the centre is screened and there is no polarization. We shall not however pursue this matter in detail here, mainly because F-centre states are normally too tightly bound and the states too strongly localized in the vacancies for the effective mass approximation to be quantitatively accurate. One qualification to this statement should, however, be inserted. Thus while it is true that the F-centre ground state and the excited state reached in a normal F-band transition are both highly localized, it is possible for the excited state after the lattice relaxes to be a diffuse one. If so this would show the strong effect of the electron–lattice coupling in determining the correct self-consistent states. The theory of excited F-centre states is a matter of topical interest and will be referred to again later.

The effective mass approximation can, of course, be qualitatively useful even outside its proper domain. For example, if the F-centre can be treated as the analogue of the hydrogen atom then the M-centre is the analogue of the hydrogen molecule (Fig. 4.1) which points immediately to the expected nature and order of the one-electron states (Fig. 4.10) and to the energy levels. In particular we observe that we expect two principal absorption bands corresponding to excitation of one of the electrons in the lowest 1sσ orbital to the 1sσ^* antibonding orbital (M-band) and to the 2pπ bonding orbital. These have transition moments respectively parallel to the M-centre axis and perpendicular to it, as observed. We can use the effective mass approximation to estimate roughly where these bands lie; for KCl with $m^* \sim m$ and $\varepsilon_\infty = 2{\cdot}13$ one obtains 1·62 eV for the ${}^1\Sigma_g^+ \rightarrow {}^1\Sigma_u^+$ transition and 2·05 eV for the ${}^1\Sigma_g^+ \rightarrow {}^1\Pi_u^+$ transition compared to experimental values of 1·55 eV and $\sim$2·3 eV respectively. Actually as the M-centre has a $\langle 110 \rangle$ orientation in crystals with the rocksalt structure and thus D_{2h} symmetry the degeneracy of the ${}^1\Pi_u$ state must be lifted by the crystal field, but in practice the splitting is small. Of course, no such splitting is to be expected in the CsCl structures

† Generally the conduction band minimum in the alkali halides lies at the zone centre (see e.g. Knox and Teegarden 1968).

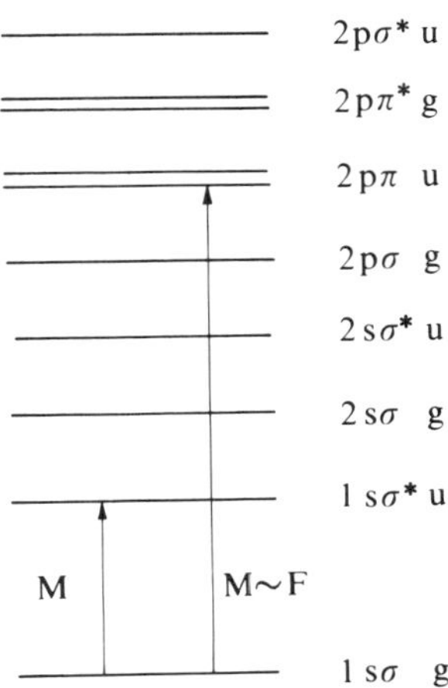

FIG. 4.10. Schematic diagram showing the one-electron energy levels of an M-centre by analogy with a diatomic molecule. The labelling uses the 'molecular' and not the 'united atom' notation.

since there the M-centres have $\langle 100 \rangle$ orientation and thus four-fold rotational symmetry about the axis. In addition to the above two principal transitions which one expects to be strong, one would also expect to see a weaker transition due to excitation of one of the ground state $1s\sigma$ electrons into an antibonding $2s\sigma^*$ orbital—weaker since the $1s \rightarrow 2s$ transition is forbidden in the free atom. This is the origin of the weak band first found by Okamoto (1961) at 1·96 eV and having a transition moment along the $\langle 110 \rangle$ axis. Thus the qualitative picture of the M-centre provided by the effective mass approximation is reasonably good. One can go further and discuss the ionized states M^+ and M^- similarly. Indeed for the M^+ (F_2^+) centre the model gives surprisingly good guidance not only for the transition energies but for the oscillator strengths as well (Aeggerter and Lüty 1971). The model has also been used for $R \equiv F_3$ centres (Silsbee 1965).

This sort of qualitative success is not too surprising; after all, the general features of Fig. 4.10 are the same in the diatomic molecules of all the elements in the first row of the periodic table. As long as the description of the F-centre is not too much in error we would expect the description of its aggregates to be roughly right. However for quantitative predictions it cannot be correct to use the effective mass approximation, at least not for the highly localized ground states—thus for the KCl F-centre ground state something like 90 per cent of the F-electron is contained within a sphere of passing through the nearest-neighbour cations. Furthermore it is clear that, as there is no explicit consideration of electron–lattice coupling, we cannot use the effective mass model to describe the widths of absorption bands nor to predict Stokes shifts and transition energies in *emission*. Additional evidence that an effective mass approach is not the right one comes from the absence of any clear correlation between F-band energies and dielectric constants, whereas the energies of F-bands and F-aggregate bands correlate quite strongly with lattice parameter a by so called Ivey laws

$$\lambda = \text{const.}\, a^{-n} \tag{4.38}$$

where n is $\simeq 1{\cdot}8$. This correlation with lattice parameter alone strongly suggests that the F-electrons are trapped primarily by Madelung fields into potential wells associated mainly with the vacant sites themselves. This observation gave rise to the purely electrostatic or point-ion model (Gourary and Adrian 1957). This very simple approach has proved extremely useful and we shall therefore devote the next few pages to it. It is however far from being the only valid description of F-type centres and many other more sophisticated calculations have been published; for details of these we refer the reader to specialist books such as those by Markham (1966) and Fowler (1968).

4.3.3. *Pseudo-potential theory and the point-ion model of ionic crystals*

In the point-ion model of Gourary and Adrian (1957) we solve the Schrödinger equation for the defect electrons moving in a potential V_{PI} obtained by replacing each ion by an equivalent point charge (e.g. $+e$ for Na^+ ions and $-e$ for Cl^- ions). For a one-electron centre, e.g. the F-centre, the Schrödinger equation is

$$\left(-\frac{\hbar^2}{2m}\nabla^2 + V_{\text{PI}}\right)\phi = E\phi. \tag{4.39}$$

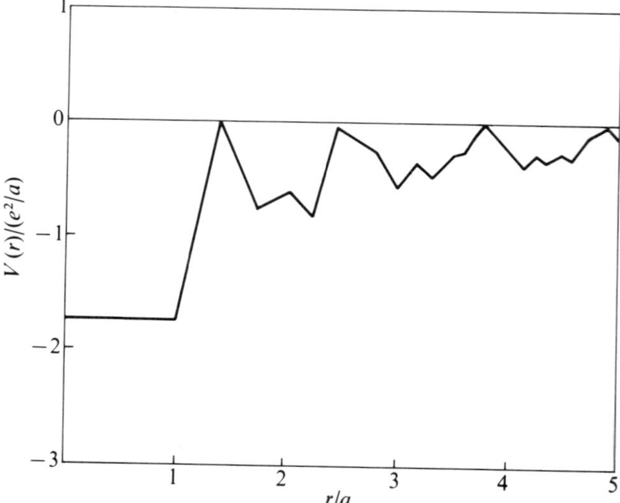

FIG. 4.11. The radial variation of the spherically symmetrical ($l = 0$) component of the point-ion potential of an F-centre. The abscissae are in units of the anion–cation nearest-neighbour distance a while the ordinates are in units of e^2/a.

Owing to the high point-symmetry of V_{PI} of the F-centre, an expansion of V_{PI} in spherical harmonics is especially convenient; thus in the rock-salt lattice, after the dominant spherically symmetric ($l = 0$) term the first angular term in V_{PI} is of $l = 4$, so that the lowest states responsible for the main observable properties of F-centres are predominantly s-like and p-like. The radial variation of $V_{PI}(l = 0)$ is as shown in Fig. 4.11. Although the eqn (4.39) with $V_{PI}(l = 0)$ has been solved exactly (Laughlin 1965) quite good energies are also obtained with relatively simple variational wavefunctions, notably those based on particle-in-a-box wavefunctions inside the vacancy ($r \leqslant a$) joined to hydrogenic wavefunctions outside ($r \geqslant a$). Examples of $1s \rightarrow 2p$ transition energies calculated in this way are given in Table 4.9 and compared with observed F-band transition energies; the correspondence on the whole is good. Also we note that although the wavefunctions ϕ which are the solutions of (4.39) are not the true wavefunctions since they are not orthogonal to the core orbitals ψ_c, when they are so orthogonalized to give ψ_v

$$|\psi_v\rangle = |\phi\rangle - \sum_c \langle\psi_c|\phi\rangle|\psi_c\rangle \tag{4.40}$$

TABLE 4.9

Theoretical and experimental F-*band energies (in eV) for several ionic crystals. The theoretical values are the results of simple point-ion calculations without allowance for ionic polarization and distortion* (*Gourary and Adrian* 1957; *Kemp and Neeley* 1963; *Bennett and Lidiard* 1965). *Even so the comparison is quite good. These point-ion states therefore form a useful starting point for more elaborate calculations* (*Stoneham and Bartram* 1970). *Explicit evaluation of the effects of distortion and polarization confirm that these corrections in many cases are relatively small.* (*For sources of experimental data see Dawson and Pooley* 1969; *Hughes and Henderson* 1972; *Cavenett, Hayes, Hunter, and Stoneham* 1969).

Substance	Point-ion 1s–2p	F-band (0K)
LiF	4·00	5·08
LiCl	2·75	3·26
NaF	3·24	3·70
NaCl	2·39	2·75
KCl	1·99	2·30
RbBr	1·71	1·85
CaF_2	3·19	3·30
SrF_2	2·91	2·85
BaF_2	2·61	2·03
MgO	4·79	4·95

they give good descriptions of electron densities as seen through hyperfine constants (Gourary and Adrian 1960; Hayes and Stoneham 1974). On the other hand it is plain that no point-ion model can describe ion-size effects, e.g. the shift of the F-centre transition caused by an alkali impurity among the neighbours of the anion vacancy (F_A-centre) or the systematic departures from the Ivey relation (Fig. 4.12).

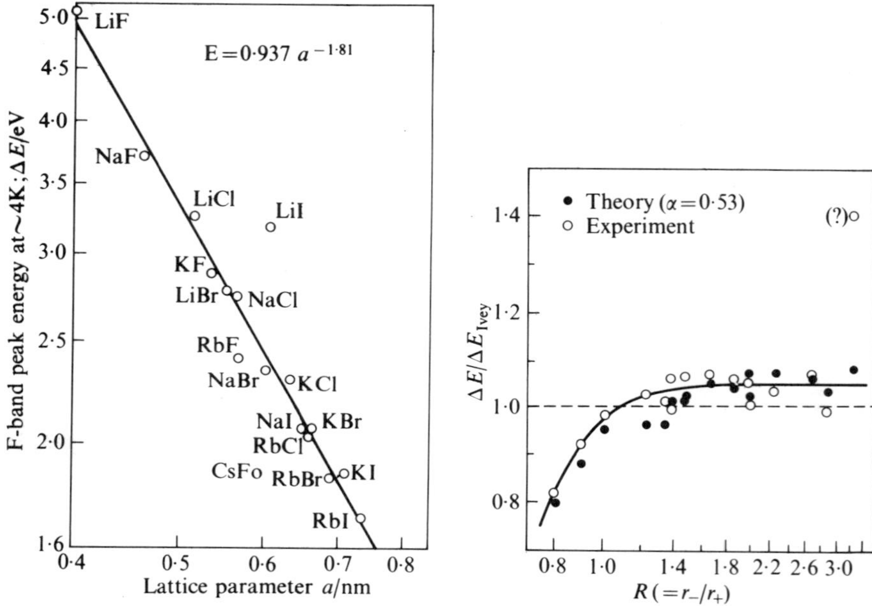

FIG. 4.12. (a) A log–log plot of the energy of the F-band maximum ΔE against the nearest neighbour lattice distance for the alkali halides having the NaCl structure (after Dawson and Pooley 1969). (b) On a larger scale the departures from the best linear relation (Ivey law) fitted as in (a) in this case plotted against the ratio of anion to cation radius. The theory of these ion-size corrections is described in Section 3.3.3. (After Bartram, Stoneham, and Gash 1968.)

A convenient general approach which allows the point-ion model to be seen as a particular mathematical approximation is provided by pseudopotential theory (see e.g. Chapter 1 for a general review). This was recognized early on in the development of pseudopotential theory (Gourary and Fein 1962). We shall here follow the later development made by Bartram, Stoneham and Gash (1968). Firstly define a projection operator P

$$P = \sum_c |\psi_c\rangle\langle\psi_c| \tag{4.41}$$

which projects out of any wavefunction that part which is orthogonal to all the core-orbitals ψ_c. We have

$$P^2 = P \tag{4.42}$$

and, since the core-states ψ_c are eigenstates of the Hamiltonian H, we also have

$$[H, P] = 0. \tag{4.43}$$

We now write the localized wavefunction ψ_v as

$$|\psi_v\rangle = (1-P)|\phi\rangle \tag{4.44}$$

with the object of finding a simplified equation for ϕ which need not itself be orthogonal to the core-orbitals as ψ_v is. If we substitute into the Schrödinger equation for the defect electron states i.e.

$$H\psi_v = E_v\psi_v \tag{4.45}$$

we obtain from (4.41) and (4.44)

$$(H+V_R)|\phi\rangle = E_v|\phi\rangle \tag{4.46}$$

where

$$V_R \equiv \sum_c (E_v - E_c)|\psi_c\rangle\langle\psi_c|. \tag{4.47}$$

Lastly, if we separate H into a kinetic energy operator T and the self-consistent crystal potential V we obtain a Schrödinger equation for the pseudowave-function ϕ

$$(T+V_p)|\phi\rangle = E_v|\phi\rangle \tag{4.48}$$

which contains the pseudopotential V_p,

$$V_p \equiv V+V_R \tag{4.49}$$

and has the same eigenvalues as (4.45). This pseudopotential V_p by (4.47) is a non-local potential but it is convenient and valid for many purposes to forget this feature. Thus a number of detailed studies show that V_R largely cancels out the true potential V in just those regions inside the ions where V is large and rapidly varying, thus giving a wavefunction ϕ which is much smoother than the true wavefunction. Of course, since V_R is derived from the localized core orbitals it cannot cancel off the long-range Coulomb field due to the net charge on any ion. We may thus expect V_p in effect to approximate a point-ion potential.

Actually there is an arbitrariness in the formalism remaining to be exploited. Thus $|\phi\rangle$ as given by the above equation is indefinite to the addition of terms such $\sum_{c'} \alpha_{c'}|\psi_{c'}\rangle$. It was proposed by Cohen and Heine (1961) that this arbitrariness be turned to advantage by choosing $|\phi\rangle$ so as to be as smooth as possible and the suggested criterion of smoothness was to be that the mean kinetic energy $\langle\phi|T|\phi\rangle/\langle\phi|\phi\rangle$ be as small as possible or equivalently by (4.48) that $\langle\phi|V_p|\phi\rangle/\langle\phi|\phi\rangle$ be as large as possible. A straightforward formal manipulation based on variations $|\delta\phi\rangle = \Sigma\,\alpha_{c'}|\psi_{c'}\rangle$ then leads to the equation

$$\langle\psi_c|(V+V_R)|\phi\rangle-\bar{V}\langle\psi_c|\phi\rangle = 0 \tag{4.50}$$

for the optimum ϕ where $\bar{V}$ is given by

$$\bar{V} = \langle\phi|(V+V_R)|\phi\rangle/\langle\phi|\phi\rangle \tag{4.51}$$

evaluated with the optimum ϕ. Further manipulation gives the equation

$$V_p|\phi\rangle = V|\phi\rangle+P(\bar{V}-V)|\phi\rangle \tag{4.52}$$

for the corresponding optimal pseudopotential V_p.

All this is standard pseudopotential formalism. Bartram *et al.* (1968) introduced two assumptions particular to colour centre problems: (1) the tight-binding or strong localization of the core-functions, and (2) the slow variation of the defect wavefunction ϕ across any particular ion. These assumptions enabled them to develop the formalism in a useful way. Matthew and Green (1971) have shown how to relax the second of these two assumptions but the implications have not yet been fully worked out. We shall therefore describe the analysis of Bartram *et al.* (1968) and only indicate the likely effects of modifying the second of their assumptions.

Firstly we set

$$V = \sum_\gamma V_\gamma \tag{4.53}$$

in which the sum is over all ions γ and where V_γ is a sum of the point-ion potential $V_{\mathrm{PI}\gamma}$ and a short-range part. Also we can write the projection operator

$$P = \sum_\gamma P_\gamma \tag{4.54}$$

where P_γ projects on to the orbitals of ion γ. From (4.53) and (4.54), (4.52) then becomes

$$V_p = V_{\mathrm{PI}}+\sum_\gamma[(1-P_\gamma)(V_\gamma-V_{\mathrm{PI}\gamma})-P_\gamma V_{\mathrm{PI}\gamma}+P_\gamma(\bar{V}-U_\gamma)] \tag{4.55}$$

where

$$U_\gamma \equiv \sum_{\gamma'\neq\gamma} V_{\gamma'} \simeq \sum_{\gamma'\neq\gamma} V_{\mathrm{PI}\gamma'} \tag{4.56}$$

by the assumption that V_γ differs negligibly from the point-ion part outside ion γ. If we now incorporate the second assumption that ϕ varies slowly over the region of any ion then the expectation value of V_p is

$$\langle\phi|V_p|\phi\rangle = \langle\phi|V_{\mathrm{PI}}|\phi\rangle+\sum_\gamma C_\gamma|\phi(r_\gamma)|^2 \tag{4.57}$$

where $\mathbf{r}_\gamma$ is the vector position of ion γ and

$$C_\gamma = A_\gamma + (\bar{V} - U_\gamma)B_\gamma, \tag{4.58}$$

with

$$A_\gamma = \int (1-P_\gamma)(V_\gamma - V_{\mathrm{PI}\gamma})\,\mathrm{d}\tau - \int P_\gamma V_{\mathrm{PI}\gamma}\,\mathrm{d}\tau, \tag{4.59}$$

and

$$B_\gamma = \int P_\gamma\,\mathrm{d}\tau = \sum_c \left| \int \psi_c(\mathbf{r})\,\mathrm{d}\tau \right|^2. \tag{4.60}$$

The quantities A_γ and B_γ are properties of the ions only and can be evaluated *a priori* when self-consistent potentials and core wavefunctions are known. Eqn (4.57) is formally equivalent to

$$V_p = V_{\mathrm{PI}} + \sum_\gamma C_\gamma \delta(r - r_\gamma) \tag{4.61}$$

so in this approximation the ion-size correction to the point-ion potential is a sum of δ-function terms. Matthew and Green (1971) have argued that better results can be obtained by relaxing the assumptions leading to (4.61) and using a pseudopotential in which the δ-function is spread out over a sphere of radius equal to the ionic radius. In any case eqn (4.61) or the Matthew–Green modifications of it provide a rational way in which to represent the point-ion model and to evaluate the corrections to it. These corrections are not large in general but can be significant, particularly in the case of heavy ions. We shall consider them in detail later. For the present, however, we note that the point-ion model can be regarded as a first approximation; its simplicity means that it is convenient for complex centres, e.g. aggregate centres such as M, R, etc. We shall therefore first review its application to such cases.

Before doing so, however, it is necessary to complete the discussion by referring to the lattice relaxation and polarization around the colour centre. In principle this is handled by one of the methods of Section 4.2.2, but to apply these we need to specify the coupling of the centre to the rest of the lattice (region II). Part of this is already included in E_v—the electronic energy of the centre for a given lattice configuration—though the lattice potential, V, and the derived pseudopotential, V_p. Even though these may be Hartree–Fock or other accurate potentials, so far in practice they have always been those appropriate to free ions i.e. they are unmodified by the presence of the defect. One can see, however, that for a defect with a net charge (e.g. F^+ in MgO) or for one with a diffuse wave-function, electrical forces will polarize the lattice ions. To describe this effect quantum-mechanically and self-consistently would scarcely be possible except in the simplest cases (e.g. neutral centres, compact defect wavefunctions, and light elements). As a result this polarization coupling is handled classically; the ions of the lattice are represented by a polarizable-ion or a shell model and their classical electrostatic coupling with the average electric field of the defect centre is evaluated i.e. a Hartree mean field approximation is used. The sources of coupling of the defect to the lattice having been specified, the actual distortion and polarization of the lattice can then be obtained by the methods of Section 4.2.2. Although both the Mott–Littleton and the Kanzaki approaches have been used in this connection, the lowest states of the F-centre in its ground-state lattice configuration have often been treated more simply. In this case the wavefunctions are compact and the centre is electrically neutral and it proves adequate to assume that only the nearest neighbours to the vacancy relax (only ~ 1 per cent outwards as it turns out; cf. the much larger relaxations around the empty vacancy given in Table 4.5). Also the polarization energy is relatively small and so has often been evaluated without allowing for the mutual interactions of the induced dipoles. In these cases then the polarization and lattice distortion introduce relatively small corrections to the point-ion predictions given in Table 4.9, so that these simplified treatments are probably adequate. This is not however true of the relaxed excited states of the F-centre. ENDOR experiments on the relaxed excited state of F-centres in KI (Mollenauer and Baldacchini 1972) and KBr (Baldacchini and Mollenauer 1974) show that this is a diffuse state. The correct representation of the polarization and lattice distortion energy is therefore certainly important in describing the relaxation of the excited state from its compact form reached by optical absorption in the ground state to this relaxed diffuse state. The theoretical problem is further complicated by the fact that the order of the nearly degenerate 2s and 2p levels apparently reverses as the lattice relaxes. The nature of these states has a direct bearing on the long lifetimes of excited F-centres ($\sim 10^{-6}$ s) and the large Stokes shifts between the energies of optical absorption and emission. For detailed calculations we refer to the review by Fowler (1968) and the subsequent articles by Bennett (1968, 1969, 1971), Öpik and Wood (1969), Wood and Öpik (1969), Stoneham and Bartram (1970), and Ham (1972).

4.3.3.1. *Application of the point-ion model to complex centres.* The simplest aggregate centres are those formed by the pairing of a single F-centre with an impurity cation, e.g. an alkali impurity to give the so-called F_A-centre (see Lüty, 1968) or with another (empty) vacancy either cation or anion. Clearly in the point-ion model F_A-centres will differ from the normal F-centre only as a result of a different and less symmetric ionic relaxation; we thus mention these particular defects again when we have considered ion-size corrections explicitly. The next simplest are the M-centres, i.e. pairs of F-centres, and the R-centres, i.e. trios of F-centres. These centres have been studied in positively and negatively ionized states as well. We have already described some of their qualitative features as inferred from continuum models. Possibly because of the success of these models in guiding and interpreting experiments rather few detailed calculations of these complex centres have been made.

The only truly point-ion calculation on an M-centre appears to be that of Evarestov (1964) who calculated the energy of a point-ion Hamiltonian using Heitler–London wavefunctions constructed from hydrogenic F-orbitals. His results for KCl are given in Table 4.10 and are in fair agreement with experiment, thus demonstrating once again the usefulness of the point-ion model as a first approximation. It will be observed that these results give a crystal field splitting of the $^1\Pi_u$ state (^1B_u state in the notation for a centre with D_{2h} symmetry). However, these results are all rigid-lattice results and an additional splitting would be expected from lattice relaxation. Even though more elaborate calculations of M-centre states have been made (Meyer and Wood 1964) the effects of such lattice relaxation have not yet been evaluated.

TABLE 4.10

Some M $\equiv$ F_2*-centre transition energies (in eV) as calculated in the point-ion approximation by Evarestov* (1964) *and as inferred from experiment. The ground state is* $^1A_{1g}$ *(or* $^1\Sigma_g$ *in the continuum model). The transition* $^1A_{1g} \rightarrow {}^1A_{2u}$ *(or* $^1\Sigma_g \rightarrow {}^1\Sigma_u$*) corresponds to the so called* M*-band while the transitions* $^1A_{1g} \rightarrow {}^1B_{1u}, {}^1B_{2u}$ *(or* $^1\Sigma_g \rightarrow {}^1\Pi_u$*) generally lie close together in the region of the* F*-band (cf Table* 4.9).

Substance		Transition energy		
		$^1A_{1g} \rightarrow {}^1A_{1u}$ ⟨110⟩	$^1A_{1g} \rightarrow {}^1B_{1u}$ ⟨100⟩	$^1A_{1g} \rightarrow {}^1B_{2u}$ ⟨1$\bar{1}$0⟩
LiF	calculated	2·85	4·11	4·03
	experimental	2·79		
LiCl	calculated	2·01	2·96	2·91
	experimental	1·91		
NaF	calculated	2·26	3·43	3·35
	experimental	2·45		
NaCl	calculated	1·82	2·67	2·56
	experimental	1·71		
KCl	calculated	1·50	2·25	2·12
	experimental	1·55	2·31	2·27
KBr	calculated	1·44	2·18	1·99
	experimental	1·35		
RbBr	calculated	1·33	2·01	1·85
	experimental	1·50		

Rather extensive point-ion model calculations have however recently been made for the R-centre in KCl by Kern and Bartram (1971). These calculations, based on single-centre orbitals and the usual configuration interaction approach, explored a large number of states; where the energy levels can be directly compared with experiment, they appear to be accurate. These calculations also demonstrated several theoretical points which are almost certain to be generally important. Firstly, it is convenient to use single-centre orbitals since one thereby avoids the computational complications caused by the need to evaluate multi-centre integrals. Less trivially, but not unexpectedly in view of a large body of similar experience in molecular theory, it was found to be very important to include enough configurations; thus even in this three-electron system correlation effects are very important. It was also found that lattice distortion was very important in fixing the relative positions of different levels. These calculations have shown that with sufficient attention to these points, it is possible to make quite accurate predictions of the levels of such several-electron centres. However, it would obviously be more difficult to repeat them for a centre of much lower symmetry, e.g. the N-centre, believed to be four F-centres arranged in neighbouring positions on a (211) plane.

An earlier calculation by Stoneham (1966) studied a postulated tetrahedral $N \equiv F_4$ aggregate in the rigid-lattice approximation. Although this again showed the convenience of using single-centre orbitals a centre of this structure and with these properties has not yet been identified experimentally. However neither has it been demonstrated that such a centre is unlikely, in the sense of being of higher energy than the alternative, observed (211) planar configuration.

4.3.3.2. *Singlet–triplet splitting of two-electron centres.* In discussing the M-centre we considered only the spin-singlet states. However, the lowest-lying spin-triplet lies only a little above the ground state even though the two component F-centres are nearest neighbours. This reflects again the strong localization of the F-functions. In MgO and CaO after fast-neutron irradiation, another two-electron centre has been found by e.s.r. with even smaller singlet-triplet splittings, only a few meV (sometimes called the F_t-centre).

For several reasons it is of interest to calculate such splittings but this is not straightforward. It cannot be done by the conventional Heitler–London method for example, since this predicts that at large enough separations of the two centres, the triplet state will lie lower, contrary to a general theorem which requires the lowest eigenstate to be nodeless.†

† The argument is as follows. The lowest eigenstate of a Sturm–Liouville equation must be nodeless. The Schrödinger equation is a member of this class. Since the complete wavefunction for any two-electron system is separable into the product of a spatial part and a spin part, the symmetric nature of the triplet spin function requires the spatial part to be antisymmetric. Thus the triplet cannot be the lowest eigenstate. This conclusion does not hold for more than two electrons since then the complete wavefunction is not separable into spin and spatial factors.

A method which can be used to calculate these small splittings was devised by Herring and Flicker (1964); it becomes more accurate the smaller the splitting, i.e. the larger the separation of the centres. It has been applied to colour centre problems by Berezin (1968, 1972) and by Norgett (1971c).

Let Φ_S and Φ_T be the singlet and triplet two-electron spatial wave-functions. Then as our starting point, we take the equation

$$(E_T - E_S)\Phi_T\Phi_S = \Phi_S H\Phi_T - \Phi_T H\Phi_S \tag{4.62}$$

where H is the two-electron Hamiltonian operator. We then define

$$\Phi_{1,2} = \frac{1}{\sqrt{2}}(\Phi_S \pm \Phi_T). \tag{4.63}$$

By the general nature of Φ_S and Φ_T for any system with two attractive centres A and B, it follows that Φ_1 corresponds to a state with electron 1 mainly in well A and electron 2 in well B while Φ_2 corresponds to a state with electron 2 in A and electron 1 in B. Referring to Fig. 4.13 we define a sub-space Ω in the hyperspace $(\mathbf{r}_1, \mathbf{r}_2)$ such that

$$\Omega \equiv \left.\begin{matrix} x_1 < 0 \\ x_2 > 0 \end{matrix}\right\} \text{all } y, z. \tag{4.64}$$

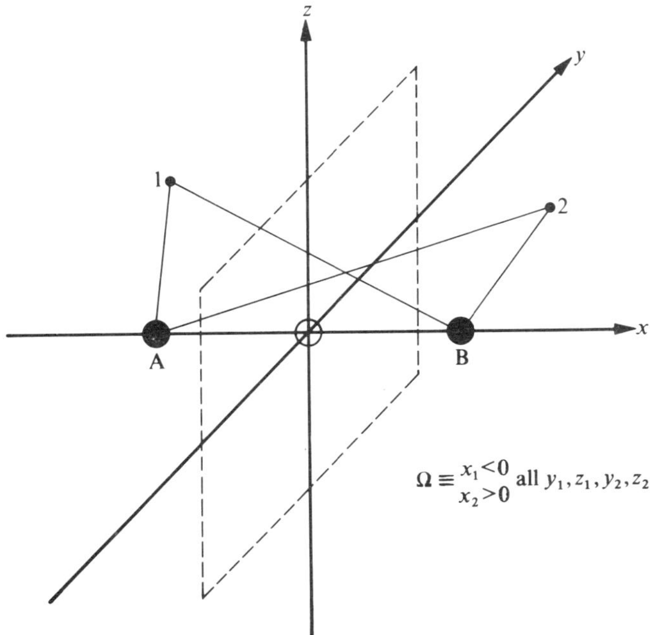

FIG. 4.13. Illustrating the sub-space Ω in the hyperspace $(\mathbf{r}_1, \mathbf{r}_2)$.

Then Φ_1 is almost entirely contained in Ω while Φ_2 is almost entirely outside Ω. We substitute from (4.63) into (4.62) and integrate both sides over the sub-space Ω. Then

$$E_T - E_S = \frac{2\int_\Omega (\Phi_2 H\Phi_1 - \Phi_1 H\Phi_2)\,d\tau_1\,d\tau_2}{\int_\Omega (\Phi_1^2 - \Phi_2^2)\,d\tau_1\,d\tau_2}. \tag{4.65}$$

The denominator, by the definition of Ω, is very close to unity. If we now insert the explicit form for H, namely

$$H = -\frac{\hbar^2}{2m}(\nabla_1^2 + \nabla_2^2) + V(\mathbf{r}_1, \mathbf{r}_2), \tag{4.66}$$

and apply Green's theorem, we have

$$E_T - E_S = \frac{\hbar^2}{m}\int_\Sigma (\Phi_2 \boldsymbol{\nabla}\Phi_1 - \Phi_1 \boldsymbol{\nabla}\Phi_2)\,.\,d\mathbf{S}, \tag{4.67}$$

in which we use vector notation $\boldsymbol{\nabla} \equiv \boldsymbol{\nabla}_1 + \boldsymbol{\nabla}_2$ and $d\boldsymbol{S}$ is a vector normal to the element of area dS on the hypersurface Σ bounding the sub-space Ω; Σ in fact is the sheet $x_1 = 0$, $x_2 > 0$ (all y, z) plus the sheet $x_2 = 0$, $x_1 < 0$ (all y, z). We have also set the denominator in (4.65) exactly equal to unity; it can be evaluated explicitly when the wavefunctions are known but in the cases we are concerned with, the error involved proves to be minute. Lastly as the two sheets give equal contributions to (4.67) we can write $E_T - E_S$ explicitly as

$$E_T - E_S = \frac{\hbar^2}{m}\int_{-\infty}^{\infty} \cdots \int \left(\Phi_1 \frac{\partial \Phi_2}{\partial x_1} - \Phi_2 \frac{\partial \Phi_1}{\partial x_1}\right)_{x_1 = 0} dy_1\,dz_1\,dx_2\,dy_2\,dz_2. \tag{4.68}$$

One can now evaluate this using given wavefunctions. The advantage of this approach is that one can use approximate wavefunctions, even Heitler–London functions, and thus obtain $E_T - E_S$ with good accuracy even though these same wavefunctions would not give this difference correctly if used to evaluate E_T and E_S separately. The symmetry of the centre about its mid-point guarantees that $E_T - E_S$ as given by (4.68) is positive. The accuracy, but not the sign of the splitting, depends on the accuracy with which the wavefunctions are known.

As a simple illustration of the usefulness of this approach we can treat the M-centre by assuming the orbitals Φ_S and Φ_T to be simple Heitler–London functions based on F-centre orbitals. If we take the F-orbitals to be simple hydrogenic functions, i.e.

$$\phi_{A,B} = \frac{\alpha^{\frac{3}{2}}}{\pi^{\frac{3}{2}}}\exp(-\alpha r_{A,B}) \tag{4.69}$$

then from (4.68)

$$\Delta \equiv E_{\mathrm{T}} - E_{\mathrm{S}} = \alpha^3 R(1 + \alpha R + \tfrac{1}{3}\alpha^2 R^2)\exp(-2\alpha R),$$

where R is the distance between the two F-centres. The parameter α can either be found variationally (as by Gourary and Adrian 1957) or empirically within the framework of the effective mass approximation by taking

$$\alpha^2 = \frac{8}{3}\left(\frac{m^*}{\hbar^2}\right)\Delta E_{\mathrm{F}}, \tag{4.70}$$

where ΔE_{F} is the energy of the F-band presumed to be a 1s–2p transition. The calculated magnitudes of Δ are roughly the same for these two choices (Table 4.11). It will be seen that the results of this rather simple calculation are in fair agreement with experiment (KCl) or other more elaborate direct calculations of the singlet and triplet state energies. We have not allowed for any lattice distortion here.

TABLE 4.11

The singlet–triplet splitting Δ(in eV) of the ground state A_{1g} of the M $\equiv$ F_2*-centre of several alkali halides as obtained from eqn* (4.68) *using the hydrogenic function* (4.69) *as described in the text.* (*After Berezin* 1972.)

Substance	Calculated from (4·68) and (4·69)		Other
	Semi-empirical, α	Variational, α	
LiF	0·249	0·569	0·18[a]
LiCl	0·201	0·411	0·18[a]
NaF	0·142	0·320	
NaCl	0·115	0·259	
KCl	0·086	0·199	0·064[b]
RbBr	0·082	0·161	

[a] Wood and Meyer (1964); [b] McCall and Grossweiner (1967).

The most elaborate use of this approach to singlet–triplet splittings however is that by Norgett (1971c) whose studied the F_t-centre previously mentioned. The e.s.r. studies of this centre showed it to have ⟨100⟩ symmetry with strong indications that it was composed of two F^+-centres at the next nearest-neighbour separation (Fig. 4.14). The question whether the intervening cation site was vacant or occupied normally could not be answered since the overwhelmingly abundant ^{24}Mg and ^{40}Ca isotopes are without nuclear moment and thus give no hyperfine structure. The object of the calculation was to answer this question from the magnitude of Δ. Norgett again took the wavefunctions to be of Heitler–London type, i.e.

$$\Phi_1 = \phi_{\mathrm{A}}(1)\phi_{\mathrm{B}}(2) \tag{4.71}$$

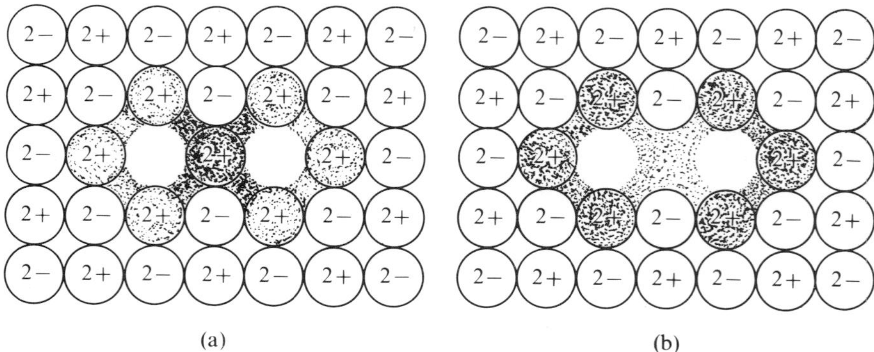

FIG. 4.14. Schematic diagram illustrating two possible structures for the F_t-centre in MgO and CaO as inferred from e.s.r. experiments.

but with more flexible single-centre orbitals ϕ than (4.69). In fact ϕ_A and ϕ_B were taken to be

$$\phi_{A,B} = (\phi_{sA,B} + \lambda\phi_{pA,B}) \tag{4.72}$$

in which ϕ_s and ϕ_p were taken to be s- and p-like *Gaussian* orbitals $N_s \exp(-\gamma_s^2 r^2)$ and $N_p \exp(-\gamma_p^2 r^2)$. The parameters γ_s, γ_p, and λ were determined in the usual way by minimization of the energy of the equivalent point-ion model. As would be expected λ is negative for the trivacancy model—the electrons are repelled by the effective negative charge $-2e$ at the centre ($\lambda \sim -0{\cdot}4$ to $-0{\cdot}5$) while for the divacancy model λ is positive, but somewhat smaller ($\lambda \sim 0{\cdot}1$–$0{\cdot}2$). The much smaller overlap of ϕ_A and ϕ_B in the trivacancy model is directly reflected in much smaller splittings Δ as shown in Table 4.12. This gives results calculated with allowance also for ionic relaxation and polarization, expressed in a Mott–Littleton approximation (Section 4.2.2). It will be seen that these effects are quantitatively important. It will also be seen that the calculated splittings Δ definitely favour the trivacancy model over that of a simple pair of F-centres. The theory thus distinguishes between two experimentally equivalent models.

It is worth commenting that the use of Gaussian orbitals in colour centre problems is not only convenient for the evaluation of the usual molecular integrals but also for evaluation of crystal field integrals, as is evident when one recalls the Ewald expression for the Madelung potentials due to point ions. However, in addition to convenience, Gaussian orbitals also appear to be quite accurate. Gaussian functions in any case would be expected to be a more accurate basis for these vacancy centres than in free molecules since the vacancy potential wells are of finite depth and are not Coulombic at the origin.

4.3.3.3. *Ion-size corrections.* We have already seen how the point-ion potential V_{PI} can be regarded as a first-order approximation to the

TABLE 4.12

Calculated and observed singlet–triplet splittings Δ (in K) of the ground state of the F_t-centre according to models (a) and (b) shown in Fig. (4.14). The calculated values are obtained from a point-ion calculation (Norgett 1971c). Results are given for a rigid unpolarized lattice as well as for one whose relaxation is described in the Mott–Littleton approximation.

Substance	Model	Δ (calculated)		Δ
		Rigid	Relaxed	(Experimental)
MgO	Trivacancy	28·4	44·4	80[a]
CaO	Trivacancy	17·4	34·1	50[b]
SrO	Trivacancy	13·7	29·3	—
BaO	Trivacancy	10·9	27·1	—
MgO	Divacancy	2448	2730	80[a]
CaO	Divacancy	1837	2095	50[b]
SrO	Divacancy	1576	1810	—
BaO	Divacancy	1369	1616	—

[a] Henderson (1966); [b] Tanimoto, Ziniker, and Kemp (1965).

pseudopotential but that the criterion of maximum smoothness for the pseudowavefunction leads to explicit expressions for the ion-size corrections to V_{PI}. In terms of the core-orbitals ψ_c the coefficients A_γ and B_γ are

$$A_\gamma = \int (V_\gamma - V_{PI\gamma})\,d\tau - \sum_c \left(\int \psi_c\,d\tau\right)\left[\int (V_\gamma - V_{PI\gamma})\psi_c\,d\tau'\right] \quad (4.73)$$

$$B_\gamma = \sum_c \left|\int \psi_c\,d\tau\right|^2. \quad (4.74)$$

These coefficients are properties of the ion γ and in principle can be calculated directly once and for all when (free) ion orbitals and self-consistent fields such as those of Slater and Clementi are known. As is evident from (4.73) and (4.74) the only contributions to A_γ and B_γ are from s-like core orbitals. This is a direct consequence of the assumption that the defect function ϕ varies slowly across any ion.

Evidently this formulation, if accurate, is extremely convenient. The coefficients A_γ and B_γ were evaluated by Bartram *et al.* (1968) using both Slater and Clementi orbitals. Tested against experimental F-band energies (by repeating the F-centre point-ion calculations with (4.61) in place V_{PI}), the A_γ coefficients turned out to be about twice too large. Arbitrary reduction of all A_γ coefficients by the same factor (0.53) however allowed an accurate description of F-band energies not only in the alkali halides but also in the three alkaline earth fluorides; see Fig. 4.12(b). Of course, empirical adjustments of atomic integrals are not unknown in molecular theory!

As we have already mentioned, the F_A-centre, i.e. an F-centre one of whose nearest cation neighbours is an alkali impurity, provides an obvious test for any theory of ion-size corrections. A considerable amount of experimental information on the splittings of the F-absorption, the nature of the relaxed excited state, reorientation activation energies, etc. now exists for these centres (Lüty 1968). Several calculations based on the above pseudopotential formulation have been made. The most complete of these are by Alig (1970) and by Ong and Vail (1973) who included lattice distortion and polarization in addition to purely ion-size effects. Unfortunately, the F_A-centre seems to pose a rather severe test for the theory and the results of these calculations are not accurate; for example the calculated splittings of the F-absorption energy can be as much as two to three times too large. However both ion-size corrections and lattice distortion and polarization contribute to these results and it is possible that the use of a polarizable-point-ion model for the lattice effects could be the source of this error (see Section 4.2.1.2).

In conclusion we may say the pseudopotential theory as given in Section 4.3.3 provides a convenient basis for calculating ion-size effects though more work evidently remains to be done. In particular, the reason why it is necessary to modify the *a priori* values of the A_γ is not yet clear though it is probably that the assumption that ϕ is effectively constant across any ion is too restrictive (Alig 1970; Matthew and Green 1971). Lastly, it should be repeated that the pseudopotential theory is only one approach to ion-size effects and that the more straightforward approach in which one evaluates the expectation value $\langle\psi_v|\mathscr{H}_{HF}|\psi_v\rangle$ of the Hartree–Fock Hamiltonian (see e.g. Wood and Joy 1964) is judged preferable by many in the field. We have however followed the pseudopotential approach here for the promise it gives of a relatively simple and transparent description.

4.3.4. *Molecular orbital theories*

There are two important areas of application which call directly upon molecular methods of calculation, namely the trapped electron–hole centres in ionic crystals and the vacancy and interstitial defects in covalent materials such as diamond and Si. Some of the trapped hole centres, e.g. so-called V_k- and H-centres, are effectively halogen molecular ions, X_2^-, and the electron hole is almost entirely localized on these two halogen atoms. The corresponding molecular orbitals of free X_2^- form the obvious starting point. They cannot be described in the same way as electron excess centres in these crystals as is shown by the fact that their optical transition energies do not satisfy any kind of Ivey relation—the lattice parameter itself appears far less important than the nature of the anion.

In the second class of defects typified by the vacancy in diamond it has seemed natural, following the early work of Coulson and Kearsley (1957), to build up defect molecular orbitals for the localized states from bond

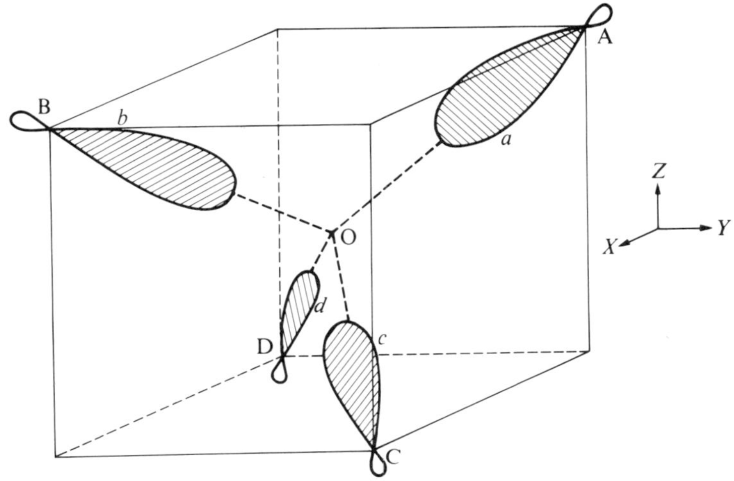

FIG. 4.15. Schematic diagram showing a vacancy in the diamond lattice with the broken-bond orbitals on the neighbouring atoms, A, B, C, and D.

orbitals on the surrounding atoms (Fig. 4.15). Recently there has been growing interest in this class of defects and there have also been attempts to apply extended Hückel and other one-electron methods to large clusters of atoms in the crystal coordination in the hope of extrapolating to the infinite lattice results. At the present time there exists some conflict between the defect molecule approach which is carried through by a full configuration interaction treatment and these one-electron results. In particular the interactions of the electrons in the defect molecule description are found to be large and to dominate the order and energies of the defect states. In this section we describe these two areas of application of molecular methods to solid-state problems. In both of them the defect-lattice coupling proves to be quantitatively and qualitatively important.

4.3.4.1. *Trapped hole centres.* A considerable amount of evidence exists which shows that the structures of so-called V_k-, H-, and V_1-centres are as shown schematically in Fig. 4.16, i.e. basically a halogen molecule ion X_2^- fixed in a surrounding ionic lattice. In the V_k-centre the two nuclei essentially occupy a pair of neighbouring anion sites while in the H-centre they are symmetrically placed about a single anion site, so that the H-centre is an interstitial centre. The molecular axes of both the V_k- and H-centres have $\langle 110 \rangle$ orientation. The V_1-centre is an H-centre perturbed by the presence of a foreign alkali ion; in this case the foreign ion perturbs the $\langle 110 \rangle$ orientation slightly (Itoh 1972).

Since all these centres are basically X_2^- molecule ions perturbed by the surrounding lattice, we expect that a good starting point for the representation

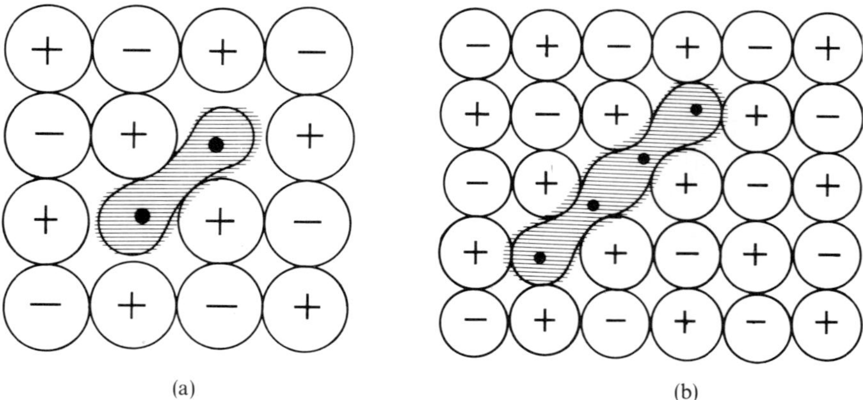

(a) (b)

FIG. 4.16. Schematic diagram showing the structure of the so-called V_k- and H-centres in alkali halides with the NaCl structure. The V_k-centre (a) is an electron–hole self-trapped on a pair of neighbouring anions, X^-, giving an X_2^- molecular ion. The H-centre (b) results from the formation of an interstitial halogen atom X^0. However, the stable structure is again an X_2^- molecular ion whose geometric centre lies on an anion site, although the electron–hole spreads to a small extent on to the neighbouring ⟨110⟩ anions (which are therefore also shown shaded here). The so-called V_1 centre is an H-centre one of whose *cation* neighbours is an alkali different from the host e.g. Na^+ in KCl. For the optical properties of these centres see Table (4.14).

of the electronic configuration of all of them will be the molecular structure;

$$(\text{closed shells})(np\sigma_g)^2(np\pi_u)^4(np\pi_g^*)^4(np\sigma_u^*)$$

formed from the bonding and antibonding orbitals generated by the atomic np orbitals. The one-electron levels corresponding to these molecular orbitals are shown schematically in Fig. 4.17. The degeneracy of the π-orbitals in the free molecule will be lifted in crystals with the NaCl structure since this structure gives only orthorhombic symmetry about the molecular

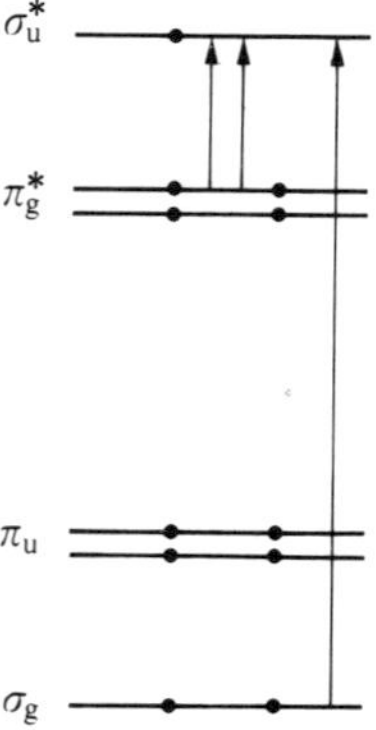

FIG. 4.17. Schematic diagram illustrating the one-electron energy levels of the molecular orbitals in the X_2^- molecule–ion and their occupancy in the $^2\Sigma_u^+$ ground state.

axis. In this structure, the symmetry axes of the states in the crystal field are ⟨001⟩ and ⟨1$\bar{1}$0⟩ when the molecular axis is ⟨110⟩.

It is clear from the Fig. 4.17 that we should expect a high energy transition corresponding to excitation of an electron in a σ_g orbital to the empty σ_u^* orbital in all these centres ($^2\Sigma_u^+ \rightarrow {}^2\Sigma_g^+$). In NaCl structures this will have an electric dipole moment along ⟨110⟩. Correspondingly we should expect two transitions ($^2\Sigma_u^+ \rightarrow {}^2\Pi_g$) of nearly equal but lower energy having electric moments along ⟨1$\bar{1}$0⟩ and ⟨001⟩. This is just what is observed although the crystal field splitting of the lower-energy transitions has not been resolved. In CaF_2 and CsCl structures the axis of the high-energy transition is along ⟨100⟩ while the low-energy transition takes place perpendicular to the molecular axis. In KBr the spin–orbit splitting of the $^2\Pi_g$ state into $^2\Pi_{g,\frac{1}{2}}$ and $^2\Pi_{g,\frac{3}{2}}$ states is however resolved. The low-energy transitions ($^2\Sigma_u^+ \rightarrow {}^2\Pi_g$) are observed in the near infra-red region while the high-energy transitions ($^2\Sigma_u^+ \rightarrow {}^2\Sigma_g^+$) fall in the near u.v. region (Table 4.13).

TABLE 4.13

Energies (in eV) of the maxima of the optical absorption bands (at 0 K) due to V_k, H *and* V_1*-centres in several alkali halides with the* NaCl *structure. In the case of* V_1 *the responsible impurity cation is also given. The high-energy transition corresponds to the* $^2\Sigma_u^+ \rightarrow {}^2\Sigma_g^+$ *transition of the free molecule while the low-energy transition which is very much weaker corresponds to* $^2\Sigma_u^+ \rightarrow {}^2\Pi_g$.

Substance	V_k		H		V_1	
	$^2\Sigma_u^+ \rightarrow {}^2\Sigma_g^+$	$^2\Sigma_u^+ \rightarrow {}^2\Pi_g$	$^2\Sigma_u^+ \rightarrow {}^2\Sigma_g^+$	$^2\Sigma_u^+ \rightarrow {}^2\Pi_g$	$^2\Sigma_u^+ \rightarrow {}^2\Sigma_g^+$	$^2\Sigma_u^+ \rightarrow {}^2\Pi_g$
LiF	3·65	1·65				
NaF	3·38					
LiCl	3·15					
NaCl	3·28		4·13		3·59	
KCl	3·40	1·65	3·69	2·38	3·47 (Na^+)	2·22 (Na^+)
RbCl	3·40					
KBr	3·22	1·65, 1·38	3·26		3·02 (Na^+)	
KI	3·10	1·55, 1·08				

Thus in qualitative terms these centres appear to be well described by the picture of a molecule ion X_2^- embedded in a crystal medium. The interatomic separation of the free molecule ions however is often considerably smaller than the anion–anion separation in the crystal so that the molecule is stretched by the Coulomb and other forces of lattice cohesion. This stretching will obviously affect the energy of the transition (*cf*. Fig. 4.18) leading us to expect the transition energies to decrease as the lattice parameter increases. Likewise it will be seen that the transition energies of the interstitial centres are higher than those of the V_k-centres, reflecting the compression of the

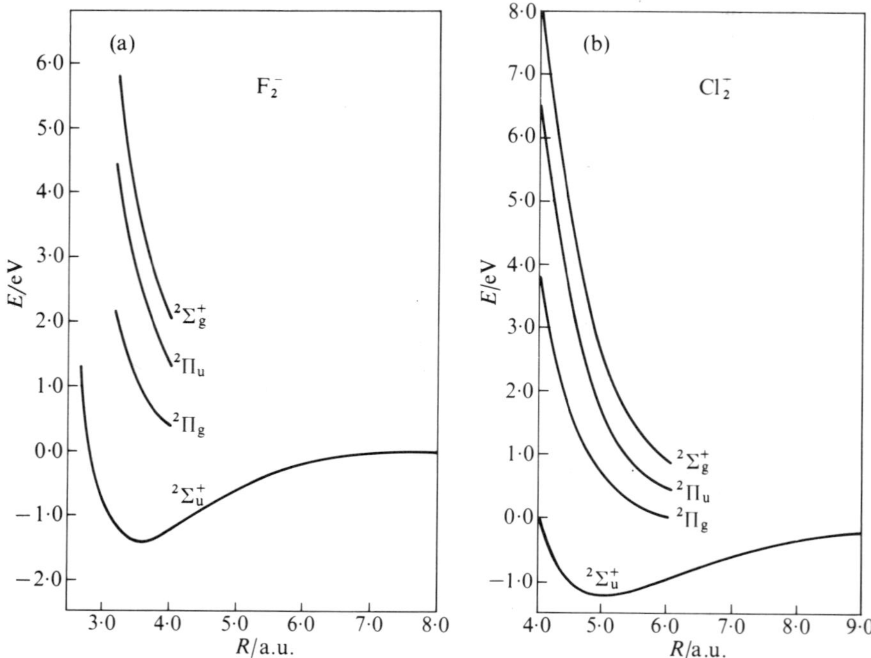

FIG. 4.18. The total energies of the free molecule–ions F_2^- and Cl_2^- as a function of internuclear distance R (in atomic units) as calculated by Gilbert and Wahl (1971) using an SCF–MO method.

interstitial molecule, the effect being the more pronounced in the smaller lattices; for example compare NaCl with KCl in Table 4.13.

To describe these effects quantitatively requires a theory of the X_2^- molecule, the specification of its interaction with the lattice and then the evaluation of the lattice energy under the influence of these forces, using the methods described in Section 4.2.2. The energies of the lowest states of free F_2^- and Cl_2^- molecules have been evaluated as a function of internuclear separation by Gilbert and Wahl (1971) by the SCF–MO method. Their basis set included a total of 15 bonding and 15 antibonding orbitals. This large number of orbitals is necessary since a previous calculation by Das, Jette, and Knox (1964) of the F_2^- levels using only the orbitals given in Fig. 4.17 plus the corresponding s-core yielded a $^2\Sigma_u^+ \rightarrow {}^2\Sigma_g^+$ transition energy of only 1·3 eV compared to the experimental value in LiF:V_k of 3·65 eV (Table 4.13). The results of Gilbert and Wahl (1971) are shown in Fig. 4.18 and have the expected form. They have been used in most of the recent calculations on trapped hole centres.

For the interaction of the X_2^- molecule with the lattice and for the lattice response we call upon the ionic model. The charge distributions of the isolated molecule show that it is reasonable to regard the electron hole as centred 50 per cent of its time on one ion and 50 per cent on the other; this allows us to specify its external electric field and thus Coulomb interactions as that due to charges $-e/2$ on each atomic centre with an effective charge $+e$ at each anion site due to absence of the normal anion from those positions. The Born–Mayer overlap repulsions with the ions surrounding the centre will, of course, be changed by the presence of the hole and at the present time one has only the simplest intuitions to use in estimating this change. Since Born–Mayer interactions in different systems are often consistent with a common hardness parameter ρ we may assume this to be unchanged. On the other hand, the radius of the atoms in the X_2^- molecule must be less than that of the X^- ion; it could be estimated from the electron distribution, e.g. by calculating the mean radius of the outer occupied s- and p-atomic orbitals, providing that such a way of estimating is also consistent with the radii used for the ion–ion interactions (Tosi 1964). The reduction is probably not very large, being perhaps a reduction of ~ 5 per cent since the reduction of $\langle r \rangle$ and $\langle r^2 \rangle$ in going from Cl^- to Cl and from F^- to F is ~ 10–12 per cent (Bagus 1965). We should, however, also expect a small reduction in the pre-exponential factor due to the reduced number of electrons contained in the outermost shell. Thus neither of the effects is very large and, in fact, in some of the explicit calculations on V_k-centres (Das, *et al.* 1964; Jette, Gilbert, and Das, 1969) the differences have been ignored altogether. Since the corresponding terms in the total energy are small relative to the Coulomb and polarization terms, this simplification would appear to be adequate. Where the assumption has been tested (as in the alkaline earth fluorides) the results for spectroscopic quantities are indeed rather insensitive to changes in the Born–Mayer interaction between the molecule and neighbouring ions (Norgett and Stoneham 1973*a*). However this is not true when the rates of thermally activated hopping of the V_k-centre are considered (Norgett and Stoneham 1973*b*). Likewise for interstitial centres (H and V_1) these interactions are much more important and the sensitivity of the results to the precise model adopted must be examined (Hatcher and Dienes 1961, 1964; Dienes, Hatcher, and Smoluchowski 1967). Lastly, it is necessary also to specify the polarizabilities of the X_2^- molecule ion. Although calculated values are now available from the work of Gilbert and Wahl (1971) explicit calculations have so far been guided by rather simple intuitions. Although all authors regard the X_2^- molecule ion as effectively a pair of hypothetical $X^{-\frac{1}{2}}$ ions, to get the polarizability of these component 'ions' some scale the polarizability of X^- ions in proportion to the reduced number of outer-shell electrons, others make different assumptions. At least in the alkaline earth fluorides it appears that many of the results are not sensitive to the precise model chosen for the molecule-ion polarizability.

Having specified the potential of the X_2^- molecule and its interactions with the ionic lattice it remains to choose the model for this lattice and to evaluate the relaxation around the defect by one of the methods of Section 4.2.2. Of the calculations for the V_k-centres those by Jette, Gilbert, and Das (1969) for alkali fluorides and chlorides and by Jette and Das (1969) for CaF_2 use a point-polarizable-ion model and a Mott–Littleton method of calculation (Section 4.2.2.1). The results for the energy levels of the Cl_2^- ion in NaCl as a function of internuclear separation R (after minimization with respect to all other region I displacements) are shown in Fig. 4.19.

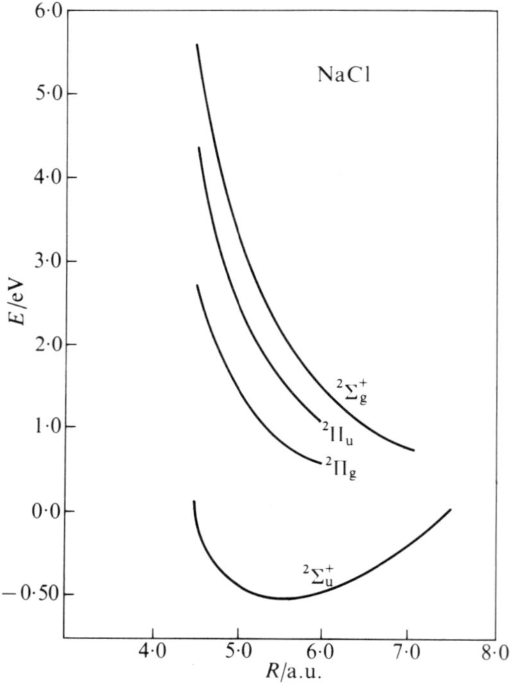

FIG. 4.19. The total energy of the V_k-centre (Cl_2^-) in NaCl as a function of the Cl–Cl separation, R, (in atomic units) in the centre (all other ions in the lattice are relaxed to be in equilibrium at each value of R). (After Jette *et al.* 1969).

Comparison with the corresponding curves for the free Cl_2^- molecule show how the molecule is extended by lattice forces in this structure. Table 4.14 compares some of the numerical results obtained by Jette *et al.* (1969) for the alkali halides with corresponding experimental values. It will be seen that the agreement is fairly satisfactory though by no means perfect. It is better for the fluorides LiF and NaF, than for the chlorides. The good agreement for the V_k-centre is also found in CaF_2 (Jette and Das 1969; Norgett and Stoneham 1973*a*). It is possible that the errors are due to the use of the polarizable-point-ion model for the lattice. Although, as we have already

TABLE 4.14

Calculated transition energies (in eV) of V_k*-centres obtained from the model of a halogen molecule-ion* X_2^- *embedded in a crystal* (*Jette* et al 1969). *Experimental energies are also given for comparison* (cf. *Table* 4.13).

	Calculated		Experimental	
Substance	$^2\Sigma_u^+ \rightarrow {}^2\Sigma_g^+$	$^2\Sigma_u^+ \rightarrow {}^2\Pi_g$	$^2\Sigma_u^+ \rightarrow {}^2\Sigma_g^+$	$^2\Sigma_u^+ \rightarrow {}^2\Pi_g$
Free F_2^-	4·74	2·23		
LiF	3·30	1·60	3·65	1·65
NaF	3·30	1·60	3·38	
KF	2·99	1·42		
RbF	3·00	1·46		
Free Cl_2^-	3·86	1·87		
LiCl	2·69	1·31	3·15	
NaCl	2·65	1·35	3·28	
KCl	2·37	1·23	3·40	1·65
RbCl	2·20	1·18	3·40	

seen (Section 4.2.1.2), this model can lead to considerable errors—particularly when used for charged defects—there is little to go on in the present connection. CaF_2 is the only case for which both polarizable-point-ion and shell models have been evaluated, but the value of the direct comparison so provided is limited because the minimum energy of V_k in both models occurs at an internuclear separation very close to that in the free F_2^- molecule ion (2·0 Å in the polarizable-point-ion model and 1·9 Å in the shell models compared to 1·95 Å in the free molecule). However the good agreement with experiment for this centre does allow us to infer that the description of the isolated F_2^- molecule provided by Gilbert and Wahl's SCF–MO calculation is a good one.

However it is possible that the generally less satisfactory picture of the V_k-centre in the chlorides is due to errors in the basic description of the free molecule Cl_2^-; in particular that the force constant for stretching in the ground state is too weak. This would result in the molecule being stretched too much in the crystal thus giving too low transition energies (Table 4.14). Support for this interpretation comes also from the corresponding study of H-centres, i.e. interstitial Cl_2^--centres, in NaCl and KCl (Dienes *et al.* 1967). These calculations gave the Cl_2^- internuclear separation in the H-centre as 2·46 Å in NaCl and 2·54 Å in KCl compared to 2·65 Å in the free molecule. Such compressions would give $^2\Sigma_u^+ \rightarrow {}^2\Sigma_g^+$ transition energies significantly *higher* than the observed H-band energies, again suggesting that the Gilbert and Wahl Cl_2^- molecule is too compressible. However the Cl_2^- model is not so good for the H-centre as for the V_k-centre, since the calculations predict the stable orientation to be ⟨111⟩ instead of ⟨110⟩ as observed. The small energy difference ($\sim 0{\cdot}2$ eV) may be accounted for by the small spread (~ 5 per cent) of the electron hole on to the two Cl^- neighbours in the $\langle 110 \rangle$ direction as observed in the e.s.r. spectra. Further work on these interesting and basic centres using the best molecular and lattice relaxation methods would be valuable.

4.3.4.2. *Defects in covalent crystals.* If it is reasonable to regard the covalent solids diamond, graphite, silicon, etc., as giant covalent molecules, then it should be equally valid to describe the states of defects using the same (hybridized) atomic orbitals as for the bonding of the perfect lattice. This idea is illustrated schematically for a vacancy in a diamond lattice in Fig. 4.15. It was first formulated mathematically by Coulson and Kearsley (1957) and the method has since been applied by a number of authors mainly to defects in diamond and Si. It should be noted at the outset, however, that the Coulson–Kearsley method followed the pattern usual in molecular theory, namely: one-electron levels and states → single configuration Slater states → configuration interaction states → lattice relaxed and Jahn–Teller distorted states. Unfortunately, quantitative calculations made by this method cannot be said to have been as successful as those defect calculations we have already described in ionic crystals. Nevertheless the underlying idea of a covalently-bonded 'defect molecule' has guided both experimental and theoretical work in useful directions. We shall review several of the uses which have been made of the basic idea including: (1) the description of empirically determined states; (2) semi-empirical calculations of defect formation and migration energies; (3) calculations of defect-molecule states; (4) other approaches to the one electron defect states.

(1) *Empirically determined defect states.* A considerable number of vacancy defects in Si have been studied in detail by e.s.r. and other spectroscopic techniques; they include isolated single vacancies (V^+ and V^-), divacancies and a number of impurity-vacancy pairs. Their ground states are generally low-spin states ($S = \frac{1}{2}$) and they are Jahn–Teller-distorted; for V^+ and V^- these distortions are along $\langle 100 \rangle$. The same appears to be true of the V^+ centre in diamond (Baldwin 1963). We can interpret these results as in Fig. 4.20. The vacancy molecular orbitals formed from the broken sp^3 band orbitals shown in Fig. 4.15, when we assume that T_d point symmetry of the site is retained, are

$$
\begin{aligned}
v &= a+b+c+d \quad (a_1\text{-symmetry}) \\
\left.\begin{aligned}
t_x &= a+b-c-d \\
t_y &= -a+b+c-d \\
t_z &= a-b+c-d.
\end{aligned}\right\} &\quad (t_2\text{-symmetry})
\end{aligned}
\tag{4.75}
$$

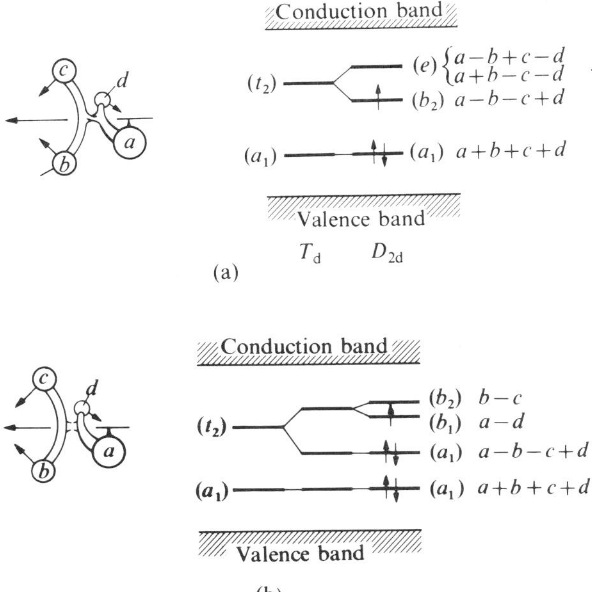

FIG. 4.20. The pattern of one-electron levels of the vacancy in Si and Jahn–Teller splittings as inferred from e.s.r. studies of (a) V^+- and (b) V^--centres.

The corresponding one-electron energies are assumed to lie within the gap between the valence and conduction bands. The lowest energy state of V^+ formed by assigning two electrons with antiparallel spin to the a_1-level and one electron to the t_2 level is a $^2T_2$ state and 3-fold orbitally degenerate. The ⟨100⟩ axis observed may thus be assumed to result from an E-type Jahn–Teller distortion which pulls the neighbours to the vacancy together in pairs.† The new one-electron levels are then as shown. For sufficiently large splitting we might then suppose that the ground state of V^0 will be that obtained by adding one-electron (of antiparallel spin) to the b_2 orbital. Experimentally V^0 in Si is a non-magnetic centre ($S = 0$) consistent with such an assignment but the symmetry is not known. If we add yet another electron then this must go into the two-fold degenerate e orbital, when a further distortion occurs. Although this leaves the axis of symmetry as ⟨100⟩, it makes the centre orthorhombic i.e. the two pairs of atoms $(a+d)$ and $(c+b)$ become non-equivalent. Experimentally V^- in Si *is* observed to be an orthorhombic centre with $S = \frac{1}{2}$, just as this model requires. We see then

† From the theory of the Jahn–Teller effect it follows that the symmetry of the undistorted state (T_2) would also allow a T-type distortion which would give the centre a ⟨111⟩ symmetry axis; but we must assume that in this case the forces exerted by the defect on the lattice are such that the E-type distortion gives the greatest energy lowering.

that this one-electron model of defect orbitals does provide a scheme for the representation of experimental results. It has been used successfully in this way not only for simple vacancy centres in Si but also for more complex defects such as divacancies and impurity–vacancy pairs. In diamond the experimental picture is far less complete, but V^+ at least appears similar to the corresponding centre in Si.

(2) *Calculation of formation and migration energies of vacancies* (V^0). An extremely simple form of the defect molecule model was first proposed by Swalin (1961) for the calculation of vacancy formation energies. It assumes that the atoms neighbouring the vacancy rebond together in pairs, e.g. $a+d$ and $b+c$. The energy of these reformed bonds is assumed to be represented by the same energy function as the normal nearest-neighbour bonds of the lattice for which a Morse function is used. When the lattice is allowed to relax this (necessarily) leads to a tetragonal $\langle 100 \rangle$ centre as above. The detailed calculations which have been made for this model lead to the values of vacancy formation energies given in Table 4.15. Some of the earlier ones

TABLE 4.15

Calculated energies of formation (in eV) of vacancies in the diamond structure obtained with Morse function bond models. Similar results have however been obtained by more sophisticated solid state methods.

Substance	Calculated formation energy
Diamond	4·2
Si	2·3–2·8
Ge	2·1–2·5

can be criticized on the grounds of restricted lattice relaxation or other internal inconsistencies. The best calculation is probably that of Seeger and Swanson (1968), but it should be recognized that since the models, in all cases, are very intuitive the principal criteria of goodness are mathematical accuracy and internal consistency. The values given in Table 4.15 are consistent with the corresponding experimental formation enthalpies. However it does seem that the present uncertainties in this simplified model are too great for it to be very reliable. This is well demonstrated by the calculations of Larkins and Stoneham (1971a,b) who evaluated accurately the responses of the valence force model (Section 4.2.1.3) to symmetrized systems of (unit) forces (E, T_2, and A_1) impressed on the neighbours of the vacancy. When the actual forces due to rebonding of the defect electrons were inserted then the calculated relaxation energies were much too great, even giving negative formation energies!

(3) *Electronic states of the defect molecule.* The method of Coulson and Kearsley (1957) attempts to calculate the rebonding of the defect electrons and the electronic states of the defect. It assumes that the basic molecular orbitals are obtained from the broken bond orbitals as in eqn (4.75), but beyond that it uses only the standard methods of molecular theory. Correctly antisymmetrized single configuration states are formed and then configuration interaction of such states of the same symmetry type is allowed by finding those linear combinations which minimize the energy functional. In this way exchange and correlation effects among the defect electrons are included. The energy levels of the centre are then obtained in terms of the usual kinds of energy integral arising in molecular theory including 3- and 4-centre integrals for which the Mulliken approximation is used. Uncertainties in the original atomic orbitals used for the sp^3 bond orbitals are directly reflected in the values of these integrals. The defect levels prove sensitive to the values of two in particular: (*i*) the self-penetration integral

$$P = \langle a(1)| T_1 + V_{A1} |a(1)\rangle \tag{4.76}$$

where T_1 is the kinetic energy operator for defect electron 1 and V_{A1} is the potential energy of this electron in the field of atom A; and (*ii*) the interatomic Coulomb interaction integral

$$Q = \left\langle a(1)a(2) \left| \frac{e^2}{r_{12}} \right| a(1)a(2) \right\rangle. \tag{4.77}$$

For these both calculated values and empirical values derived from molecular data (on the assumption that the bonding orbitals are independent of environment) have been used. The most recent calculated values obtained using self-consistent crystal wavefunctions seem to agree quite closely with the empirical values and are thus to be preferred to earlier values based on Slater and Clementi free-atom functions.

The original formulation of this approach (Coulson and Kearsley 1957) assumed the lattice to be rigid. However the method readily gives the forces $\mathbf{F}^{(0)}$ (eqn 4.4) and later developments have thus evaluated the lattice distortion in this approximation (Friedel, Lannoo, and Leman 1967; Lannoo and Stoneham 1968; Coulson and Larkins (1971) Larkins 1971*a*,*b*; Larkins and Stoneham 1971*a*,*b*). As the rebonding involves only the nearest neighbours of the vacancy in this case (Fig. 4.15) the forces $F^{(0)}$ only act upon these four atoms. From the T_d symmetry of the vacancy the forces can be classified by symmetry as A_1 (symmetric), E-type (2 independent choices θ, ε) and T_2-type (3 independent choices ξ, η, ζ). The response of the lattice to any one of these three types will belong to the same irreducible representation (of T_d) and can thus be represented in terms of a single effective force constant or equivalent normal mode frequency ω_Γ for the neighbours of the vacancy. The valence-force model of diamond and Si (Section 4.2.1.3) is the only one to have been accurately evaluated in this way (Larkins and Stoneham 1971*a*); the frequencies ω_A, ω_E, and ω_T were all considerably less than the Raman frequency whereas previous simpler estimates had given values comparable with it. Since the energy lowering is inversely proportional to ω_Γ^2 such differences are reflected as large differences in the energies of relaxation. Careful evaluations of alternative models using the methods of Section 4.2.2 are called for. *All* electronic levels may be lowered in energy from interaction with the totally symmetric distortion mode, Q_A. The corresponding symmetric force term is

$$F_A^{(0)} = -\left(\frac{\partial}{\partial Q_A}\langle\Phi|H|\Phi\rangle\right)_{Q_A=0} \tag{4.78}$$

where Φ is the electronic state of the centre and H the defect electron Hamiltonian. However only *degenerate* levels can couple with asymmetric distortion modes giving Jahn–Teller splittings of the degeneracy (Öpik and Pryce 1957). In the present connection a (two-fold degenerate) E level can couple to E modes—but not to T_2 modes—giving tetragonal distortions. A (three-fold degenerate) T_1 or T_2 level may couple to either E or T_2 modes, depending on which gives the greatest energy lowering, so that either a tetragonal or a trigonal distortion results. Mixed trigonal and tetragonal distortions do not occur for isolated levels but only when there is an accidental degeneracy (Stoneham and Lannoo 1969).

Lastly, it should be noted that these calculations give absolute energy levels of the various defect states but do not in themselves evaluate the energy of polarization of the rest of the lattice caused by any net charge on the defect (e.g. V^+, V^-, etc.). For this reason and also because neither the valence nor conduction states are calculated on the same basis it is not possible at present to infer the corresponding electrical donor and acceptor levels relative to the band edges at all accurately. However location of those electrical levels corresponding to change in the charge state of the vacancy from $V^{(x)}$ to $V^{(x-1)}$ (by addition of an electron) relative to the valence band maximum (Γ'_{25}) is given in principle by

$$E(V^{(x)}/V^{(x-1)}) = I + E_{DM}(V^{(x-1)}) - E_{DM}(V^{(x)}) + \Delta E_{pol}, \tag{4.79}$$

where I is the energy required to ionize an electron from the top of the valence band to 'a state of rest at infinity' and E_{DM} are the absolute ground-state energies of the defect molecule as usually calculated without allowance for the electrical polarization of the rest of the lattice by the charge on the defect. The change in this polarization energy ΔE_{pol} could be estimated by the Mott–Littleton method (Section 4.2.2) as was done by Weiser (1962) for interstitial defects. A rough guide to the magnitude and importance of this term can be obtained from the Jost continuum formula, i.e.

$$\Delta E_{\text{pol}} = (2x-1)\frac{e^2}{2R}\left(1-\frac{1}{\varepsilon_0}\right) \tag{4.80}$$

where R is the radius of the vacancy and ε_0 is the static dielectric constant. With the existing calculations it would seem sensible to take R as the nearest-neighbour bond distance since no change in bond orbitals has been assumed even on nearest neighbours to the vacancy.

The predictions of this defect molecule model show a rather wide spread resulting from the various uncertainties already mentioned and we shall therefore not go into details of the calculations further but instead list several relevant general observations on the successes and limitations of this approach.

(1) If only relaxations which retain the T_d symmetry of the defect are allowed then level schemes for the vacancy in diamond such as those shown in Fig. 4.21 result. The relative order of closely spaced levels may, however, be sensitive to the choice of atomic orbitals and to the values of the integrals P and Q in particular.

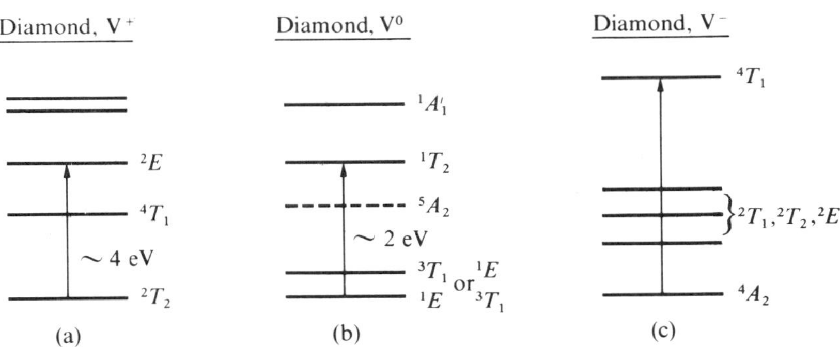

FIG. 4.21. Diagram summarizing some common features of the results of several calculations of the energy levels and states of vacancies (V^+, V, and V^-) in diamond. Exact energy separations and the order of closely spaced levels may depend on details of the calculation. Only symmetric (A_1) relaxations have been allowed and further Jahn–Teller relaxations splitting the degenerate E, T_1, and T_2 states will occur. Except in cases of near degeneracy, E-levels undergo E-type distortion (giving ⟨100⟩ orthorhombic centres) while T_1 and T_2 levels may undergo either an E-type distortion or a T_1-type distortion (giving ⟨111⟩ trigonal centres).

(2) Electronic interaction terms are important in leading to these patterns. In each case the levels shown largely derive from the same configuration $a_1^2 t_2^n$.

(3) The forces exerted by the vacancy defects on the rest of the lattice are large. This applies particularly to the symmetric forces but those leading to Jahn–Teller distortions can also be large. Their accurate estimation is difficult as a consequence, and present calculations

appear to overestimate them substantially, possibly because only $F^{(0)}$ is evaluated i.e. their dependence on relaxation ($F^{(1)}$, etc.) is not evaluated. The calculation of the corresponding lattice distortion, energy lowering and Jahn–Teller splittings requires in addition a good model for the lattice response. While the valence-force model is consistent with the bond approach as presently applied, it does represent a separate model assumption.

(4) Comparison with experiment remains difficult for diamond owing to the rather limited amount of definitive experimental information (Clark and Mitchell 1971). The ground state of V^+ appears to be confirmed (Baldwin 1963) but it would be surprising if it were not $^2T_2$ in almost any scheme based on an a_1 orbital lying below a t_2 orbital. The pattern of levels shown for V^0 and having the ^1E level lowest seems to correspond well to the optical properties of a centre seen in irradiated diamond and generally denoted by GR1 (Lannoo and Stoneham 1968). However other centres having a similar level scheme (e.g. the interstitial, Yamaguchi 1963) would give similar predictions, and since the level is calculated to be lowest in some cases of V^0 it remains possible that the GR1 centre is not V^0 after all.

(5) Within the limitations already noted, the check upon the absolute ground-state energies provided by (4.79) is reasonably satisfactory if we use (4.80) for ΔE_{pol} and the band calculations of Painter, Ellis, and Lubinsky (1971) to give I. The levels $E(V^+/V^0)$ and $E(V^0/V^-)$ lie within the gap when the semi-empirical integral values (P and Q) are used, although not otherwise. Of course uncertainties in the exact values of lattice relaxations are directly important here.

(6) For vacancies in Si the range of possibilities associated with present uncertainties in orbitals and integral values is rather wider than in diamond. However the position is summarized schematically in Fig. 4.22. The pattern of levels of V^+ is very similar although the separations are smaller as would be expected. For V^0 the order of levels is changed when the (symmetric) lattice distortion is included. One of the reasonable predicted possibilities gives the ground state as a $^1A_1$ with the next highest level $^1T_2$. For V^-, configuration mixing and lattice distortion are both important in determining the level scheme. One reasonable possibility gives the lowest state as $^2T_2$ followed by $^4T_1$ and ^2E within about 1 eV.

(7) The experimental information on the ground levels of V^+ and V^- is very detailed in Si. The defect molecule prediction about the nature of the ground state of V^+ is correct, but as we have previously remarked it would be rather surprising if it were otherwise. The ground state of V^- is a spin doublet Jahn–Teller distorted mainly in an E-mode. This is consistent with the defect-molecule prediction of a $^2T_2$ ground state

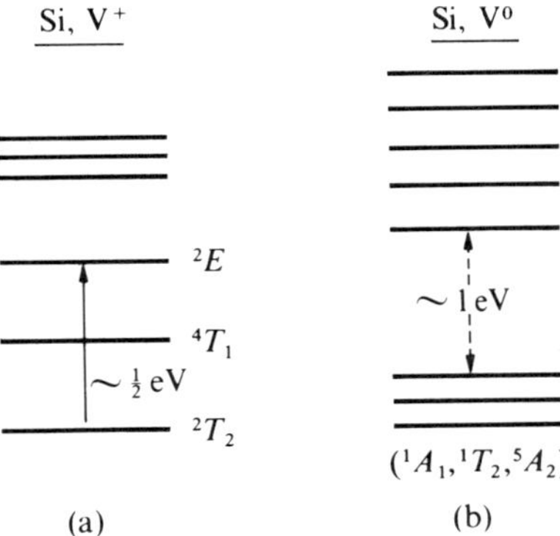

FIG. 4.22. Diagram summarizing some common features of the results of several calculations of the energy levels and states of vacancies (V^+, V) in Si. As for diamond the exact energy separations and the order of closely spaced levels may depend on details of the calculation. For V^- the predicted level structure is too sensitive to these to be usefully put on a single diagram. As in Fig. (4.21) only symmetric distortions have been taken into account. The same remarks about Jahn–Teller splittings apply as there.

since this will undergo either T_2 or E-mode Jahn–Teller distortions, although the detailed estimation of these distortions actually favours a T_2-type (trigonal) distortion. There is no evidence on excited states of V^+ or V^-. Furthermore V^0 is not seen in e.s.r. and is thus presumed to be non-magnetic, again consistent with those defect-molecule predictions which give a ${}^1A_1$ ground state.

For Si then, the defect molecule calculations again show the importance of configuration interaction and of lattice distortion in determining states and energy levels but unfortunately they show a wider spectrum of predictions than for diamond. The predictions made using a semi-empirical value for the one-centre Coulomb integral Q are in general agreement with experiment; however, for V^0 this agreement also rests on the correctness of the lattice distortion.

In conclusion it seems fair to summarize these defect-molecule calculations as follows. They correctly embody many of the basic chemical intuitions about centres in covalent solids such as diamond and Si and permit the calculation of the electronic states and derived quantities of the defect. However the predictions are sensitive to several uncertainties and are only partially in agreement with experiment. Separate model assumptions are necessary to calculate lattice distortion and the location of levels relative to the edges of the conduction and valence bands. The method does not provide any internal check on the assumption that the defect states involved are truly localized states as distinct from resonance states lying within the continuous bands of valence or conduction states. Perhaps its most serious limitation is its starting point, namely the assumption, expressed in eqn

(4.75), that the one-electron defect orbitals are localized on first neighbours of the defect.

(4) A priori *calculations of one-electron states.* Several authors have therefore attempted to calculate the one-electron defect orbitals by using various one-electron methods. These include specifically solid-state methods based on the states of the perfect crystal as well as applications of semi-empirical molecular methods such as the extended Hückel method and most recently, the so-called X_α method (Slater and Johnson 1972). The extended Hückel calculations have attracted interest recently particularly when they have been employed to suggest new physical hypotheses. However all these methods although they provide information about the limitations of the basic defect-molecule assumption (4.75), at present remain one-electron calculations and have not been used to assess theoretically the more important qualitative conclusions of the defect-molecule model, e.g. the importance of exchange and correlation effects among the localized defect electrons. They too provide a fairly wide spectrum of predictions. For these reasons we shall again not go into detail but list a few pertinent observations.

(1) The extended-Hückel method (EH) has mainly been applied to finite clusters† containing a few dozen atoms although recently with insertion of periodic boundary conditions it has in effect also been applied to an infinite solid containing a superlattice of vacancy defects (Messmer and Watkins, 1973b). In this last application of course any localized *levels* will spread into a defect *band*, the width of which reflects the delocalization of the state. The studies of perfect *clusters* disclose a banding of states corresponding to the valence and conduction bands of the infinite solid with a forbidden gap between. However, in magnitude this band gap does not converge uniformly or at all rapidly with increasing cluster size to the limit obtained for the infinite solid and furthermore is sensitive to the assumed boundary conditions, e.g. dangling or paired surface bonds. For the infinite perfect solid the EH *valence* band structure of diamond is in reasonable quantitative agreement with some *ab initio* calculations (e.g. Painter *et al.* 1971) but in only qualitative agreement with semi-empirical calculations (e.g. Saslow, Bergstrasser, and Cohen 1966; Hemstreet, Fong, and Cohen 1970). The EH *conduction* band is not qualitatively correct.

(2) When a vacancy is introduced at the centre of the cluster the energy levels of a_1- and t_2-symmetry are affected and new levels may appear inside the gap. These we might expect to identify with localized levels in the infinite solid, and in particular to compare with the a_1 and t_2

† See Messmer and Watkins (1971, 1973*a*) for the vacancy in diamond and Larkins (1971*d*) for vacancies in diamond and in Si. For the diamond interstitial see Watkins, Messmer, Weigel, Peak, and Corbett (1971). For nitrogen impurities in diamond see Messmer and Watkins (1970) and Larkins (1971*c*).

orbitals of the defect-molecule method. However, the results obtained are sensitive to cluster size, boundary conditions, and lattice relaxation (Messmer and Watkins 1971; Larkins 1971*c,d*) and the possibilities range from localized a_1 and t_2 orbitals to no localized states at all.

(3) Previous work on the infinite solid by other one-electron methods also indicates a considerable sensitivity to the assumptions made. Thus Benneman's (1967) calculation, which gave four localized levels in diamond, contrasts with the pseudopotential Slater–Koster calculation of Callaway and Hughes (1967) which gave no localized state (unless the perturbing potential was arbitrarily increased over the pseudopotential value). Similarly Rouhani, Lannoo, and Lenglart (1971) reported that they obtained a localized a_1 and a localized t_2 orbital in diamond but only resonances in Si.

(4) The assumption of Hückel theory that the energy can be taken simply as the sum of the one-electron energies of the occupied orbitals permits the economic calculation of atomic relaxation within clusters. The uses of the method in this way for defects in diamond have thrown up interesting heuristic ideas on defect structure; for example the interstitial configuration, which has previously been assumed to be an atom in either a tetrahedral or hexagonal interstice, is suggested by these calculations to be either a dumbbell or a bond-centre configuration. However, it would be naive to accept the Huckle energy function as providing more than an initial and rough guide to the possible stable configurations.

In concluding this section on defects in covalent solids we note that ideas based on chemical intuition and evaluated by methods often directly taken over from chemical theory have provided useful and suggestive results but that these predictions are not yet as reliable as those in ionic crystals. This is to be expected partly because in the covalent solid one is dealing generally with a centre containing several defect electrons and partly because in an ionic crystal by contrast the strong electrostatic forces have a constraining effect on the defect electrons. Existing theories of covalent solids are clearly incomplete but it seems likely that the present indications of the importance of interelectronic interactions and of lattice distortion in determining defect states and levels will also be found when a more complete theory is developed.

4.4. Conclusion

In this chapter we have discussed rather selectively a number of topics in the theory of point defects in crystalline solids. To be comprehensive would be impossible in the space available and we do not claim to have done more than touched on some of the important ideas and results in this diverse and flourishing field. Certainly our illustrations have been limited only to the simplest defect structures. Neither have we been able to look at the usefulness of the knowledge described here, e.g. in understanding radiation damage processes, nor to see how it is used in identifying defect structures experimentally. Our object instead as part of this volume has been to bring out the analogies and interplay between the ideas of theoretical chemistry and the theory of defects in solid-state physics. We hope to have established how close the interaction often is, the very simplest ideas of molecular levels and symmetries often giving clear guidance towards the identification of defects. The solid-state physicists' 'molecule' however is always firmly embedded in a crystal structure and this almost always exerts a strong quantitative influence on its properties, even allowing the observation of 'molecules' which have no analogue in the free state (e.g. the R- or F_3-centre). In detailed quantitative calculations the necessity to evaluate the effects of lattice structure and its relaxation upon electronic levels adds another dimension of complexity to the theoretician's task. It is not surprising, therefore that the sophistication of defect calculations is often not comparable with that of *a priori* molecular calculations. However as in that field general methods and computer programs are emerging which will permit these calculations to be carried out much more widely.

4.5. Bibliography

For those interested in following the subject further we have included the following bibliography in addition to the specific references cited in the text.

An introduction to defect solid state physics covering basic aspects of the study is provided by the recent book:

(1) *Defects in crystalline solids* by B. Henderson (Arnold, London 1972).

As already mentioned in the text a comprehensive treatment from the solid-state viewpoint is provided by

(2) *Point defects and diffusion by* C. P. Flynn (Clarendon Press, Oxford 1972).

For works emphasising colour centres we refer to:

(3) *Colour centres in solids* by J. H. Schulman and W. D. Compton (Pergamon Press, Oxford 1962).

(4) *Physics of color centres*, ed. W. B. Fowler (Academic Press, New York 1968), (for alkali halides).

(5) *Crystals with the fluorite structure*. Clarendon Press, Oxford, (ed. W. Hayes) to be published.

(6) 'Colour Centres in Simple Oxides' by A. E. Hughes and B. Henderson in *Point defects in solids*, eds. J. W. Crawford and L. Slifkin (Plenum, New York, 1972).

(7) The proceedings of the Albany N.Y. and Reading Conferences on defects in semiconductors published as (i) *Radiation Effects in Semiconductors*, eds. J. W. Corbett and G. D. Watkins (Gordon and Breach, London 1971); and (ii) *Defects in Semiconductors*, ed. J. H. Whitehouse (Institute of Physics, London 1973), (for covalent materials).

REFERENCES

ABRAHAMSON, A. A. (1964). *Phys. Rev.* **133**, *A*990.

—— (1969). *Phys. Rev.* **178**, 76.

AEGERTER, M. A. and LÜTY, F. (1971). *Phys. Status Solidi.* **43**, 245.

ALIG, R. C. (1970). *Phys. Rev. B***2**, 2108.

AXILROD, B. M. (1951). *J. chem. Phys.* **19**, 719.

AXILROD, B. M. and TELLER, E. (1943). *J. chem. Phys.* **11**, 299.

BAGUS, P. S. (1965). *Phys. Rev.* **139**, *A*619.

BALDACCHINI, G. and MOLLENAUER, L. F. (1974). *J. Phys.* (*Paris*) **34**, Colloque C-9, 141.

BALDWIN, J. A. (1963). *Phys. Rev. Letts.* **10**, 220.

BARKER, J. A., FISHER, R. A., and WATTS, R. O. (1971). *Mol. Phys.* **21**, 657.

BARR, L. W. and LIDIARD, A. B. (1970). In *Physical chemistry—an advanced treatise*. vol. X, p. 151. Academic Press, New York.

BARTRAM, R. H., STONEHAM, A. M., and GASH, P. (1968). *Phys. Rev.* **176**, 1014.

BASSANI, F. (1971). In *Theory of imperfect crystalline solids* p. 265. International Atomic Energy Agency, Vienna.

BENNETT, H. (1968). *Phys. Rev.* **132**, 215.

—— (1969). *Phys. Rev.* **169**, 729.

—— (1971). *Phys. Rev. B*3, 2763.

—— and LIDIARD, A. B. (1965). *Phys. Letts.* **18**, 253.

BENNEMAN, K. (1967). In *Calculation of the properties of vacancies and interstitials* (ed. A. D. Franklin). National Bureau of Standards Miscellaneous Publication No. 287, p. 127.

BEREZIN, A. A. (1968). *Fiz. tverd. Tela* **10**, 2882 (1969 English version *Sov. Phys.: Sol. State* **10**, 2880).

—— (1972). *Phys. Status Solidi.* **49**, 51.

BERNE, A., BOATO, G., and de PAZ, M. (1966). *Nuovo Cim.* **46***B*, 182.

BILZ, H. (1972). In *Computational solid state physics* (ed. F. Herman, N. W. Dalton, and T. R. Koehler), p. 309. Plenum Press, New York.

BORN, M. and HUANG, K. (1954). *Dynamical theory of crystal lattices*, Clarendon Press, Oxford.

—— and MAYER, J. E. (1932). *Z. Phys.* **75**, 1.

BURTON, J. J. (1969). *Phys. Rev.* **182**, 885.

—— and JURA, G. (1966). *J. Phys. Chem. Solids* **27**, 961.

CALLAWAY, J. and HUGHES, A. J. (1967). *Phys. Rev.* **156**, 860.

CATLOW, C. R. A. and HAYNS, M. R. (1972). *J. Phys. C* **5**, L 237.

—— and NORGETT, M. J. (1973). *J. Phys. C* **6**, 1325.

CAVENETT, B. C., HAYES, W., HUNTER, I. C., and STONEHAM, A. M. (1969). *Proc. R. Soc. A* **309**, 53.

CHADWICK, A. V. and MORRISON, J. A. (1970). *Phys. Rev. B* **1**, 2748.
CLARK, C. D. and MITCHELL, E. W. J. (1971). In *Radiation effects in semiconductors* (ed. J. W. Corbett and G. D. Watkins), p. 257. Gordon and Breach, London.
COCHRAN, W. (1971). *CRC Crit. rev. solid st. sci.* **2**, 1.
COHEN, M. H. and HEINE, V. (1961). *Phys. Rev.* **122**, 1821.
CORISH, J. and JACOBS, P. W. M. (1973). In *Surface and defect properties of solids* **2**, p. 160. Chemical Society Specialist Periodical Report.
COULSON, C. A. (1961). *Valence*, (2nd edn). Clarendon Press, Oxford.
—— and KEARSLEY, M. J. (1957). *Proc. R. Soc. A* **241**, 433.
—— and LARKINS, F. P. (1971). *J. Phys. Chem. Solids* **32**, 2245.
DAS, T. P., JETTE, A. N., and KNOX, R. S. (1964). *Phys. Rev.* **134**, *A*1079.
DAWSON, R. K. and POOLEY, D. (1969). *Phys. Status solidi* **35**, 95.
DIENES, G. J., HATCHER, R. D., and SMOLUCHOWSKI, R. (1967). *Phys. Rev.* **157**, 692.
DRUGER, S. D. (1971). *Phys. Rev. B* **3**, 1391.
DYMOND, J. H. (1968). *J. chem. Phys.* **49**, 3673.
ELCOMBE, M. E. and PRYOR, A. W. (1970). *J. Phys. C.* **3**, 492.
ENGLERT, A., TOMPA, H., and BULLOUGH, R. (1970). In *Fundamental aspects of dislocation theory* (ed. J. A. Simmons, R. de Wit, and R. Bullough). National Bureau of Standards Special Publication No. 317, vol. 1, p. 273.
EVARESTOV, R. A. (1964). *Optika Spektroskp.* **16**, 361 (1964 English version: *Opt. Spectrosp.* (*USSR*) **16**, 198).
FAUX, I. D. (1971*a*). *Ph.D. Thesis*. University of London.
—— (1971*b*). *J. Phys. C* **4**, L 211.
—— and LIDIARD, A. B. (1971). *Z. Naturf.* **26a**, 62.
FELIX, F. and MEIER, K. (1969). *Phys. Status solidi* **32**, *K*139.
FLETCHER, R. (1972). *Comput. Phys. Commun.* **3**, 159.
—— and POWELL, M. J. D. (1963). *Comput. J.* **6**, 163.
—— and REEVES, C. M. (1964). *Comput. J.* **7**, 149.
FLOCKEN, J. W. and HARDY, J. R. (1970). *Phys. Rev. B* **1**, 2447.
FLYNN, C. P. (1972). *Point defects and diffusion*. Clarendon Press, Oxford.
FRIEDEL, J., LANNOO, M., and LEMAN, G. (1967). *Phys. Rev.* **164**, 1056.
FOWLER, W. B. (1968). In *Physics of color centers* (ed. W. B. Fowler), p. 54. Academic Press, New York.
GILBERT, T. L. and WAHL, A. C. (1967). *J. Chem. Phys.* **47**, 3425.
—— —— (1971). *J. Chem. Phys.* **55**, 5247.
GLYDE, H. R. and VENABLES, J. A. (1968). *J. Phys. Chem. Solids* **29**, 1093.
GOURARY, B. S. and ADRIAN, F. J. (1957). *Phys. Rev.* **105**, 1180.
—— —— (1960). *Solid St. Phys.* **10**, 127.
GOURARY, B. S. and FEIN, A. E. (1962). *J. appl. Phys. Suppl.* **33**, 331.
HAM, F. S. (1972). *Phys. Rev. Lett.* **28**, 1048.
HARDY, J. R. (1962). *Phil. Mag.* **7**, 315.
—— and LIDIARD, A. B. (1967). *Phil. Mag.* **15**, 825: for corrections to the part of this paper dealing with the elastic strain field around a defect see Faux and Lidiard (1971).
HATCHER, R. D. and DIENES, G. J. (1961). *Phys. Rev.* **124**, 726.
—— —— (1964). *Phys. Rev.* **134**, *A*214.
HAYES, W. and STONEHAM, A. M. (1974). In *Crystals with the fluorite structure*, ch. 4, (ed. W. Hayes). Clarendon Press, Oxford.

HEMSTREET, L. A., FONG, C. Y., and COHEN, M. L. (1970). *Phys. Rev. B***2**, 2054.
HENDERSON, B. (1966). *Brit. J. Appl. Phys.* **17**, 851.
HENRY, R. and NORBERG, E. A. (1972). *Phys. Rev. B* **6**, 16 45.
—— —— (1972). *Phys. Rev. B* **6**, 1645.
HERMAN, R. C., WALLIS, M. C., and WALLIS, R. F. (1956). *Phys. Rev.* **103**, 87.
HERRING, C. and FLICKER, M. (1964). *Phys. Rev.* **134**, *A*362.
HORTON, G. K. (1968). *Am. J. Phys.* **36**, 93.
HUGHES, A. E. and HENDERSON, B. (1972). In *Point Defects in Solids* (ed. J. W. Crawford and L. M. Slifkin), p. 381. Plenum Press, New York.
ITOH, N. (1972). *Crystal Lattice Defects* **3**, 115.
JANSEN, L. (1964). *Phys. Rev.* **135**, *A*1292.
JETTE, A. N. and DAS, T. P. (1969). *Phys. Rev.* **186**, 919.
——, GILBERT, T. L., and DAS, T. P. (1969). *Phys. Rev.* **184**, 884.
KANZAKI, H. (1957). *J. Phys. Chem. Solids* **2**, 24, 37.
KARO, A. M. and HARDY, J. R. (1963). *Phys. Rev.* **129**, 2024.
—— —— (1968). *J. Chem. Phys.* **48**, 3795.
—— —— (1971). *Phys. Rev. B* **3**, 3418.
KEELER, G. J. and BATCHELDER, D. N. (1970). *J. Phys. C* **3**, 510.
KEETON, S. C. and WILSON, W. D. (1973). *Phys. Rev. B* **7**, 834.
KEMP, J. C. and NEELEY, V. I. (1963). *Phys. Rev.* **132**, 215.
KERN, R. C. and BARTRAM, R. H. (1971). Paper presented to the International Conference on Colour Centres in Ionic Crystals held at Reading in September 1971: to be published.
KITTEL, C. (1971). *Introduction to Solid State Physics* (4th edn.) especially ch. 19. Wiley, New York.
KNOX, R. S. and TEEGARDEN, K. J. (1968). In *Physics of color centers* (ed. W. B. Fowler), p. 1. Academic Press, New York.
KOHN, W. and LUTTINGER, J. M. (1955). *Phys. Rev.* **98**, 915.
KOSTER, G. F. and SLATER, J. C. (1954). *Phys. Rev.* **96**, 439.
LANNOO, M. and STONEHAM, A. M. (1968). *J. Phys. Chem. Solids* **29**, 1987.
LARKINS, F. P. (1971*a*). *J. Phys. Chem. Solids* **32**, 2123.
—— (1971*b*). *J. Phys. Chem. Solids* **32**, 965.
—— (1971*c*). *J. Phys. C.* **4**, 3065.
—— (1971*d*). *J. Phys. C* **4**, 3077.
—— and STONEHAM, A. M. (1971*a*). *J. Phys. C* **4**, 143.
—— —— (1971*b*). *J. Phys. C* **4**, 154.
LAUGHLIN, C. (1965). *Solid St. Commun.* **3**, 55.
LOMBARDI, E. and JANSEN, L. (1968). *Phys. Rev.* **167**, 822.
LOSEE, D. L. and SIMMONS, R. O. (1968). *Phys. Rev.* **172**, 944.
LÖWDIN, P. O. (1956). *Adv. Phys.* **5**, 3.
LUNDQVIST, S. (1957). *Ark. Fysik* **12**, 263.
LÜTY, F. (1968). In *Physics of color centers* (ed. W. B. Fowler), p. 181. Academic Press, New York.
MCCALL, R. T. and GROSSWEINER, L. I. (1967). *J. appl. Phys.* **38**, 284.
MACDONALD, R. F. (1972). *J. Phys. F* **2**, 209.
MCMURRY, H. L., SOLBRIG, A. W., BOYTER, J. K., and NOBLE, C. (1967). *J. Phys. Chem. Solids* **28**, 2359.

MARKHAM, J. J. (1966). *The F-center in Alkali Halides*. Academic Press, New York.
MATTHEW, J. A. D. and GREEN, B. (1971). *J. Phys. C* **4**, *L*101.
MESSMER, R. P. and WATKINS, G. D. (1970). *Phys. Rev. Letts* **25**, 656.
—— —— (1971). In *Radiation effects in semiconductors* (ed. J. W. Corbett and G. D. Watkins), p. 23. Gordon and Breach, London.
—— —— (1973a). *Phys. Rev. B* **7**, 2568.
—— —— (1973b). In *defects in semiconductors* (ed. J. H. Whitehouse) p 228. Institute of Physics, London.
MEYER, A. and WOOD, R. F. (1964). *Phys. Rev.* **133**, *A*1436.
MOLLENAUER, L. F. and BALDACCHINI, G. (1972). *Phys. Rev. Lett.* **29**, 465.
MOTT, N. F. and GURNEY, R. W. (1940). *Electronic processes in ionic crystals*. Clarendon Press, Oxford.
—— and LITTLETON, M. J. (1938). *Trans. Faraday Soc.* **34**, 485.
MÜLLER, M. and NORGETT, M. J. (1972). *J. Phys. C* **5**, *L*256.
—— —— (1973). *Z. Naturf.* **28**a, 311.
NARDELLI, G. F. and REPANAI-CHIAROTTI, A. (1960). *Nuovo Cim.* **18**, 1053.
NORGETT, M. J. (1971*a*). *J. Phys. C* **4**, 298.
—— (1971*b*). *J. Phys. C* **4**, 1284.
—— (1971*c*). *J. Phys. C* **4**, 1289.
—— and FLETCHER, R. (1970). *J. Phys. C* **3**, *L*190.
—— and LIDIARD, A. B. (1968). *Phil. Mag.* **18**, 1193.
—— and STONEHAM, A. M. (1973*a*). *J. Phys. C* **6**, 223.
—— —— (1973*b*). *J. Phys. C* **6**, 229.
——, PERRIN, R. C., and SAVINO, E. J. (1972). *J. Phys. F* **2**, *L*73.
OKAMOTO, F. (1961). *Phys. Rev.* **124**, 1090.
ONG, C. K. and VAIL, J. M. (1973). *Phys. Rev. B* **8**, 1636.
ÖPIK, U. and PRYCE, M. H. L. (1957). *Proc. Roy. Soc. A* **238**, 425.
—— and WOOD, R. F. (1969). *Phys. Rev.* **179**, 772.
PAINTER, G. S., ELLIS, D. E., and LUBINSKY, A. R. (1971). *Phys. Rev. B* 4, 3610.
PARKER, E. H. C., GLYDE, H. R., and SMITH, B. L. (1968). *Phys. Rev.* **176**, 1107.
PASKIN, A. and RAHMAN, A. (1966). *Phys. Rev. Letts* **16**, 300.
PAULING, L. (1960). *The Nature of the chemical bond* (*3rd edn.*) Cornell University Press, Ithaca, New York.
PIRENNE, J. and KARTHEUSER, E. (1964). *Physica* **30**, 2005.
POLLACK, G. L. (1964). *Rev. Mod. Phys.* **36**, 748.
POWELL, M. J. D. (1964). *Comput. J.* **7**, 155.
ROUHANI, D., LANNOO, M., and LENGLART, P. (1971). In *Radiation effects in semiconductors* (ed. J. W. Corbett and G. D. Watkins), p. 15. Gordon and Breach, London.
SANGSTER, M. J. (1973). *J. Phys. Chem. Solids.* **34**, 355.
——, PECKHAM, G. E., and SAUNDERSON, D. H. (1970). *J. Phys. C* **3**, 1026.
SASLOW, W., BERGSTRASSER, T. K., and COHEN, M. L. (1966). *Phys. Rev. Letts* **16**, 354.
SAUNDERSON, D. H. and PECKHAM, G. E. (1971). *J. Phys. C* **4**, 2009.
SCHOKNECHT, W. E. (1971). Ph.D. Thesis, University of Illinois, Champaign-Urbana.
—— (1965). *Phys. Rev.* **138**, *A*180.
SCHULZE, P. D. and HARDY, J. R. (1972*a*). *Phys. Rev. B* **5**, 3270.
—— —— (1972*b*). *Phys. Rev. B* **6**, 1580.

SEEGER, A. and SWANSON, M. L. (1968). In *Lattice defects in semiconductors* (ed. R. R. Hasiguti), p. 93. University of Tokyo and Pennsylvania State University Presses.
SILSBEE, R. H. (1965). *Phys. Rev.* **138**, *A*180.
SINCLAIR, J. E., NIEBEL, K. F., and VENABLES, J. A. (1973). A.E.R.E. Report T.P.564; to be published.
SLATER, J. C. (1939). *Introduction to chemical physics.* McGraw Hill, New York.
—— and JOHNSON, K. H. (1972). *Phys. Rev. B* **5**, 844.
SOLBRIG, A. W. (1971). *J. Phys. Chem. Solids* **32**, 1761.
STONEHAM, A. M. (1966). *Proc. Phys. Soc.* **88**, 135.
—— and BARTRAM, R. H. (1970). *Phys. Rev. B* **2**, 3403.
—— and LANNOO, M. (1969). *J. Phys. Chem. Solids* **30**, 1769.
SWALIN, R. A. (1961). *J. Phys. Chem. Solids* **18**, 290.
SWENBERG, C. E. (1967). *Phys. Lett.* **24***A*, 163.
SZIGETI, B. (1950). *Proc. R. Soc. A* **204**, 51.
TANIMOTO, D. H., ZINIKER, W. M., and KEMP, J. C. (1965). *Phys. Rev. Letts.* **14**, 645.
TESSMAN, J. R., KAHN, A. H., and SHOCKLEY, W. (1953). *Phys. Rev.* **92**, 890.
TEWARY, V. K. (1969). A.E.R.E. Report T.P. 388.
—— (1973). *Adv. Phys.* **22**, 757.
—— and Bullough, R. (1971). *J. Phys. F* **1**, 554.
TOSI, M. P. (1964). *Solid St. Phys.* **16**, 1.
WATKINS, G. D., MESSMER, R. P., WEIGEL, C., PEAK, D., and CORBETT, J. W. (1971). *Phys. Rev. Letts* **27**, 1573.
WEDEPOHL, P. T. (1967). *Proc. Phys. Soc.* **92**, 79.
WEISER, K. (1962). *Phys. Rev.* **126**, 1427.
WILSON, W. D. and JOHNSON, R. A. (1972). In *Interatomic Potentials and simulation of lattice defects* (ed. P. C. Gehlen, J. R. Beeler, and R. I. Jaffee), p. 375. Plenum Press, New York.
WOOD, R. F. and JOY, H. W. (1964). *Phys. Rev.* **136**, *A*451.
WOOD, R. F. and MEYER, A. (1964). *Solid St. Commun.* **2**, 225.
WOOD, R. F. and ÖPIK, U. (1969). *Phys. Rev.* **179**, 783.
YAMAGUCHI, T. (1963). *J. Phys. Soc. Japan*, **18**, 368.
YEN, W. L. and NORBERG, E. R. (1963). *Phys. Rev.* **131**, 269.

PART II

ISOLATED MOLECULES AND MOLECULAR CRYSTALS

5

ON THE ORIGIN OF ELECTRONIC PROPERTIES OF MOLECULES

R. MCWEENY

5.1. Introduction

SOME 35 years ago, C. A. Coulson introduced two types of numerical index —the 'charges' and the 'bond orders'—to characterize, using simple molecular orbital (MO) theory, the distribution of the π-electrons in a conjugated molecule. It could hardly have been foreseen at the time that the introduction of such indices was to be the starting point for so many developments in molecular quantum mechanics, particularly those concerned with the physical interpretation of molecular properties, and that many years later† whole conferences would be devoted to the subject of the electron distribution and its description in terms of charge density, spin density, and other experimentally-accessible distribution functions.

This chapter is concerned with the distribution of electrons in a molecule, in fixed nucleus approximation, and its relationship to observable electronic properties. Density matrix theory, referred to earlier in Section 1.1 and in Chapter 3, provides a useful formalism and source of ideas for this purpose; however, the aim will not be to discuss density matrix theory *per se*, but rather to use its concepts as a means of improving our understanding of molecular properties. We wish to do this on the most general level, transcending as far as possible any particular computational approach, and therefore work for most of the time in terms of a general many-electron wavefunction $\Psi(\mathbf{x}_1, \mathbf{x}_2, \ldots, \mathbf{x}_N)$ which may be exact or approximate, and whose precise form will not, for the most part, be considered except as a means of illustrating the theory. The variables $\mathbf{x}_i$ refer to the ith electron, and include $\mathbf{r}_i$ (space) and s_i (spin); thus $\mathbf{x}_i = \mathbf{r}_i,\ s_i$ and similarly a volume element $\mathrm{d}\mathbf{x}_i = \mathrm{d}\mathbf{r}_i\,\mathrm{d}s_i$. We recall that the spin variable s is introduced only for formal convenience. An electron has two spin eigenstates, represented by vectors α and β of a two-dimensional spin space: the orthogonality and unit length properties are

$$\langle\alpha|\alpha\rangle = \langle\beta|\beta\rangle = 1, \qquad \langle\alpha|\beta\rangle = \langle\beta|\alpha\rangle^* = 0$$

but it is often convenient to use a functional notation analogous to that used

† First Sagamore Conference on Charge and Spin Density (1964).

in defining the scalar product of two functions, writing

$$\langle\alpha|\alpha\rangle = \int \alpha^*(s)\alpha(s)\,ds = 1, \qquad \langle\alpha|\beta\rangle = \int \alpha^*(s)\beta(s)\,ds = 0. \tag{5.1}$$

If $\{\phi_i(\mathbf{r})\}$ denotes a complete set of spatial functions, then the set of products $\{\phi_i(\mathbf{r})\alpha(s),\ \phi_i(\mathbf{r})\beta(s)\}$ is complete for all functions of space and spin variables and a general space–spin state of a one-electron system is represented by the two-component function

$$\psi(\mathbf{r}, s) = \phi_\alpha(\mathbf{r})\alpha(s) + \phi_\beta(\mathbf{r})\beta(s) \tag{5.2}$$

which is a *general spin-orbital*. Space and spin variables are then handled in a formally identical manner; for example, using (5.1),

$$\langle\psi|\psi\rangle = \int \psi^*(\mathbf{r}, s)\psi(\mathbf{r}, s)\,d\mathbf{r}\,ds = \int \phi_\alpha^*(\mathbf{r})\phi_\alpha(\mathbf{r})\,d\mathbf{r} + \int \phi_\beta^*(\mathbf{r})\phi_\beta(\mathbf{r})\,d\mathbf{r}$$

and 'integration' over spin conveniently introduces the summation over discrete components. The statistical interpretation is most direct if s is identified with the spin z component (S_z), in which case $\alpha(s)$ and $\beta(s)$ are delta functions at $s = \pm\frac{1}{2}$ respectively; (5.2) may then be consistently interpreted in the usual way

$$|\psi(\mathbf{r}, s)|^2\,d\mathbf{r}\,ds = \begin{cases} \text{probability of finding the electron} \\ \text{with coordinates in the range } (\mathbf{r}, \mathbf{r}+d\mathbf{r}) \\ \text{and spin in the range } (s, s+ds). \end{cases} \tag{5.3}$$

The probability of finding the electron in $d\mathbf{r}$ with spin $s \sim \frac{1}{2}$ is then clearly $|\phi_\alpha(\mathbf{r})|^2\,d\mathbf{r}$ (integration around $s = \frac{1}{2}$ giving unity by (5.1)), and the factors ϕ_α, ϕ_β therefore determine individually the probability densities for 'up-spin' and 'down-spin' electrons. This statistical picture is developed extensively in Section 5.4.

For many-electron systems it is customary to build up the wavefunction Ψ from antisymmetrized products (Slater determinants) of *pure* spin-orbitals, such as

$$\Phi_\kappa(\mathbf{x}_1, \mathbf{x}_2, \dots, \mathbf{x}_N) = (N!)^{-\frac{1}{2}}|\psi_R(\mathbf{x}_1)\psi_S(\mathbf{x}_2)\dots\psi_U(\mathbf{x}_N)| \tag{5.4}$$

where $\psi_R(\mathbf{x})$ is any member of the set $\phi_1(\mathbf{r})\alpha(s)$, $\phi_1(\mathbf{r})\beta(s)$, $\phi_2(\mathbf{r})\alpha(s)$, ... and κ indicates a particular selection of spin-orbitals $\psi_R, \psi_S, \dots, \psi_U$. The set of all possible antisymmetrized products is in principle complete for many-electron functions, in the sense that

$$\Psi(\mathbf{x}_1, \mathbf{x}_2, \dots, \mathbf{x}_N) = \sum_\kappa c_\kappa \Phi_\kappa(\mathbf{x}_1, \mathbf{x}_2, \dots, \mathbf{x}_N) \tag{5.5}$$

with arbitrarily high accuracy if sufficient terms are taken. However, such expansions of the wavefunction are invariably severely truncated for practical reasons, and *ab initio* calculations of properties may then be seriously in error. It is for such reasons that we frequently wish to shift attention away from the precise calculation of numerical values and towards the main qualitative features of the electron distribution: if we cannot hope to calculate some physical property accurately, at least we may hope to obtain some qualitative picture of its origin in terms of more primitive concepts.

5.2. Charges and bond orders

First we recall the definition of charges and bond orders (Coulson 1939) in the original context of π-electron theory. The (ground state) wavefunction describing the π-electrons of a conjugated molecule may be written, assuming an even number of electrons,

$$\Psi(\mathbf{x}_1, \mathbf{x}_2, \ldots, \mathbf{x}_N) = (N!)^{-\frac{1}{2}}|\psi_A(\mathbf{x}_1)\bar{\psi}_A(\mathbf{x}_2) \ldots \bar{\psi}_U(\mathbf{x}_N)| \quad (5.6)$$

where it is convenient to use ψ_R and $\bar{\psi}_R$ for α- and β-type spin-orbitals:

$$\psi_R(\mathbf{x}) = \phi_R(\mathbf{r})\alpha(s), \qquad \bar{\psi}_R(\mathbf{x}) = \phi_R(\mathbf{r})\beta(s). \quad (5.7)$$

In principle (5.6) may be a Hartree–Fock function for the 'closed-shell' of doubly-occupied molecular orbitals (MOs): in practice, each MO is approximated as a linear combination of atomic orbitals (the LCAO–MO approximation) in the form

$$\phi_R = \sum_i c_{Ri}\chi_i \qquad (R = A, B, \ldots, U) \quad (5.8)$$

where χ_i is a π-type 2p orbital on centre i. Since $|\phi_R(\mathbf{r})|^2$ indicates the probability per unit volume of finding an electron in ϕ_R at point $\mathbf{r}$, which may be pictured as a 'charge density' (i.e. the average electronic content per unit volume), it is clear that the sum over all doubly occupied orbitals

$$P(\mathbf{r}) = 2 \sum_{R(\mathrm{occ})} |\phi_R(\mathbf{r})|^2 \quad (5.9)$$

will have an interesting physical significance as the total charge density contributed by the π-electrons. In some ways this quantity is more fundamental than the MOs themselves since it is an *invariant*; the MOs in (5.6) may be replaced by any new set of orthonormal linear combinations $(\phi'_A, \phi'_B, \ldots, \phi'_U)$ without changing the wavefunction Ψ and therefore have no intrinsic physical significance, but the charge density is unaffected by such a change of description and has a more direct connection with physical observables. The definition of P is not limited by the one-determinant form of wavefunction (5.6), and its wider significance is explored in later sections.

With the LCAO-type approximation (5.8), the charge density function (5.9) at once reduces to

$$P(\mathbf{r}) = \sum_{i,j} P_{ij}\chi_i(\mathbf{r})\chi_j^*(\mathbf{r}) \quad (5.10)$$

in which the coefficients are defined by

$$P_{ii} = 2 \sum_{R(\text{occ})} |c_{Ri}|^2, \qquad P_{ij} = 2 \sum_{R(\text{occ})} c_{Ri}c^*_{Rj}. \tag{5.11}$$

Since P_{ii} is the 'weight factor' with which the 'orbital density' $|\chi_i(\mathbf{r})|^2$ appears in the charge density it indicates the amount of charge associated with centre i; while P_{ij} similarly indicates the importance of the 'overlap density' $\chi_i(\mathbf{r})\chi_j^*(\mathbf{r})$. For an important class of conjugated hydrocarbons, using the simple form of MO theory due to Hückel, it may be shown (Coulson and Rushbrooke 1940) that $P_{ii} = 1$ for every conjugated centre, showing that each atom formally retains the one π-electron which it contributes to the π-electron distribution, i.e. no appreciable shifts of charge occur. On the other hand, P_{ij} is an important indicator of the degree of 'double bonding' in link i–j: in the simple MO description of the ethylene molecule (with *one* π-bond) this quantity takes the value unity, while in other situations a good correlation is found between values of P_{ij} and other measures of double-bond character (e.g. the shortening of the C—C single bond by the presence of π-bonding). In later sections we shall find there are good physical reasons for the validity of such correlations. A rather complete theory of the properties of conjugated molecules was subsequently developed by Coulson and Longuet-Higgins (1949) and this was the starting point for many further developments and applications.

Some years after its introduction, the charge-and-bond-order concept was re-examined by various authors. Löwdin (1950) pointed out the implications of the orthogonality assumption $\langle\chi_i|\chi_j\rangle \simeq 0 (i \neq j)$, which is not well satisfied by the AOs on neighbouring atoms, and showed that the formal features of simple MO theory could be retained provided the χs were reinterpreted as orthogonal linear combinations of the original AOs. Inclusion of overlap has, however, a profound effect on the interpretation of the coefficients in (5.10) and the resolution of the charge density into orbital and overlap terms then requires more careful consideration: if (McWeeny 1951) we introduce *normalized* densities, putting $\langle\chi_i|\chi_j\rangle = S_{ij}$ (overlap integral),

$$d_i(\mathbf{r}) = |\chi_i(\mathbf{r})|^2, \qquad d_{ij}(\mathbf{r}) = \chi_i(\mathbf{r})\chi_j^*(\mathbf{r})/S_{ij} \tag{5.12}$$

and assume for simplicity real functions, then (5.10) may be written

$$P(\mathbf{r}) = \sum_i q_i d_i(\mathbf{r}) + \sum_{i<j} q_{ij}\, d_{ij}(\mathbf{r}) \tag{5.13}$$

where

$$q_i = P_{ii}, \qquad q_{ij} = 2S_{ij}P_{ij} \qquad (i \neq j). \tag{5.14}$$

These quantities, first referred to as 'atom and bond charges', indicate numerically the electronic content of the regions of space associated with the density functions (5.12); for integration of (5.13) must yield the total charge in the distribution (i.e. N electrons) of which q_i comes from the term d_i and q_{ij} from d_{ij}. There is thus a conservation equation

$$\sum_i q_i + \sum_{i<j} q_{ij} = N. \tag{5.15}$$

This formal resolution of the electron density into orbital and overlap contributions extends far beyond the context in which it was first introduced and has many applications. For example (McWeeny 1952–4) by summing the orbital charges on any atom and adding half of every overlap charge connecting it with neighbouring atoms, it is possible to define the 'gross electronic charge' formally associated with the atom in its environment, and to compute properties of the *bonded* atoms in a molecule or solid. Such charges are nowadays referred to as 'populations' and 'electron population analysis' (Mulliken 1955) is widely employed in the interpretation of electronic structures computed from elaborate non-empirical wavefunctions.

Before turning to the more general analysis of the electron distribution, it is worth emphasizing the intimate connection between the elements P_{ij} and the total electronic energy of the system E. In Hückel theory, electron interactions are assumed to be absorbed into an effective or 'model' Hamiltonian h, such that the contribution to the energy arising from an electron in MO ϕ_R is $\varepsilon_R = \langle\phi_R|h|\phi_R\rangle$; the sum of the orbital energies then yields (using (5.8) and the definitions (5.11))

$$E = \sum_{i,j} P_{ij}\langle\chi_j|h|\chi_i\rangle = tr\mathbf{Ph} \tag{5.16}$$

where $\mathbf{h}$ is the matrix of the Hamiltonian h with respect to the AOs $\{\chi_i\}$. The charges and bond orders are thus the coefficients of the diagonal and off-diagonal elements, respectively, of the matrix of the Hamiltonian, quantities known in this context as the 'coulomb' and 'resonance' (or 'bond') integrals, and usually denoted by α_i and β_{ij}. Thus (5.16) may be written†

$$E = \sum_i q_i\alpha_i + 2\sum_{i<j} P_{ij}\beta_{ij} \tag{5.17}$$

and the charges and bond-orders acquire a new significance: they determine the first order response of the system to changes in its coulomb and bond integrals, respectively, in the sense

$$q_i = P_{ii} = \left(\frac{\partial E}{\partial \alpha_i}\right), \qquad P_{ij} = \frac{1}{2}\left(\frac{\partial E}{\partial \beta_{ij}}\right). \tag{5.18}$$

These quantities play a leading role in theories of chemical reactivity, when the variations $\delta\alpha_i$, $\delta\beta_{ij}$ may arise from the approach of charged reactants, the twisting of bonds, etc.; but for present purposes their main interest lies

† Note that, with the assumption of real functions, $\beta_{ji} = \beta_{ij}$.

in their capacity for generalization and the richness of the concepts to which they lead. Before making these generalizations it is useful to consider in some detail the nature of the Hamiltonian (always in fixed-nucleus approximation) for a general many-electron molecule.

5.3. The Hamiltonian

The usual non-relativistic Hamiltonian may be written in the form

$$H = \sum_{i=1}^{N} h(i) + \tfrac{1}{2} \sum_{i,j=1}^{N}{}' g(i,j), \tag{5.19}$$

where the prime on the second sum denotes omission of the term with $i = j$ and†

$$h(i) = \frac{1}{2m} p_i^2 + V(i) \tag{5.20}$$

$$g(i,j) = e^2/\kappa_0 r_{ij}. \tag{5.21}$$

Here, $\mathbf{p}_i$ is the momentum operator and $V(i)$ the potential energy of electron i in the field of the fixed nuclei. Thus, $p_i^2 = \mathbf{p}_i^2 = p_{xi}^2 + p_{yi}^2 + p_{zi}^2 = -\hbar^2\nabla^2(i)$.

In considering spin- and field-dependent properties it is necessary to add the 'small terms' to the Hamiltonian; these arise in principle from the relativistic generalization of the Schrödinger equation, with external fields included. For one electron the correct equation is the Dirac equation, which involves a four-component wavefunction; two components are small, however, and development in powers of the fine structure constant ($\alpha = 1/137{\cdot}030$) permits their elimination in terms of the two large components, (cf. the discussion in Section 1.2) yielding the equation

$$h\psi = \mathrm{i}\hbar(\partial\psi/\partial t), \tag{5.22}$$

in which the effective Hamiltonian (Dirac–Pauli), up to terms in α^2 and omitting the rest energy mc^2, is

$$h = \left[\frac{1}{2m}\pi^2 + V\right] + \left[-\frac{\pi^4}{8m^3c^2} + \frac{e\hbar}{m}\mathbf{B}\,.\,\mathbf{S} + \frac{e\hbar}{4m^2c^2}(\mathbf{S}\,.\,\mathbf{F}\times\boldsymbol{\pi} - \mathbf{S}\,.\,\boldsymbol{\pi}\times\mathbf{F}) + \frac{e\hbar^2}{8m^2c^2}\operatorname{div}\mathbf{F}\right], \tag{5.23}$$

and the wavefunction has only two components:

$$\psi = \phi_\alpha\alpha + \phi_\beta\beta. \tag{5.24}$$

† In this chapter we use SI units throughout, with $\kappa_0 = 4\pi\varepsilon_0$ (ε_0 being the permittivity of free space). However, F is used for electric field intensity (E for energy).

The α and β in (5.24), prefixed by spatial functions ϕ_α and ϕ_β, correspond exactly to the spin states ($m_s = \pm\frac{1}{2}$) in Pauli's phenomenological introduction of spin, and the vector operator **S** in (5.23) has components S_x, S_y, S_z with properties exactly the same as those of the Pauli operators†

$$\begin{array}{lll} S_x\alpha = \frac{1}{2}\beta & S_y\alpha = \frac{1}{2}\mathrm{i}\beta & S_z\alpha = \frac{1}{2}\alpha \\ S_x\beta = \frac{1}{2}\alpha & S_y\beta = -\frac{1}{2}\mathrm{i}\beta & S_z\beta = -\frac{1}{2}\beta. \end{array} \tag{5.25}$$

The other quantities in (5.23) are the flux density **B** of the applied magnetic field, the electric field intensity **F**, and the 'gauge invariant' momentum operator

$$\boldsymbol{\pi} = \mathbf{p} + e\mathbf{A} \tag{5.26}$$

where **A** is the magnetic vector potential associated with the field **B**. The expectation value of π gives in fact the mass times the expectation velocity and the presence of the term $e\mathbf{A}$ ensures that a change in **A** (which is not uniquely defined for a given field) does not affect computed expectation values. The fields and the electric and magnetic (vector) potentials (ϕ and **A**) are related by

$$F_x = -\frac{\partial\phi}{\partial x} = e^{-1}\frac{\partial V}{\partial x}, \qquad B_x = \frac{\partial}{\partial y}A_z - \frac{\partial}{\partial z}A_y, \tag{5.27}$$

with similar equations for the other components. With the definitions (5.23) *et seq.*, eqn (5.22), and its stationary state counterpart

$$h\psi = \varepsilon\psi, \tag{5.28}$$

are the usual Schrodinger equations, as modified by Pauli to include relativistic effects to order α^2.

For two interacting particles the appropriate relativistic generalization is the Breit equation. This is a 16-component equation (with solutions expressible in terms of products of four-component Dirac functions), and elimination of the small components then leads to a Breit–Pauli equation whose status is similar to that of (5.22) and (5.23). The solution is expressible in terms of products of two 2-component functions of the form (5.24) or, to use the usual terminology, products of *spin-orbitals*. The status of the Breit equation is less secure than that of the Dirac equation, and the generalization from two to N particles even less so. The general N-electron Hamiltonian is taken to be of the form (5.19) with $h(i)$ defined by (5.23) (one term for each electron i) but the fields now including those due to other electrons. We shall retain $V(i)$ for the potential energy of electron i in the field of fixed nuclei, etc., and include interactions in $g(i,j)$; additional interactions,

† It has been convenient to remove a factor $\hbar$ so that the operators are dimensionless: thus $\hbar S_x, \hbar S_y, \hbar S_z$ have the dimensions of angular momentum.

involving the *spins* of pairs of particles, are taken from the Breit–Pauli (two-particle) equation, with the assumption of pairwise additivity. Some of the most important terms are included in the $h(i)$ and $g(i,j)$ given below, in which $\mathbf{r}_{ij}$ denotes the vector $i \rightarrow j$;

$$h(i) = \frac{1}{2m}\pi^2(i) + V(i) + g\beta\mathbf{B}\,.\,\mathbf{S} + \frac{g\beta}{4mc^2}[\mathbf{S}(i)\,.\,\mathbf{F}(i)\times\boldsymbol{\pi}(i) - \mathbf{S}(i)\,.\,\boldsymbol{\pi}(i)\times\mathbf{F}(i)] \tag{5.29}$$

$$g(i,j) = \frac{e^2}{\kappa_0 r_{ij}} - \frac{g\beta^2}{\kappa_0 c^2\hbar}\left(\frac{[2\mathbf{S}(i)\,.\,\mathbf{r}_{ij}\times\boldsymbol{\pi}(j) + \mathbf{S}(j)\,.\,\mathbf{r}_{ij}\times\boldsymbol{\pi}(j)]}{r_{ij}^3}\right) - \frac{g^2\beta^2}{\kappa_0 c^2}\left(\frac{[3(\mathbf{S}(j)\,.\,\mathbf{r}_{ij})(\mathbf{S}(i)\,.\,\mathbf{r}_{ij}) - r_{ij}^2\mathbf{S}(i)\,.\,\mathbf{S}(j)]}{r_{ij}^5} + \frac{8\pi}{3}\delta(\mathbf{r}_{ij})\mathbf{S}(i)\,.\,\mathbf{S}(j)\right). \tag{5.30}$$

The first two terms in (5.29) with $\boldsymbol{\pi}$ defined by (3.7) and $\mathbf{A}$ referring to the externally applied magnetic field, represent the generalization of (5.20) to admit the presence of the magnetic field; the third is the spin-only part of the Zeeman term, with $\beta = e\hbar/2m$ (the Bohr magneton) and $g = 2{\cdot}0030$;† and the last describes the spin–orbit interaction. $\mathbf{F}(i)$ is here 'external' to the electronic system, including the field at electron i due to the nuclei—but not that of other electrons. The first term in (5.30) is the usual coulomb interaction; the second describes spin–other-orbit interaction; and the last is a dipole–dipole interaction between the spins of different electrons, corrected by a 'contact' interaction associated with the singularity in the potential.

Additional terms arise when the nuclei are recognized as structured particles, possessing spins, magnetic dipole and electric quadrupole moments. The electron–nuclear spin interactions follow from the Breit equation for two unlike particles, and other terms arise from the modification of the electric and magnetic vector potentials describing the field in which the electrons move, and from the direct interaction between nuclear spins and the applied fields.

When the above terms are summed over all electrons (assuming only pair-wise interactions), and the vector potential terms in $\pi^2(i)$ are separated from the rest, we obtain a many-electron Hamiltonian which may be written

$$H = H_0 + H_{\text{elec}} + H_{\text{mag}} + H_{\text{SL}} + H_{\text{Z}} + H_{\text{SS}} + H_{\text{N}} \tag{5.31}$$

where H_0 is the usual non-relativistic Hamiltonian defined in (5.19), (5.20) and (5.21). The other terms are readily obtained on separating the fields

† In the Dirac–Pauli Hamiltonian (3.4) $g = 2$; by introducing g explicitly we may allow a small quantum-electrodynamic correction (see, for example, Bethe and Salpeter, 1957).

and potentials into 'internal' and 'external' parts and giving the 'internal' terms the forms appropriate to a set of fixed nuclei. On assuming a uniform external field **B** (with $\mathbf{A} = \frac{1}{2}\mathbf{B}\times\mathbf{r}$) we obtain

$$H_{\text{elec}} = \sum_i V_{\text{ext}}(i) \tag{5.32a}$$

$$H_{\text{mag}} = \beta \sum_i \mathbf{B}\,.\,\mathbf{L}(i) + (e^2/8m)\sum_i (\mathbf{B}\times\mathbf{r}_i)^2 = H_{\text{P}} + H_{\text{D}} \tag{5.32b}$$

$$H_{\text{SL}} = \frac{g\beta^2}{\kappa_0 c^2}\left(\sum_{n,i} Z_n \frac{\mathbf{S}(i)\,.\,\mathbf{M}^n(i)}{r_{ni}^3} - \sum_{i,j} \frac{[\mathbf{S}(i)\,.\,\mathbf{M}^j(i) + 2\mathbf{S}(j)\,.\,\mathbf{M}^j(i)]}{r_{ji}^3}\right)$$
$$+\frac{g\beta}{4mc^2}[\mathbf{S}(i)\,.\,\mathbf{F}(i)\times\boldsymbol{\pi}(j) - \mathbf{S}(i)\,.\,\boldsymbol{\pi}(i)\times\mathbf{F}(i)] \tag{5.32c}$$

$$H_{\text{Z}} = g\beta \sum_i \mathbf{B}\,.\,\mathbf{S}(i) \tag{5.32d}$$

$$H_{\text{SS}} = -\frac{1}{2}\frac{g^2\beta^2}{\kappa_0 c^2}\sum_{i,j}\left(\frac{[3(\mathbf{S}(j)\,.\,\mathbf{r}_{ij})(\mathbf{S}(i)\,.\,\mathbf{r}_{ij}) - r_{ij}^2\mathbf{S}(i)\,.\,\mathbf{S}(j)]}{r_{ij}^5}\right.$$
$$\left.+\frac{8\pi}{3}\delta(\mathbf{r}_{ij})\mathbf{S}(i)\,.\,\mathbf{S}(j)\right) \tag{5.32e}$$

$$H_{\text{N}} = -\sum_n g_n\beta_p\mathbf{B}\,.\,\mathbf{I}(n) + \frac{2\beta\beta_p}{\kappa_0 c^2}\sum_{n,i}\frac{g_n\mathbf{I}(n)\,.\,\mathbf{M}^n(i)}{r_{ni}^3}$$
$$+\frac{g\beta\beta_p}{\kappa_0 c^2}\sum_{n,i} g_n\left(\frac{3(\mathbf{S}(i)\,.\,\mathbf{r}_{ni})(\mathbf{I}(n)\,.\,\mathbf{r}_{ni}) - r_{ni}^2\mathbf{I}(n)\,.\,\mathbf{S}(i)}{r_{ni}^5}\right.$$
$$\left.+\frac{8\pi}{3}\delta(\mathbf{r}_{ni})\mathbf{I}(n)\,.\,\mathbf{S}(i)\right). \tag{5.32f}$$

The nomenclature is self-explanatory: the potential V_{ext} and fields **F**, **B** are now truly external (i.e. applied from outside the molecule), the magnetic perturbation H_{mag} being written as a sum of two terms which it is customary to refer to as its 'paramagnetic' and 'diamagnetic' parts; Z_n is the atomic number of nucleus n, and all the terms involving nuclear spins have been absorbed into H_{N} which is still incomplete in so far as electric quadrupole and direct nuclear dipole interactions have been omitted. It should be noted that $\mathbf{L}^n(i)$ and $\mathbf{M}^n(i)$, for example, denote angular momentum operators for electron i about the position of nucleus n, the latter containing a gauge-invariant momentum operator, defined by

$$\hbar\mathbf{L}^n(i) = \mathbf{r}_{ni}\times\mathbf{p}(i), \qquad \hbar\mathbf{M}^n(i) = \mathbf{r}_{ni}\times\boldsymbol{\pi}(i). \tag{5.33}$$

Nuclear spin operators have been denoted by $\mathbf{I}(n)$, the fine structure constant $\alpha = e^2/\kappa_0\hbar c$, and $\beta_p = e\hbar/2m_p$ is the *proton* magneton. Both the electron–

electron and electron–nuclear spin–spin interactions contain, as well as the dipole–dipole part, a 'contact' term which contains a δ-function; this short-range interaction effectively vanishes unless its argument (the separation of the particles) vanish, but is nevertheless important. The second term in H_N arises from the nuclear contribution to the vector potential in the $\boldsymbol{\pi}$ operator.

All the terms listed above lead to observable effects, many of which may be measured with high precision, as for example in n.m.r. and e.s.r. spectroscopy. Our eventual aim is to relate such effects to the form of the electron distribution.

5.4. The basic density functions

Before considering the electron distribution in its full generality, it is useful to start from a one-electron system and to look for simple semi-classical interpretations of the wealth of information contained in the wavefunction. This will lead to the definition of various density functions, which are the prototypes of those required in the many-electron case.

Let us consider, then, a single electron described by a general two-component wavefunction of the form (5.2), noting first that $|\psi|^2$ has the standard interpretation as a probability density ρ: just as in (5.3), but with the abbreviation $\mathbf{r}, s = \mathbf{x}$:

$$\rho(\mathbf{x})\,\mathrm{d}\mathbf{x} = |\psi|^2\,\mathrm{d}\mathbf{x} = \begin{cases} \text{probability of finding an} \\ \text{electron in } \mathrm{d}\mathbf{x}. \end{cases} \tag{5.34}$$

The term 'in $\mathrm{d}\mathbf{x}$' means with spatial variables between $\mathbf{r}$ and $\mathbf{r}+\mathrm{d}\mathbf{r}$, and spin between s and $s+\mathrm{d}s$ (i.e. in $\mathrm{d}\mathbf{r}\,\mathrm{d}s$). If we are not interested in the spin, only the position of the electron, we may integrate over all spin possibilities and denote the result by $P(\mathbf{r})\,\mathrm{d}\mathbf{r}$:

$$P(\mathbf{r})\,\mathrm{d}\mathbf{r} = \left[\int |\psi(\mathbf{r}, s)|^2\,\mathrm{d}s\right]\mathrm{d}\mathbf{r} = \begin{cases} \text{probability of finding an} \\ \text{electron in } \mathrm{d}\mathbf{r} \text{ (irrespective} \\ \text{of spin).} \end{cases} \tag{5.35}$$

The probability density functions with and without spin are thus related by

$$P(\mathbf{r}) = \int \rho(\mathbf{x})\,\mathrm{d}s. \tag{5.36}$$

For an electron described by the two-component function (1.2) it follows at once that

$$P(\mathbf{r}) = |\phi_\alpha(\mathbf{r})|^2 + |\phi_\beta(\mathbf{r})|^2 \tag{5.37}$$

which means simply that the spatial electron density is a sum of 'up-spin' and 'down-spin' densities. This result is consistent with the fact that the expectation value of any quantity $V(\mathbf{r})$ depending solely on electronic position is given by

$$\langle V\rangle = \int \psi^*(\mathbf{x})V(\mathbf{r})\psi(\mathbf{x})\,\mathrm{d}\mathbf{x} = \int V(\mathbf{r})\psi(\mathbf{x})\psi^*(\mathbf{x})\,\mathrm{d}\mathbf{x} = \int V(\mathbf{r})P(\mathbf{r})\,\mathrm{d}\mathbf{r} \quad (5.38)$$

which is simply the classical average of $V(\mathbf{r})$.

The expectation value of a general operator A (say) cannot be expressed as a classical average over $P(\mathbf{r})$ (or over $\rho(\mathbf{x})$ if A is spin-dependent) because the operator comes *between* ψ^* and ψ and works only on the latter. The re-arrangement in (5.38) was possible only because $V(\mathbf{r})$ was chosen to be simply a multiplicative factor in the integrand, not for example a differential operator. We can however make a similar reduction at the expense of a very slight complication; we take the complex conjugate function over to the right, as in (5.38), but change its variable from $\mathbf{x}$ to $\mathbf{x}'$ to protect it from the operator (which by convention works on functions of $\mathbf{x}$); then *after* the operation we put $\mathbf{x}' = \mathbf{x}$ and complete the integration. In this way

$$\langle A\rangle = \int \psi^*(\mathbf{x})A\psi(\mathbf{x})\,\mathrm{d}\mathbf{x} = \int \{A\psi(\mathbf{x})\psi^*(\mathbf{x}')\}_{\mathbf{x}'=\mathbf{x}}\,\mathrm{d}\mathbf{x}.$$

It is convenient to write

$$\rho(\mathbf{x};\mathbf{x}') = \psi(\mathbf{x})\psi^*(\mathbf{x}') \quad (5.39)$$

as a generalization of the $\rho(\mathbf{x})$ appearing in (5.34), which arises when $\mathbf{x}' = \mathbf{x}$. This quantity is the 'density matrix' ($\mathbf{x}$ and $\mathbf{x}'$ corresponding formally to row and column indices in the matrix representation of an operator) for a system in state ψ. It is also convenient to retain $\rho(\mathbf{x})$ to indicate the 'diagonal element' with $\mathbf{x}' = \mathbf{x}$:

$$\rho(\mathbf{x}) = \rho(\mathbf{x};\mathbf{x}) \quad (5.40)$$

The expectation value of *any* operator may now be written

$$\langle A\rangle = \int \{A\rho(\mathbf{x};\mathbf{x}')\}_{\mathbf{x}'=\mathbf{x}}\,\mathrm{d}\mathbf{x} \quad (5.41)$$

where $\rho(\mathbf{x};\mathbf{x}')$ is a generalized density function, whose diagonal element gives the usual probability density.

5.4.1. *Charge, current, and spin density*

From $\rho(\mathbf{x};\mathbf{x}')$ it is possible to derive expectation value formulae which give considerable physical insight into the origin of certain properties. The first of these is rather trivial, and by now very well known, being essentially contained in (5.38). If A is a *spinless* operator, we may put $s' = s$ in (5.41) and integrate over spin at once to obtain

$$\langle A \rangle = \int \{AP(\mathbf{r};\mathbf{r}')\}_{\mathbf{r}'=\mathbf{r}}\,\mathrm{d}\mathbf{r} \tag{5.42}$$

where

$$P(\mathbf{r};\mathbf{r}') = \int \{\rho(\mathbf{x};\mathbf{x}')\}_{s'=s}\,\mathrm{d}s. \tag{5.43}$$

This quantity is the *spinless* density matrix, whose diagonal element $P(\mathbf{r})$ is the probability density introduced in (5.36); (5.42) is the generalization of (5.38) to all types of spinless operator, and (5.43) is the generalization of (5.36). A consequence of (5.38), which arises from (5.42) when $A = V(\mathbf{r})$, is simply that the change of energy of the system described by ψ, due to a perturbing potential $V(\mathbf{r})$, may be computed classically, as if the system were a 'smeared out' charge of density $P(\mathbf{r})$. For $P(\mathbf{r})\,\mathrm{d}\mathbf{r}$ would be the amount of charge (in electrons) in element $\mathrm{d}\mathbf{r}$; $V(\mathbf{r})$ is the potential energy of unit charge (one electron) at $\mathbf{r}$; and the integral is thus the potential energy of the 'charge cloud' in the perturbing field. For this reason $P(\mathbf{r})$, although strictly speaking it is a probability density, is often called the 'charge density' or 'electron density'. It consists of two parts, according to (5.37), and in general, since

$$\rho(\mathbf{x};\mathbf{x}') = \phi_\alpha(\mathbf{r})\phi_\alpha^*(\mathbf{r}')\alpha(s)\alpha^*(s') + \phi_\beta(\mathbf{r})\phi_\beta(\mathbf{r}')\beta(s)\beta^*(s')$$
$$+\phi_\alpha(\mathbf{r})\phi_\beta^*(\mathbf{r}')\alpha(s)\beta^*(s') + \phi_\beta(\mathbf{r})\phi_\alpha(\mathbf{r}')\beta(s)\alpha^*(s')$$

we have by integration over spin

$$P(\mathbf{r};\mathbf{r}') = P^{\alpha,\alpha}(\mathbf{r};\mathbf{r}') + P^{\beta,\beta}(\mathbf{r};\mathbf{r}'), \tag{5.44}$$

where

$$P^{\alpha,\alpha}(\mathbf{r};\mathbf{r}') = \phi_\alpha(\mathbf{r})\phi_\alpha^*(\mathbf{r}'), \qquad P^{\beta,\beta}(\mathbf{r};\mathbf{r}') = \phi_\beta(\mathbf{r})\phi_\beta^*(\mathbf{r}') \tag{5.45}$$

are 'up-spin and 'down-spin' components. On putting $\mathbf{r}' = \mathbf{r}$ and using the contracted notation (*cf.* 5.40)) for diagonal elements, we recover (5.37) in the form

$$P(\mathbf{r}) = P^\alpha(\mathbf{r}) + P^\beta(\mathbf{r}). \tag{5.46}$$

As we shall see presently, exactly similar quantities may be defined for many-electron systems and similar results hold.

Now let us consider the possibility of current flow in the charge distribution. The presence of currents may be inferred by observing the magnetic field to which they give rise and comparing it with the classical Biot–Savart expression

$$B_\lambda = \frac{\mu_0}{4\pi}\int \frac{[\mathbf{r}\times\mathbf{j}]_\lambda}{r^3}\,\mathrm{d}r \qquad (\lambda = x, y, z) \tag{5.47}$$

for the λ-component of the field at the origin due to current flow (current density $\mathbf{j}$) in a continuous medium (permeability μ_0). For states described by real wavefunctions the current density is everywhere zero; for generality, therefore, we should consider systems in the presence of an applied magnetic field, where the wavefunction is essentially complex and serves to describe induced currents. This means an applied field $\mathbf{B}$ should be admitted from the start, and the simplest Hamiltonian we can use is

$$h = \frac{1}{2m}\pi^2 + V(r) \qquad (\boldsymbol{\pi} = \mathbf{p} + e\mathbf{A}) \tag{5.48}$$

which follows from (5.23) with neglect of spin and other small terms. The field due to currents induced in the electron distribution may be computed by inserting an infinitesimal 'test dipole' $\boldsymbol{\mu}$ at an arbitrary point (taken as origin) and computing the expectation value of the corresponding perturbation operator due to the dipole field arising from vector potential

$$\mathbf{A}^{(\mu)} = \frac{\mu_0}{4\pi}\frac{\boldsymbol{\mu}\times\mathbf{r}}{r^3}. \tag{5.49}$$

The energy change is then written $E_{\text{dip}} = -\boldsymbol{\mu} \cdot \mathbf{B}^{\text{ind}}$ and $\mathbf{B}^{\text{ind}}$ is compared with (5.47) to identify $\mathbf{j}$. In this way we find easily $\mathbf{j} = -e\mathbf{J}$ where†

$$J_\lambda(\mathbf{r}) = Rl\left[m^{-1}\int \psi^*\pi_\lambda\psi\, \mathrm{d}s\right] \qquad (\lambda = x, y, z).$$

This result may be expressed in terms of $\rho(\mathbf{x};\mathbf{x}')$ by the usual device of introducing primed variables. The result is

$$J_\lambda(\mathbf{r}) = Rl[m^{-1}\pi_\lambda P(\mathbf{r};\mathbf{r}')]_{\mathbf{r}'=\mathbf{r}} \tag{5.50}$$

which may also be written as a sum of two parts, according to (5.44), namely

$$J^\alpha_\lambda(\mathbf{r}) = Rl[m^{-1}\pi_\lambda P^{\alpha,\alpha}(\mathbf{r};\mathbf{r}')]_{\mathbf{r}'=\mathbf{r}}, \qquad J^\beta_\lambda(\mathbf{r}) = Rl[m^{-1}\pi_\lambda P^{\beta,\beta}(\mathbf{r};\mathbf{r}')]_{\mathbf{r}'=\mathbf{r}}. \tag{5.51}$$

These λ-components of the total current represent currents in the up-spin and down-spin distributions, respectively.

It is easily verified that the current components (5.50) reduce to the usual expression for the 'probability current density'. Thus, taking a spin-up electron with wavefunction $\phi_\alpha(\mathbf{r})\alpha(s)$, $P^{\alpha,\alpha}(\mathbf{r};\mathbf{r}') = \phi_\alpha(\mathbf{r})\phi^*_\alpha(\mathbf{r}')$ and

$$J^\alpha_x(\mathbf{r}) = \frac{1}{m}Rl\left[\left(\frac{\hbar}{\mathrm{i}}\frac{\partial}{\partial x} + \mathrm{e}A_x\right)\phi_\alpha(\mathbf{r})\phi^*_\alpha(\mathbf{r}')\right]_{\mathbf{r}=\mathbf{r}'}$$

† It should be noted that only the real part of the quantity in square brackets is significant, owing to the reality of the expectation value of the perturbation energy. The imaginary part may be non-zero but gives no contribution to the velocity dependent interaction energy arising from any applied field with a vector potential satisfying div $\mathbf{A} = 0$ (i.e. commuting with $\boldsymbol{\pi}$).

This yields at once the familiar result

$$J_x^\alpha(\mathbf{r}) = \frac{\hbar}{2mi}\left[\phi_\alpha^*(\mathbf{r})\frac{\partial}{\partial x}\phi_\alpha(\mathbf{r}) - \phi_\alpha(\mathbf{r})\frac{\partial}{\partial x}\phi_\alpha^*(\mathbf{r})\right] + \frac{e}{m}A_x\phi_\alpha(\mathbf{r})\phi_\alpha^*(\mathbf{r}). \quad (5.52)$$

It should be noted that the present derivation does not depend on the assumption that the wavefunction is an exact solution of the time-dependent Schrödinger equation: thus with an approximate wave function it gives an approximate current density expression in the sense that it yields an appropriate interaction energy with any externally applied field.

Finally, we ask whether the magnetic effects of electron *spin* can also be visualized in a semi-classical way. To do this we again start from the unperturbed Hamiltonian (5.48), which is the first term in the Dirac–Pauli Hamiltonian (5.23), and calculate the first-order interaction energy with the test dipole; but now we include the $g\beta\mathbf{B}\,.\,\mathbf{S}$, noting that in addition to the direct spin–field interaction, due to the uniform applied field, there will be a term

$$E_{\text{dip}}^{(\text{spin})} = g\beta\left\langle\psi\left|\sum_\lambda B_\lambda^{(\mu)}S_\lambda\right|\psi\right\rangle \qquad (\lambda = x, y, z). \quad (5.53)$$

This represents the interaction between the test dipole $\boldsymbol{\mu}$, producing the field $\mathbf{B}^{(\mu)}$, and the electron spin. For a general one-electron state (5.2) we obtain, after inserting field components derived from the vector potential (5.49) and making suitable re-arrangements,

$$E_{\text{dip}}^{(\text{spin})} = \frac{\mu_0}{4\pi}\int\frac{r^2(\boldsymbol{\mu}\,.\,\mathbf{M}) - 3(\boldsymbol{\mu}\,.\,\mathbf{r})(\mathbf{r}\,.\,\mathbf{M})}{r^5}\,\mathrm{d}\mathbf{r} \quad (5.54)$$

where $\mathbf{M}$ is a continuous *density of magnetization*, with components

$$M_\lambda = -g\beta\int(\phi_\alpha\alpha + \phi_\beta\beta)^*S_\lambda(\phi_\alpha\alpha + \phi_\beta\beta)\,\mathrm{d}s \qquad (\lambda = x, y, z). \quad (5.55)$$

This interpretation is appropriate because (5.54) has been cast in classical form, corresponding to the interaction between dipole $\boldsymbol{\mu}$ and a magnetized medium in which $\mathbf{M}\,\mathbf{dr}$ is the magnetic dipole moment associated with volume element $\mathbf{dr}$. The electron in its 'orbit' is thus magnetically equivalent not only to a current distribution defined by (5.50) but to a continuous distribution of magnetization. The components follow at once from the properties of the spin operators (5.25):

$$\begin{aligned} M_x(\mathbf{r}) &= -\tfrac{1}{2}g\beta(\phi_\alpha^*(\mathbf{r})\phi_\beta(\mathbf{r}) + \phi_\beta^*(\mathbf{r})\phi_\alpha(\mathbf{r})) = -g\beta Q_x(\mathbf{r}) \\ M_y(\mathbf{r}) &= -\tfrac{1}{2}\mathrm{i}g\beta(\phi_\beta^*(\mathbf{r})\phi_\alpha(\mathbf{r}) - \phi_\alpha^*(\mathbf{r})\phi_\beta(\mathbf{r})) = -g\beta Q_y(\mathbf{r}) \\ M_z(\mathbf{r}) &= -\tfrac{1}{2}g\beta(\phi_\alpha^*(\mathbf{r})\phi_\alpha(\mathbf{r}) - \phi_\beta^*(\mathbf{r})\phi_\beta(\mathbf{r})) = -g\beta Q_z(\mathbf{r}). \end{aligned} \quad (5.56)$$

Since, for an electron, $-g\beta$ is the factor relating magnetic moment to angular momentum (as a multiple of $\hbar$) it is clear that $Q_x(\mathbf{r})$, $Q_y(\mathbf{r})$, and $Q_z(\mathbf{r})$ may be interpreted as the density of *spin angular momentum*, in units of $\hbar$, at point $\mathbf{r}$ in the electron distribution. These expressions agree with those which can be derived directly from the four-component Dirac equation (see, for example, Slater 1960; vol. II, Sections 23–26), and the Pauli approximation is therefore consistent in giving correct expressions for properties other than the energy.

The results (5.56) are easily expressed in terms of the density function $\rho(\mathbf{x};\mathbf{x}')$ defined in (5.39). Thus from (5.55)

$$M_\lambda(\mathbf{r}) = -g\beta \int [S_\lambda \rho(\mathbf{x};\mathbf{x}')]_{\mathbf{x}'=\mathbf{x}}\,\mathrm{d}\mathbf{x} = -g\beta Q_S(\mathbf{r})_\lambda,$$

where $Q_S(\mathbf{r})_\lambda$ ($\lambda = x, y, z$) is the density function occurring in (5.56) and is the diagonal element ($\mathbf{r}' = \mathbf{r}$) of the more general quantity

$$Q_S(\mathbf{r};\mathbf{r}')_\lambda = \int [S_\lambda \rho(\mathbf{x};\mathbf{x}')]_{s'=s}\,\mathrm{d}s. \tag{5.57}$$

This expression should be compared with (5.43); it is the spin density matrix, which plays an important role in determining a wide range of spin-dependent properties. The diagonal element $Q_S(\mathbf{r})_\lambda$ has a very direct physical interpretation, for from (5.41), the expectation value of S_λ is given by

$$\langle S_\lambda \rangle = \langle \psi | S_\lambda | \psi \rangle = \int [S_\lambda \rho(\mathbf{x};\mathbf{x}')]_{\mathbf{x}'=\mathbf{x}}\,\mathrm{d}\mathbf{r}\,\mathrm{d}s = \int Q_S(\mathbf{r})_\lambda\,\mathrm{d}\mathbf{r}. \tag{5.58}$$

In other words, the contributions to the λ-component of the spin angular momentum may be regarded as smeared out in space with a density $Q_S(\mathbf{r})_\lambda$, the contribution from volume element $\mathrm{d}\mathbf{r}$ being $Q_S(\mathbf{r})_\lambda\,\mathrm{d}\mathbf{r}$.

To summarize: the expectation values of all observables, for a one-electron system described by a Pauli-type (two-component) wavefunction, are determined by the density matrix $\rho(\mathbf{x};\mathbf{x}')$ according to eqn (5.41). In particular, the electric and magnetic fields at all points in space may be computed as if the electron were replaced by a continuous distribution of matter with charge density $-eP(\mathbf{r})$, current density $-e\mathbf{J}(\mathbf{r})$, and magnetization density $\mathbf{M}(\mathbf{r}) = -g\beta\mathbf{Q}(\mathbf{r})$, where the three densities (one scalar and two vector densities) are defined by (5.43), (5.50) and (5.57). Thus the electric potential at any point (taken as origin) is

$$\phi = -\frac{e}{4\pi\varepsilon_0}\int \frac{P(\mathbf{r})}{r}\,\mathrm{d}\mathbf{r} \tag{5.59}$$

while the magnetic field (due to both velocity and spin effects) has components

$$B_\lambda = \frac{\mu_0}{4\pi}\int \frac{\{\mathbf{r}\times(-e\mathbf{J})\}_\lambda}{r^3}\,\mathrm{d}\mathbf{r} + \frac{\mu_0}{4\pi}\int \frac{\{r^2\mathbf{M}_\lambda - 3(\mathbf{M}\,.\,\mathbf{r})r_\lambda\}}{r^5}\,\mathrm{d}\mathbf{r}. \tag{5.60}$$

Such densities, which we shall find may be defined equally well for many-electron systems, often provide qualitative insight into the origins of many observable properties. Before proceeding further we consider one example in some detail.

5.4.2. *Example. One-electron atom with spin–orbit coupling*

For a single electron in a p-type orbital, L–S coupling leads to $^2P_{\frac{3}{2}}$ and $^2P_{\frac{1}{2}}$ states. For example, the state with $j = m = \frac{1}{2}$ has wavefunction

$$\psi(\mathbf{r}, s) = \tfrac{1}{3}[p_0(\mathbf{r})\alpha(s) - \sqrt{2}p_{+1}(\mathbf{r})\beta(s)].$$

On identifying the components ϕ_α and ϕ_β, and substituting in (5.45) and (5.56), we obtain at once

$$\begin{aligned}
P(\mathbf{r}\,;\mathbf{r}') &= \tfrac{1}{3}\{p_0(\mathbf{r})p_0^*(\mathbf{r}') + 2p_{+1}(\mathbf{r})p_{+1}^*(\mathbf{r}')\}\\
Q(\mathbf{r}\,;\mathbf{r}')_x &= -(\sqrt{2}/6)\{p_0(\mathbf{r})p_{+1}^*(\mathbf{r}') + p_{+1}(\mathbf{r})p_0^*(\mathbf{r}')\}\\
Q(\mathbf{r}\,;\mathbf{r}')_y &= -\mathrm{i}(\sqrt{2}/6)[p_0(\mathbf{r})p_{+1}^*(\mathbf{r}') - p_{+1}(\mathbf{r})p_0^*(\mathbf{r}')]\\
Q(\mathbf{r}\,;\mathbf{r}')_z &= (1/6)[p_0(\mathbf{r})p_0^*(\mathbf{r}') - 2p_{+1}(\mathbf{r})p_{+1}^*(\mathbf{r}')].
\end{aligned}$$

In terms of cartesian coordinates the p-functions (Condon–Shortley phase conventions, and with $f(r)$ for the normalized radial factor) are

$$p_{+1} = -N_p(x+\mathrm{i}y)f(r), \qquad p_0 = \sqrt{2}N_pzf(r), \qquad p_{-1} = N_p(x-\mathrm{i}y)f(r)$$

in which $N_p = \sqrt{(3/8\pi)}$ and it is then easy to derive the following density functions:

$$\begin{aligned}
P(\mathbf{r}) &= (4\pi)^{-1}r^2f^2(r) & Q_x(\mathbf{r}) &= (4\pi^{-1})f^2(r)\times zx\\
Q_y(\mathbf{r}) &= (4\pi)^{-1}f^2(r)\times zy & Q_z(\mathbf{r}) &= (4\pi^{-1})f^2(r)\times\tfrac{1}{2}(3z^2-r^2).
\end{aligned}$$

Thus, although the charge density has full spherical symmetry, the spin density exhibits a curious polarization around the axis of quantization (Fig. 5.1). The current density is also readily obtained from (5.50): its components are

$$J_x(\mathbf{r}) = -\,y(3\hbar/16\pi m)f^2(\mathbf{r}), \qquad J_y(\mathbf{r}) = x(3\hbar/16\pi m)f^2(\mathbf{r}), \qquad J_z(\mathbf{r}) = 0$$

and the current thus flows in circles around the z-axis, as would be expected. The spin distribution is reminiscent of the static spin density waves introduced in solid-state theory by Overhauser (1959).

5.5. Generalization to many-electron systems

In introducing the basic density functions we have considered a single electron, characterized by a Pauli–Dirac Hamiltonian and a two-component

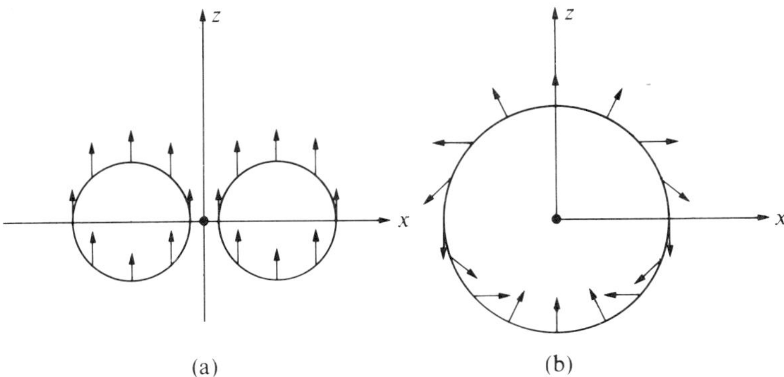

FIG. 5.1. Schematic indication of spin distribution for one electron in a p-state, with spin–orbit coupling: contours indicate form of the charge density, arrows indicate the spin density. (a) $^2P_{\frac{3}{2}}$ state ($j = m = \frac{3}{2}$). Charge distribution has cylindrical symmetry (torus around z axis), spin polarization uniformly along z direction. (b) $^2P_{\frac{1}{2}}$ state ($j = m = \frac{1}{2}$). Charge distribution has spherical symmetry, spin polarization non-uniform.

wavefunction or spin–orbit. For a many-electron system, we use the Breit–Pauli Hamiltonian—some of whose principal terms have been described in Section 5.2—and a 2^N-component wavefunction Ψ which may be expanded in terms of spin-products $\alpha_1\alpha_2 \ldots \alpha_N$, $\beta_1\alpha_2 \ldots \alpha_N$, $\alpha_1\beta_2 \ldots \alpha_N, \ldots$, the expansion coefficients being functions of the spatial variables $\mathbf{r}_1, \mathbf{r}_2, \ldots, \mathbf{r}_N$. Again it is convenient to introduce spin 'functions' $\alpha(s_i)$, $\beta(s_i)$ and to use $\mathbf{x}_i$ to indicate collectively the space–spin variables $\mathbf{r}_i, s_i$. The state is then characterized by a wavefunction $\Psi(\mathbf{x}_1, \mathbf{x}_2, \ldots, \mathbf{x}_N)$ with the interpretation

$$|\Psi(\mathbf{x}_1, \mathbf{x}_2, \ldots, \mathbf{x}_N)|^2 \, \mathrm{d}\mathbf{x}_1 \, \mathrm{d}\mathbf{x}_2 \ldots \mathrm{d}\mathbf{x}_N = \begin{cases} \text{probability of finding electron} \\ \text{1 in } \mathrm{d}\mathbf{x}_1\text{, electron 2 simul-} \\ \text{taneously in } \mathrm{d}\mathbf{x}_2\text{, etc.} \end{cases} \tag{5.61}$$

which is the generalization of (5.3) with the abbreviation $(\mathbf{r}, s) = \mathbf{x}$.

It is well known, however, that $\Psi(\mathbf{x}_1, \mathbf{x}_2, \ldots, \mathbf{x}_N)$ (and even $|\Psi(\mathbf{x}_1, \mathbf{x}_2, \ldots, \mathbf{x}_N)|^2$) contains much redundant information; for the most general Hamiltonian we have considered contains only one- and two-electron operators and in evaluating corresponding expectation values it is always necessary to eliminate $N-1$ or $N-2$ variables by trivial integrations. To eliminate this superfluous information from the start we introduce certain 'reduced' density functions, analogous to those used in Section 5.4. Thus, from (5.61), the probability of finding electron 1 in a volume element $\mathrm{d}\mathbf{x}$ at the point $\mathbf{x}$, and the other electrons *anywhere*, is obtained by putting $\mathbf{x}_1 = \mathbf{x}$, $\mathrm{d}\mathbf{x}_1 = \mathrm{d}\mathbf{x}$ and integrating over all possible positions of the remaining electrons (2, 3, ..., N): it is

$$\mathrm{d}\mathbf{x} \int \Psi(\mathbf{x}, \mathbf{x}_2, \ldots, \mathbf{x}_N)\Psi^*(\mathbf{x}, \mathbf{x}_2, \ldots, \mathbf{x}_N) \, \mathrm{d}\mathbf{x}_2 \ldots \mathrm{d}\mathbf{x}_N. \tag{5.62}$$

The wavefunction is however antisymmetric under any permutation of variables and the product in the integrand is therefore symmetric: the probability of finding electron 2 in $\mathbf{dx}$ at point $\mathbf{x}$ is thus

$$\mathrm{d}\mathbf{x} \int \Psi(\mathbf{x}_1, \mathbf{x}, \mathbf{x}_3, \ldots, \mathbf{x}_N)\Psi^*(\mathbf{x}_1, \mathbf{x}, \mathbf{x}_3, \ldots, \mathbf{x}_N)\, \mathrm{d}\mathbf{x}_1\, \mathrm{d}\mathbf{x}_3 \ldots \mathrm{d}\mathbf{x}_N$$

and, by interchanging $\mathbf{x}$ and $\mathbf{x}_1$ and replacing the dummy variable $\mathbf{x}_1$ by $\mathbf{x}_2$, this is seen to be identical with (5.62). The probability of finding *any one* of the N-electrons in $\mathrm{d}\mathbf{x}$ at $\mathbf{x}$ is given by exactly the same expression and hence the probability of finding an electron, no matter which, in $\mathrm{d}\mathbf{x}$ at $\mathbf{x}$ is

$$N\, \mathrm{d}\mathbf{x} \int \Psi(\mathbf{x}, \mathbf{x}_2, \ldots, x_N)\Psi^*(\mathbf{x}, \mathbf{x}_2, \ldots, \mathbf{x}_N)\, \mathrm{d}\mathbf{x}_2 \ldots \mathrm{d}\mathbf{x}_N = \rho_1(\mathbf{x})\, \mathrm{d}\mathbf{x}$$

where we have introduced a *one*-electron density function $\rho_1(\mathbf{x})$ analogous to that appearing in (5.34). Since the name of the point at which the function is evaluated is irrelevant, we can define

$$\rho_1(\mathbf{x}_1) = N \int \Psi(\mathbf{x}_1, \mathbf{x}_2, \ldots, \mathbf{x}_N)\Psi^*(\mathbf{x}_1, \mathbf{x}_2, \ldots, \mathbf{x}_N)\, \mathrm{d}\mathbf{x}_2 \ldots \mathrm{d}\mathbf{x}_N. \qquad (5.63)$$

But there has now been a subtle change in the interpretation of $\mathbf{x}_1$: when it appears as the argument in a density function $\mathbf{x}_1$ $(=\mathbf{r}_1, s_1)$ is the set of co-ordinates of an *arbitrary point* in configuration space—it does not indicate the space–spin coordinates of 'electron 1'. The indistinguishability of of electrons (reflected in the antisymmetry of the wavefunction) is thus incorporated, from the start, and $\rho_1(\mathbf{x}_1)\, d\mathbf{x}_1$ refers to the probability of a volume element $\mathrm{d}\mathbf{x}_1$ at point $\mathbf{x}_1$ being occupied by *any* electron.

In exactly the same way, we can pick two volume elements, $\mathrm{d}\mathbf{x}_1$ and $\mathrm{d}\mathbf{x}_2$, and ask for the probability that they be simultaneously occupied by any two electrons. We write the result as $\rho_2(\mathbf{x}_1, \mathbf{x}_2)\, \mathrm{d}\mathbf{x}_1\, \mathrm{d}\mathbf{x}_2$ and find

$$\rho_2(\mathbf{x}_1,\mathbf{x}_2) = N(N-1)\int \Psi(\mathbf{x}_1,\mathbf{x}_2,\mathbf{x}_3,\ldots,\mathbf{x}_N)\Psi^*(\mathbf{x}_1,\mathbf{x}_2,\mathbf{x}_3,\ldots,\mathbf{x}_N)\, \mathrm{d}\mathbf{x}_3 \ldots \mathrm{d}\mathbf{x}_N. \qquad (5.64)$$

The factor $N(N-1)$ arises because, having chosen the volume elements at $\mathbf{x}_1$ and $\mathbf{x}_2$, there are N electrons available to put in the first one, but only $(N-1)$ for the second; there are thus $N(N-1)$ equally probable situations, the typical probability (referring to electron 1 in $\mathrm{d}\mathbf{x}_1$ and 2 in $\mathrm{d}\mathbf{x}_2$) being given by the integral.

The arguments above have been given in detail to avoid confusion over the statistical interpretation: when they appear in the density functions,

$\mathbf{x}_1, \mathbf{x}_2, \ldots$ refer to 'generic' situations, *not* to particular electrons 1, 2, ... but rather to the points at which they may be found. We note also that, since Ψ is assumed to be a normalized wavefunction, the density functions have the normalization properties

$$\int \rho_1(\mathbf{x}_1)\,\mathrm{d}\mathbf{x}_1 = N, \qquad \int \rho_2(\mathbf{x}_1, \mathbf{x}_2)\,\mathrm{d}\mathbf{x}_1\,\mathrm{d}\mathbf{x}_2 = N(N-1), \tag{5.65}$$

i.e. they are not normalized to *unity*. The normalization adopted here goes back to Husimi (1940) and has been adopted by McWeeny (1954, 1955), Born and Green (1947), Yang (1962) and others; but alternative choices have been used by Löwdin (1955), Ter Haar (1961), and Coleman (1963).

It is a trivial matter to repeat the same arguments in evaluating matrix elements of typical operators. Thus one- and two-electron operators always appear in the N-electron Hamiltonian as symmetric sums,

$$\sum_{i=1}^{N} A(i), \qquad \sum_{i,j=1}^{N} B(i,j)$$

and corresponding expectation values consist of N or $N(N-1)$ identical terms. For the one-electron operator, we proceed as in the derivation of (5.41), writing

$$\left\langle \Psi \left| \sum_{i=1}^{N} \right| A(i)\Psi \right\rangle$$

$$= N \int \Psi^*(\mathbf{x}_1, \mathbf{x}_2, \ldots, \mathbf{x}_N) A(1)(\mathbf{x}_1, \mathbf{x}_2, \ldots, \mathbf{x}_N)\,\mathrm{d}\mathbf{x}_1\,\mathrm{d}\mathbf{x}_2 \ldots \mathrm{d}\mathbf{x}_N$$

$$= N \int [A(1)\Psi(\mathbf{x}_1, \mathbf{x}_2, \ldots, \mathbf{x}_N)\Psi^*(\mathbf{x}_1', \mathbf{x}_2, \ldots, \mathbf{x}_N)]_{\mathbf{x}_1'=\mathbf{x}_1}\,\mathrm{d}\mathbf{x}_1\,\mathrm{d}\mathbf{x}_2 \ldots \mathrm{d}\mathbf{x}_N$$

and hence, if we introduce

$$\rho_1(\mathbf{x}_1; \mathbf{x}_1') = N \int \Psi(\mathbf{x}_1, \mathbf{x}_2, \ldots, \mathbf{x}_N)\Psi^*(\mathbf{x}_1', \mathbf{x}_2, \ldots, \mathbf{x}_N)\,\mathrm{d}\mathbf{x}_2 \ldots \mathrm{d}\mathbf{x}_N \tag{5.66}$$

as the N-electron analogue of (5.39), it follows that

$$\left\langle \Psi \left| \sum_{i=1}^{N} A(i) \right| \Psi \right\rangle = \int \{A(1)\rho_1(\mathbf{x}_1; \mathbf{x}_1')\}_{\mathbf{x}_1'=\mathbf{x}_1}\,\mathrm{d}\mathbf{x}_1 \tag{5.67}$$

which is formally identical with the one-electron result (5.41). In the same way, on defining

$$\rho_2(\mathbf{x}_1, \mathbf{x}_2; \mathbf{x}_1', \mathbf{x}_2') = N(N-1) \int \Psi(\mathbf{x}_1, \mathbf{x}_2, \ldots, \mathbf{x}_N) \times \Psi^*(\mathbf{x}_1', \mathbf{x}_2', \ldots, \mathbf{x}_N)\,\mathrm{d}\mathbf{x}_3, \ldots, \mathrm{d}\mathbf{x}_N \tag{5.68}$$

we find at once

$$\left\langle \Psi \middle| \sum_{i,j=1}^{N} B(i,j) \middle| \Psi \right\rangle = \int \{B(1,2)\rho_2(\mathbf{x}_1, \mathbf{x}_2; \mathbf{x}'_1, \mathbf{x}'_2)\}_{\substack{\mathbf{x}'_1=\mathbf{x}_1 \\ \mathbf{x}'_2=\mathbf{x}_2}} \mathrm{d}\mathbf{x}_1\, \mathrm{d}\mathbf{x}_2. \qquad (5.69)$$

The *many*-electron expectation values are in this way written in terms of *one*- and *two*-electron density functions; ρ_1 and ρ_2 are the one- and two-body 'reduced density matrices' (Husimi 1940) for a system described by wave function Ψ. The 'diagonal elements', obtained by identifying primed and unprimed variables, give the probability functions defined earlier and we retain the notation (4.7) for these elements:

$$\rho_1(\mathbf{x}_1) = \rho_1(\mathbf{x}_1; \mathbf{x}_1), \qquad \rho_2(\mathbf{x}_1, \mathbf{x}_2) = \rho_2(\mathbf{x}_1, \mathbf{x}_2; \mathbf{x}_1, \mathbf{x}_2).$$

It is now possible to derive other useful density functions as in Section 5.4.

5.5.1. *Charge, current and spin densities*

In the case of spinless operators, we may put $s'_1 = s_1$ in (5.67) and complete the spin integration immediately to obtain (*cf.* (5.42))

$$\left\langle \Psi \middle| \sum_i A(i) \middle| \Psi \right\rangle = \int \{A(1)P_1(\mathbf{r}_1; \mathbf{r}'_1)\}_{\mathbf{r}'_1=\mathbf{r}_1} \mathrm{d}\mathbf{r}_1 \qquad (5.70)$$

where

$$P_1(\mathbf{r}_1; \mathbf{r}'_1) = \int \{\rho_1(\mathbf{x}_1; \mathbf{x}'_1)\}_{s'_1=s_1} \mathrm{d}s \qquad (5.71)$$

is the spinless one-electron density matrix, analogous to (5.43). The diagonal element is the charge density, essentially as observed by X-ray crystallographers. More precisely, as in (5.35),

$$P_1(\mathbf{r}_1)\,\mathrm{d}\mathbf{r}_1 = \begin{cases} \text{probability of finding an electron} \\ \text{(either spin) in volume element } \mathrm{d}\mathbf{r}_1 \end{cases} \qquad (5.72)$$

The validity of a 'charge cloud' interpretation rests upon the generalization of (5.38): if $A(1) = V(\mathbf{r}_1)$ is simply a function of electronic position—let us say potential energy of the electron in some applied field—the prime may be removed immediately in (5.70) and the expectation value of $V = \sum_i V(i)$ becomes

$$\langle V \rangle = \int V(\mathbf{r}_1)P_1(\mathbf{r}_1)\,\mathrm{d}\mathbf{r}_1, \qquad (5.73)$$

which is simply a classical average—in this example, the potential energy of a charge cloud (density P_1) in the applied field.

Next we consider the current density, using exactly the same procedure as in the one-electron case. The presence of a test dipole μ perturbs the operator $\pi(i)$ for each of the N electrons, and the total perturbation energy becomes the expectation value of a *sum* of one-electron contributions instead of a single term. The result may thus be written, using (5.67), in terms of ρ_1. We find a coupling energy of the same form $E_{\text{dip}} = -\boldsymbol{\mu} \cdot \mathbf{B}^{\text{ind}}$ where $\mathbf{B}^{\text{ind}}$ (including any terms induced by application of an external field) is again given by the Biot–Savart expression (5.47), but the current density has components $j_\lambda = -eJ_\lambda$ given by

$$J_\lambda(\mathbf{r}_1) = Rl\left[\frac{1}{m}\int [\pi_\lambda(1)\rho_1(\mathbf{x}_1\,;\mathbf{x}'_1)]_{s'_1=s_1}\,\mathrm{d}s_1\right].$$

Since π_λ is a spatial operator, the integration over spin is immediate and yields

$$J_\lambda(\mathbf{r}_1) = Rl\left[\frac{1}{m}\pi_\lambda(1)P_1(\mathbf{r}_1\,;\mathbf{r}'_1)\right]_{\mathbf{r}'_1=\mathbf{r}_1}. \tag{5.74}$$

As in the one-electron case, the total current density may be written as a sum of up-spin and down-spin parts; for the most general form of ρ_1 is

$$\rho_1(\mathbf{x}_1\,;\mathbf{x}'_1) = P_1^{\alpha,\alpha}(\mathbf{r}_1\,;\mathbf{r}'_1)\alpha(s_1)\alpha^*(s'_1)+P_1^{\alpha,\beta}(\mathbf{r}_1\,;\mathbf{r}'_1)\alpha(s_1)\beta^*(s'_1)$$
$$+P_1^{\beta,\alpha}(\mathbf{r}_1\,;\mathbf{r}'_1)\beta(s_1)\alpha^*(s'_1)+P_1^{\beta,\beta}(\mathbf{r}_1\,;\mathbf{r}'_1)\beta(s_1)\beta^*(s'_1), \tag{5.75}$$

where the spatial factors are labelled according to the four possible spin products. The integration leading from ρ_1 to P_1 then shows that

$$P_1(\mathbf{r}_1\,;\mathbf{r}'_1) = P_1^{\alpha,\alpha}(\mathbf{r}_1\,;\mathbf{r}'_1)+P_1^{\beta,\beta}(\mathbf{r}_1\,;\mathbf{r}'_1)$$

and hence

$$J_\lambda(\mathbf{r}_1) = J_\lambda^\alpha(\mathbf{r}_1)+J_\lambda^\beta(\mathbf{r}_1), \tag{5.76}$$

where the two parts are of the form (5.51).

Finally, we turn to the spin density. The derivation indicated in Section 5.4 again applies, single terms being replaced by sums of terms—whose expectation values are then expressible in terms of ρ_1 according to (5.67). The final conclusion is that the electron distribution behaves like a magnetized medium, the density of magnetization at point $\mathbf{r}_1$ being $\mathbf{M}(\mathbf{r}_1) = -g\beta\mathbf{Q}_S(\mathbf{r}_1)$ where $\mathbf{Q}_S(\mathbf{r}_1)$ is the (vector) density of spin angular momentum at point $\mathbf{r}_1$ and has components which are diagonal elements ($\mathbf{r}'_1 = \mathbf{r}_1$) of

$$Q_S(\mathbf{r}_1\,;\mathbf{r}'_1)_\lambda = \int \{S_\lambda(1)\rho_1(\mathbf{x}_1\,;\mathbf{x}'_1)\}_{s'_1=s_1}\,\mathrm{d}s_1. \tag{5.77}$$

This result coincides (apart from a trivial change of notation) with the one-electron result (5.57). Again, it follows that the expectation values of *total* spin components are given by

$$\langle S_\lambda\rangle = \int Q_S(\mathbf{r}_1)_\lambda\,\mathrm{d}\mathbf{r}_1 \qquad (\lambda = x, y, z) \tag{5.78}$$

as if the angular momentum were smeared out in space with density $Q_S(\mathbf{r}_1)_\lambda$. One important result follows immediately: if we put (5.75) in (5.77) and note the properties of the spin operators, we obtain

$$\begin{aligned} Q_S(\mathbf{r}_1)_x &= \tfrac{1}{2}\{P_1^{\alpha,\beta}(\mathbf{r}_1\,;\mathbf{r}_1)+P_1^{\beta,\alpha}(\mathbf{r}_1\,;\mathbf{r}_1)\} \\ Q_S(\mathbf{r}_1)_y &= \tfrac{1}{2}\mathrm{i}\{P_1^{\alpha,\beta}(\mathbf{r}_1\,;\mathbf{r}_1)-P_1^{\beta,\alpha}(\mathbf{r}_1\,;\mathbf{r}_1)\} \\ Q_S(\mathbf{r}_1)_z &= \tfrac{1}{2}\{P_1^{\alpha,\alpha}(\mathbf{r}_1\,;\mathbf{r}_1)-P_1^{\beta,\beta}(\mathbf{r}_1\,;\mathbf{r}_1)\}. \end{aligned} \tag{5.79}$$

Thus $Q_S(\mathbf{r}_1)_z$ is proportional simply to the difference of up-spin and down-spin densities, $P_1^\alpha(\mathbf{r}_1)$ and $P_1^\beta(\mathbf{r}_1)$, and measures the *resultant* up-spin component; but the other components depend on the cross-terms in the general expression (5.75). Now it may be shown by symmetry arguments (McWeeny and Mizuno, 1961) that for any spin eigenstate, with the z direction as the axis of quantization, the cross terms must vanish. It follows that in such a state $Q_S(\mathbf{r}_1)_x = Q_S(\mathbf{r}_1)_y = 0$ everywhere and the distribution has uniform magnetization along the z-axis, as in the one-electron example Fig. 5.1(a) (p. 215). It is the remaining z-component of the spin angular momentum density which is usually referred to as *the* spin density: this concept was introduced by Weissman (1956) and McConnell (1958), the function often being normalized so that $D_S(\mathbf{r}) = Q_S(\mathbf{r})_z/M_S$ integrates to give unity. It may also be shown that the spin angular momentum densities for all states of a multiplet ($M_S = S, S-1, \ldots, -S$) are identical in form, differing only through a proportionality factor M_S: thus, if we define a 'standard' density for the state with $M_S = S$, the density for any other component of the multiplet is

$$Q_S(\mathbf{r}_1)_z = (M_S/S)Q_S^{\mathrm{st}}(\mathbf{r}_1)_z. \tag{5.80}$$

This result is true only for a pure spin multiplet and does not hold if there is any L–S coupling (as in the Example of Section 5.4, where the distributions for different M_J differ widely in form). The following example illustrates the calculation of density functions for a many-electron system.

5.5.2. *Example. Spin density in ^2P-states of the lithium atom*

Let us consider the states formally associated with the configuration $1s^22p$, with spin–orbit coupling to $J = \frac{1}{2}, \frac{3}{2}$, as in an earlier example (p. 214) but with the addition of the K shell. The $J = M = \frac{3}{2}$ state is fairly well represented by the single-determinant wavefunction $|s\bar{s}p_{+1}|$; but the Hamiltonian (5.19) has non-zero matrix elements between other functions with similar quantum numbers and mixing therefore occurs. A suitable two-configuration function has the form

$$\Psi = A|s\bar{s}p_{+1}|+B\{|s\bar{p}_0p_{+1}|+|sp_0\bar{p}_{+1}|\}+C|\bar{s}p_0p_{+1}|$$

and the density functions are obtained by integrating $\Psi\Psi^*$. Each pair of determinants gives a contribution which is easily obtained using Slater's rules: thus $|s\bar{s}p_{+1}|$ with itself gives a 'diagonal' contribution $(ss^* + \bar{s}\bar{s}^* + p_{+1}p^*_{+1})$ with coefficient $|A|^2$, while $(s\bar{s}p_{+1})$ with $(s\bar{p}_0 p_{+1})$ gives a term $\bar{s}\bar{p}_0^*$ with coefficient AB^* (plus a complex conjugate term). On inserting explicit forms of the orbitals as in the previous example (p. 214) we obtain charge and spin densities of the form

$$P = as^2 + b(x^2 + y^2)f^2 + cz^2f^2 + dszf$$

$$Q_z = a's^2 + b'(x^2 + y^2)f^2 + c'z^2f^2 + d'szf.$$

The individual density terms are localized in different regions of space as depicted schematically in Fig. 5.2, which indicates a spin distribution

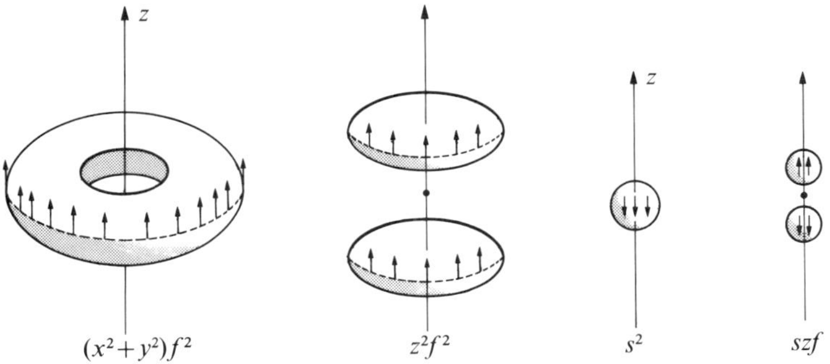

FIG. 5.2. Contributions to spin density in $^2P_{\frac{3}{2}}$ state of lithium atom. With $j = m = \frac{3}{2}$ the main component has the form of a torus of electron density with spin polarization uniformly along the z direction (as in Fig. 5.1a). The other contributions arise when configuration interaction is admitted, and give a 'core polarization' effect with non-zero spin density at the nucleus.

identical with that for a single 2p-electron (Fig. 5.1a) supplemented by small polarization terms near the nucleus. The reversal of spin direction at the nucleus results from a negative value of a', and it should be noted that the polarization is uniform in the sense $Q_x = Q_y = 0$. A similar treatment of the $^2P_{\frac{1}{2}}$ state gives a spin distribution like that in Fig. 5.1b, but with an inner-shell polarization reflecting the curious form of the density in the 2p region. The reversed spin density at the nucleus can be detected in n.m.r. experiments (see p. 233).

5.5.3. *Two-electron density functions*

For a many-electron system, it is also important to know the relative probabilities of different situations involving two or more particles. In particular, since the Hamiltonian normally contains only one- and two-electron operators, the function $\rho_2(\mathbf{x}_1, \mathbf{x}_2; \mathbf{x}'_1, \mathbf{x}'_2)$ will have a special importance. The question then arises whether other important *spatial* densities, involving a *pair* of points in ordinary three-dimensional space, may usefully be defined. Clearly, one such function will be important, namely,

$$P_2(\mathbf{r}_1,\mathbf{r}_2;\mathbf{r}'_1,\mathbf{r}'_2) = \int \{\rho_2(\mathbf{x}_1,\mathbf{x}_2;\mathbf{x}'_1,\mathbf{x}'_2)\}_{\substack{s'_1=s_1\\ s'_2=s_2}} \mathrm{d}s_1\,\mathrm{d}s_2 \tag{5.81}$$

since this will determine the expectation values of all *spinless* two-electron operators. From (5.69) we obtain at once, when B is spinless,

$$\left\langle \Psi \middle| \sum_{i,j} B(i,j) \middle| \Psi \right\rangle = \int B(1,2)P_2(\mathbf{r}_1,\mathbf{r}_2;\mathbf{r}'_1,\mathbf{r}'_2)\,\mathrm{d}\mathbf{r}_1\,\mathrm{d}\mathbf{r}_2 \tag{5.82}$$

which compares with the one-electron result (5.70).

The (spinless) 'pair-function' $P_2(\mathbf{r}_1,\mathbf{r}_2)$ is the diagonal element of $P_2(\mathbf{r}_1,\mathbf{r}_2;\mathbf{r}'_1,\mathbf{r}'_2)$ and is essentially the quantity used in the classical theory of fluids for discussing the correlation between the motions of different particles: $P_2(\mathbf{r}_1,\mathbf{r}_2)\,\mathrm{d}\mathbf{r}_1\,\mathrm{d}\mathbf{r}_2$ is the probability of finding volume elements $\mathrm{d}\mathbf{r}_1$ and $\mathrm{d}\mathbf{r}_2$, at spatial points $\mathbf{r}_1$ and $\mathbf{r}_2$, simultaneously occupied by particles. This quantity appears whenever we deal with expectation values of functions of position such as $g(i,j) = 1/r_{ij}$; for then the primes are redundant and (5.82) becomes

$$\left\langle \Psi \middle| \sum_{i,j} g(i,j) \middle| \Psi \right\rangle = \int g(1,2)P_2(\mathbf{r}_1,\mathbf{r}_2)\,\mathrm{d}\mathbf{r}_1\,\mathrm{d}\mathbf{r}_2 \tag{5.83}$$

which is a simple classical average for particles distributed with pair function $P_2(\mathbf{r}_1,\mathbf{r}_2)$.

As in the one-electron case, the pair function may be resolved into components referring to different spin situations. There are up to 16 components, from which various important density functions may be derived. Thus, the pair function $P_2(\mathbf{r}_1,\mathbf{r}_2)$ may be written (McWeeny 1954)

$$P_2(\mathbf{r}_1,\mathbf{r}_2) = P_2^{\alpha\alpha}(\mathbf{r}_1,\mathbf{r}_2)+P_2^{\alpha\beta}(\mathbf{r}_1,\mathbf{r}_2)+P_2^{\beta\alpha}(\mathbf{r}_1,\mathbf{r}_2)+P_2^{\beta\beta}(\mathbf{r}_1,\mathbf{r}_2) \tag{5.84}$$

where, for example, $P_2^{\alpha\beta}(\mathbf{r}_1,\mathbf{r}_2)\,\mathrm{d}\mathbf{r}_1\,\mathrm{d}\mathbf{r}_2$ is the probability of finding electrons in $\mathrm{d}\mathbf{r}_1$ and $\mathrm{d}\mathbf{r}_2$, with spin up and spin down, simultaneously. The physical significance and basic properties of such functions have been widely discussed elsewhere (e.g. McWeeny 1967; Kutzelnigg, Del Re, and Berthier 1968) in particular, they completely determine the expectation values of all the spin-dependent terms in the Breit–Pauli Hamiltonian.

The most familiar spin-dependent two-particle function is the spin–spin coupling function $Q_{\mathrm{SS}}(\mathbf{r}_1,\mathbf{r}_2;\mathbf{r}'_1,\mathbf{r}'_2)_m$ introduced by McWeeny and Mizuno (1961) and further discussed by McLachlan (1963) and McWeeny (1965): this is actually a five-component function ($m = 0, \pm 1, \pm 2$) but for a spin eigenstate only one component is non-zero, namely

$$Q_{SS}(\mathbf{r}_1, \mathbf{r}_2; \mathbf{r}'_1, \mathbf{r}'_2)_0$$
$$= \int [\{3S_z(1)S_z(2) - \mathbf{S}(1) . \mathbf{S}(2)\}\rho_2(\mathbf{x}_1, \mathbf{x}_2; \mathbf{x}'_1, \mathbf{x}'_2)]_{\substack{s'_1 = s_1 \\ s'_2 = s_2}} \, ds_1 \, ds_2. \quad (5.85)$$

The diagonal element has a simple interpretation: $Q_{SS}(\mathbf{r}_1, \mathbf{r}_2)_0$ measures the 'local' anisotropy in the coupling of the spins of electrons in volume elements $d\mathbf{r}_1$ and $d\mathbf{r}_2$, due to quantization of total spin along the given z-axis. The total anisotropy, obtained by integration over $\mathbf{r}_1$ and $\mathbf{r}_2$, would vanish in a completely isotropic situation with $\langle S_x^2 \rangle = \langle S_y^2 \rangle = \langle S_z^2 \rangle$ but in an (S, M_S) eigenstate has the value $3M_S^2 - S(S+1)$.

As the discussion of two-electron properties closely parallels that of the one-electron properties, and has been presented elsewhere in some detail (McWeeny and Sutcliffe 1969; McWeeny 1971), the theory will not be developed further in this review.

5.5.4. *Transition densities*

Before proceeding to the theory of electronic properties, one straightforward generalization is necessary: the density functions for a given state determine expectation values, but for many purposes it is also necessary to know off-diagonal matrix elements connecting two different states. Obvious examples occur in the evaluation of transition probabilities (where the perturbing operator may be, for example, a time-dependent electric-dipole term, or a spin magnetic moment) and in the theory of the splitting of degenerate levels by a static field. To this end we define

$$\rho_1(AB|\mathbf{x}_1; \mathbf{x}'_1) = N \int \Psi_A(\mathbf{x}_1, \mathbf{x}_2, \dots, \mathbf{x}_N)\Psi_B^*(\mathbf{x}'_1, \mathbf{x}_2, \dots, \mathbf{x}_N) \, d\mathbf{x}_2 \dots d\mathbf{x}_N \quad (5.86)$$

as the transition density associated with transition from state Ψ_A to state Ψ_B, and note that all the other density functions so far defined may be generalized in a precisely similar way. The functions employed so far (without explicit labelling of the states) simply correspond to the special case $A = B$. Just as expectation values are determined by formulae such as (5.67), off-diagonal matrix elements follow from

$$\left\langle \Psi_B \left| \sum_i A(i) \right| \Psi_A \right\rangle = \int \{A(1)\rho_1(AB|\mathbf{x}_1; \mathbf{x}'_1)\}_{\mathbf{x}'_1 = \mathbf{x}_1} \, d\mathbf{x}_1 \quad (5.87)$$

and similar expressions. In particular, corresponding to (5.73) we obtain for any function of electronic position ($A(i) = A(\mathbf{r}_i)$)

$$\left\langle \Psi_B \left| \sum A(i) \right| \Psi_A \right\rangle = \int A(\mathbf{r}_1)P_1(AB|\mathbf{r}_1) \, d\mathbf{r}_1. \quad (5.88)$$

The matrix element may thus be obtained 'classically', simply by averaging the function over the transition density. This result makes it easy to visualize pictorially the origin of many effects: for example the intensity of an electric dipole induced transition, with electric vector in the x-direction, depends on the x-moment of a transition charge density $P_1(AB|\mathbf{r})$ connecting the states concerned, and a transition is forbidden when this density vanishes identically or when the appropriate moment is zero. Similarly, $Q_S(AB|\mathbf{r})_x$ represents a distribution of magnetization density in the x-direction, associated with the transition $A \to B$, and determines an e.s.r. transition probability under the influence of an oscillating microwave field in the x-direction.

The transition densities connecting states belonging to a degenerate spin multiplet, or to two different multiplets, are normally inter-related by equations such as (5.80). This means that only one 'standard' transition density need be calculated for a given pair of multiplets. In fact (5.80) is a special case of an important result which we need later in discussing the resolution of degeneracies by the spin terms in the Hamiltonian.

The relationships we require arise from the Wigner–Eckart theorem, which applies to matrix elements of sets of operators (e.g. S_x, S_y, S_z) between wavefunctions (e.g. spin eigenfunctions) with standard transformation properties under rotation of axes. Thus the set of $2S+1$-functions Ψ_{A,M_A} with $M_A = S_A, S_A-1, \ldots, -S_A$ provides a standard irreducible representation of the rotation group, in the sense that corresponding functions $\{\Psi'_{A,M_A}\}$ with rotated axes of spin quantization are related to the original set by the rule

$$\Psi'_{A,M_A} = \sum_{M'_A} \Psi_{A,M'_A} D_{S_A}(R)_{M'_A M_A} \tag{5.89}$$

where $\mathbf{D}_{S_A}(R)$ is the matrix associated with rotation $\mathbf{R}$ in irreducible representation D_{S_A}. The theorem states that if $\{T_m^{(s)}\}$ is a set of *operators* (so-called 'irreducible tensor operators of rank s') transforming like angular momentum eigenfunctions with quantum numbers s, m, then

$$\langle \Psi_{B,M_B}| T_m^{(s)} |\Psi_{A,M_A}\rangle = \begin{pmatrix} S_A & s & | & S_B \\ M_A & m & | & M_B \end{pmatrix} \times \text{'constant'} \tag{5.90}$$

where the 'constant' is independent of the values of M_A, M_B, and m, and its coefficient is the Clebsch–Gordon coefficient† for coupling angular momenta S_A and s to a resultant S_B.

The 'vector operators' S_x, S_y, S_z, used in defining the components of the spin density according to (5.77), do not in fact possess the standard transformation properties. But operators of the type $T_m^{(s)}$ with $s = 1$ may be set up as follows (dropping the superscript):

$$S_{+1} = -(S_x + \mathrm{i}S_y)/\sqrt{2}, \qquad S_0 = S_z, \qquad S_{-1} = (S_x - \mathrm{i}S_y)/\sqrt{2}. \tag{5.91}$$

† Many notations are in use: the one used here parallels that for the $3j$ symbols.

These transform, under rotation of axes, exactly like a set of p-functions (with Condon–Shortley phase conventions). We may now define the transition spin density connecting states Ψ_{A,M_A} and Ψ_{B,M_B} by taking the matrix element of a typical spin-dependent term in the Hamiltonian $\sum_i f(i)S_m(i)$ where $f(i)$ is a spatial operator. By (5.67) we obtain

$$\left\langle \Psi_{B,M_B} \left| \sum_i f(i)S_m(i) \right| \Psi_{A,M_A} \right\rangle = \int \{f(1)Q_S(A, M_A; B, M_B|\mathbf{r}_1; \mathbf{r}'_1)_m\}_{\mathbf{r}'_1=\mathbf{r}_1}\, d\mathbf{r}_1 \tag{5.92}$$

where the transition spin density arises from the spin integration:

$$Q_S(A, M_A; B, M_B|\mathbf{r}_1; \mathbf{r}'_1)_m = \int [S_m(1)\rho_1(A, M_A; B, M_B|\mathbf{x}_1; \mathbf{x}'_1)_{s'_1=s_1}\, ds_1. \tag{5.93}$$

Comparison of (5.92) and (5.90), however, shows that the transition densities must contain a common density function (independent of M_A, M_B, m) multiplied by a Clebsch–Gordon coefficient. As in (5.80) it is convenient to define a standard density by considering the case $M_A = S_A$, $M_B = S_B$, noting that the vanishing of the Clebsch–Gordon coefficient unless $M_A + m = M_B$ implies that there is only one non-vanishing component between the standard states, that with $m = S_B - S_A$. The final result is then

$$Q_S(A, M_A; B, M_B|\mathbf{r}_1; \mathbf{r}'_1)_m = \begin{bmatrix} S_A & 1 & | & S_B \\ M_A & m & | & M_B \end{bmatrix} Q_S^{\text{st}}(AB|\mathbf{r}_1; \mathbf{r}'_1) \tag{5.94}$$

where the standard density is

$$Q_S^{\text{st}}(AB|\mathbf{r}_1; \mathbf{r}'_1) = Q_S(A, S_A; B, S_B|\mathbf{r}_1; \mathbf{r}'_1)_{S_B - S_A} \tag{5.95}$$

and determines all other transition densities connecting multiplets A and B. The square-bracket coefficient in (5.94) is simply the ratio of two Clebsch–Gordon coefficients

$$\begin{bmatrix} S_A & s & | & S_B \\ M_A & m & | & M_B \end{bmatrix} = \begin{pmatrix} S_A & s & | & S_B \\ M_A & m & | & M_B \end{pmatrix} \begin{pmatrix} S_A & s & | & S_B \\ S_A & (S_B - S_A) & | & S_B \end{pmatrix}^{-1} \tag{5.96}$$

and may be obtained from tables. In particular, for the individual states of a given multiplet ($B = A$ and $M_B = M_A$) the coefficients reduce to (M_A/S_A) and we obtain the result (5.80) for the spin densities. A convenient formula for the *transition* spin densities ($M_B \neq M_A$) within a multiplet may be obtained by noting that the Clebsch–Gordon coefficients can be eliminated in favour of matrix elements of simple (total) spin operators (see, for example, McWeeny 1971, Chapter 5). For any pair of multiplets of the same multiplicity ($S_A = S_B$)

we obtain

$$Q_S(A, M_A; B, M_B|\mathbf{r}_1; \mathbf{r}_1')_m = S^{-1}\langle S, M_B|S_m|S, M_A\rangle Q_S^{\text{st}}(AB|\mathbf{r}_1; \mathbf{r}_1') \qquad (S_A = S_B = S) \quad (5.97)$$

where S_m is the m component of total spin, whose eigenstates are the kets $|S, M\rangle$. This obviously yields (5.80) in the case $M_A = M_B = M$ since $\langle S, M|S_z|S, M\rangle = M$.

Similar considerations apply to the two-electron functions, such as Q_{SS} which has five components, but the generalizations present no new features and have been fully discussed elsewhere (McWeeny 1965, 1971).

5.6. Perturbation theory and molecular properties. First-order properties

The experimentally observable properties of a molecule may in general be related to its response to some kind of perturbation; the response may be calculated by perturbation theory and different properties may be classified according to the order of perturbation theory in which they arise. Thus, when a uniform electric field is applied, the first-order perturbation energy (linear in field strength) determines the dipole moment, while the second-order term (quadratic in the field) determines the electric polarizability tensor. In this section we show how a wide range of properties, including those associated with all the small terms in the Hamiltonian (Section 5.3), may be discussed in terms of perturbation theory, and how these properties are related to the basic density functions of the unperturbed system.

In order to obtain a unified treatment, including both degenerate and non-degenerate states, it is convenient to use a matrix partitioning approach (Löwdin 1959). In general the wavefunction Ψ may be expanded in terms of any complete orthonormal set of N-electron functions $\{\Phi_\kappa\}$ in the form

$$\Psi = \sum_\kappa c_\kappa \Phi_\kappa. \tag{5.98}$$

We assume that the set may be broken into two subsets; a truncated set $\{\Phi_{\kappa_a}\}$ suitable for obtaining a first approximation to the states of interest; and $\{\Phi_{\kappa_b}\}$, the orthogonal complement of $\{\Phi_{\kappa_a}\}$. The Schrödinger equation then takes a matrix form

$$\mathbf{Hc} = E\mathbf{c} \tag{5.99a}$$

and may be partitioned to conform to the choice of subsets:

$$\left(\begin{array}{c|c} \mathbf{H}^{AA} & \mathbf{H}^{AB} \\ \hline \mathbf{H}^{BA} & \mathbf{H}^{BB} \end{array}\right)\left(\begin{array}{c} \mathbf{a} \\ \hline \mathbf{b} \end{array}\right) = E\left(\begin{array}{c} \mathbf{a} \\ \hline \mathbf{b} \end{array}\right). \tag{5.99b}$$

† If the complete set includes a continuum, the summation is understood to include integration over the corresponding range.

It is then well known that the solutions of (5.99) may be obtained iteratively by solution of the equation

$$\mathbf{H}_{\text{eff}}\mathbf{a} = E\mathbf{a} \tag{5.100}$$

in which $\mathbf{H}_{\text{eff}}$ is a matrix of n_a rows and columns (n_a functions in set A) and $\mathbf{H}_{\text{eff}}$ is the matrix of an 'effective Hamiltonian' which depends implicitly on the eigenvalue parameter E. Clearly (5.100) must be solved iteratively, a given approximate eigenvalue being used to define an $\mathbf{H}_{\text{eff}}$ from which an improved approximation may be obtained. The effective Hamiltonian is defined by its matrix elements

$$\mathbf{H}_{\text{eff}} = \mathbf{H}^{AA}+\mathbf{H}^{AB}(\mathbf{E}-\mathbf{H}^{BB})^{-1}\mathbf{H}^{BA} \tag{5.101}$$

where $\mathbf{E}$ is E times the n_B-dimensional unit matrix. Expansion of the inverse shows that $\mathbf{H}_{\text{eff}}$ has elements

$$\langle\kappa'_a|H_{\text{eff}}|\kappa_a\rangle = \langle\kappa'_a|H|\kappa_a\rangle+\sum_{\kappa_b}\frac{\langle\kappa'_a|H|\kappa_b\rangle\langle\kappa_b|H|\kappa_a\rangle}{E-E_{\kappa_b}}+\dots. \tag{5.102}$$

This expansion may be used whenever the off-diagonal elements are small enough to ensure convergence. In dealing with a particular state Φ_{κ_a}, the energy in the denominator is replaced by a first approximation $E \sim E_{\kappa_a} = \langle\kappa_a|H|\kappa_a\rangle$ and, using the resultant $\mathbf{H}_{\text{eff}}$, (5.100) is solved to obtain an improved estimate.† After the iteration, the final values of the expansion coefficients in (5.98) can be obtained from the partitioned equation (5.99b); the coefficients c_{κ_a} are contained in the column $\mathbf{a}$, while the corresponding c_{κ_b} follow from the column

$$\mathbf{b} = (\mathbf{E}-\mathbf{H}^{BB})^{-1}\mathbf{H}^{BA}\mathbf{a}. \tag{5.103}$$

This result may be expanded, as in (5.102), to give

$$c_{\kappa_b} = \frac{\langle\kappa_b|H|\kappa_a\rangle}{E-E_{\kappa_b}}+\sum_{\kappa'_b(\neq\kappa_b)}\frac{\langle\kappa_b|H|\kappa'_b\rangle\langle\kappa'_b|H|\kappa_a\rangle}{(E-E_{\kappa_b})(E-E_{\kappa'_b})}+\dots. \tag{5.104}$$

The connection between these results and those of the Brillouin–Wigner and Rayleigh–Schrödinger perturbation theories, which they closely resemble, becomes clear when H is separated into an unperturbed part and a perturbation.

In the present context we write

$$H = H_0+H' = H_0+\lambda H^{\lambda}+\mu H^{\mu}+\dots \tag{5.105}$$

where $\lambda, \mu \to 1$ switches on the two perturbations H^{λ} and H^{μ}, which represent typical small terms (e.g. from spins and fields) in the Hamiltonian (5.31).

† Normally by taking the root nearest to E_{κ_a} (or the eigenvector with the greatest Φ_{κ_a} component). Care must be taken, in cases of poor convergence, to select the appropriate solution (see, for example, Löwdin 1959).

First, the Φ_κ are taken to be exact eigenfunctions of the 'model' Hamiltonian H_0 (in which such effects are ignored) with eigenvalues $E^0_{\kappa_a}$. The precise division of H into H_0 and H' will of course depend on the situation considered. We then have

$$\begin{aligned}\langle\kappa'_a|H|\kappa_a\rangle &= \delta_{\kappa'_a\kappa_a}E^0_{\kappa_a}+\langle\kappa'_a|H'|\kappa_a\rangle \\ \langle\kappa_b|H|\kappa_a\rangle &= \langle\kappa_b|H'|\kappa_a\rangle,\end{aligned} \tag{5.106}$$

where the first result implies a choice of 'correct zero-order eigenfunctions' in the usual sense. The first approximation to each perturbed energy is then given by

$$E_{\kappa_a} = E^0_{\kappa_a}+\langle\kappa_a|H'|\kappa_a\rangle, \qquad E_{\kappa_b} = E^0_{\kappa_b}+\langle\kappa_b|H'|\kappa_b\rangle. \tag{5.107}$$

The elements of the effective Hamiltonian then become to first order (i.e. with the first-order energy E_{κ_a} substituted in place of the unknown E),

$$\langle\kappa'_a|H_{\text{eff}}|\kappa_a\rangle = \delta_{\kappa'_a\kappa_a}E^0_{\kappa_a}+\langle\kappa'_a|H'|\kappa_a\rangle+\sum_{\kappa_b}\frac{\langle\kappa'_a|H'|\kappa_b\rangle\langle\kappa_b|H'|\kappa_a\rangle}{(E_{\kappa_a}-E_{\kappa_b})}+\ \dots. \tag{5.108}$$

If the A-set comprises a single non-degenerate function Φ_{κ_a}, the single element $\langle\kappa_a|H_{\text{eff}}|\kappa_a\rangle$ becomes an approximation to the perturbed energy,

$$E_{\kappa_a} = E^0_{\kappa_a}+\langle\kappa_a|H'|\kappa_a\rangle+\sum_{\kappa_b}\frac{\langle\kappa_a|H'|\kappa_b\rangle\langle\kappa_b|H'|\kappa_a\rangle}{E_{\kappa_a}-E_{\kappa_b}}+\ \dots \tag{5.109}$$

and agrees with the Rayleigh–Schrödinger result except that the energy denominators contain 'shifted' energies (i.e. first-order corrected) instead of unperturbed energies; such denominators are characteristic of 'improved perturbation formulas' (Morse and Feshbach, 1954; Section 9.1). In fact, use of $E \sim E_{\kappa_a}$ in the effective Hamiltonian reproduces the Rayleigh–Schrödinger terms to *third* order plus parts of higher order terms, and for most purposes an extended iteration is quite unnecessary.

If the perturbation H' is a sum of parts as in (5.105) we may collect terms of given degree in the perturbation parameters λ, μ, ... and write the secular matrix $\mathbf{H}_{\text{eff}}$ as a sum of first-, second-, and higher-order terms. Thus

$$\mathbf{H}_{\text{eff}} = \mathbf{H}^{(1)}+\mathbf{H}^{(2)}+\ \dots \tag{5.110}$$

where, using (5.107) in the denominators,

$$H^{(1)}_{\kappa'_a\kappa_a} = \lambda\langle\kappa'_a|H^\lambda|\kappa_a\rangle+\mu\langle\kappa'_a|H^\mu|\kappa_a\rangle+\ \dots \tag{5.111a}$$

$$H^{(2)}_{\kappa'_a\kappa_a} = \lambda^2\sum_{\kappa_b}\frac{\langle\kappa'_a|H^\lambda|\kappa_b\rangle\langle\kappa_b|H^\lambda|\kappa_a\rangle}{E^0_{\kappa_a}-E^0_{\kappa_b}}+\ \dots \tag{5.111b}$$

$$+\lambda\mu\sum_{\kappa_b}\frac{\langle\kappa'_a|H^\lambda|\kappa_b\rangle\langle\kappa_b|H^\mu|\kappa_a\rangle+\langle\kappa'_a|H^\mu|\kappa_b\rangle\langle\kappa_b|H^\lambda|\kappa_a\rangle}{E^0_{\kappa_a}-E^0_{\kappa_b}}+\ \dots.$$

The perturbations therefore have additive effects in first order, and may be considered one at a time, but in second order there are cross-terms (in $\lambda\mu$) which lead to pairwise couplings among the various perturbations. To investigate any particular property it is only necessary to pick out the associated term, or pair of terms, and evaluate the associated perturbation energies. We now consider a number of examples, showing how the results may be interpreted pictorially in terms of charge and spin density functions.

5.6.1. *First-order effects. Non-degenerate state*

In the applications we shall make, the perturbation H' in (5.105) contains the small terms discussed in Section 5.3 which may include both spin interactions and applied fields. The unperturbed Hamiltonian H_0 is normally that defined in (5.19), (5.20) and (5.21), in which $V(i)$ represents the potential energy of electron i in the field of the fixed nuclei. The perturbation H' then contains sums of one- and two-electron terms, $\sum_i \delta h(i)$ and $\frac{1}{2}\sum_{i,j}\delta g(i,j)$, in which a typical one-electron perturbation would be $\delta h(i) = \delta V(i)$, a change in potential energy of electron i due to an applied electric field. Some of the effects of a magnetic field have been anticipated already in Sections 5.4 and 5.5.

Before turning to specific examples it is useful to consider generally the first-order effect of a one-electron perturbation in which the operator $h(i)$ of (5.19) is broken into two parts

$$h(i) = h_0(i) + \delta h(i)$$

where $\delta h(i)$ may stand for any of the small terms (in (5.29), for example) or for a small change in the potential energy function $V(i)$.

The change in the many-electron Hamiltonian is simply $H' = \sum_i \delta h(i)$ and the first-order energy changes in the levels of a degenerate set follow on solving (5.100) with elements given by (5.102). For a non-degenerate level, dropping the state label κ_a, the energy change is simply the expectation value of H' (i.e. of $\sum_i \delta h(i)$) and on using (5.67) we obtain

$$\delta E = \int \{\delta h(1)\rho_1(\mathbf{x}_1;\mathbf{x}'_1)\}_{\mathbf{x}'_1=\mathbf{x}_1}\,\mathrm{d}\mathbf{x}_1. \tag{5.112}$$

This is essentially a statement of the result of first-order perturbation theory; the energy change is the expectation value of the perturbation, computed using the density function of the unperturbed system. But it shows very clearly the importance of the 'charge-cloud' picture of the electron distribution; for if $\delta h(1) = \delta V(\mathbf{r}_1)$, a change in electronic potential energy (e.g. due to an externally applied field), then as in (5.73)

$$\delta E = \int \delta V(\mathbf{r}_1)P_1(\mathbf{r}_1)\,\mathrm{d}\mathbf{r}_1 \tag{5.113}$$

which is simply the change in potential energy of a charge cloud of density $P_1(\mathbf{r}_1)$. This interpretation is consistent because $P_1(\mathbf{r}_1)\,d\mathbf{r}_1$ then represents the *fraction* of an electron in volume element $d\mathbf{r}_1$ while $\delta V(\mathbf{r}_1)$ is the change of potential energy that would be suffered by *one* electron at point $\mathbf{r}_1$.

The following examples, all referring to a non-degenerate unperturbed state, indicate the value of the charge-cloud picture:

(1) *Dipole and multipole moments.* A uniform applied field F, along the z-axis, produce a perturbation $\delta V(i) = Fez_i$ and δE may be written in the classical form

$$\delta E = Fe\int zP_1(\mathbf{r})\,d\mathbf{r} = -FM_z,$$

where M_z is the z-component of the electric moment of a charge-cloud of density $-eP_1(\mathbf{r})$. More generally, $\delta E = (F_xM_x+F_yM_y+F_zM_z) = -\mathbf{F}\,.\,\mathbf{M}$ where the x- and y-components are similarly defined. Quadrupole and higher moments may be obtained by assuming δV to be produced by an external point charge and writing the interaction energy in a corresponding classical form; the multipole moments in this classical form then again turn out to be just those of a static distribution of charge of density $P_1(\mathbf{r})$ (electrons/unit volume).

There are countless other applications of the same general type: for example, interaction with a nuclear quadrupole moment is determined by a field gradient computed classically from the charge density $P_1(\mathbf{r})$, and is revealed in nuclear quadrupole resonance experiments (degenerate nuclear spin states being resolved by interaction with the electronic environment).

(2) *Intermolecular forces.* At infinite separation, the wavefunction for two interacting molecules, A and B (in non-degenerate ground states), may be written $\Psi = \Psi_A\Psi_B$. At moderate distances, this function should be antisymmetrized to allow for exchange and then defines an unperturbed system to which the interactions (electron–nuclear and electron–electron) may be added as a perturbation. The interaction energy was first given, without exchange, by Longuet–Higgins (1956) and, with exchange, by McWeeny (1959) in the form

$$\begin{aligned} E_{\text{int}} = &\int V_A(1)P_1^B(\mathbf{r}_1)\,d\mathbf{r}_1 + \int V_B(1)P_1^A(\mathbf{r}_1)\,d\mathbf{r}_1 \\ &+ \int g(1,2)P_1^A(\mathbf{r}_1)P_1^B(\mathbf{r}_2)\,d\mathbf{r}_1\,d\mathbf{r}_2 \\ &- \int g(1,2)P_1^A(\mathbf{r}_2;\mathbf{r}_1)P_1^B(\mathbf{r}_1;\mathbf{r}_2)\,d\mathbf{r}_1\,d\mathbf{r}_2 \end{aligned} \tag{5.114}$$

where the terms represent, respectively, the potential energy of the charge cloud of B in the field due to the nuclei of A and vice versa, the repulsion energy between the two charge clouds (again classically computed), and finally an 'exchange correction' which is small until A and B begin to interpenetrate. The more usual expressions for interaction energies involve dipole–dipole, dipole–quadrupole, and similar interactions, but such terms correspond merely to a multipole expansion (which is quite unnecessary) of the integrals in the exact expression (5.114). Intermolecular energies are discussed more fully by Professor Murrell in Chapter 7 of this volume.

(3) *Force on a nucleus. Hellmann–Feynman theorem.* Displacement of nucleus n along the x-axis, by an amount δX_n, produces a certain perturbation $\delta V(\mathbf{r}_i)$ in the potential energy of electron i. Let us divide both sides of (5.113) by δX_n and consider the limit $\delta X_n \to 0$: we obtain

$$F_{nx} = \int F_{nx}(\mathbf{r}) P_1(\mathbf{r})\, d\mathbf{r} \qquad (5.115)$$

where $F_{nx} = -\partial E/\partial X_n$ is the x-component of force on nucleus n due to interaction with the electron distribution (i.e. rate of change of total electronic energy with displacement), and $F_{nx}(\mathbf{r}) = -\partial V(\mathbf{r})/\partial X_n$ is the x-component due to a single electron at point $\mathbf{r}$. Since there are $P(\mathbf{r})$ electrons per unit volume at point $\mathbf{r}$, this result means that each element of the charge cloud exerts a force on each nucleus and the total force acting may be computed by summing the contributions. This is the famous Hellmann–Feynman theorem (for a recent review see Epstein, Hurley, Wyatt, Parr (1967), which states that the forces holding the nuclei together in a molecule (against their mutual repulsion) may be computed classically, just as if they were embedded in a continuous distribution of negative charge.

(4) *Geometry changes in excited states.* For a molecule in its ground-state equilibrium configuration the Hellmann–Feynman forces are exactly balanced by the nuclear repulsions. After transition to an excited state, however, this is no longer so, and a change of geometry normally occurs. The forces causing such changes may be computed from (5.115), using $P_1(\mathbf{r})$ for the excited state: inspection of the form of the new density function will often indicate the probable nature of the geometry changes (Coulson and Luz 1968). It is surprising that so little effort has been put into the calculation of charge densities in excited states, in view of their potential importance in the discussion of dissociation and reaction paths.

(5) *The Jahn–Teller effect.* Another familiar instance of a spontaneous geometry change occurs in the Jahn–Teller effect (Jahn and Teller 1937), which may be discussed alternatively (Clinton and Rice 1959) in terms of the force field associated with the charge density according to (5.115). If the electronic ground state of a polyatomic molecule, calculated with some assumed geometry, turns out to be degenerate, the theorem asserts that a distortion will occur in such a way that the degeneracy will be resolved as far as possible.† The reason for this result, from the present point of view, is that whereas the product $\Psi\Psi^*$ is invariant under all symmetry operations when Ψ belongs to a unidimensional irreducible representation of the molecular point group it is *not* invariant if Ψ is any member of a degenerate set (providing a many-dimensional irreducible representation): this simply means that the charge density has the full symmetry of the assumed nuclear configuration in the non-degenerate case (i.e. is invariant under all symmetry operations) but has a lower symmetry in the degenerate case —and hence gives rise to a distribution of forces tending to reduce the symmetry. In addition to its conceptual transparency, the force-field approach can in principle allow the prediction of the type of distortion most likely to occur. An interesting application to the molecule VCl_4 has been considered by Coulson and Deb (1969). Other interesting applications, to the dynamic Jahn–Teller effect and to vibronic coupling in general, have been suggested by Clinton (1960) but still await further development.

5.6.2. *First–order effects. Degenerate states*

In a general polyatomic molecule, the degeneracy of the unperturbed levels is normally a *spin* degeneracy which is resolved when fields are applied and spin terms are taken into account. In the previous section, with no such degeneracy, the charge density played the major role; when spin degeneracy is admitted, interest turns towards the various spin density functions (which vanish in a singlet state).

In order to include the hyperfine coupling terms (5.32f) we need first to make a slight extension of the perturbation theory. Instead of the basis functions Φ_κ, it is necessary to employ the electron-nuclear product functions $\Phi_{\kappa\lambda}\Theta_\lambda$ in which Θ_λ describes a nuclear spin state: thus for two protons, n and n', Θ_λ will be a linear combination of the spin products $\alpha(n)\alpha(n')$, $\alpha(n)\beta(n')$, $\beta(n)\alpha(n')$, $\beta(n)\beta(n')$. The ground-state manifold of electronically degenerate

† With an odd number of electrons a spin degeneracy (Kramers degeneracy) will persist until a magnetic field is applied; in the present context spin effects are unimportant. It should be noted also that this is a case where, for completeness, degenerate perturbation theory should be used; we may, however, consider each state of the set $\{\Psi_{\kappa a}\}$ separately if we use the 'correct zero order states' with the assumed distortion as the perturbation.

states $\{\Phi_{\kappa_a}\}$ then acquires an additional nuclear spin degeneracy, and the aim of the perturbation theory is to discuss the resolution of degeneracies in the set $\{\Phi_{\kappa_a}\Theta_\lambda\}$ due to the small terms in the Hamiltonian. The theory developed earlier may be taken over in its entirety on replacing $\langle\kappa'_a|H|\kappa_a\rangle$ by $\langle\kappa'_a\lambda'|H'|\kappa_a\lambda\rangle$, etc., and setting up the effective Hamiltonian with matrix elements $\langle\kappa'_a\lambda'|H_{\text{eff}}|\kappa_a\lambda\rangle$; solution of the corresponding secular equations gives the effect of electron–nuclear spin interactions. Examples are:

5.6.2.1. *Hyperfine coupling constants in e.s.r.* Here we consider the contact coupling term in (5.32f) and the mechanism by which it produces a hyperfine splitting of the electronic Zeeman levels of a molecule in a magnetic field. We assume the electronic level possesses spin degeneracy only, comprising $2S+1$ states labelled by the quantum number M_S, and treat the spin–field and spin–spin interactions as the perturbation. The elements of the matrix $\mathbf{H}^{(1)}$ in (5.110), which gives the first-order splitting, are then $\langle M'_S\lambda'|H'|M_S\lambda\rangle$ where we use M_S in place of κ_a to label the degenerate states, and where (setting $(8\pi/3c^2\kappa_0)g\beta g_n\beta_p = \mu_n$):

$$H' = g\beta\sum_i \mathbf{B}\,.\,\mathbf{S}(i) - \sum_n g_n\beta_p\mathbf{B}\,.\,\mathbf{I}(n) + \sum_{i,n}\mu_n\delta(\mathbf{r}_{ni})\mathbf{I}(n)\,.\,\mathbf{S}(i). \qquad (5.116)$$

Matrix elements of the first two sums are trivial: the first may be written $g\beta\mathbf{B}\,.\,\mathbf{S}$ (*total* spin operator) while the second is independent of electronic variables. To deal with the third we write the scalar product in standard form (Racah 1942) in terms of the irreducible components defined in (5.91): simple rearrangement shows that

$$\mathbf{I}(n)\,.\,\mathbf{S}(i) = \sum_m (-1)^m I_{-m}(n)S_m(i) \qquad (m = 0, \pm 1). \qquad (5.117)$$

We then use (5.92) to obtain, putting $\mathbf{r}'_1 = \mathbf{r}_1$ since the δ-function is simply a multiplicative factor,

$$\left\langle M'_S\lambda'\left|\sum_i \delta(\mathbf{r}_{ni})\mathbf{I}(n)\,.\,\mathbf{S}(i)\right|M_S\lambda\right\rangle = \sum_m (-1)^m I_{-m}(n)\int \delta(\mathbf{r}_{n1})Q_S(M_SM'_S|\mathbf{r}_1)_m\,\mathrm{d}\mathbf{r}_1$$
$$= \sum_m (-1)^m I_{-m}(n)Q_S(M_SM'_S|\mathbf{R}_n)_m$$

since the effect of the δ-function is to identify $\mathbf{r}_1$ with $\mathbf{R}_n$, the position of nucleus n. Finally, since the transition densities for all choices of M_S and M'_S are determined by the standard density Q_S^{st} (for $M'_S = M_S = S$) according to (5.97), we obtain

$$\left\langle M'_S\lambda'\left|\sum_{i,n}\mu_n\delta(\mathbf{r}_{ni})\mathbf{I}(n)\,.\,\mathbf{S}(i)\right|M_S\lambda\right\rangle$$
$$= \sum_n \mu_n S^{-1}Q_S^{\text{st}}(\mathbf{R}_n)\left\langle M'_S\lambda'\left|\sum_m (-1)^m I_{-m}(n)S_m\right|M_S\lambda\right\rangle.$$

According to (5.117), the matrix element on the right is that of an electron–nuclear spin scalar product $\mathbf{I}(n).\mathbf{S}$, where $\mathbf{S}$ refers to *total* electron spin: it is therefore quite independent of the orbital form of the wavefunction, which merely determines the numerical factor outside. This is a key result and leads to the concept of the *spin Hamiltonian*: if we define

$$H_S = g\beta\mathbf{B}.\mathbf{S} - \sum_n g_n\beta_p\mathbf{B}.\mathbf{I}(n) + \sum_n a_n\mathbf{I}(n).\mathbf{S} \tag{5.118}$$

then this spin Hamiltonian, which contains only (total) electron and nuclear spin operators and numerical parameters, will have exactly the same matrix elements as H' provided (inserting the appropriate $\mu_n = g\beta g_n\beta_p \times (8\pi/3\kappa_0 c^2)$ we identify the 'coupling constant' a_n as

$$a_n = \frac{8\pi}{3}\frac{1}{c^2\kappa_0}g\beta\beta_p g_n S^{-1}Q_S^{\text{st}}(\mathbf{R}_n) = \frac{8\pi g\beta\beta_p}{3\kappa_0 c^2}g_n D_S(\mathbf{R}_n) \tag{5.119}$$

where $D_S(\mathbf{R}_n)$ is the *normalized* spin density (p. 220) evaluated at nucleus n. In other words, to first order in the perturbation calculation, the actual molecule will behave exactly like a fictitious spin-only system with Hamiltonian (5.118); setting up the secular equations, using a basis of electron–nuclear spin functions, the resultant matrix $\mathbf{H}_S$ will be numerically identical with $\mathbf{H}^{(1)}$ (the first order approximation to $\mathbf{H}_{\text{eff}}$) which gives the perturbed levels according to degenerate perturbation theory, through solution of (5.100).

It is also worth noting that the electron–nuclear dipole–dipole interaction in (5.32f) may also be described by a simple spin-Hamiltonian term, this time of the form

$$\sum_n \sum_{\lambda,\mu} a_{n,\lambda\mu}^{\text{dip}} S_\lambda I_\mu(n) \qquad (\lambda, \mu = x, y, z) \tag{5.120}$$

where, as is customary, the result is expressed in Cartesian components. The elements of the coupling tensor are then, for example,

$$\begin{aligned} a_{n,xx}^{\text{dip}} &= (3g\beta g_n\beta_P)\int r_{n1}^{-5}(x_{n1}^2 - \tfrac{1}{2}r_{n1}^2)S^{-1}Q_S^{\text{st}}(\mathbf{r}_1)\,\mathrm{d}\mathbf{r}_1 \\ a_{n,xy}^{\text{dip}} &= (3g\beta g_n\beta_P)\int r_{n1}^{-5}x_{n1}y_{n1}S^{-1}Q_S^{\text{st}}(\mathbf{r}_1)\,\mathrm{d}\mathbf{r}_1 \end{aligned} \tag{5.121}$$

and thus measure second moments of the spin distribution in the vicinity of the nucleus.

5.6.2.2. *Spin-orbit coupling in diatomic molecules.* In the absence of the spin–orbit coupling terms in (5.29) and (5.30), the ground state of a diatomic molecule or radical may possess both orbital and spin degeneracy. The kets of the degenerate manifold (i.e. the $|\kappa_a\rangle$) may then be denoted by $|\Lambda, \Sigma\rangle$ where Λ and Σ are quantum numbers for the axial component of total orbital and spin angular momenta, respectively. The analysis is similar to that in Section (5.6.2.1), with omission of nuclear spin effects, and will only be sketched. The main spin–orbit terms appear in (5.32c), which for $\mathbf{B} = 0$ becomes

$$H_{\text{SL}} = \frac{g\beta^2}{\kappa_0 c^2}\left[\sum_{n,i} Z_n \frac{\mathbf{S}(i)\,.\,\mathbf{L}^n(i)}{r_{ni}^3} - \frac{1}{2}\sum_{i,j}\frac{\mathbf{S}(i)\,.\,\mathbf{L}^j(i)}{r_{ji}^3} - \sum_{i,j}\frac{\mathbf{S}(j)\,.\,\mathbf{L}^j(i)}{r_{ji}^3}\right] \quad (5.122)$$

where Z_n is the atomic number of nucleus n, and $\mathbf{L}^p(q)$ indicates the angular momentum of particle q about the position of particle p. The dominant term is the first, which becomes increasingly important for heavy nuclei.

It follows easily that the first-order contribution to the matrix $\mathbf{H}_{\text{eff}}$ is diagonal, and the diagonal elements thus give directly the resolved levels. The matrix elements again coincide with those of a simple 'equivalent operator' of the form

$$H_{\text{SL}} = AL_zS_z \quad (5.123)$$

where L_z and S_z are orbital and spin angular momentum operators (for components along the molecular axis) and the main contribution to the coupling constant A is

$$A = \frac{g\beta^2}{\kappa_0 c^2}\sum_n Z_n \int [r_{n1}^{-3}L_z^n(1)S^{-1}Q_S^{st}(\mathbf{r}_1\,;\mathbf{r}_1')]_{\mathbf{r}_1'=\mathbf{r}_1}\,\mathrm{d}\mathbf{r}_1\,. \quad (5.124)$$

In this case the off-diagonal element of the standard spin density is required, owing to the presence of the angular momentum operator. Thus A depends on the angular character of the spin density in the immediate vicinity of the nuclei. The second and third terms in (5.122) introduce 'spin–other-orbit' effects and therefore depend on a two-electron function (Q_{SL}) akin to the Q_{SS} introduced in (5.85); their inclusion leads simply to two extra terms in A.

The value of calculations of A, as an aid to the assignment of spectra, has been stressed by Walker and Richards (1968); thus for a molecule such as NO, with a $^2\Pi$ ground state, A measures directly the separation of the $^2\Pi_{\frac{1}{2}}$ and $^2\Pi_{\frac{3}{2}}$ states. Calculations along the present lines are fully reported elsewhere (McWeeny and Moores 1972).

5.6.2.3. *Spin–spin splitting of Zeeman levels.* In describing the results of e.s.r. experiments in terms of a spin Hamiltonian, as in Section (5.6.2.1) above, it is sometimes necessary to add a term of the form

$$H_{\text{S}}(\text{spin–spin}) = \sum_{\mu,\nu} d_{\mu\nu}\,S_\mu S_\nu \qquad (\mu, \nu = x, y, z) \quad (5.125)$$

attributable to the electron spin–spin dipolar interaction appearing in (5.32e). This term leads to a splitting of molecular Zeeman levels in the zero-field limit, first observed by Hutchison and Mangum (1958) and van der

Waals and de Groot (1959), and now well documented for the triplet states of many organic molecules and radicals. This is a pure two-electron effect and depends entirely on the spin–spin coupling function introduced in (5.85). It may then be shown (McWeeny and Mizuno 1961) that the elements of the coupling tensor are given by

$$d_{xx} = 3g^2\beta^2 \int r_{12}^{-5}(x_{12}^2 - \tfrac{1}{3}r_{12}^2)D_{SS}(\mathbf{r}_1, \mathbf{r}_2)\,d\mathbf{r}_1\,d\mathbf{r}_2$$

and similar expressions (*cf.* (5.121)), where $D_{SS} = Q_{SS}^{st}/S(2S-1)$ is the normalized density obtained from Q_{SS}^{st} for the state in question (e.g. excited triplet), and only the diagonal element is required.

5.7. Second-order properties

The second-order effects of a perturbation may be obtained in all cases (manifold $|\kappa_a\rangle$ degenerate or non-degenerate) by setting up the matrix $\mathbf{H}^{(2)}$ defined in eqn (5.111b), and solving the secular equations (5.100), in the degenerate case, with $\mathbf{H}_{\text{eff}} = \mathbf{H}^{(1)} + \mathbf{H}^{(2)}$. A vast range of molecular properties may be discussed in this way and we indicate only a few typical examples. The density functions which determine such properties are now clearly the *transition* densities, which connect the state of interest with all other states. The calculation of second-order effects is thus much more difficult, requiring inclusion of the 'excited' functions $|\kappa_b\rangle$, whose admixture describes the perturbation of the wavefunction.

There are two main types of second-order effect, those which are quadratic in a single perturbation (λ^2 terms in (5.111b)), and those which are linear in each of two perturbations ($\lambda\mu$ terms). Electric polarizability belongs to the first type, λ being the field strength; the orbital modification of free-electron g values is of the second type, the two relevant perturbations being the electron-field interaction (orbital Zeeman term; H_P in (5.32*b*)) and the spin–orbit coupling, respectively. Before discussing either type, however, we note that the perturbation calculation may often be effected in two stages with a simple physical interpretation. For example, the field of a point charge polarizes the electron distribution (first-order change of wavefunction); the interaction between the charge and the polarization then gives the polarization energy (second-order energy). Similarly, a coupling between two nuclear spins results from the fact that one produces a spin polarization (first-order change of spin density) which is then 'detected' by the other. It is therefore sometimes useful to switch on one perturbation, calculate its first-order effect on the electron distribution, and finally calculate a related second-order energy change. This procedure has recently acquired some popularity in 'finite perturbation calculations' (Pople, McIver, and Ostlund 1968). There are two particularly important cases.

Method A *Non-degenerate level, single perturbation*

If μH^{μ} represents, say, a perturbation by a point charge, with μ as a strength parameter, the perturbed state to first order is

$$|\tilde{\kappa}_a\rangle = |\kappa_a\rangle + \mu|\kappa_a^{\mu}\rangle \tag{5.126}$$

where

$$|\kappa_a^{\mu}\rangle = \sum_{\kappa_b(\neq\kappa_a)} \frac{|\kappa_b\rangle\langle\kappa_b|H^{\mu}|\kappa_a\rangle}{(E_{\kappa_b}^0 - E_{\kappa_a}^0)}. \tag{5.127}$$

The energy to *second* order may then be computed correctly as the expectation value of μH^{μ} (exactly as in first-order theory) over the perturbed state $|\tilde{\kappa}_a\rangle$, provided the perturbed function is re-normalized including μ^2 terms. Simple rearrangement shows that, with a normalizing factor M, the second-order energy is

$$E^{\mu\mu} = M^2\langle\tilde{\kappa}_a|H^{\mu}|\tilde{\kappa}_a\rangle = \langle\kappa_a|H^{\mu}|\kappa_a^{\mu}\rangle = \langle\kappa_a^{\mu}|H^{\mu}|\kappa_a\rangle \tag{5.128}$$

which corresponds to the μ^2 term in (5.111b). This establishes the validity of a procedure in which the (correctly normalized) *density functions* are calculated using first-order theory, and then employed in obtaining the second-order energy.

Method B *Degenerate level, double perturbation*

We consider the case where λH^{λ} resolves the levels but μH^{μ}, though important, does not. Thus an applied magnetic field interacting with electronic *orbital* motion does not resolve a set of *spin*-degenerate levels, having vanishing matrix elements between different component states, but may lead to important effects in conjunction with a second perturbation (e.g. spin–orbit coupling). Such cases occur rather frequently. We assume then that

$$H' = \lambda H^{\lambda} + \mu H^{\mu}$$

with $\langle\kappa_a'|H^{\mu}|\kappa_a\rangle = 0\ (\kappa_a' \neq \kappa_a)$, and that the effects of interest are those which involve both perturbations jointly ($\lambda\mu$ terms in eqn (5.111b). First we switch on μH^{μ} and obtain the perturbed states (which are non-mixing) according to (5.98) and (5.104). Then we set up the secular problem using for the matrix $\mathbf{H}_{\text{eff}}$ the matrix of the second perturbation λH^{λ} referred to the perturbed states $|\tilde{\kappa}_a\rangle$, i.e. we take

$$\langle\kappa_a'|H_{\text{eff}}|\kappa_a\rangle \simeq \langle\tilde{\kappa}_a'|\lambda H^{\lambda}|\tilde{\kappa}_a\rangle \tag{5.129}$$

It is easily verified that this secular matrix correctly reproduces all the terms linear in λ in $\mathbf{H}_{\text{eff}}$ given by (5.110) and (5.111). This means that instead of employing double perturbation theory, we may use first-order (single) perturbation theory to calculate the $|\tilde{\kappa}_a\rangle$ and the perturbed distribution functions; and then deal with the 'interesting' perturbation (which resolves the levels) within the framework of first-order theory for a degenerate level, using the basis of states $|\tilde{\kappa}_a\rangle$.

Various other special perturbation formulae may be derived but will not be discussed here. We now consider concrete examples, in which Methods A and B will both be useful.

5.7.2.1. *Polarization by a point charge.* A classic problem in the theory of chemical reactivity (see, for example, Coulson and Longuet–Higgins, 1949) is that of predicting the response of a molecule towards a perturbing point charge, representing for example an approaching ion. We have seen (p. 229) that for any perturbation described by a change δV in the potential energy function the first-order energy change (non-degenerate case) is given by (5.113); it therefore represents the classical interaction energy of the charge cloud with the potential source—in the present case a point charge.

If we suppose $\delta V(i) = -qe/r_{0i}$, for a point charge q at point 0, and take this as the perturbation μH^{μ} (using q instead of μ), then (5.127) determines the effect on the electron distribution. The electron density deriving from (5.126) is obtained by integrating $\tilde{\Psi}_{\kappa_a}\tilde{\Psi}^*_{\kappa_a}$ over all variables except $\mathbf{r}_1$ (see (5.66) and (5.71)) and may thus be written $P_1(\tilde{\kappa}_a\tilde{\kappa}_a|\mathbf{r}_1) = P_1(\kappa_a\kappa_a|\mathbf{r}_1)+qP_1^{(q)}(\kappa_a\kappa_a|\mathbf{r}_1)$ where the polarization per unit charge is

$$P_1^{(q)}(\kappa_a\kappa_a|\mathbf{r}_1) = \sum_{\kappa_b}\{\pi_{\kappa_a\kappa_b}P_1(\kappa_a\kappa_b|\mathbf{r}_1)+(\text{c.c.})\} \tag{5.130}$$

and (c.c.) indicates the complex conjugate of the preceeding term. All the densities and transition densities refer to the unperturbed system. The coefficients $\pi_{\kappa_a\kappa_b}$ are determined by, using (5.88),

$$\pi_{\kappa_a\kappa_b} = \frac{\left\langle\kappa_a\left|\sum_i -e/r_{0i}\right|\kappa_b\right\rangle}{(E_{\kappa_a}-E_{\kappa_b})} = \frac{e\int r_{01}^{-1}P_1(\kappa_b\kappa_a|\mathbf{r}_1)\,\mathrm{d}\mathbf{r}_1}{\Delta E(\kappa_a\rightarrow\kappa_b)}. \tag{5.131}$$

The meaning of (5.130) and (5.131) may be stated as follows: the point charge q produces a contamination of the unperturbed charge density in which each transition density (for a $\kappa_a \rightarrow \kappa_b$ transition) occurs with a weight proportional to its electrostatic interaction with a $\kappa_b \rightarrow \kappa_a$ transition density, and inversely proportional to the energy difference $\Delta E(\kappa_a \rightarrow \kappa_b)$. This result gives insight into the connection between spectroscopic transition probabilities—which are likewise determined by the transition densities (p. 224)—and the perturbing effect of a point charge presented in a given position; the most effective terms in the sum are largely determined by geometry and simple electrostatics.

It follows also from the discussion of Method A that the second-order energy change is

$$q^2E^{qq} = q^2 \sum_{\kappa_b} \pi_{\kappa_a\kappa_b} \int r_{01}^{-1} P_1(\kappa_a\kappa_b|\mathbf{r}_1)\mathrm{d}\mathbf{r}_1$$

$$= \tfrac{1}{2}q^2 \int r_{01}^{-1} P_1^{(q)}(\kappa_a\kappa_a|\mathbf{r}_1)\,\mathrm{d}\mathbf{r}_1 \qquad (5.132)$$

and therefore represents the interaction of the point charge with the total polarization it produces, as would be computed classically.

5.7.2.2. *Intermolecular forces, dispersion interactions.* In first order, the interaction between two molecules in non-degenerate ground states may be computed classically (p. 230) from their charge-density functions: for neutral molecules such interactions are very small unless highly polar groups are present. In second order, however, significant long-range forces always occur, due to the response of each molecule to the presence of the other. These effects may be calculated by using a single product $\Psi_{ab} = \Psi_{Aa}\Psi_{Bb}$ as the unperturbed function (Ψ_{κ_a}), corresponding to molecules A and B in states a and b, and allowing admixture with functions $\Psi_{a'b'} = \Psi_{Aa'}\Psi_{Bb'}$ in which one or both of the two molecules have been excited. The dispersion interactions arise from the double excitations and the second-order energy is given by (Longuet–Higgins 1956; McWeeny 1959)

$$E_{\mathrm{disp}} = -\sum_{a',b'} \frac{\left|\int r_{12}^{-1} P_1^A(aa'|\mathbf{r}_1) P_1^B(bb'|\mathbf{r}_2)\,\mathrm{d}\mathbf{r}_1\,\mathrm{d}\mathbf{r}_2\right|^2}{\Delta E(a \to a', b \to b')}. \qquad (5.133)$$

The denominator is positive, being an excitation energy, and the interaction is thus an attraction. The integral in the numerator is the electrostatic interaction of two transition charge clouds, and this is an ‘exact’ first-order result provided the molecular wavefunctions do not overlap; if there is a slight overlap the wavefunction must be antisymmetrized and exchange terms appear (McWeeny 1959); if there is considerable overlap the analysis is more complicated (Dacre and McWeeny 1970) but the interaction energy can still be expressed in terms of density functions for the individual molecules. The important point about (5.133) is that no approximation of the integral is necessary: the traditional expression (see, for example, Margenau and Kestner 1969) results as the first term of a multipole expansion, each density being approximated by a transition *dipole*, but such approximations are unnecessary and are often rather poor.

5.7.2.3. *Electronic g values.* As a final example, we choose one to illustrate the occurrence of transition *spin* densities and show the value of Method B (p. 230). We consider a polyatomic molecule or radical with a non-singlet ground state (spin degeneracy only), and try to discover the origin of the spin-Hamiltonian term

$$H_S(\text{Zeeman}) = \beta \sum_{\mu,\nu} g_{\mu\nu} B_\mu S_\nu \tag{5.134}$$

which is known to give a good description of field–spin interaction. For a free electron, the g-tensor is diagonal with $g_{xx} = g_{yy} = g_{zz} = g$ and this term is merely the spin-only Zeeman term (5.32d):

$$H_z = \sum_i g\beta \mathbf{B} \,.\, \mathbf{S}(i) = g\beta \mathbf{B} \,.\, \mathbf{S}. \tag{5.135}$$

The first-order splitting of the multiplet is obtained by taking the matrix elements of this perturbation to obtain the matrix $\mathbf{H}^{(1)}$ in (5.110). Any modifications must occur in second order and must be bilinear in spin and field components: the obvious choice of two perturbations (λH^λ and μH^μ in the general theory) is

$$H_1 = \sum_i \beta \mathbf{B} \,.\, \mathbf{L}(i) = \beta \mathbf{B} \,.\, \mathbf{L} \tag{5.136}$$

$$H_2 = \frac{g\beta^2}{\kappa_0 c^2} \sum_{n,i} Z_n r_{ni}^{-3} \mathbf{S}(i) \,.\, \mathbf{M}^n(i) \tag{5.137}$$

where H_1 denotes the 'paramagnetic' term in (5.32*b*), and H_2 is the spin–orbit coupling, with neglect of spin–other-orbit effects. It should be noted that $\mathbf{L}(i)$ $(= \hbar^{-1}\mathbf{r}_i \times \mathbf{p}(i))$ is the electronic angular momentum about the (arbitrary) origin, but that $\mathbf{M}^n(i) = \hbar^{-1}\mathbf{r}_{ni} \times \boldsymbol{\pi}(i)$ is about nucleus n and in the presence of the field contains the gauge invariant operator $\boldsymbol{\pi}$ instead of $\mathbf{p}$.

The physical mechanism which leads from these perturbations is simple: H_1 turns on the field and this imparts angular momentum to the electron distribution, in the form of induced currents; the induced currents produce a magnetic moment (proportional to B) and this couples to the electron spin through the spin-orbit interaction H_2; the coupling energy will be linear in field and spin components and must occur as the corresponding $\lambda\mu$ cross-term in (5.111*b*).

General expressions for the g-tensor components have been given elsewhere (McWeeny 1955, 1970) and detailed non-empirical calculations, including spin–other-orbit effects, have been successfully performed (McWeeny and Moores 1972). Here, however, we obtain alternative expressions in terms of the current density induced by H_1. This perturbation is spinless and therefore produces no mixing between functions which differ in their S, M_S values. Up to terms linear in $\mathbf{B}$, the density functions for the perturbed system are obtained as in the derivation of (5.130). Thus

$$\rho_1(\tilde{\kappa}_a\tilde{\kappa}_a|\mathbf{x}_1\,;\mathbf{x}'_1) = \rho_1(\kappa_a\kappa_a|\mathbf{x}_1\,;\mathbf{x}'_1) + \left[\sum_m (-1)^m B_{-m}\sum_{\kappa_b}\pi^m_{\kappa_a\kappa_b}\rho_1(\kappa_a\kappa_b|\mathbf{x}_1\,;\mathbf{x}'_1)+(\text{c.c.})\right], \tag{5.138}$$

where the coefficients $\pi^m_{\kappa_a\kappa_b}$ are this time determined by

$$\pi^m_{\kappa_a\kappa_b} = \frac{-\beta\int\{L_m(1)P_1(\kappa_b\kappa_a|\mathbf{r}_1\,;\mathbf{r}'_1)\}_{\mathbf{r}'_1=\mathbf{r}_1}\,\mathrm{d}\mathbf{r}_1}{\Delta E(\kappa_a\to\kappa_b)} \tag{5.139}$$

(after an integration over spin), and vanish unless $S_b = S_a$, $M_b = M_a$. Let us write this result in the Cartesian form

$$\rho_1(\tilde{\kappa}_a\tilde{\kappa}_a|\mathbf{x}_1\,;\mathbf{x}'_1) = \rho_1(\kappa_a\kappa_a|\mathbf{x}_1\quad\mathbf{x}'_1)+\left\{\sum_{\mu=x,y,z} B_\mu\delta_\mu(\kappa_a\kappa_a|\mathbf{x}_1\,;\mathbf{x}'_1)+(\text{c.c.})\right\} \tag{5.140}$$

so that δ_μ determines the polarization produced by the μ component of the applied field. From this result we may obtain the induced current density by integrating over spin and then using (5.74).

In the present application, however, the perturbation H_2 is spin-dependent, and we must therefore retain the first-order result in the form (5.140). The term bilinear in flux density and spin components arises when, according to (5.129), we evaluate $\langle\tilde{\kappa}'_a|H_2|\tilde{\kappa}_a\rangle$ to obtain the matrix of the spin–orbit coupling within the degenerate manifold. In this way we obtain

$$\begin{aligned}\langle\kappa'_a|H_{\text{eff}}|\kappa_a\rangle &= \frac{g\beta^2}{\kappa_0c^2}\sum_n Z_n\int\{r_{n1}^{-3}[\mathbf{S}(1)\,.\,\mathbf{M}^n(1)]\rho_1(\tilde{\kappa}_a\tilde{\kappa}'_a|\mathbf{x}_1\,;\mathbf{x}'_1)\}_{\mathbf{x}'_1=\mathbf{x}_1}\,\mathrm{d}\mathbf{x}_1\\ &= \frac{g\beta^2}{\kappa_0c^2}\sum_n Z_n\sum_m\int\{(-1)^m r_{n1}^{-3}M^n_{-m}(1)Q_S(\tilde{\kappa}_a\tilde{\kappa}'_a|\mathbf{r}_1\,;\mathbf{r}'_1)_m\}_{\mathbf{r}'_1=\mathbf{r}_1}\,\mathrm{d}\mathbf{r}_1,\end{aligned}$$

in which $Q(\tilde{\kappa}_a\tilde{\kappa}'_a|\mathbf{r}_1\,;\mathbf{r}'_1)_m$ is a transition spin density, as defined in (5.93). This result is reminiscent of the derivation of the formula (5.74) for the current density, but now in place of P_1 there appears Q_S. Analogously however we can define a 'spin' current density. First of all we use (5.97) to obtain

$$\langle\kappa'_a|H_{\text{eff}}|\kappa_a\rangle = \frac{g\beta^2}{\kappa_0c^2}\sum_n Z_nS^{-1}\sum_m(-1)^m\langle M'_a|S_m|M_a\rangle \times\int[r_{n1}^{-3}\hbar^{-1}(\mathbf{r}_{n1}\times\boldsymbol{\pi}(1)\}_{-m}Q_S^{\text{st}}(\tilde{a}\tilde{a}|\mathbf{r}_1\,;\mathbf{r}'_1)]_{\mathbf{r}'_1=\mathbf{r}_1}\,\mathrm{d}\mathbf{r}_1,$$

which may be written (*cf*. (5.117))

$$\langle\kappa'_a|H_{\text{eff}}|\kappa_a\rangle = g\beta\langle M'_a|\mathbf{Y}\,.\,\mathbf{S}|M_a\rangle. \tag{5.141}$$

The vector **Y**, so defined, has Cartesian components

$$Y_\lambda = -\frac{1}{\kappa_0 c^2}\sum_n Z_n \int r_{n1}^{-3}[\mathbf{r}_{n1}\times(-e\mathbf{J}^S)]_\lambda \, \mathrm{d}\mathbf{r}_1, \tag{5.142}$$

where the *spin current density* $\mathbf{J}^S$ has components given by

$$J_\lambda^S(\mathbf{r}_1) = m^{-1}Rl[\pi_\lambda Q_S^{\mathrm{st}}(\tilde{a}\tilde{a}|\mathbf{r}_1\,;\,\mathbf{r}_1')]_{\mathbf{r}_1'=\mathbf{r}_1}. \tag{5.143}$$

This is exactly like the current density (5.74), but contains the spin density $\frac{1}{2}(P_1^\alpha - P_1^\beta)$ in place of the charge density $P_1 = P_1^\alpha + P_1^\beta$: it measures the flow of spin density, induced by the magnetic field, rather than the flow of charge density.

Finally, we note that Q_S^{st} follows from (5.140) and (5.73) and that (5.143) then gives

$$J_\lambda^S(\mathbf{r}_1) = \sum_\mu B_\mu J_{\mu\lambda}^S(\mathbf{r}_1) \tag{5.144}$$

where $J_{\mu\lambda}^S(\mathbf{r}_1)$ is the λ component of the spin current due to unit applied field in the μ direction and is defined by

$$J_{\mu\lambda}^S(\mathbf{r}_1) = m^{-1}Rl\int [\pi_\lambda \delta_\mu(aa|\mathbf{x}_1\,;\,\mathbf{x}_1')]_{\mathbf{x}_1'=\mathbf{x}_1}\,\mathrm{d}s_1. \tag{5.145}$$

This introduces tensor character into the coupling between field and spin; it follows readily that

$$\langle \kappa_a'|H_{\mathrm{eff}}|\kappa_a\rangle = \langle M_a'|g\beta \sum_{\mu,\nu} B_\mu G_{\mu\nu} S_\nu|M_a\rangle \tag{5.146}$$

where, with $\mathbf{J}_\mu(\mathbf{r}_1)$ as the spin current with components (5.145),

$$G_{\mu\nu} = \sum_n \frac{-Z_n}{\kappa_0 c^2}\int r_{n1}^{-3}\{\mathbf{r}_{n1}\times -e\mathbf{J}_\mu\}_\nu \,\mathrm{d}\mathbf{r}_1. \tag{5.147}$$

Eqn (5.146) shows that the part of the perturbation matrix which corresponds to a coupling bilinear in field and spin is correctly reproduced by a spin Hamiltonian term

$$g\beta\mathbf{B}\,.\,\mathbf{GS}$$

where the g-tensor elements are simply numerical parameters determined by (5.147). There is, of course, also a direct *first*-order effect due to the spin Zeeman term (5.135): what we have now shown is that there is a second-order modification arising from induced currents and spin– orbit coupling. To this order, the complete Zeeman term in the spin Hamiltonian may be written

$$H_S(\text{Zeeman}) = \beta\sum_{\mu\,\nu} B_\mu(g\delta_{\mu\nu}+G_{\mu\nu})S_\nu = g\beta\sum_{\mu\nu} B_\mu g_{\mu\nu} S_\nu. \tag{5.148}$$

There is no difficulty in extending this result to include the spin–other-orbit interaction, given in (5.138), further terms then being added to the expression for $G_{\mu\nu}$.

It is interesting to note that, physically, the g-tensor connects induced current components with the applied field components—but that, instead of the total current, it is only the *difference* between its up-spin and down-spin parts that is relevant.

This case has been considered in some detail as the same analysis may be adapted for many other purposes. Thus, a similar analysis gives the second-order effects which arise if we take for H_2 the second nuclear spin term in H_N (5.32f): the result is a field-proportional term in $\mathbf{I}(n)$ which modifies the first-order interaction $(-\sum_n g_n \beta_p \mathbf{B} \cdot \mathbf{I}(n))$ occurring in (5.118) and leads to the form

$$H_S^{\text{nuc}}(\text{Zeeman}) = -\beta_p \sum_{\mu,\nu} g_n B_\mu (\delta_{\mu\nu} - \sigma_{\mu\nu}) I_\nu(n) \tag{5.149}$$

where the numerical coefficients $\sigma_{\mu\nu}$ define the *chemical shift tensor*. In this case, however, the tensor components are determined by the *total* induced current, rather than the difference of its up-spin and down-spin parts. General expressions for the tensor components, in terms of the currents, have been given elsewhere (e.g. McWeeny 1970).

5.8. Conclusion

Throughout this chapter it has been unnecessary to specify in any detail the actual form of the wavefunction; we have assumed only that, in principle, the wavefunctions for the unperturbed molecule are available, and have shown that a vast range of properties may be discussed in terms of a few density functions connecting the unperturbed states. In terms of the densities and currents it is possible to get considerable insight into the physical origin of molecular properties and behaviour, without becoming pre-occupied with the numerical computations. Finally, however, it is necessary to make a few comments on practicalities.

First we note that in *all* orbital theories, where Ψ is constructed from some complete set of spin-orbitals $\{\psi_R\}$, ρ_1 and ρ_2 must appear as linear combinations of products of two spin-orbitals and four spin-orbitals, respectively:

$$\rho_1(\mathbf{x}_1; \mathbf{x}_1') = \sum_{R,S} \rho_{1RS} \psi_R(\mathbf{x}_1) \psi_S^*(\mathbf{x}_1') \tag{5.150}$$

$$\rho_2(\mathbf{x}_1, \mathbf{x}_2; \mathbf{x}_1', \mathbf{x}_2') = \sum_{R,S,T,U} \rho_{2RS,TU} \psi_R(\mathbf{x}_1) \psi_S(\mathbf{x}_2) \psi_T^*(\mathbf{x}_1') \psi_U^*(\mathbf{x}_2'). \tag{5.151}$$

The sets of coefficients provide matrix representations of the density *operators* ρ_1 and ρ_2, obtained by regarding the above functions as integral kernels. Thus the integral operator description of the effect of ρ_1 on an arbitrary (space–spin) function $\psi(\mathbf{x}_1)$ is

$$\rho_1\psi(\mathbf{x}_1) = \int \rho_1(\mathbf{x}_1\,;\mathbf{x}'_1)\psi(\mathbf{x}'_1)\,\mathrm{d}\mathbf{x}'_1. \tag{5.152}$$

It is then apparent that the coefficients ρ_{1RS} are simply matrix elements of the operator in the usual sense, provided the spin-orbitals are assumed orthonormal so that $\langle\psi_R|\psi_S\rangle = \delta_{RS}$:

$$\rho_{1RS} = \langle\psi_R|\rho_1|\psi_S\rangle. \tag{5.153}$$

Similar considerations apply to ρ_2, the spin-orbital products $\psi_R(\mathbf{x}_1)\psi_S(\mathbf{x}_2)$ providing a discrete basis for all two-electron (space–spin) functions and the elements $\rho_{2RS,TU}$ providing a corresponding matrix representation of the operator ρ_2 (rows and columns labelled by index pairs).

The basic spatial density functions defined in Section 5.5 arise by elimination of spin and take similar forms; thus the charge and spin density functions appear as linear combinations of the orbital products $\phi_R(\mathbf{r}_1)\phi_S^*(\mathbf{r}'_1)$. On writing each orbital as a linear combination of basis functions $\{\chi_i\}$ we obtain finally

$$P_1(\mathbf{r}_1\,;\mathbf{r}'_1) = \sum_{i,j} P_{1ij}\chi_i(\mathbf{r}_1)\chi_j^*(\mathbf{r}'_1) \tag{5.154}$$

$$Q_S(\mathbf{r}_1\,;\mathbf{r}'_1) = \sum_{i,j} Q_{Sij}\chi_i(\mathbf{r}_1)\chi_j^*(\mathbf{r}'_1) \tag{5.155}$$

in which the coefficients may be collected into matrices $\mathbf{P}_1$ and $\mathbf{Q}_S$ associated with charge and spin density operators P_1, Q_S. Although the properties of the operators are of considerable interest (see, for example McWeeny and Kutzelnigg 1968) we are here concerned more with the individual elements of $\mathbf{P}_1$ and $\mathbf{Q}_S$, as a means of characterizing the distribution of charge and spin in a molecule. In its original context (Coulson 1935), $\mathbf{P}_1$ was derived for a one-determinant wavefunction, describing the π-electrons of a conjugated molecule in terms of 2p atomic orbitals (the χs); $\mathbf{P}_1$ is then the 'charge-and-bond-order' matrix, with elements whose significance was discussed in Section 5.2. It should be noted that similar matrices may be defined for all the density functions and that it is therefore possible to speak of 'electron populations', 'transition spin populations' etc. at any level of accuracy. The use of a finite basis set $\{\chi_i\}$ simply provides a useful (though in no sense unique) way of resolving a density function into 'orbital' and 'overlap' contributions as defined in Section 5.2, and of formally associating amounts of charge or spin with corresponding regions in space.

Secondly, we note that, although the perturbation theory used in Sections 5.5 and 5.6 appears to require a knowledge of exact molecular wavefunctions, the formulation is general enough to admit systematic refinement of computed properties. For example, if H_0 in (5.31) is re-written

$$H_0 = H_M + (H_0 - H_M) \tag{5.156}$$

H_M being the Hamiltonian of a 'Hartree–Fock model', for which accurate occupied and virtual orbitals are obtainable, then the same treatment applies exactly with an extra perturbation term $\Delta H_M = (H_0 - H_M)$. The effect of ΔH_M on quantities computed using a Hartree–Fock function may then be obtained systematically by proceeding to higher orders of perturbation theory. If ΔH_M is large compared with the other perturbations in (5.31) this correction may have to be carried to high order to obtain good values of computed quantities (see, for example, Green (1971) for a discussion of dipole moments) and for many purposes it may be preferable to estimate the effect of ΔH_0 by an initial configuration interaction calculation. In this connection, it is necessary to consider *which* configurations should be admitted before proceding to calculate the effects of the 'real' perturbations in (5.31). The importance of this question is immediately evident from the example (p. 220) in which the configuration interaction required in order to take some account of electron correlation (i.e. of the interactions contained in ΔH_M) may be absolutely essential to the correct description of nuclear hyperfine coupling: for the spin density (which determines the coupling constant according to (5.119)) is *zero at the nucleus*, in the one-configuration approximation, and non-zero density only appears through the 'polarization' terms.† This problem will not be considered further; it has been discussed by various authors in terms of double or multiple perturbation theory of the type used in Section 5.6 (see, for example, Hall 1964) and it is clear that much care and effort is needed in obtaining good 'unperturbed' functions before attempting the accurate calculation of properties; in particular the functions should provide sufficient flexibility in the distribution function on which the properties of interest may depend.

It seems appropriate to end this review by turning once again to the early work of Coulson and his collaborators; for the concepts they introduced have proved remarkably independent of refinements in the theory. It is in fact easy to show that charges, bond orders, and the 'atom and bond polarizabilities' (as defined by Coulson and Longuet-Higgins in 1949) all re-emerge from the general theory—based in principle on exact wavefunctions—on making the appropriate specializations and approximations. To confirm that this is so, we use LCAO–MO approximations to reformulate the results of earlier sections.

† It is also noteworthy that the required terms arise only in the *second*-order correction of the one-configuration wavefunction; ΔH_{HF} is in fact very much larger than H_{N} and it is not surprising that its effect must be included to higher order.

We assume that the ground state (κ_a) is represented in terms of molecular orbitals $\{\phi_R\}$, expressed in terms of atomic orbitals $\{\chi_i\}$ according to

$$\phi_R = \sum_i c_{Ri}\chi_i \tag{5.157}$$

and (5.154) becomes, with a more explicit notation,

$$P_1(\kappa_a\kappa_a|\mathbf{r}_1) = \sum_{i,j} P_{ij}\chi_i(\mathbf{r}_1)\chi_j^*(\mathbf{r}_1). \tag{5.158}$$

The first-order energy change, under any one-electron perturbation, $h \to h+\delta V$, then follows from (5.113). Let us consider, following Coulson and Longuet-Higgins, an ideally 'localized' perturbation, which affects only one diagonal matrix element of the Hamiltonian h, and is thus defined by

$$\langle\chi_i|\delta V|\chi_j\rangle = \delta\alpha_r\delta_{ir}\delta_{rj}, \tag{5.159}$$

the Kronecker deltas ensuring that all elements except the rth vanish. The required energy change, from (5.113), is then $\delta E = \delta\alpha_r P_{rr}$ essentially as in Hückel theory (Section 5.2). A similar result follows for a perturbation $\delta\beta_{rs}$ of one off-diagonal element, namely (assuming all quantities real) $\delta E = 2\delta\beta_{rs}P_{rs}$, and the **P**-matrix elements thus rather generally measure first-order response to 'atom' and 'bond' perturbations in the sense

$$P_{rs} = \frac{\partial E}{\partial \alpha_r} \qquad P_{rs} = \frac{1}{2}\frac{\partial E}{\partial \beta_{rs}}. \tag{5.160}$$

This result is formally identical with (5.18) but holds under far less restrictive conditions: although it has been necessary to introduce a finite basis $\{\chi_i\}$, E is the *total* electronic energy and may, for example, be calculated from an *ab initio* wavefunction with full configuration interaction. The validity of this result for variationally determined approximate wavefunctions has been discussed elsewhere (McWeeny 1955).

The second-order response, which depends on first-order change of the electron density, may be cast in the form due to Coulson and Longuet-Higgins on making slightly more restrictive assumptions. If the ground state (κ_a) is represented by one determinant of doubly occupied MOs, then the 'excited' functions (κ_b) which (through configuration interaction) give a first-order modification of the charge density are those which differ by promotion of *one* electron to an unoccupied MO. With this choice of functions, $\Phi_{\kappa_b} = \Phi(R \to S)$ with singlet spin coupling, we obtain easily

$$P_1(\kappa_a\kappa_b|\mathbf{r}_1) = \sqrt{2}\phi_R(\mathbf{r}_1)\phi_S^*(\mathbf{r}_1), \quad P_1(\kappa_b\kappa_a|\mathbf{r}_1) = \sqrt{2}\phi_S(\mathbf{r}_1)\phi_R^*(\mathbf{r}_1) \tag{5.161}$$

and in place of (5.131), using (5.159) and (5.161),

$$\pi_{\kappa_a\kappa_b} = \frac{\int \delta V(\mathbf{r}_1)P_1(\kappa_b\kappa_a|\mathbf{r}_1)\,\mathrm{d}\mathbf{r}_1}{\Delta E(\kappa_a \to \kappa_b)} = \frac{\sqrt{2}c_{Rr}^*c_{Sr}\,\delta\alpha_r}{\Delta E(R \to S)}. \tag{5.162}$$

In place of (5.130), the change in charge density becomes,

$$\delta P_1(\kappa_a\kappa_a|\mathbf{r}_1) = \delta\alpha_r \sum_{i,j} \pi_{r,ij}\chi_i(\mathbf{r}_1)\chi_j^*(\mathbf{r}_1), \tag{5.163}$$

where, on approximating the denominator in (5.162) by a difference of orbital energies,

$$\pi_{r,ij} = 4 \sum_{\substack{R(\text{occ})\\ S(\text{unocc})}} \frac{c_{Rr}^* c_{Sr} c_{Ri} c_{Sj}^*}{\varepsilon_S - \varepsilon_R}. \tag{5.164}$$

This is the quantity defined by Coulson and Longuet-Higgins as an 'atom–bond polarizability': comparison of (5.163) and (5.158) shows that it gives the change in bond order P_{ij} due to unit change ($\delta\alpha_r = 1$) in the matrix element $\alpha_r = \langle\chi_r|h|\chi_r\rangle$. In particular, for $i = j = r$, we obtain the 'self-polarizability' ($\pi_{r,r}$) of atom r and it follows as in the derivation of (5.132) that the energy change up to second order is

$$\delta E = \delta\alpha_r P_{rr} + \tfrac{1}{2}(\delta\alpha_r)^2\pi_{r,r}. \tag{5.165}$$

The self-polarizability may thus be defined either as the first derivative of a bond order, or the second derivative of an energy, with respect to change of the matrix element α_r:

$$\pi_{r,r} = \left(\frac{\partial P_{rr}}{\partial\alpha_r}\right) = \left(\frac{\partial^2 E}{\partial\alpha_r^2}\right). \tag{5.166}$$

Similar results may be obtained for variation of an off-diagonal element ($\beta_{rs} = \langle\chi_r|h|\chi_s\rangle$) and it is clear that the approximations of Hückel theory are in no way crucial to the analysis; in fact the approach of Section 5.7 applies equally well in non-empirical, all-electron theory.

Coulson was certainly among the first to direct attention towards the charge density and its profound importance in the discussion of molecular properties; 35 years later it is clear that, however much more rigorous molecular quantum mechanics has become, however much more refined its approximations and computational techniques, his original ideas retain a validity and significance extending far beyond their original context.

REFERENCES

BETHE, H. and SALPETER, E. (1957). *Quantum mechanics of one- and two-electron atoms.* Springer–Verlag, Berlin.

BORN, M. and GREEN, H. S. (1947). *Proc. R. Soc. A* **191**, 168; *see also* Born, M. (1964). *Natural philosophy of cause and chance.* Dover, New York.

CLINTON, W. L. and RICE, R. (1959). *J. Chem. Phys.* **30**, 542.

COLEMAN, A. J. (1963). *Rev. Mod. Phys.* **35**, 668.

COULSON, C. A. (1939). *Proc. R. Soc. A* **169**, 413.
—— and RUSHBROOKE, G. S. (1940). *Proc. Cambridge Phil. Soc.* **36**, 193.
—— and LONGUET-HIGGINS, H. C. (1949). *Proc. R. Soc. A* **191**, 39; *see also, ibid. A* **192**, 16; *ibid. A* **193**, 447, 456; *ibid. A* **195**, 188.
—— and LUZ, Z. (1968). *Trans. Farad. Soc.* **64**, 2884.
DACRE, P. D. and MCWEENY, R. (1970). *Proc. R. Soc. A* **317**, 435; *see also*, McWeeny, R. (1970). *Spins in chemistry. Appendix* 1. Academic Press, New York and London.
EPSTEIN, S. T., HURLEY, A. C., WYATT, R. E., and PARR, R. G. (1967). *J. Chem. Phys.* **47**, 1275.
GREEN, S. (1971). *J. Chem. Phys.* **53**, 827.
TER HAAR, D. (1961). *Rept. Prog. Phys.* **24**, 304.
HALL, G. G. (1964). *Adv. Quantum Chem.* **1**, 241.
HUSIMI, K. (1940). *Proc. phys.-math. Soc. Japan* **22**, 264.
HUTCHISON, C. A. and MANGUM, B. W. (1958). *J. chem. Phys.* **29**, 952.
JAHN, H. A. and TELLER, E. (1937). *Proc. R. Soc. A* **161**, 220.
KUTZELNIGG, W., DEL RE, G., and BERTHIER, G. (1968). *Phys. Rev.* **172**, 49.
LONGUET-HIGGINS, H. C. (1956). *Proc. R. Soc. A* **235**, 537.
LÖWDIN, P. -O. (1950). *J. chem. Phys.* **18**, 365.
—— (1955). *Phys. Rev.* **97**, 1474.
—— (1959). *Adv. chem. Phys.* **2**,
MCCONNELL, H. M. (1958). *J. chem. Phys.* **28**, 1188.
MCLACHLAN, A. D. (1962). *Mol. Phys.* **5**, 51.
MCWEENY, R. (1951). *J. chem. Phys.* **19**, 1614; *see also*; *ibid.* **20**, 920 (errata).
—— (1952). *Acta Crystallogr.* **5**, 463: *ibid.* (1953), **6**, 631; *ibid.* (1954). **7**, 180.
—— (1954). *Proc. R. Soc. A* **223**, 63.
—— (1955). *Proc. R. Soc. A* **232**, 114.
—— (1959). *Proc. R. Soc. A* **235**, 242.
—— (1960). *Rev. mod. Phys.* **32**, 335.
—— (1965). *J. chem. Phys.* **42**, 1717.
—— (1967). *Int. J. Quantum. Chem.* **1***S*, 351.
—— (1971). *Spins in chemistry*. Academic Press, New York and London.
—— and MIZUNO, Y. (1961). *Proc. R. Soc. A* **259**, 554.
—— and KUTZELNIGG, W. (1968). *Int. J. Quantum. Chem.* **2**, 187.
—— and SUTCLIFFE, B. T. (1969). *Methods of molecular quantum mechanics*. Academic Press, London.
MARGENAU, H. and KESTNER, N. R. (1969). *Theory of Intermolecular forces*. Pergamon Press, Oxford.
MOORES, W. H. and MCWEENY, R. (1973). *Proc. R. Soc. A*. **332**, 365–384.
MORSE, P. M. and FESHBACH, H. (1954). *Methods of theoretical physics, vol.* II. McGraw-Hill, New York.
MULLIKEN, R. S. (1955). *J. Chem. Phys.* **23**, 1833, 2343.
OVERHAUSER, A. W. (1959). *Phys. Rev. Letts* **3**, 414; *see also*; *ibid.* (1960), **4**, 462.
POPLE, J. A., MCIVER, J. W., and OSTLUND, N. S. (1968). *J. chem. Phys.* **49**, 2960; *see also* Blizzard, A. C. and Santry, D. P. (1971). *J. chem. Phys.* **55**, 950.
RACAH, G. (1942). *Phys. Rev.* **62**, 438.
SLATER, J. C. (1960). *Quantum theory of atomic structure. vol.* 2, ch. 23. McGraw-Hill, New York.

VAN DER WAALS, J. H. and DE GROOT, M. S. (1959). *Mol. Phys.* **2**, 333.
WALKER, R. E. H. and RICHARDS, W. G. (1968). *Symp. Farad. Soc.* **2**, 64.
WEISSMAN, S. I. (1956). *J. chem. Phys.* **25**, 890.
YANG, C. N. (1962). *Rev. mod. Phys.* **34**, 694.

6

ATOMS IN MOLECULES

G. G. BALINT-KURTI AND M. KARPLUS

CHEMISTS have traditionally regarded molecules as being built up from constituent atoms (Coulson 1961; Pauling 1960), with the atoms retaining many of their characteristics in the molecular environment. It was from such an atomic viewpoint that the first theory of valence was constructed by Heitler and London (1927). Moffitt (1951*a*) noted that the electronic energy of a molecule is almost equal to that of its separated atoms, so that the energy required for dissociation is only a small fraction of the total energy. It seemed logical, therefore, to treat the process of molecule formation as a perturbation, with the unperturbed states corresponding to those of the isolated atoms and their ions. Even in the absence of a rigorous perturbation development (see, for example, Hirschfelder 1967), knowledge of the constituent atoms can serve to facilitate the description of the molecule. Moffitt showed how to isolate the atomic contribution to the energy of a molecule and proposed a procedure by which experimental atomic term values can be used to estimate this contribution. Application of Moffitt's atoms-in-molecules method to the test case of H_2 (Hurley 1955; Pauncz 1954; Batana and Cohan 1962) demonstrated, however, that it possessed some undesirable properties and led to the introduction of two alternative schemes (Hurley 1956*a*, Arai 1957*a*). In spite of these various possibilities, the difficulties were such that at the 1959 Boulder Conference on Molecular Quantum Mechanics, Coulson remarked 'It is not true to say that in this last week we have killed the theory of atoms in molecules, but it is true to say that it has been very seriously wounded.' (Coulson 1960).

We have reconsidered Moffitt's atoms-in-molecules method and proposed an alternative approach called the Orthogonalized Moffitt method. (Balint-Kurti and Karplus 1969) This method overcomes some unsatisfactory attributes of the previous atoms-in-molecules treatments. The equations of the method correct the Hamiltonian matrix elements calculated *ab initio* by use of the observed energies of the atoms and ions used to build up the molecule. The introduction of such corrections enables one to obtain a better, and more reliable result for a given size of basis set than would otherwise be possible. As the size of the basis set is increased, the magnitude of the corrections decreases and, in the limit of a complete basis set, the Orthogonalized Moffitt method becomes identical with the *ab initio* result; that is, both give the exact solution.

Our interest will be centred mainly on calculating potential energy curves and surfaces. In order to obtain accurate surfaces the wavefunction must dissociate correctly. A Hartree–Fock type wavefunction is not always capable of giving a reasonable description of the dissociation products (see Coulson 1961). By contrast, multistructure valence-bond type wavefunctions, which form the basis of the atoms-in-molecules methods, can be specifically constructed so that they provide approximate descriptions of the eigenstates of the dissociation products in the limit of large internuclear separations (Moffitt 1954). Moreover, potential energy surfaces calculated with the Orthogonalized Moffitt method converge to the experimental energies of the dissociation products. This property is crucial for use of the surfaces in scattering calculations or for the analysis of curve crossing processes.

As an illustration of the reasoning employed in the atoms-in-molecules method, we consider the example of LiF. To obtain the so-called *ab initio* binding energy for this molecule the calculated molecular energy at the equilibrium internuclear separation is compared with the corresponding Hartree–Fock atomic energies. From the very accurate Hartree–Fock calculation of McLean (1963) there results a binding energy of about 0·15 a.u.,† which is not in good agreement with the experimental value of 0·22 a.u. The origin of the discrepancy of 0·07 a.u. is a simple one. Near its equilibrium, LiF is best described as Li^+F^-. The experimental electron affinity of the fluorine atom (Berry and Reinmann 1963) is approximately 0·13 a.u., while the Hartree–Fock value is only 0·05 a.u. (Clementi and McLean 1964). Lowering the molecular energy by this difference of 0·08 a.u., yields nearly the correct binding energy; the error in describing Li^+ relative to Li is an order of magnitude less. Thus, the inaccuracy in the Hartree–Fock binding energy for LiF is almost entirely atomic in origin (McLean 1963). We summarize this discussion by the diagram on the following page.

The objective of all the atoms-in-molecules methods is to provide a systematic framework for applying atomic corrections, corresponding to those introduced for LiF, to the general case.

6.1. Heuristic introduction to atom-in-molecules method

Let us now consider the calculation of a potential energy curve for a diatomic molecule AB in a little more detail. The first step in an atoms-in-molecules calculation is to write down approximate eigenfunctions for the ground and excited states of the atoms and ions involved. These atomic eigenfunctions are then multiplied together in pairs and antisymmetrized to form basis functions for the expansion of the total molecular wavefunction. Moffitt showed that if exact atomic eigenfunctions were used then the

† 1 a.u. of energy $= 4{\cdot}359\,828 \times 10^{-18}$ J.

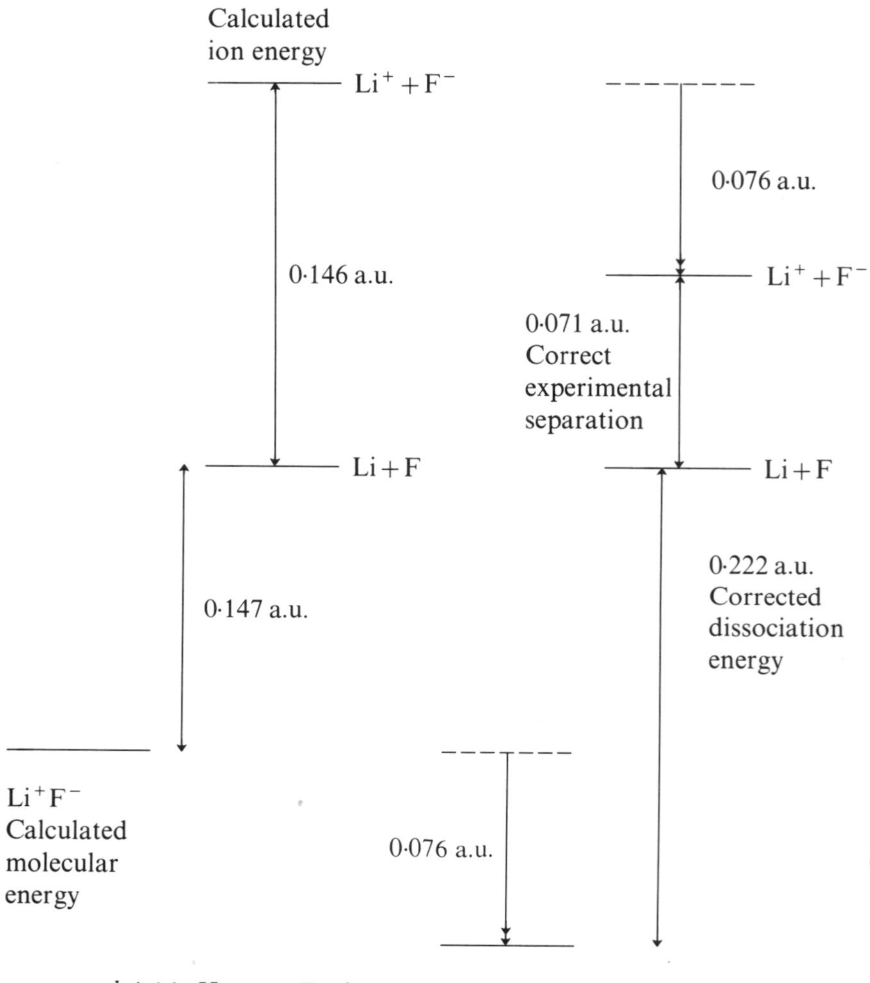

Hamiltonian matrix elements between molecular basis functions can be written as

$$H_{ij,kl} = (E_k^A + E_l^B)S_{ij,kl} + V_{ij,kl}^{AB} \tag{6.1}$$

where the indices ij refer to a molecular basis function built up from atomic eigenfunctions for the ith state of atom A and the jth state of atom B, E_k^A and E_l^B are the experimental energies of the kth state of atom A and the lth state of atom B, $S_{ij,kl}$ is the overlap of the two basis functions and $V_{ij,kl}^{AB}$ is an interatomic term, which is in general much smaller than the first intra-atomic term.

For a simple calculation on H_2, we can form two molecular basis functions of the above type using only 1s orbitals on the two hydrogen atoms

$$\begin{aligned}\Phi_{12} &= A\{a(\mathbf{r}_1)\alpha(1)b(\mathbf{r}_2)\beta(2)\}\\ &= a(\mathbf{r}_1)\alpha(1)b(\mathbf{r}_2)\beta(2)-a(\mathbf{r}_2)\alpha(2)b(\mathbf{r}_1)\beta(1) \qquad (6.2)\end{aligned}$$

$$\begin{aligned}\Phi_{21} &= A\{a(\mathbf{r}_1)\beta(1)b(\mathbf{r}_2)\alpha(2)\}\\ &= a(\mathbf{r}_1)\beta(1)b(\mathbf{r}_2)\alpha(2)-a(\mathbf{r}_2)\beta(2)b(\mathbf{r}_1)\alpha(1) \qquad (6.3)\end{aligned}$$

where a and b are 1s-functions on the two nuclei, $\mathbf{r}_1$ and $\mathbf{r}_2$ are the spatial coordinates of the two electrons, α and β are z-component spin eigenfunctions, and A is the antisymmetrizer. The two basis functions Φ_{12} and Φ_{21} are not spin eigenfunctions. If we take a linear combination of them, we can form the (unnormalized) Heitler–London function,

$$\Psi^{\mathrm{HL}} = 2^{-\frac{1}{2}}[\Phi_{12}-\Phi_{21}] = [a(\mathbf{r}_1)b(\mathbf{r}_2)+a(\mathbf{r}_2)b(\mathbf{r}_1)]2^{-\frac{1}{2}}[\alpha(1)\beta(2)-\beta(1)\alpha(2)] \qquad (6.4)$$

which has ${}^1\Sigma_g^+$ symmetry.

It is entirely equivalent to use the two basis functions Φ_{12} and Φ_{21} or the Heitler–London function as far as the ground state of H_2 is concerned.

The Hamiltonian for H_2 is (in atomic units)

$$H = -\frac{1}{2}\nabla_1^2-\frac{1}{2}\nabla_2^2-\frac{1}{R_{A_1}}-\frac{1}{R_{A_2}}-\frac{1}{R_{B_1}}-\frac{1}{R_{B_2}}+\frac{1}{r_{12}}+\frac{1}{R} \qquad (6.5)$$

where for example R_{A_1} is the distance from nucleus A to electron 1 and r_{12} is the distance between the two electrons. If we assume for the moment that we can assign electron 1 to nucleus A and electron 2 to nucleus B then the Hamiltonian for the molecule can be rewritten as

$$H = H_A+H_B-\frac{1}{R_{B_1}}-\frac{1}{R_{A_2}}+\frac{1}{r_{12}}+\frac{1}{R} \qquad (6.6)$$

where H_A and H_B are Hamiltonians for hydrogen atoms A and B. By use of the facts that $a(\mathbf{r}_1)$ is an eigenfunction of H_A,

$$H_A a(\mathbf{r}_1) = E_H a(\mathbf{r}_1) \qquad (6.7)$$

where E_H is the ground-state electronic energy of a hydrogen atom, and that the antisymmetrizer A commutes with the molecular Hamiltonian H, it follows that

$$\begin{aligned}H\Phi_{12} &= HA[a(\mathbf{r}_1)\alpha(1)b(\mathbf{r}_2)\beta(2)]\\ &= AH[a(\mathbf{r}_1)\alpha(1)b(\mathbf{r}_2)\beta(2)]\\ &= \left(2E_H+\frac{1}{R}\right)\Phi_{12}+A\left\{\left(-\frac{1}{R_{B_1}}-\frac{1}{R_{A_2}}+\frac{1}{r_{12}}\right)[a(\mathbf{r}_1)\alpha(1)b(\mathbf{r}_2)\beta(2)]\right\}. \qquad (6.8)\end{aligned}$$

With eqn (6.8) the matrix element $H_{12,12}$ can be written as

$$H_{12,12} = \int \Phi_{12} H \Phi_{12}\, d\tau = \left(2E_H + \frac{1}{R}\right) S_{12,12} + V_{12,12} \tag{6.9}$$

where $S_{12,12} = \int \Phi_{12}\Phi_{12}\, d\tau = 2$, and

$$V_{12,12} = \int \Phi_{12} \left\{ A \left[\left(-\frac{1}{R_{B_1}} - \frac{1}{R_{A_2}} + \frac{1}{r_{12}} \right) (a(\mathbf{r}_1)\alpha(1) b(\mathbf{r}_2)\beta(2)) \right] \right\} d\tau$$

$$= -4 \int \frac{a(\mathbf{r}_1)^2}{R_{B_1}} d\mathbf{r}_1 + 2 \int \frac{a(\mathbf{r}_1)^2 b(\mathbf{r}_2)^2}{r_{12}} d\mathbf{r}_1\, d\mathbf{r}_2 .$$

The matrix element $H_{21,12}$ is given by

$$H_{21,12} = \left(2E_H + \frac{1}{R}\right) S_{21,12} + V_{21,12} \tag{6.10}$$

where

$$S_{21,12} = -2S^2$$

$$V_{21,12} = 4S \int \frac{a(\mathbf{r}_1) b(\mathbf{r}_1)}{R_{B_1}} d\mathbf{r}_1 - 2 \int \frac{a(\mathbf{r}_1) b(\mathbf{r}_1) a(\mathbf{r}_2) b(\mathbf{r}_2)}{r_{12}} d\mathbf{r}_1\, d\mathbf{r}_2$$

and $S = \int a(\mathbf{r}_1) b(\mathbf{r}_1)\, d\mathbf{r}_1$. As $H_{12,12} = H_{21,21}$ and $H_{12,21} = H_{21,12}$, the calculated molecular energy corresponding to the Heitler–London wavefunction is

$$E^{HL} = \frac{\int \Psi^{HL} H \Psi^{HL}\, d\tau}{\int (\Psi^{HL})^2\, d\tau} = \frac{H_{12,12} - H_{21,12}}{S_{12,12} - S_{21,12}}$$

$$= 2E_H + \frac{1}{R} + \left\{ -2 \int \frac{a(\mathbf{r}_1)^2}{R_{B_1}} d\mathbf{r}_1 - 2S \int \frac{a(\mathbf{r}_1) b(\mathbf{r}_1)}{R_{B_1}} d\mathbf{r}_1 \right.$$

$$\left. + \int \frac{a(\mathbf{r}_1)^2 b(\mathbf{r}_2)^2}{r_{12}} d\mathbf{r}_1\, d\mathbf{r}_2 + \int \frac{a(\mathbf{r}_1) b(\mathbf{r}_1) a(\mathbf{r}_2) b(\mathbf{r}_2)}{r_{12}} d\mathbf{r}_1\, d\mathbf{r}_2 \right\} \frac{1}{1+S^2} . \tag{6.11}$$

(See Coulson 1961, page 119.)

Comparing eqn (6.1) with eqns (6.9) and (6.10) we see that the latter are special cases of the former. The molecule H_2 is, however, an atypical case in that the exact atomic wavefunctions of its component atoms (H atoms) are known. For more complex atoms, only approximate eigenfunctions are available and eqn (6.1) is not exact. Nevertheless, Moffitt suggested that eqn (6.1) could be employed if the approximate atomic eigenfunctions were used to evaluate $S_{ij,kl}$ and $V^{AB}_{ij,kl}$ and experimental atomic energies were introduced for E^A_k and E^B_l. This is equivalent in the HL energy expression (eqn 6.11), to using atomic orbitals to calculate only the term in curly brackets. The essential point in such a procedure is that the interatomic term $V_{ij,kl}^{AB}$ is generally much smaller than the intra-atomic energy terms E_k^A and E_l^B, and that by use of experimental atomic energies for the latter the large intra-atomic errors in a molecular calculation would be eliminated. Further, the eigenvalues of the corrected Hamiltonian in eqn (6.1) coincide with the experimental energies of the atoms at large internuclear separations. As we improve the description of the states of the atoms, the errors in eqn (6.1) are reduced and the atoms-in-molecules calculation approaches an *ab initio* calculation.

The atoms-in-molecules method views the process of molecule formation as the coming together of atoms. As the internuclear distances decrease, the excited states of the atoms are mixed with their ground states. The extent of this mixing, which is responsible for such processes as hybridization and ionic–covalent resonance, is very sensitive to intra-atomic errors in a calculation. By using experimental atomic energies in an attempt to eliminate intra-atomic errors, the degree of mixing of the different atomic states is improved. Thus, one might hope that the atoms-in-molecules method will give an improved potential function, relative to the *ab initio* result for an equivalent basis set.

While the basic idea of Moffitt's atoms-in-molecules method seems very reasonable, actual application of the method often leads to erroneously large molecular binding energies. The origins of these errors, and the various modified atoms-in-molecules methods which have been formulated to overcome them are discussed in Section 6.2 using the H_2 molecule as an illustrative example. This discussion is preceded by a more rigorous derivation of Moffitt's basic equations.

In Section 6.3, the mechanics of performing multistructure valence-bond calculations are outlined. Since these calculations involve the evaluation of Hamiltonian matrix elements between Slater determinants constructed from a non-orthogonal orbital basis set, they present a number of problems (Van Vleck 1936). Several different schemes for overcoming the 'non-orthogonality problem' have been proposed (Löwdin 1955; Hurley 1960), and some of them have been used in molecular calculations. We describe one of these that provides a tractable solution to the problem of evaluating the required matrix elements.

Some results for diatomic molecules are presented in Section 6.4. These calculations were performed with sufficiently small basis sets so that they could serve as a starting point for building up basis sets to describe potential surfaces for triatomic systems. In Section 6.5, the results of some potential energy surface calculations for LiF_2, Li_2F and LiHF are described. These systems are simple prototypes for reactions that have been extensively studied in crossed molecular beam experiments (Herschbach 1970).

6.2. Formulation

The basis set for the atoms-in-molecules approach consists of the eigenfunctions of the Hamiltonian for the system with the atoms infinitely separated from each other. For simplicity, we fix our attention on a hypothetical diatomic AB; the results can be generalized to a polyatomic system without difficulty. For such a diatomic, the basis set is constructed from a complete set of eigenfunctions, in principle including continuum functions, of the atoms A and B.† The resulting product functions are

$$\chi_{ij}^{AB} = \Phi_i^A \Phi_j^B \tag{6.12}$$

where Φ_i^A and Φ_j^B are atomic eigenfunctions and are antisymmetric with respect to interchange of the electrons assigned to them. A perturbation or variation approach using the χ_{ij}^{AB} as the basis set would not satisfy the Pauli exclusion principle without additional constraints. The most straightforward way of satisfying the exclusion principle is to antisymmetrize the functions χ_{ij}^{AB}. The antisymmetrized product functions so formed are called *composite functions*‡ (CFs) (Moffitt 1954) and we denote them by

$$\Phi_{ij}^{AB} = N_{ij} A'[\Phi_i^A \Phi_j^B] \tag{6.13}$$

where N_{ij} is a normalization factor (i.e. $N_{ij} = 1/\langle A'[\Phi_i^A\Phi_j^B] | A'[\Phi_i^A\Phi_j^B]\rangle^{\frac{1}{2}}$) and A' is the partial antisymmetrizer,

$$A' = \left[\frac{N_A!N_B!}{(N_A+N_B)!}\right]^{\frac{1}{2}} \sum_P (-1)^P P.$$

The summation runs over all permutation P of the electrons which interchange electrons originally assigned to the two different atoms.

If we were to use a complete set of atomic functions on both atoms to form the CF basis set, it would be overcomplete (Arai 1957a). With the finite set of atomic eigenfunctions which are employed in practice, however, no overcompleteness problem arises. Indeed, it is customary to augment the CF basis set by including CFs built up from eigenfunctions of the ionic species (i.e. Li^+F^-). While in principle this introduces another possible source of overcompleteness, in practice it rarely leads to any problem. For the moment we assume that we know the exact atomic eigenfunctions and use them to construct both the product functions of eqn (6.12) and the composite functions of eqn (6.13). At finite internuclear separations the composite functions are not orthogonal. As the internuclear distance tends to infinity, however,

† The atomic eigenfunctions are taken to be real. They are, therefore, not eigenfunctions of the z component of the angular momentum operator, but only of the total angular momentum operator.

‡ The composite functions defined in eqn (6.13) are normalized for convenience of later manipulation. This differs from Moffitt's original definition.

they become orthogonal due to the orthogonality of the atomic eigenfunctions which were used to construct them.

Let us now examine the Hamiltonian of the diatomic system. If we use the product functions of eqn (6.12) we can assign certain electrons to atom A and others to atom B. The electronic Hamiltonian can then be written (in atomic units)

$$H = \left\{-\frac{1}{2}\sum_{i\in A}\nabla_i^2 - \sum_{i\in A}\frac{Z_A}{R_{Ai}} + \frac{1}{2}\sum_{\substack{i\in A\\ j\in A}}\frac{1}{r_{ij}} - \frac{1}{2}\sum_{j\in B}\nabla_j^2 - \sum_{j\in B}\frac{Z_B}{R_{Bj}} + \frac{1}{2}\sum_{i\in B}\frac{1}{r_{ij}}\right\}$$

$$+\left\{-\sum_{i\in B}\frac{Z_A}{R_{Ai}} - \sum_{j\in A}\frac{Z_B}{R_{Bj}} + \sum_{\substack{i\in A\\ j\in B}}\frac{1}{r_{ij}}\right\} \tag{6.14}$$

where R_{Ai} is the distance from atom A to electron i and r_{ij} is the distance between electrons i and j. The terms in the first curly bracket are intra-atomic terms, which are regarded as of zeroth order, and are denoted by $\{H_A + H_B\}$. Those in the second curly bracket, which we denote by V^{AB}, are interatomic terms and in the present approach they correspond to the 'perturbation.' Thus, we decompose the Hamiltonian operator in the form

$$H = H_A + H_B + V^{AB}. \tag{6.15}$$

This decomposition is valid only as long as specific electrons are assigned to each of the atoms; that is, using the basis functions of eqn (6.12), we can write

$$H\chi_{ij}^{AB} = (E_i^A + E_j^B + V^{AB})\chi_{ij}^{AB} \tag{6.16}$$

where E_i^A and E_j^B are experimental atomic energies. The use of the product basis set of eqn (6.12) and of the above decomposition of the Hamiltonian would enable us to set up a rigorous perturbation theory. The problem of satisfying the Pauli exclusion principle would, however, have to be handled by constraining the wavefunction. This procedure, which introduces considerable complications (Hirschfelder 1967), is not used here. Instead, we employ the more conventional type of antisymmetrized basis set given in eqn (6.13). The Hamiltonian, consequently, cannot be separated into a zeroth-order and a perturbation term, as it is now impossible to assign specific electrons to each of the two atoms. It is, therefore, difficult to develop a perturbation theory (see, for example, Murrell and Shaw 1967) and we resort to a variational approach. As the major contribution to the molecular energy is atomic in origin, we will, however, attempt to identify the atomic contributions to the Hamiltonian matrix elements in the molecule and will use experimental data to estimate them.

The approach of the two atoms, in their respective ground states at infinite separation, leads to a mixing of their excited states with the ground states.

To obtain the correct contribution from the excited states, it is imperative that the relative energies of the atomic states be accurate. Zeroth-order orbital wavefunctions for atoms (i.e. ones in which there is no intra-atomic configuration interaction) often give very poor results for the spacing of the atomic levels (Moffitt 1951*a*) and for the electron affinities of atoms. The resulting 'atomic' errors are reflected in most simple molecular calculations and often lead to inaccurate dissociation energies (Moffitt 1951*b*; Hurley 1958*a*). The 'atoms-in-molecules' approach was designed to prevent such atomic errors from propagating into the molecular problem.

A general Hamiltonian matrix element between composite functions is

$$H_{ij,kl} = \langle \Phi_{ij}^{\mathrm{AB}} | H | \Phi_{kl}^{\mathrm{AB}} \rangle. \tag{6.17}$$

Making use of the fact that the Hamiltonian is totally symmetric with respect to interchange of electrons and therefore commutes with the partial antisymmetrizer A' (or with any operator which merely permutes electrons), we obtain

$$H_{ij,kl} = \langle \Phi_{ij}^{\mathrm{AB}} | N_{kl} A' [H \Phi_k^{\mathrm{A}} \Phi_l^{\mathrm{B}}] \rangle. \tag{6.18}$$

We can now partition the Hamiltonian in the same manner as in eqns (6.14) and (6.15) since it is possible to assign specific electrons to each of the two atoms in the term in the square brackets. This yields

$$\begin{aligned} H_{ij,kl} &= \langle \Phi_{ij}^{\mathrm{AB}} | N_{kl} A' [(E_k^{\mathrm{A}} + E_l^{\mathrm{B}} + V^{\mathrm{AB}}) \Phi_k^{\mathrm{A}} \Phi_l^{\mathrm{B}}] \rangle \\ &= (E_k^{\mathrm{A}} + E_l^{\mathrm{B}}) \langle \Phi_{ij}^{\mathrm{AB}} | \Phi_{kl}^{\mathrm{AB}} \rangle + \langle \Phi_{ij}^{\mathrm{AB}} | N_{kl} A' [V^{\mathrm{AB}} \Phi_k^{\mathrm{A}} \Phi_l^{\mathrm{B}}] \rangle \end{aligned} \tag{6.19}$$

or

$$H_{ij,kl} = (E_k^{\mathrm{A}} + E_l^{\mathrm{B}}) S_{ij,kl} + V_{ij,kl}^{\mathrm{AB}} \tag{6.20}$$

It should be noted that the partial antisymmetrizer does not commute with V^{AB}. Since the Hamiltonian operator is self-adjoint, we have (with $S_{ij,kl}$ and $V_{ij,kl}^{\mathrm{AB}}$ real)

$$\begin{aligned} H_{ij,kl} &= (E_k^{\mathrm{A}} + E_l^{\mathrm{B}}) S_{ij,kl} + V_{ij,kl}^{\mathrm{AB}} \\ &= (E_i^{\mathrm{A}} + E_j^{\mathrm{B}}) S_{ij,kl} + V_{kl,ij}^{\mathrm{AB}} \end{aligned} \tag{6.21}$$

where the two parts of the right-hand side of eqns (6.20) and (6.21) are not separately Hermitian.

Moffitt (1951*a*) proposed two distinct schemes for performing a molecular calculation in which corrections are applied to eliminate the intra-atomic errors. The first, and in principle more straightforward, procedure stems directly from eqn (6.21). Moffitt argued that a fairly crude orbital approximation to the composite functions Φ_{ij}^{AB} will reproduce the true charge density reasonably well and will, therefore, yield satisfactory results for $S_{ij,kl}$ and for the small perturbation term $V_{ij,kl}^{\mathrm{AB}}$. Since such an orbital approximation may be inadequate for calculating the atomic eigenvalues E_i^{A} and E_j^{B}, he suggested that these should be taken from experiment or from more extensive atomic calculation. If all quantities which are calculated using orbital-type wavefunctions are denoted by a tilde, the Moffitt proposal can be written as

$$H_{ij,kl} \sim (E_k^{\mathrm{A}}+E_l^{\mathrm{B}})\tilde{S}_{ij,kl}+\tilde{V}_{ij,kl}^{\mathrm{AB}}. \tag{6.22}$$

With approximate atomic eigenfunctions and exact atomic energies, the right-hand side of eqn (6.22) is not necessarily Hermitian. It is clearly not physically realistic to use a non-Hermitian Hamiltonian matrix since the eigenvalues of such a matrix are generally complex. There is no unique way of making the Hamiltonian matrix in eqn (6.22) Hermitian. Since

$$H_{kl,ij} = H_{ij,kl} \sim (E_k^{\mathrm{A}}+E_l^{\mathrm{B}})\tilde{S}_{ij,kl}+\tilde{V}_{ij,kl}^{\mathrm{AB}} \tag{6.23}$$

and

$$H_{ij,kl} = H_{kl,ij} \sim (E_i^{\mathrm{A}}+E_j^{\mathrm{B}})\tilde{S}_{ij,kl}+\tilde{V}_{kl,ij}^{\mathrm{AB}} \tag{6.24}$$

it has been customary to take the average of eqns (6.23) and (6.24). Thus, one obtains the Hamiltonian matrix

$$H_{kl\,ij} = H_{ij\,kl} \sim \tfrac{1}{2}(E_i^{\mathrm{A}}+E_j^{\mathrm{B}}+E_k^{\mathrm{A}}+E_l^{\mathrm{B}})\tilde{S}_{ij,kl}+\tfrac{1}{2}(\tilde{V}_{ij,kl}^{\mathrm{AB}}+\tilde{V}_{kl,ij}^{\mathrm{AB}}). \tag{6.25}$$

The expression for $\tilde{V}_{ij,kl}^{\mathrm{AB}}$ involves the partial antisymmetrizer,

$$\tilde{V}_{ij,kl}^{\mathrm{AB}} = \langle\tilde{\Phi}_{ij}^{\mathrm{AB}}|N_{kl}A'(V^{\mathrm{AB}}\tilde{\Phi}_k^{\mathrm{A}}\tilde{\Phi}_l^{\mathrm{B}})\rangle \tag{6.26}$$

Such a term is the sum of very many integrals and the algebra involved in its evaluation is somewhat complicated. To simplify the expression and obtain Moffitt's second scheme, we make use of eqn (6.20) to write

$$V_{ij,kl}^{\mathrm{AB}} = H_{ij,kl}-(E_k^{\mathrm{A}}+E_l^{\mathrm{B}})S_{ij,kl}. \tag{6.27}$$

Moffitt argued that the use of calculated quantities in the right-hand side of eqn (6.27) would provide an adequate estimate of $\tilde{V}_{ij,kl}^{\mathrm{AB}}$; that is,

$$\tilde{V}_{ij,kl}^{\mathrm{AB}} = \tilde{H}_{ij,kl}-(\tilde{E}_k^{\mathrm{A}}+\tilde{E}_l^{\mathrm{B}})\tilde{S}_{ij,kl}, \tag{6.28}$$

where $\tilde{E}_k^{\mathrm{A}}$ and $\tilde{E}_l^{\mathrm{B}}$ are the atomic energies calculated with the approximate atomic eigenfunctions; e.g.

$$\tilde{E}_k^{\mathrm{A}} = \langle\tilde{\Phi}_k^{\mathrm{A}}|H_{\mathrm{A}}|\tilde{\Phi}_k^{\mathrm{A}}\rangle. \tag{6.29}$$

Substituting eqn (6.28) into (6.25), we obtain

$$\begin{aligned} H_{ij,kl} \sim \tilde{H}_{ij,kl}&+\tfrac{1}{2}\{(E_i^{\mathrm{A}}-\tilde{E}_i^{\mathrm{A}})+(E_j^{\mathrm{B}}-\tilde{E}_j^{\mathrm{B}}) \\ &+(E_k^{\mathrm{A}}-\tilde{E}_k^{\mathrm{A}})+(E_l^{\mathrm{B}}-\tilde{E}_l^{\mathrm{B}})\}\,\tilde{S}_{ij,kl}. \end{aligned} \tag{6.30}$$

The use of eqn (6.30) will be referred to as the AIM method. Two major approximations have been made in deriving it. First, there is the approximation involved in substituting approximate composite functions for exact ones in obtaining eqn (6. 22)from eqn (6.21) and second, there is the approximation involved in obtaining eqn (6.28) from (6.27). These approximations appear reasonable, but we will not attempt to justify them in detail. Instead we adopt the attitude that the value of eqn (6.30) should be judged on the basis of the results obtained by its use.

The AIM Hamiltonian eqn (6.30) can be broken up into two parts, the first part being the *ab initio* Hamiltonian $\tilde{H}_{ij,kl}$ and the second part being the term in curly brackets. This second term may be regarded as a correction term. It contains corrections for the discrepancies between the calculated and experimental energies of the states of the isolated atoms and ions. In the limit as the $\tilde{\Phi}_i^{\mathrm{A}}$ and $\tilde{\Phi}_j^{\mathrm{B}}$ approach exact atomic eigenfunctions, the AIM method becomes equivalent to an *ab initio* calculation; it is also identical with the customary '*ab initio*' procedure if the corrections for all atomic states are equal (see below).

In most applications, the approximate CFs are very crude and the corrections appearing in the $H_{ij,kl}$ matrix elements are so large and so different from each other that erroneously large dissociation energies result. This is particularly true if a single set of atomic orbitals appropriate to a few given composite functions (usually those best suited to the ground-state dissociation products) is used in forming the CFs for excited and ionic states, as well. Hurley (1956*a*) has suggested that for such crude treatments, which are the type most feasible for polyatomics and potential surfaces, the AIM corrections are too large and that instead of the $\tilde{E}_i^{\mathrm{A}}$ and $\tilde{E}_j^{\mathrm{B}}$ defined in eqn (6.29), functions with exponents optimized for the particular states under consideration should be used in evaluating the atomic energies (ICC method). Hurley's modification of Moffitt's method has the desired effect of reducing the correction energies appearing in eqn (6.30). For a number of diatomics it has yielded more reasonable dissociation energies than the AIM method itself (Hurley 1955; Balint–Kurti and Karplus 1969; Hurley 1956*b*), and generally it seemed to give good results for most systems to which it has been applied (Hurley 1958*b*; 1956*c*, Krauss and Wehner 1958; Ohno 1957). However, the ICC procedure introduces other difficulties. Although the basic AIM method leads to the correct experimental energy at large internuclear distances ($R \to \infty$) for all CFs, the ICC approach does so only for those built up from approximate eigenfunctions corresponding to the ground states of the atoms (or whatever atomic states the orbitals best approximate, Hurley 1958*a*). Moreover, in homonuclear diatomic ions (e.g., Li_2^+, F_2^-) not even the ground-state potential energy curve tends to the correct asymptotic value if a minimal atomic orbital basis set is used (Balint–Kurti and Karplus 1969).

It has long been recognized that, if an atomic orbital basis set is used to describe a molecule, the energies obtained from a given calculation are much improved by allowing the sizes of the atomic orbitals to vary (i.e., by optimizing orbital exponents) (Coulson 1937; Wang 1928). The possibility of such deformations of the atomic orbitals was not included in Moffitt's original method. Arai (1957*a*) has stressed the importance of allowing for these distortions within the framework of the atoms-in-molecules method. He has proposed a method based on eqn (6.25) in which the atomic orbitals are allowed to distort so as to minimize the electronic energy of the molecule (Arai 1957*a*) His 'deformed atoms-in-molecules' method includes a term which estimates the energy needed to deform the atomic eigenfunctions. The estimates of the deformation energy involve the calculation of new types of integrals which are not normally encountered in molecular calculations. This, together with the need to calculate the awkward quantities $V^{AB}_{ij,kl}$, makes the method difficult to apply. While Arai's deformed atoms-in-molecules method has been used with reasonable success for a few small molecules (Arai 1957*b*), application to larger systems is likely to prove too complicated.

We will not discuss either of the two above modifications of Moffitt's atoms-in-molecules methods in detail. Reviews of Hurley's ICC method (Hurley 1958*a*, 1963) and Arai's deformed atoms-in-molecules method (Arai, 1960) are available and some short but useful summaries of atoms-in-molecules methods have been published (Kotani 1961; Parr, 1964).

6.2.1. *The Orthogonalized Moffitt Method*

All the atoms-in-molecules methods depend on the identification of individual composite functions at finite internuclear separations with the asymptotic atomic states to which they correspond at large distances. At finite internuclear separations, the different composite functions are not orthogonal. Thus, identifying individual functions with the asymptotic states may no longer be appropriate; that is, as the overlap increases with decreasing distance, some procedure for separating out the atomic corrections is required. Although there is no *unique* procedure for doing this, it appears reasonable to sequentially project out from the higher CFs their overlap with the lower ones. This can be done by Schmidt orthogonalization of the CFs arranged in order of increasing AIM correction, $\{(E^{A}_{i}-\tilde{E}^{A}_{i})+(E^{B}_{j}-\tilde{E}^{B}_{j})\}$. It is the orthogonalized functions that are associated with the corresponding asymptotic limits and the AIM correction is applied to them (Orthogonalized Moffitt or OM method). Since the functions are now orthogonal, eqn (6.30) is replaced by the expression

$$H^{0}_{ij,kl} \sim \tilde{H}^{0}_{ij,kl}+\{(E^{A}_{i}-\tilde{E}^{A}_{i})+(E^{B}_{j}-\tilde{E}^{B}_{j})\}\delta_{ik}\delta_{jl}, \qquad (6.31)$$

where $\tilde{H}^{0}_{ij,kl}$ is the Hamiltonian matrix element over the orthogonalized CF's.

The calculated atomic energies appearing in the correction terms of eqn (6.31) are obtained with the approximate atomic eigenfunctions used to construct the approximate composite functions. This ensures that the energies resulting from the calculation at large internuclear distances will correspond to the experimental ones. It should be noted that the Hamiltonian matrix used in the OM method is automatically Hermitian and that it is not necessary to adopt any arbitrary symmetrization procedure in order to ensure this property.

We denote the corrections to the Hamiltonian matrix elements by

$$\Delta_{ij} = (E_i^{\mathrm{A}} - \tilde{E}_i^{\mathrm{A}}) + (E_j^{\mathrm{B}} - \tilde{E}_j^{\mathrm{B}}). \tag{6.32}$$

In terms of the non-orthogonal composite functions the OM Hamiltonian (eqn 6.31) can then be written, for a problem involving only two CFs, as

OM Hamiltonian–Non-orthogonal CF Basis

$$\begin{aligned} H_{ij,ij} &= \tilde{H}_{ij,ij} + \Delta_{ij} \\ H_{ij,kl} \equiv H_{kl,ij} &= \tilde{H}_{ij,kl} + \tilde{S}_{ij,kl}\Delta_{ij} \\ H_{kl,kl} &= \tilde{H}_{kl,kl} + \Delta_{kl} + \tilde{S}^2_{ij,kl}(\Delta_{ij} - \Delta_{kl}) \end{aligned} \tag{6.33}$$

This is to be compared with the AIM Hamiltonian in the non-orthogonal basis,

Moffitt's AIM Hamiltonian–Non-orthogonal CF Basis

$$\begin{aligned} H_{ij,ij} &= \tilde{H}_{ij,ij} + \Delta_{ij} \\ H_{ij,kl} \equiv H_{kl,ij} &= \tilde{H}_{ij,kl} + \tfrac{1}{2}\tilde{S}_{ij,kl}(\Delta_{ij} + \Delta_{kl}) \\ H_{kl,kl} &= \tilde{H}_{kl,kl} + \Delta_{kl} \end{aligned} \tag{6.34}$$

The relative magnitude of the corrections in eqn (6.33) and (6.34) clearly depends on the ordering of the non-orthogonal CFs. Since the atomic corrections Δ_{ij} are by definition negative, the chosen order of the non-orthogonal CFs with

$$|\Delta_{ij}| < |\Delta_{kl}| \tag{6.35}$$

leads to the result that the magnitudes of the corrections in the OM method are less than or equal to those in the AIM method. As experience has shown that the corrections in the AIM method are too large in the case of H_2 (Hurley 1955; Paunсz, 1954; Batana and Cohan 1962), as well as in many other molecules, this ordering, which is physically most reasonable, yields satisfactory results.

Intuitively we would expect the negative corrections which are applied to the Hamiltonian matrix elements in both the AIM and OM methods to lead to a lowering of the calculated molecular energy, and this is almost invariably found to be the result. The expected lowering of energy cannot however be proven, in general, and for unusual relationships of magnitude and sign between the Hamiltonian and overlap matrix elements and the atomic corrections, it may not be realized (Batana and Cohan 1962). In a similar manner, the smaller corrections applied in the OM method are expected to result in higher calculated molecular energies for the OM method than the AIM one. This has been found to be true for every system examined; however, there is no proof that this is a general result.

6.2.2. *Properties of atoms-in-molecules methods*

We consider what happens in Moffitt's atoms-in-molecules method if the corrections for all the states of a given atom are the same. Eqn (6.30) becomes

$$\begin{aligned} H_{ij,kl} &= \tilde{H}_{ij,kl} + \tilde{S}_{ij,kl}\{(E^{\mathrm{A}} - \tilde{E}^{\mathrm{A}}) + (E^{\mathrm{B}} - \tilde{E}^{\mathrm{B}})\} \\ &= \langle \tilde{\Phi}_{ij}^{\mathrm{AB}} | H + (E^{\mathrm{A}} - \tilde{E}^{\mathrm{A}}) + (E^{\mathrm{B}} - \tilde{E}^{\mathrm{B}}) | \tilde{\Phi}_{kl}^{\mathrm{AB}} \rangle . \end{aligned} \tag{6.36}$$

In this special case, the only effect of the AIM correction is to lower the calculated molecular energy by the atomic error, while the *shape* of the potential energy curve remains identical to that which would have been obtained without correction. In the standard *ab initio* method, the dissociation energy is determined by comparing the calculated molecular energy with the calculated energies of the atoms and assuming that the errors in the molecular calculation are equal to those in the atomic calculations. Thus, if the corrections for all the states of a given atom are the same, Moffitt's AIM method gives the same results as the standard *ab initio* procedure. In general, the atoms-in-molecules methods apply a separate correction for each of the atomic states. In this sense, therefore, the *ab initio* procedure can be regarded as a crude atoms-in-molecules method (Hurley 1958*a*).

By comparing eqn (6.33) and (6.34), we see that, if the corrections for all the states of a given atom are the same (i.e. $\Delta_{ij} = \Delta_{kl}$) the OM and AIM Hamiltonian matrices are identical. In this case, therefore, the OM method also gives results identical with the *ab initio* calculation.

The importance of the limit of large internuclear separations has been stressed during the development of the atoms-in-molecules method. As the atoms separate, both $H_{ij,kl}$ and $S_{ij,kl}$ become diagonal, and eqn (6.30) reduces to

$$\begin{aligned} H_{ij,kl} &= (\tilde{E}_i^{\mathrm{A}} + \tilde{E}_j^{\mathrm{B}})\delta_{ij,kl} + [(E_i^{\mathrm{A}} - \tilde{E}_i^{\mathrm{A}}) + (E_j^{\mathrm{B}} - \tilde{E}_j^{\mathrm{B}})]\delta_{ij,kl} \\ &= (E_i^{\mathrm{A}} + E_j^{\mathrm{B}})\delta_{ij,kl} \end{aligned} \tag{6.37}$$

We see, therefore, that at large internuclear separations, Moffitt's atoms-in-molecules method yields the experimental energies of the constituent atoms. This is just what we would expect of a reasonable empirical scheme. The OM method becomes equivalent to the AIM method in the large separation limit (this may be seen by comparing eqns (6.33) and (6.34) in this limit) and therefore has the same behaviour.

As stated above, the ICC method destroys this limiting behaviour for all but a few molecular states, because the corrections which are used in eqn (6.37) are, in general, too small to lower the calculated atomic energies to the experimental values. If we were to introduce atomic orbitals which were chosen by exponent optimization of the energies E_i^{A} and E_j^{B} in the molecular calculation, the corresponding molecular states would asymptotically approach the correct experimental energies in the ICC method. For homonuclear diatomic ions with unit charge, there are no states which satisfy this condition (if a minimum orbital basis set is used), and, therefore, none of the molecular states dissociate to the experimental atomic energies in the ICC method for these systems.

6.2.3. *Improving an atoms-in-molecules calculation*

The crudest type of calculation we envisage is one in which a minimum atomic orbital basis set is used to construct the approximate atomic eigenfunctions, and the minimum number of CFs needed to correctly describe the symmetry and dissociation products of the lowest potential energy curve is included in the molecular calculation. For HF as an example, this would correspond to using only a 1s orbital on H and 1s, 2s, and 2p orbitals on F and constructing from these the two composite functions for $\mathrm{H}(^2\mathrm{S})+\mathrm{F}(^2\mathrm{P}^0)$ combined to give a $^1\Sigma^+$ molecular state. This description of HF could be improved by successively introducing certain refinements into the calculation:

(1) Using only the minimum orbital basis set, we form all possible composite functions which have the correct symmetry to contribute to the molecular state. For HF we could construct two composite functions corresponding to the atomic states $\mathrm{H}(^2\mathrm{S})+\mathrm{F}(^2\mathrm{S})$, and composite functions corresponding to the ions $\mathrm{H}^+\mathrm{F}^-(^1\mathrm{S})$, $\mathrm{H}^-(^1\mathrm{S})\mathrm{F}^+(^1\mathrm{D})$, $\mathrm{H}^-(^1\mathrm{S})\mathrm{F}^+(^1\mathrm{S})$ and $\mathrm{H}^-\mathrm{F}^+(^1\mathrm{P}^0)$. On chemical grounds, the composite function corresponding to $\mathrm{H}^+\mathrm{F}^-(^1\mathrm{S})$ is expected to be by far the most important of these. We would expect the omission of the other CFs to have little effect on the calculated potential energy curve.

(2) The atomic orbital basis set could be increased so as to permit a description of other low-lying atomic states. For HF this would mean adding 2s and 2p orbitals on H and 3s, 3p, and 3d orbitals on F. In keeping with the spirit of the atoms-in-molecules method, these

orbitals would be those of isolated H and F atoms. The additional approximate atomic eigenfunctions would be used to construct additional CFs.

(3) The approximate atomic eigenfunctions could be improved by including the interelectron distance r_{12} explicitly in their functional form or by including intra-atomic configuration interaction. A limited amount of intra-atomic configuration interaction could be included in the orbital basis set of stage (2) (e.g., for $H^-(^1S)$). To obtain a still better description of the atomic states, the orbital basis set must be further enlarged. As the description of the atomic states improves, the calculated atomic energies tend towards the experimental values and the corrections to the Hamiltonian matrix elements tend to zero. In the limit, the atoms-in-molecules methods give the same results as the *ab initio* method.

(4) The description of the molecule could be improved by adding additional composite functions. To some extent, these new composite functions could be built up from the same orbital basis set that is used to improve the description of the atomic states. On the other hand, we might depart slightly from the atoms-in-molecules philosophy and introduce orbitals suitable for describing the distortion of the low-lying atomic states in the molecular environment.

The increase in size of the composite function basis set will eventually give rise to problems. There may be approximate composite functions which are built up from approximate atomic eigenfunctions corresponding to atomic states for which no experimental data is available. Alternatively, composite functions that allow for the distortion of the atoms in the molecule may not form suitable approximations to any atomic states. Since it is impossible to determine a reliable atomic correction for these functions, it is proposed that no correction be applied. In the OM method, where the ordering of the composite functions matters, the approximate composite functions containing such atomic functions are placed at the end. In this procedure, therefore, the size of the CF basis set can be increased indefinitely and the exact solution may be approached in the limit of an infinite CF basis set. The situation often arises that there are one or more groups of CFs which are associated with the same correction. It can be shown that the ordering, among themselves, of any group of CFs associated with the same correction does not affect the outcome of an OM calculation (Balint–Kurti 1969).

We have seen that to approach the exact molecular energy and eigenfunction we must strive to attain the dual limit of exact atomic functions and infinite configuration interaction. The atoms-in-molecules methods attempt to estimate what the results of attaining the first of these two limits would be without actually performing the calculation. The second limit must be approached in the normal brute-force manner.

6.2.4. *Illustrative example*

To illustrate the various atoms-in-molecules methods we consider the H_2 molecule, which has served as the model system in previous discussions (Hurley 1955; Pauncz 1954; Batana and Cohan 1962). In keeping with our interest in more complex systems, we use the approximation in which the covalent and ionic valence-bond structures† are both formed from 1s orbitals on the hydrogen atoms, the same orbital exponent ($\zeta = 1$) being employed throughout. We have

$$\begin{aligned}\Phi_{\text{cov}} &= \tfrac{1}{2}\{a(1)b(2)+b(1)a(2)\}\{\alpha(1)\beta(2)-\beta(1)\alpha(2)\}/(1+S_{ab}^2)^{\frac{1}{2}}\\ \Phi_{\text{ion}} &= \tfrac{1}{2}\{a(1)a(2)+b(1)b(2)\}\{\alpha(1)\beta(2)-\beta(1)\alpha(2)\}/(1+S_{ab}^2)^{\frac{1}{2}}\end{aligned} \tag{6.38}$$

with $(E_{\text{cov}}-\tilde{E}_{\text{cov}}) = 0$ and $(E_{\text{ion}}-\tilde{E}_{\text{ion}}) = -0{\cdot}1528$ a.u. (Stewart 1964); in Hurley's approach $\tilde{E}_{\text{ion}}$ is calculated with the exponent optimized for H^- ($\zeta = 0{\cdot}687$) and $(E_{\text{ion}}-\tilde{E}_{\text{ion}}) = -0{\cdot}05518$ a.u. The orthogonalized Moffitt method replaces the functions Φ_{cov} and Φ_{ion} by the orthogonalized set

$$\begin{aligned}\Psi_{\text{cov}} &= \Phi_{\text{cov}}\\ \Psi_{\text{ion}} &= (\Phi_{\text{ion}}-S\Phi_{\text{cov}})(1-S^2)^{-\frac{1}{2}}\end{aligned} \tag{6.39}$$

where $S = \int \Phi_{\text{ion}}\Phi_{\text{cov}}\,d\tau$. For this case the AIM Hamiltonian matrix is

$$\begin{aligned}H_{\text{cov,cov}} &= \tilde{H}_{\text{cov,cov}}+(E_{\text{cov}}-\tilde{E}_{\text{cov}})\\ H_{\text{ion,cov}} &= \tilde{H}_{\text{ion,cov}}+\frac{S}{2}\{(E_{\text{ion}}-\tilde{E}_{\text{ion}})+(E_{\text{cov}}-\tilde{E}_{\text{cov}})\}\\ H_{\text{ion,ion}} &= \tilde{H}_{\text{ion,ion}}+(E_{\text{ion}}-\tilde{E}_{\text{ion}})\end{aligned} \tag{6.40}$$

where we have included the covalent correction terms which are zero for $\zeta = 1$, to show the general form. The ICC expression is of the same form and differs only in the numerical value of the correction. To compare the OM method with eqn (6.40), we write the OM Hamiltonian in the non-orthogonal CF basis (see eqn (6.33)).

$$\begin{aligned}H_{\text{cov,cov}} &= \tilde{H}_{\text{cov,cov}}+(E_{\text{cov}}-\tilde{E}_{\text{cov}})\\ H_{\text{ion,cov}} &= \tilde{H}_{\text{ion,cov}}+S(E_{\text{cov}}-\tilde{E}_{\text{cov}})\\ H_{\text{ion.ion}} &= \tilde{H}_{\text{ion.ion}}+(E_{\text{ion}}-\tilde{E}_{\text{ion}})+S^2\{(E_{\text{cov}}-\tilde{E}_{\text{cov}})-(E_{\text{ion}}-\tilde{E}_{\text{ion}})\}.\end{aligned} \tag{6.41}$$

As the two hydrogen atoms approach each other, the 1s orbitals a and b in eqn (6.38) become progressively more alike, as do also the covalent and

† Valence-bond structure and CF basis sets yield identical results for this case.

ionic valence-bond structures. In the limit of zero internuclear separation, Φ_{cov} and Φ_{ion} are identical. The diagonal Hamiltonian matrix elements in eqn (6.40) do not, however, become equal in this limit. Thus, two identical functions lead to different diagonal Hamiltonian matrix elements in the AIM (or ICC) method. This is clearly an undesirable attribute of the methods. In the same limit (i.e. $S = 1$), the diagonal Hamiltonian matrix elements in the OM expression eqn (6.41) behave correctly; i.e. they become equal.

TABLE 6.1

Ab initio and empirical calculations for ground state of H_2.[a,b]

R	*ab initio*	OM	AIM	ICC	Kolos and Wolnicwicz (1964)
0·3	2·01305	2·01305	1·28022	1·91237	—
0·4	1·22431	1·22431	0·22577	1·08518	0·87980
0·42	1·11396	1·11395	0·03471	0·96151	—
0·44	1·01418	1·01415	−0·11117	0·85214	—
0·46	0·92365	0·92357	−0·20278	0·75896	—
0·48	0·84137	0·84124	0·24433	0·68224	—
0·5	0·76653	0·76636	0·24884	0·61969	0·47337
0·6	0·47895	0·47879	−0·12629	0·40307	0·23037
0·7	0·28865	0·28857	−0·08270	0·24519	0·07797
0·8	0·15830	0·15825	−0·09429	0·12883	−0·02006
1·0	0·00355	0·00355	−0·13978	−0·01489	−0·12454
1·2	−0·07223	−0·07226	−0·17003	−0·08701	−0·16493
1·3	−0·09314	−0·09321	−0·17803	−0·10703	−0·17235
1·4009	−0·10665	−0·10680	−0·18209	−0·11993	(−0·17447)
1·5	−0·11438	−0·11463	−0·18284	−0·12721	−0·17285
1·6	−0·11802	−0·11839	−0·18099	−0·13047	−0·16858
1·8	−0·11670	−0·11738	−0·17155	−0·12848	−0·15507
2·0	−0·10840	−0·10944	−0·15723	−0·11947	−0·13813
2·2	−0·09659	−0·09800	−0·14041	−0·10687	−0·12012
3·0	−0·04672	−0·04872	−0·07292	−0·05293	−0·05726
4·0	−0·01278	−0·01356	−0·02168	−0·01471	—
5·0	−0·00228	−0·00291	−0·00475	−0·00318	—

[a] Atomic units are used. 1 a.u. of energy = 4·359 828 10^{-18} J. 1 a.u. of length = 0·529 177 × 10^{-10} m.

[b] All numbers are accurate to at least one unit in the fifth decimal place. The zero of energy is taken to be the energy of the separated ground state atoms.

The calculated values for H_2 obtained with no correction, with the AIM, ICC and OM methods are shown in Table 6.1 and Fig. 6.1, which also includes the accurate Kolos–Wolniewicz (1964) results for comparison. The Moffitt calculation overcorrects for this case, yielding too large a dissociation energy and an overall potential curve below the true one. Hurley's approximation reduces the correction and yields not unreasonable results. The orthogonalized Moffitt method leads to energy values that are only very slightly below the non-empirical valence-bond calculation; that is,

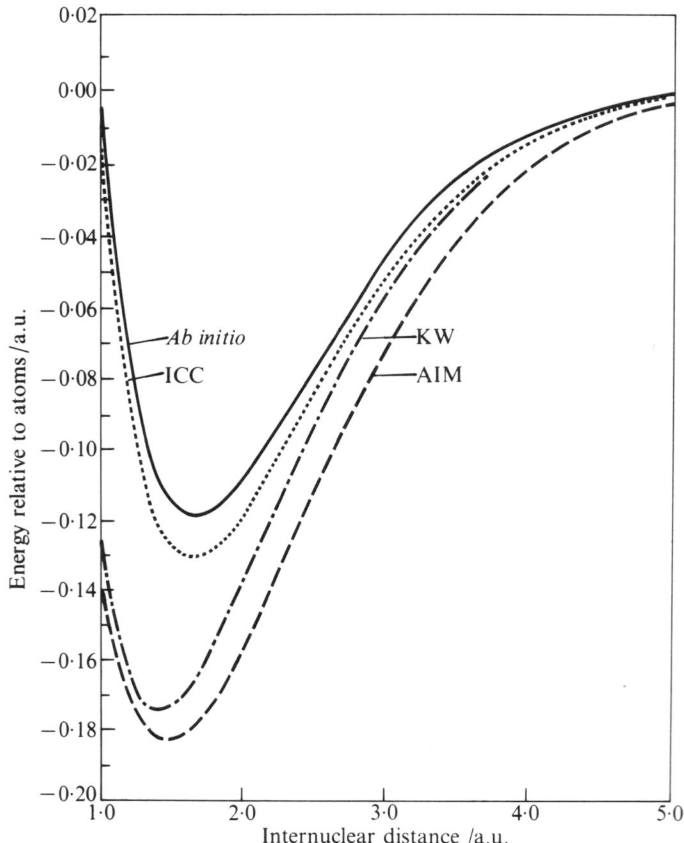

FIG. 6.1. Potential energy curves for H_2, KW refers to Kolos and Wolniewicz (1964). The OM curve is not shown as it lies almost on top of the *ab initio* one. All curves tend to zero at large internuclear distances.

the overlap between the two functions is sufficiently large that the correction appearing in eqn (6.41) due to the ionic wavefunction is very small. For the excited singlet state resulting from the calculation, the AIM and OM methods yield the correct energy at infinitely large internuclear separations (0·4722 a.u., corresponding to H^+, H^- relative to H), while the ICC method gives 0·5682 a.u. and the *ab initio* calculation gives 0·6709 a.u.

If one examines the ground-state energies at small internuclear distances ($R < 1{\cdot}0$ a.u.) one finds the results shown in Fig. 6.2. It is clear that the AIM method has a spurious inner minimum at 0·5 a.u., which is not present in any of the other calculations. The minimum arises from the large correction to the off-diagonal Hamiltonian matrix element in the AIM method.

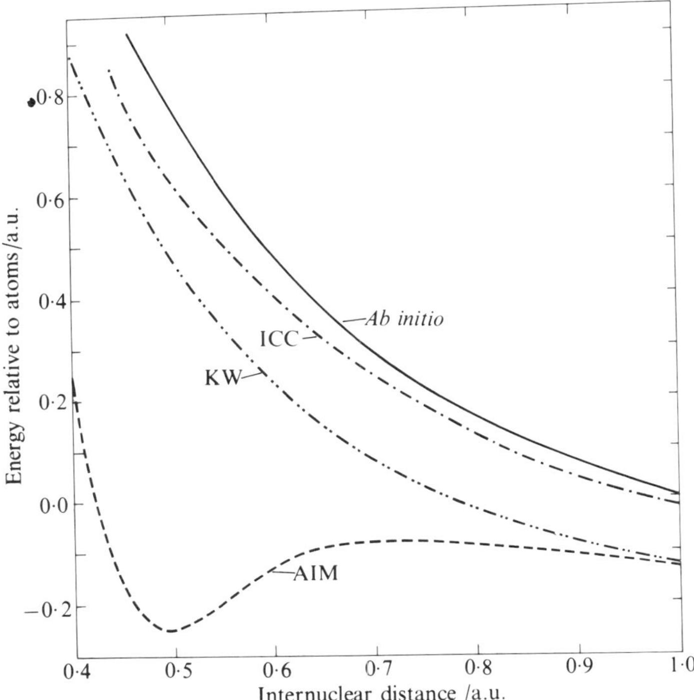

FIG. 6.2. Potential energy curves for H_2 at small internuclear separations: KW refers to Kolos and Wolniewicz (1964). The OM curve is not shown as it lies almost on top of the *ab initio* one.

6.3. The nonorthogonality problem

Both the composite functions, which form the basis set for atoms-in-molecules calculations, and valence-bond functions are, in general, linear combinations of Slater determinants built up from non-orthogonal orbitals. In this section the problem of evaluating the relevant matrix elements between two such non-orthogonal Slater determinants is considered. Several procedures for calculations of this type have been proposed (Löwdin, 1955) and some calculations have been performed with them (Hurley 1960; Copsey, Murrell and Stamper 1971). We outline here a procedure suggested by Hurley (1960). It involves transforming the non-orthogonal orbital basis set to an orthogonal one, then evaluating the required matrix elements between Slater determinants built from this orthogonal basis, and finally transforming the matrix elements between the orthogonal Slater determinants to matrix elements between the non-orthogonal determinants. This procedure is not the most efficient in every case. It is particularly suitable for problems in which there are doubly-occupied orbitals that do not participate in the bonding. Moreover, its implementation does not involve any major problems. The formalism used for describing the method is based on that given by Moffitt (1953).

The non-orthogonal Slater determinants are constructed from a non-orthogonal, real, spin-orbital basis set ϕ_j. If the ϕ_j are expanded in terms of an orthogonal set ψ_i, we can write

$$\phi_j = \sum_{i=1}^{N_{\text{so}}} \psi_i \gamma_j^i \qquad j = 1, 2, \ldots, N_{\text{so}} \tag{6.42}$$

where γ_j^i are the expansion coefficients of the non-orthogonal orbitals in terms of the orthogonal orbitals, and there are N_{so} spin-orbitals in both basis sets. Using the Einstein summation convention, which implies a summation over all repeated indices, we can write eqn (6.42) in the form

$$\phi_j = \psi_i \gamma_j^i. \tag{6.43}$$

The non-orthogonal Slater determinants are written as

$$\begin{aligned} \Phi_{\text{A}} &= (N!)^{-\frac{1}{2}} \det|\phi_{a_1}(1)\phi_{a_2}(2) \ldots \phi_{a_N}(N)| \\ &= (N!)^{-\frac{1}{2}} \Phi_{a_1 \ldots a_N}^{1 \ldots N} \end{aligned} \tag{6.44}$$

where the ϕ_{a_i} are spin-orbitals and the symbol $\Phi_{a_1 \ldots a_N}^{1 \ldots N}$ denotes the Slater determinant of the preceding line. Slater determinants constructed from the orthogonal orbital basis set ('orthogonal' Slater determinants) are denoted in an identical manner except that the function $\Phi_{a_1 \ldots a_N}^{1 \ldots N}$ is replaced by $\Psi_{a_1 \ldots a_N}^{1 \ldots N}$. The expansion in eqn (6.43) and the properties of determinants (Aitken 1959) may be used to express the non-orthogonal Slater determinant of eqn (6.44) in terms of orthogonal determinants; that is,

$$\Phi_{a_1 \ldots a_N}^{1 \ldots N} = \gamma_{a_1}^{i_1} \gamma_{a_2}^{i_2} \ldots \gamma_{a_N}^{i_N} \Psi_{i_1 \ldots i_N}^{1 \ldots N}. \tag{6.45}$$

In eqn (6.45) there is a summation over all possible values of the indices $i_1, i_2, \ldots, i_N$, each of which can take on values from 1 and N_{so}. To simplify eqn (6.45), we introduce the generalized Kronecker delta $\delta_{i_1 i_2 \ldots i_N}^{j_1 j_2 \ldots j_N}$ (Eisenhart, 1926). This quantity is equal to zero if either the bottom or the top rows of numbers contain any number more than once, or if the two rows are composed of different numbers; otherwise, it is equal to ± 1 depending on whether it requires an even or an odd number of permutations to bring the two sets of numbers into coincidence ($+1$ for even, -1 for odd). With the generalized Kronecker delta, determinants can be written in the expanded form; e.g.

$$\Psi_{i_1 \ldots i_N}^{1 \ldots N} = \delta_{i_1 i_2 \ldots i_N}^{j_1 j_1 \ldots j_N} \psi_{j_1}^1 \psi_{j_2}^2 \ldots \psi_{j_N}^N \tag{6.46}$$

where $\psi_j^k = \psi_j(k)$. Using eqn (6.46), we can re-express eqn (6.45) in the form

$$\Phi_{a_1 \ldots a_N}^{1 \ldots N} = \gamma_{a_1}^{i_1} \gamma_{a_2}^{i_2} \ldots \gamma_{a_N}^{i_N} \delta_{i_1 \ldots i_N}^{j_1 \ldots j_N} \psi_{j_1}^1 \psi_{j_2}^2 \ldots \psi_{j_N}^N \tag{6.47}$$

The Kronecker delta can be expanded by utilizing the property (Eisenhart 1926)

$$\delta^{j_1 \dots j_N}_{i_1 \dots i_N} = \frac{1}{N!} \delta^{k_1 \dots k_N}_{i_1 \dots i_N} \delta^{j_1 \dots j_N}_{k_1 \dots k_N} \tag{6.48}$$

If this is introduced into eqn (6.47), we obtain

$$\Phi^{1 \dots N}_{a_1 \dots a_N} = \frac{1}{N!} \gamma^{i_1}_{a_1} \gamma^{i_2}_{a_2} \dots \gamma^{i_N}_{a_N} \delta^{k_1 \dots k_N}_{i_1 \dots i_N} \delta^{j_1 \dots j_N}_{k_1 \dots k_N} \psi^1_{j_1} \psi^2_{j_2} \dots \psi^N_{j_N} \tag{6.49}$$

The properties of the generalized Kronecker delta enable us to rewrite eqn (6.49) as the product of two determinants (compare with eqn (6.46)

$$\Phi^{1 \dots N}_{a_1 \dots a_N} = \frac{1}{N!} \Gamma^{k_1 \dots k_N}_{a_1 \dots a_N} \Psi^{1 \dots N}_{k_1 \dots k_N} \tag{6.50}$$

where $\Gamma^{k_1 \dots k_N}_{a_1 \dots a_N}$ denotes the $N \times N$ determinant, $\det[\gamma^{k_i}_{a_j}]$, which is a minor of the determinant of the matrix $\mathbf{\Gamma}$. (It should be noted that N is the number of electrons, and is generally less than the number of spin-orbitals N_{so}; the $\mathbf{\Gamma}$ matrix is of dimensionality $N_{so} \times N_{so}$, and its elements are the expansion coefficients γ^i_j). The notation $A^{k_1 \dots k_N}_{a_1 \dots a_N}$ will be used to denote minors of the determinant of the general matrix $\mathbf{A}$ throughout this section.

In eqn (6.50) there is a summation over the indices $k_1 \dots k_N$. None of these indices can be the same, because if they are, the corresponding determinant is zero. There is an $N!$-fold repetition of terms in the summation as any given set of numbers $k_1 \dots k_N$ occurs with all possible orderings. The work involved in performing the summation in eqn (6.50) can be reduced by insisting that the indices $k_1 \dots k_N$ always be ordered in the sense $k_1 < k_2 \dots < k_N$ and omitting the $1/N!$ Thus, the final expression relating the non-orthogonal to the orthogonal Slater determinants is

$$\Phi_{\mathrm{A}} = T_{\mathrm{AK}} \Psi_{\mathrm{K}} \tag{6.51}$$

where

$$\Phi_{\mathrm{A}} = \left(\frac{1}{N!}\right)^{\frac{1}{2}} \det[\phi_{a_1}(1) \dots \phi_{a_N}(N)]$$

$$\Psi_{\mathrm{K}} = \left(\frac{1}{N!}\right)^{\frac{1}{2}} \det[\psi_{k_1}(1) \dots \psi_{k_N}(N)]$$

$$T_{\mathrm{AK}} = \Gamma^{k_1 \dots k_N}_{a_1 \dots a_N}.$$

The implied summation over K in eqn (6.51) runs over all distinct combinations of the indices $k_1 \dots k_N$ arranged in ascending order.

In applying eqn (6.51) several simplifications can be made. The most obvious of these is to re-express the minor $\Gamma^{k_1 \dots k_N}_{a_1 \dots a_N}$ as a product of minors of smaller order. Since the matrix elements of $\mathbf{\Gamma}$ between two spin orbitals with different spins are zero, T_{AK} can be rewritten as

$$T_{\mathrm{AK}} = \Gamma^{k_1 \ldots k_m}_{a_1 \ldots a_m} \Gamma^{k_{m+1} \ldots k_N}_{a_{m+1} \ldots a_N} \tag{6.52}$$

if the orbitals with sub-indices 1 to m have alpha spin, and those with $m+1$ to N have beta spin. Further simplifications are possible when the orbitals ϕ_i separate into groups, each of which span mutually orthogonal subspaces. If the orbitals ψ_i are chosen so that they fall into the same groups, the $\mathbf{\Gamma}$ matrix becomes block diagonal; that is,

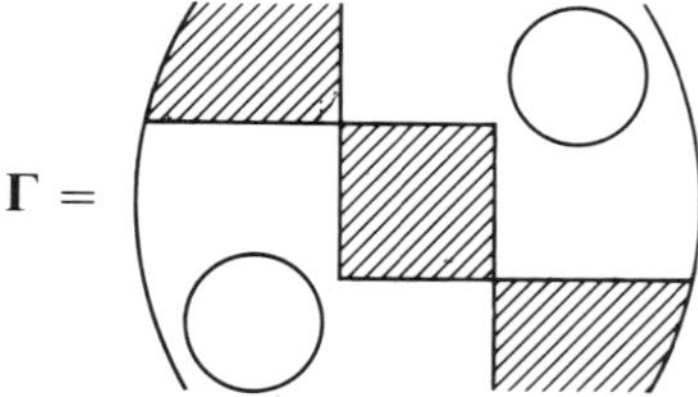

and the minors $\Gamma^{k_1 \ldots k_m}_{a_1 \ldots a_m}$ reduce to products of minors in the individual subspaces. Since in a problem involving many composite functions the matrix $\mathbf{T}$ is very large and very many minors must be evaluated, it is useful to be able to reduce their dimensionality.

The choice of the orthogonal orbitals ψ_i is important in further simplifying the $\mathbf{\Gamma}$ matrix and the required transformations; that is, the number of Ψ_{K} appearing in eqn (6.51) can be reduced by choosing the ψ_i such that $\mathbf{\Gamma}$ has the maximum possible number of zero elements. This is accomplished by using the Schmidt orthogonalization procedure to define the orbitals ψ_i and, thereby, reducing the $\mathbf{\Gamma}$ matrix to triangular form,

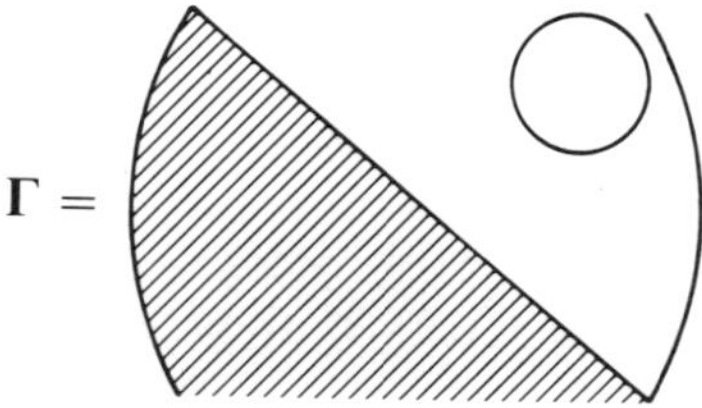

Finally, the ordering of the non-orthogonal basis ϕ_i can be utilized to simplify the transformation by reducing the number of Ψ_{K} arising in the expansion of the set Φ_{A}. This becomes clear once we realize that ϕ_i gives rise only to ψ_j with $j < i$, and recall that a determinant is equal to zero if a given ψ_i occurs more than once with a given spin. Thus, the number of Ψ_{K} in eqn (6.51) can be minimized by choosing the order of the ϕ_i such that the ones which are most often occupied have the lowest indices. It should be pointed out that the matrix **T** in eqn (6.51) is not necessarily square. The only requirement on it is that it must have at least as many columns as rows (i.e., there must be at least as many Ψ_K as Φ_A).

In summary, calculation of the Hamiltonian matrix in the basis set of the non-orthogonal Slater determinants Φ_A involves the following steps:

(1) Find the matrix $\mathbf{\Gamma}$ which relates the non-orthogonal to the Schmidt orthogonalized orbital basis set eqn (6.43). The non-orthogonal orbitals should be arranged in the optimal order for the problem at hand, as described above.

(2) From each non-orthogonal Slater determinant Φ_A generate all the orthogonal Slater determinants Ψ_K to which it gives rise. This is accomplished by expanding each of the non-orthogonal orbitals in Φ_A in terms of the orthogonal orbitals and multiplying out the determinant formed in this manner to generate determinants built up from the orthogonal orbitals. When this has been done to all the non-orthogonal Slater determinants Φ_A, a complete non-redundant set of orthogonal Slater determinants Ψ_K is assembled.

(3) Calculate the transformation matrix **T** which relates the non-orthogonal and orthogonal Slater determinants eqn (6.51).

(4) Calculate the Hamiltonian matrix between the orthogonal Slater determinants. This is done using standard and simple procedures (Condon and Shortley 1935).

(5) Transform the Hamiltonian and overlap matrices back from the orthogonal to the non-orthogonal Slater determinant basis sets by means of the matrix **T**.

Before the Hamiltonian matrix can be calculated in the basis set of the orthogonal Slater determinants, the two-electron integrals must be transformed from the non-orthogonal orbital basis set, in which they were calculated, to the orthogonalized orbital basis set. A similar transformation is also needed to perform a configuration intereaction calculation within the molecular orbital framework (Nesbet 1963). The transformation is always time-consuming and care has to be exercised to choose the best alogarithm for performing it (Nesbet 1963; Kim and Edminston 1970). Alternatively, a dummy calculation may be performed to discover which transformed integrals will be required in the final calculation and then the transformation need only be carried through for these integrals (Morokuma and Pedersen 1967). As normally only a very small fraction of the total possible number of two-electron integrals are needed in a calculation, this procedure is often useful.

The orthogonalized orbitals which occur in the above scheme are determined purely by convenience and have no physical significance. As an example of the relative speeds of performing various parts of the calculation, an approximate breakdown of computer time used on an IBM 7094 computer for a calculation on LiF_2 is given below (Clementi and Davis 1965). The calculation used 39 uncontracted Gaussian orbitals, 15 contracted orbitals, 123 CFs, 135 non-orthogonal Slater determinants, and 177 orthogonal Slater determinants.

Stage of computation	Time in minutes	
	Linear	Non-linear
Calculation of integrals	8·3	11·5
Transformation of integrals	1·8	8·8
Evaluation of Hamiltonian and overlap matrices in CF basis set	4·6	5·5
Solution of secular equations	1·0	1·3
Total time	15·7	27·1

6.4. Diatomic calculations

Several calculations using the Orthogonalized Moffitt (OM) method have now been performed on both diatomic (Balint-Kurti and Karplus 1969; Balint-Kurti 1969; 1971) and triatomic (Balint-Kurti 1969; Balint-Kurti and Karplus 1971) systems. In all cases a small atomic-orbital basis set was used. The majority of the calculations were preliminary studies for the determination of potential energy surfaces for chemical reactions (see Section 6.5), Thus, the objective, in the case of the diatomics, was not to obtain the best possible results for the diatomic at hand, but rather to find an adequate description of the diatomic which was sufficiently simple to permit its extension to the triatomic calculations.

In the calculations, the atomic orbitals were chosen to be suitable for the ground-state neutral atoms. Ions (e.g. F^-) were built up from the same atomic orbitals. In general, 1s, 2s, and 2p atomic orbitals were included in the calculations. The atomic orbitals themselves were contracted Gaussian-type orbitals (Shavitt 1963), the 1s and 2s atomic orbitals being composed of a single set of 4 or 5 1s type Gaussians and the 2p orbitals of between 1 and 3 Gaussian orbitals (Whitman *et al.*, 1969).

For every diatomic considered, the *ab initio*, multistructure valence-bond and the three atoms-in-molecules (AIM, OM, and ICC) calculations were performed. The time-consuming part of the calculation was always the evaluation of the Hamiltonian matrix needed in the *ab initio* calculation.

TABLE 6.2

Calculated and experimental dissociation energies (D_e) for some diatomic molecules.[a]

Molecule	*ab initio*	OM	AIM	ICC	Experimental	References
LiF	0·110	0·225	0·302	0·196	0·221	Balnit–Kurti and Karplus (1969)
F_2	0·051	0·075	0·139	0·068	0·061	Balnit–Kurti and Karplus (1969)
F_2^-	0·087	0·088	0·089	−0·050[b]	—	Balnit–Kurti and Karplus (1969)[c]
Li_2	0·023	0·023	—[d]	—[d]	0·039 Herzberg (1950)	Balnit–Kurti (1969)
Li_2^+	0·039	0·039	0·039	0·026[b]	—	Balint–Kurti (1969)
HF	0·157	0·191	0·287	0·223	0·225 Johns and Barrow (1959)	Balint–Kurti (1969)
LiH	0·064	0·064	—[d]	0·074	0·093 Velasco (1957)	Balint–Kurti (1969)
LiH^+	0·001	0·001	0·001	−0·009[b]	—	Balint–Kurti (1969)
$F_2^+(^2\Pi_g)$	0·110	0·110	0·113	0·018[b]	0·127 Potts and Price (1971)	Balint–Kurti (1971)

[a] All energies are in atomic units (see Table 6.1). The results quoted are for the same CF basis sets as were used to build up the CF basis sets for the triatomic calculations (see text).

[b] The ICC method does not dissociate correctly in this case. The quoted dissociation energies are obtained by comparing the lowest point on the ICC potential energy curve with experimental energies of the atoms (see text).

[c] In Balint–Kurti and Karplus (1969) a typographical error was made in all the ICC numbers of Table VI; Fig. 5 is, however, correct.

[d] The potential-energy curves obtained in these cases are unreasonable (see text).

Once this had been calculated, it was a relatively simple matter to modify the Hamiltonian matrix by introducing the various atoms-in-molecules corrections.

In Table 6.2 the calculated and experimental dissociations energies are given for the lowest states of some diatomics of interest. If the *ab initio* and OM results in Table 6.2 are compared, it is seen that for LiF and HF there is a large difference between the predictions of the two methods. In the ground states of these molecules at their equilibrium separation, ionic valence-bond structures make an important contribution and the relative energies of the ionic and covalent structures play a crucial role in determining the dissociation energy. In such a situation the atoms-in-molecules methods are expected to correct for the large atomic errors present in the *ab initio* calculation (see Section 6.2). The results indeed show that the OM method is successful in supplying the necessary corrections and leads to considerably improved results. The AIM method overcorrects for both LiF and HF and yields too great a binding energy, while the ICC method gives reasonable results for both molecules.

In other cases there is little or no difference between the *ab initio* and OM methods predictions. This correspondence can arise from two distinct causes. The first is that the corrections associated with all the CFs are identical or very similar, since the *ab initio* OM and AIM methods then give the same dissociation energies (see Section 6.2). This is the explanation for F_2^-, Li_2^+, LiH^+, and F_2^+. It should be noted that, although the *ab initio* and OM (and AIM) methods yield identical dissociation energies for these cases, they give very different predictions as to their absolute energies. For example, the *ab initio* method predicts that F_2 has a lower energy than F_2^- and that the electron affinity of F_2 is negative (Balint-Kurti and Karplus 1969). The OM method, on the other hand, predicts a positive value for the electron affinity, which goes to the correct limiting value for the separated atoms.

The *ab initio* and OM methods also give very similar results if the overlaps between CFs which are associated with the smallest correction and those associated with corrections differing appreciably from these, are very large. This is the case for Li_2 and LiH, as for the model calculations on H_2 (see Section 6.2.4) where the ionic and covalent CFs overlapped very strongly. When the overlap between ionic and covalent structures became large, the Hamiltonian matrix element $H_{\text{ion,cov}}$ from the AIM method became more negative than the element $H_{\text{cov,cov}}$ and the AIM method gave rise to a spurious minimum in the potential energy curve (see Fig. 6.2). This unreasonable behaviour is repeated, for much the same reasons, for Li_2 and LiH (Batana and Cohan 1962). In Fig. 6.3, potential energy curves calculated for the $^1\Sigma_g^+$ ground state of the Li_2 diatomic molecule are shown. The CF basis set consisted of 18 CF's, 12 of them covalent and 6 ionic. Both the AIM and ICC results are seen to be unsatisfactory. This demonstrates that the ICC method

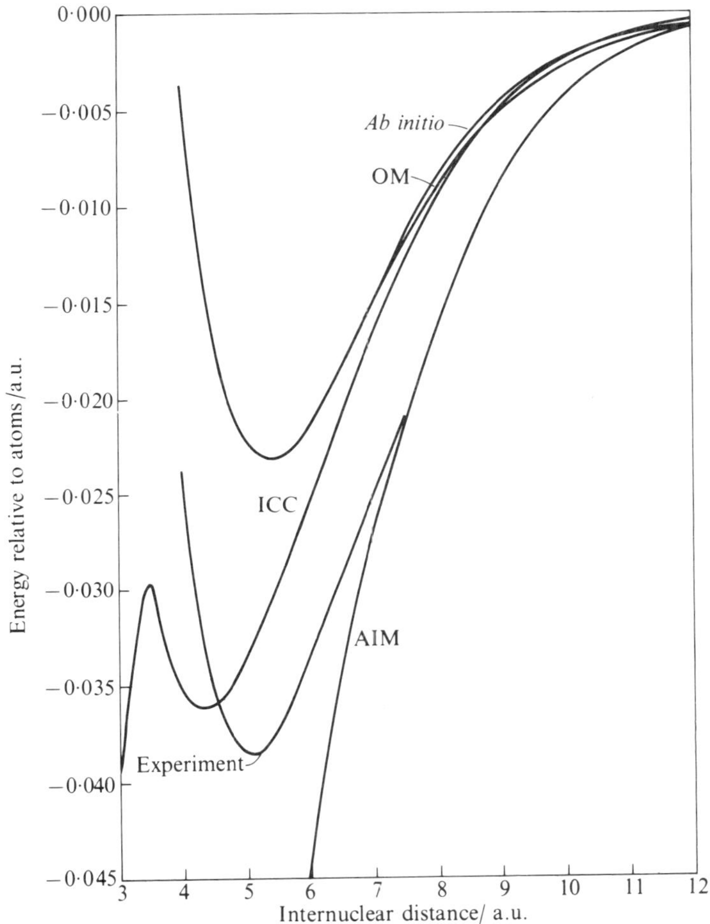

FIG. 6.3. Potential energy curves for the ground state of Li_2; for experimental parameters see Herzberg (1950).

suffers from the same basic inconsistencies as the AIM approach. If the 6 ionic CFs are omitted from the basis set, all the atoms-in-molecules methods give results similar to the *ab initio* values (Rahman 1954).

For F_2 the *ab initio*, OM, and ICC methods all yield acceptable results. This contrasts with the SCF treatment, which does not predict F_2 to be bound (Wahl 1964). For F_2^-, Li_2^+, and F_2^+, the ICC method does not dissociate to the correct experimental energies, since the same atomic orbitals were used to represent the states of the atom and its ion. In the case of LiH^+, the ICC method also fails to dissociate to the correct experimental energy because atomic orbitals optimized to describe $Li(^2S)$ are used to represent

$Li^+(^1S)$; consequently, the correction applied for $Li^+(^1S)$ in the ICC method is insufficient to lower the calculated energy to the experimental value.

6.4.1. *The ground state of* HF

To illustrate the details of the OM method, we consider some calculations on the ground state of HF. For both atoms, basis sets consisting of 1s, 2s, and 2p orbitals are used. For fluorine the orbitals are chosen to give a good description of the $^2P^0$ ground state, while for hydrogen they are chosen to correspond to the 1s, 2s, and 2p orbitals of the neutral atoms (see Appendix 6.1, which includes the various calculated and experimental energies of the atomic and ionic states involved).

On chemical grounds, we expect the binding in HF to arise largely from a single σ bond with a sizeable admixture of an ionic H^+F^- term (Coulson 1961, pp. 131–138). For the basis set used here, there is only one CF with ionicity H^+F^- and this is $H^+F^-(^1S)$. There are 16 additional CFs that can be constructed from states of the neutral atoms with the correct symmetry. They are made up from combinations of the atomic states H(1s)F($^2P^0$), H(2s)F($^2P^0$), H(2p)F($^2P^0$), H(1s)F(^2S), H(2s)F(^2S) and H(2p)F(^2S).

TABLE 6.3

Potential energy curves for $^1\Sigma^+$ ground state of HF.[a]

	17 CF basis set				100 CF basis set		
R	*ab initio*[b]	OM[c]	AIM[c]	ICC[c,e]	*ab initio*[b]	OM[c]	Experimental[c,d]
1·5	−0·087	−0·122	−0·240	−0·167	−0·092	−0·126	−0·203
1·7	−0·143	−0·178	−0·281	−0·215	−0·152	−0·183	−0·224
1·8	−0·153	−0·188	−0·287	−0·222	−0·165	−0·195	−0·223
1·9	−0·157	−0·191	−0·285	−0·222	−0·170	−0·200	−0·218
2·0	−0·155	−0·189	−0·278	−0·217	−0·169	−0·199	−0·208
2·1	−0·149	−0·184	−0·268	−0·208	−0·165	−0·195	−0·197
2·4	−0·120	−0·154	−0·227	−0·171	−0·140	−0·168	−0·158
2·8	−0·076	−0·107	−0·165	−0·116	−0·100	−0·124	−0·110
4·0	−0·011	−0·018	−0·035	−0·018	−0·032	−0·038	−0·030
4·5	−0·004	−0·007	−0·014	−0·007	−0·023	−0·025	−0·017

[a] All quantities are in atomic units (see Table 6.1).
[b] The zero of energy is taken to be the calculated energy of H(^2S)+F($^2P^0$).
[c] The zero of energy is taken to be the experimental energy of H(^2S)+F($^2P^0$).
[d] The experimental curve is taken to be a Morse curve with parameters $R_e = 1{\cdot}7325$ a.u., $D_e = 0{\cdot}2247$ a.u. and $\omega_e = 0{\cdot}01886$ a.u. See Johns and Barrow (1959).
[e] The corrections used in the ICC calculations were not strictly correct (see text).

The potential-energy curves calculated with the 17 CFs of ionicities HF and H^+F^- are tabulated in Table 6.3 and are displayed graphically in Fig. 6.4. The experimental results (Johns and Barrow 1959) are included in the table and the figure for comparison.

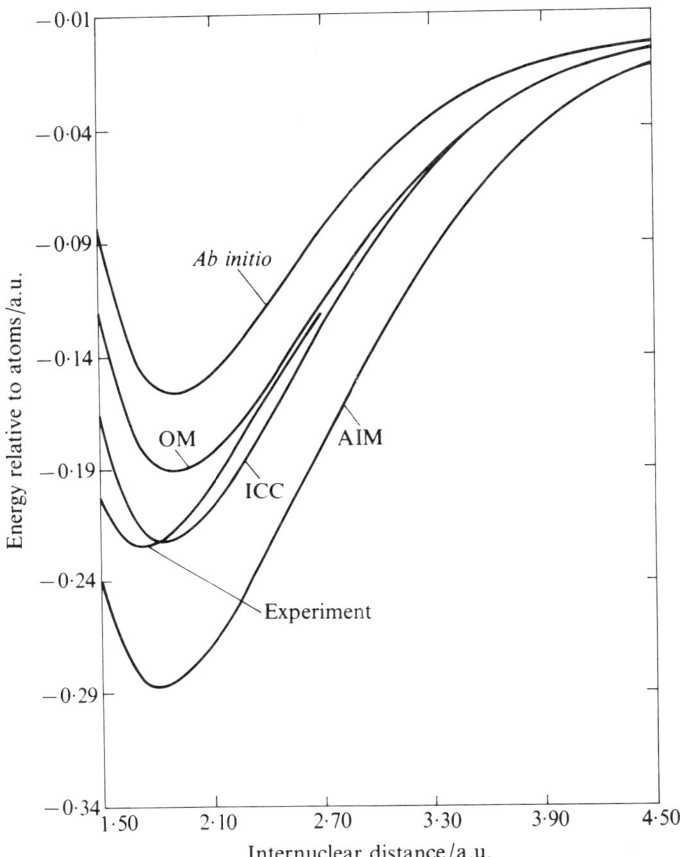

FIG. 6.4. Potential energy curves for the ground state of HF using 17 CF basis; for experimental parameters see footnote *d* of Table 6.3.

The *ab initio* method gives the smallest binding energy, and the AIM method, as is often the case, gives too deep a well. The ICC method† gives very good agreement with experiment, while the binding energy predicted by the OM method (0·191 a.u.) is somewhat smaller than the experimental value (0·225 a.u.).

The next most important CFs are those of ionicity H^-F^+. There are 83 such CFs which have the correct symmetry to mix with the $^1\Sigma^+$ wavefunction. They are made up from combinations of the atomic states $H^-(^1S)$, $H^-(^1D)$,

† The ICC calculations described here are not strictly valid as the correlations used for the hydrogen atom and ion were the same as those used in the AIM method and were not determined from exponent optimized SCF calculations. As the corrections for the F atom ions are two orders of magnitude greater than those for the H atom, the present results should not differ significantly from a proper ICC calculation.

$H^-(^1P^0)$, $H^-(^3P)$, $H^-(^3P^0)$, $H^-(^3S)$ and $F^+(^1D)$, $F^+(^1S)$, $F^+(^1P^0)$, $F^+(^3P)$, $F^+(^3P^0)$. An intra-atomic configuration interaction calculation is used for $H^-(^1S)$; this lowers the calculated energy considerably and reduces the corrections needed for CFs containing the $H^-(^1S)$ state (see Appendix). The other H^- 'states' lie in the continuum and no corrections are applied for these. They can, however, be included in the calculation, as discussed in Section 6.2.

The *ab initio* and OM potential-energy curves obtained with the 100 CF basis set, which includes all CFs of ionicities HF, H^+F^- and H^-F^+ are also displayed in Table 6.3. The results of both methods have been improved by inclusion of the extra CF basis functions. The error in the OM calculation has now been reduced to 0·023 a.u., as compared with 0·034 a.u. in the 17 CF calculation. As may be seen from Table 6.3, the improvement in the *ab initio* results is considerably greater than that in the OM results. The additional flexibility in the wavefunction is used to compensate, in part, for the poor description of the F^- ion in the *ab initio* method. Since the OM method includes a correction for this atomic error, the inclusion of the H^-F^+ CFs has not changed the qualitative description.

To investigate the nature of the bonding, several calculations were performed using only a selected number of CFs; the results of some of these are given in Table 6.4. From the last five calculations in the table it is clear that the essential factors contributing to the binding of HF are the resonance between the two $H(1s)F(^2P^0)$ σ-bonding CFs and the H^+F^- ionic CF. As long as these three CFs are present, a qualitatively reasonable potential-energy curve results. If either of these two types of CFs are omitted, the potential-energy curve is drastically changed. These conclusions are in complete agreement with chemical intuition (Coulson 1961, pp. 131–138).

The omission of any of the physically distinguishable types of CFs leads to a noticeable decrease of the binding energy. The following, physically distinct, types of CFs can be identified as having a small quantiative effect on the binding, but their omission does not lead to a qualitative change in the potential energy curve:

(1) 2s and 2p orbitals on the H atom; these orbitals allow for radial distortion and polarization of the 1s orbital on H;

(2) $F(^2S)$ states; these states allow for hybridization of the 2p orbitals on F;

(3) π-bonding CFs; these may provide a mechanism for back donation (Balint-Kurti and Karplus 1969).

In Table 6.5 we list the dipole moments corresponding to the different wavefunctions. The experimentally observed dipole moment of HF in its ground vibrational state is 1·820 Debye (Weiss 1963). We see that all the calculations predict somewhat too large dipole moments. The fact that the atoms-in-molecules methods predict larger dipole moments than the *ab initio* treatment confirms their well-established tendency of increasing the importance of ionic structures in the wavefunction. This trend is due to the

TABLE 6.4
HF *Calculations with selected CFs.*[a]

Basis of calculation		*ab initio*	OM	AIM	ICC
Full 100 CF basis set, includes all CFs of ionicities HF, H^+F^- and H^-F^+	D_e	0·170	0·200	0·296	0·227
	R_e	1·94	1·94	1·84	1·87
All HF, H^+F^- CFs plus all $H^-(1s2p)F^+(^3P)$ and $H^-(1s2p)F^+(^3P^0)$ (47 CF2)	D_e	0·169	0·199	0·295	0·227
	R_e	1·94	1·93	1·83	1·87
All HF, H^+F^- CFs plus all $H^-(^1S)$ F^+ (29 CFs)	D_e	0·161	0·194	0·290	0·224
	R_e	1·92	1·92	1·83	1·86
All HF and H^+F^- CFs (17 CFs)	D_e	0·157	0·191	0·287	0·223
	R_e	1·91	1·91	1·82	1·85
Omit 4 π-bonding CFs from 17 CF basis	D_e	0·138	0·181	0·278	0·209
	R_e	1·92	1·92	1·82	1·85
Omit 12 CFs containing 2s and 2p orbitals on H from 17 CF basis	D_e	0·114	0·169	0·263	0·188
	R_e	0·94	1·93	1·82	1·85
Omit 6 CFs containing $F(^2S)$ states from 17 CF basis	D_e	0·123	0·170	0·259	0·186
	R_e	1·96	1·93	1·83	1·88
Only neutral HF CFs (16 CFs)	D_e	0·088	0·096	0·105	0·104
	R_e	1·97	1·94	1·92	1·92
Omit 2 $H(1s)F(^2P^0)$ σ-bonding CFs from 17 CF basis	D_e	0·005	0·104	0·206	0·115
	R_e	1·73	1·77	1·71	1·72
$H(1s)F(^2P^0)$ plus H^+F^- CFs only (3 CFs)	D_e	0·093	0·164	0·243	0·161
	R_e	2·01	1·92	1·82	1·89
H^+F^- CF only[b]	D_e	−0·163	0·134	0·134	−0·004
	R_e	1·67	1·67	1·67	1·66
$H(1s)F(^2P^0)$ CFs only (2 CFs)[c]	D_e	0·010	0·010	0·010	0·010
	R_e	2·77	2·77	2·77	2·77

[a] All quantities are in atomic units (see Table 6.1). The D_e and R_e values are calculated by fitting a quadratic to the three lowest calculated points on the potential energy curves.
[b] The potential energy curves are parallel for this case.
[c] All four potential energy curves are identical as the correction matrix is a constant matrix.

TABLE 6.5
Dipole moment of $^1\Sigma^+$ *ground state of* HF.[a,b,c]

	17 CF Basis				100 CF Basis		
R	*ab initio*	OM	AIM	ICC	*ab initio*	OM	Point charge
1·5	2·072	2·395	2·761	2·476	1·977	2·328	3·812
1·7	2·100	2·506	2·937	2·585	1·913	2·361	4·321
1·8	2·110	2·568	*3·030*	2·640	1·875	2·378	4·575
1·9	*2·114*	*2·630*	3·124	*2·692*	*1·831*	*2·395*	4·830
2·0	2·111	2·692	3·219	2·740	1·780	2·411	5·083
2·1	2·099	2·751	3·313	2·782	1·724	2·425	5·337
2·4	1·990	2·894	3·576	2·852	1·510	2·437	6·100
2·8	1·652	2·914	3·821	2·738	1·124	2·314	7·116
3·2					0·690	1·934	8·133
3·6					0·330	1·339	9·150
4·0	0·380	1·216	2·560	0·965	0·098	0·744	10·166
4·5	0·166	0·547	1·328	0·428	−0·045	0·254	11·437

[a] The dipole moment is given in Debyes and the internuclear separation in atomic units.
[b] The experimental dipole moment in the ground vibrational state is 1·820 Debye.
[c] The value of the calculated dipole moment corresponding to the energy minimum (see Table 6.3) is in italic numerals.

fact that the ionic structures are almost invariably associated with larger corrections than the covalent structures. Of all the atoms-in-molecules calculations, the OM method yields the smallest dipole moments, demonstrating that the tendency to overemphasize the ionic structures is least pronounced in this method. It should also be noted that the 100 CF calculations yield smaller dipole moments than the corresponding 17 CF values. Thus, in both the *ab initio* and atoms-in-molecules methods, increasing the flexibility of the basis set brings the dipole moment closer to its true value.

In Fig. 6.5 the dipole moment vs. internuclear separation for the 100 CFs *ab initio* and OM methods are compared with the calculations of Bender and Davidson (1968). The shape of the OM curve agrees quite well with that of Bender and Davidson. The dipole moment goes through a maximum value in the region 2·5 to 2·7 a.u. and then decreases with increasing internuclear separation. This decrease corresponds to that of the ionic nature of the wavefunction. To illustrate this, we plot in Fig. 6.6 the weight† of the ionic H^+F^- CF for the *ab initio* and OM methods with the 100 CF basis. The H^+F^- contribution is seen to decrease monotonically with increasing nuclear separation. This is in sharp contrast to LiF, where the ionic contribution increases with internuclear distance until the electron-jump radius is reached (Balint–Kurti 1968). We note also that the calculated

† If the wavelength is $\Psi = \sum_i c_i\Phi_i$ the weight or occupation number of the ith CF is defined as $v_i = c_i \sum_j c_j S_{ji}$ (see Hurley, 1958).

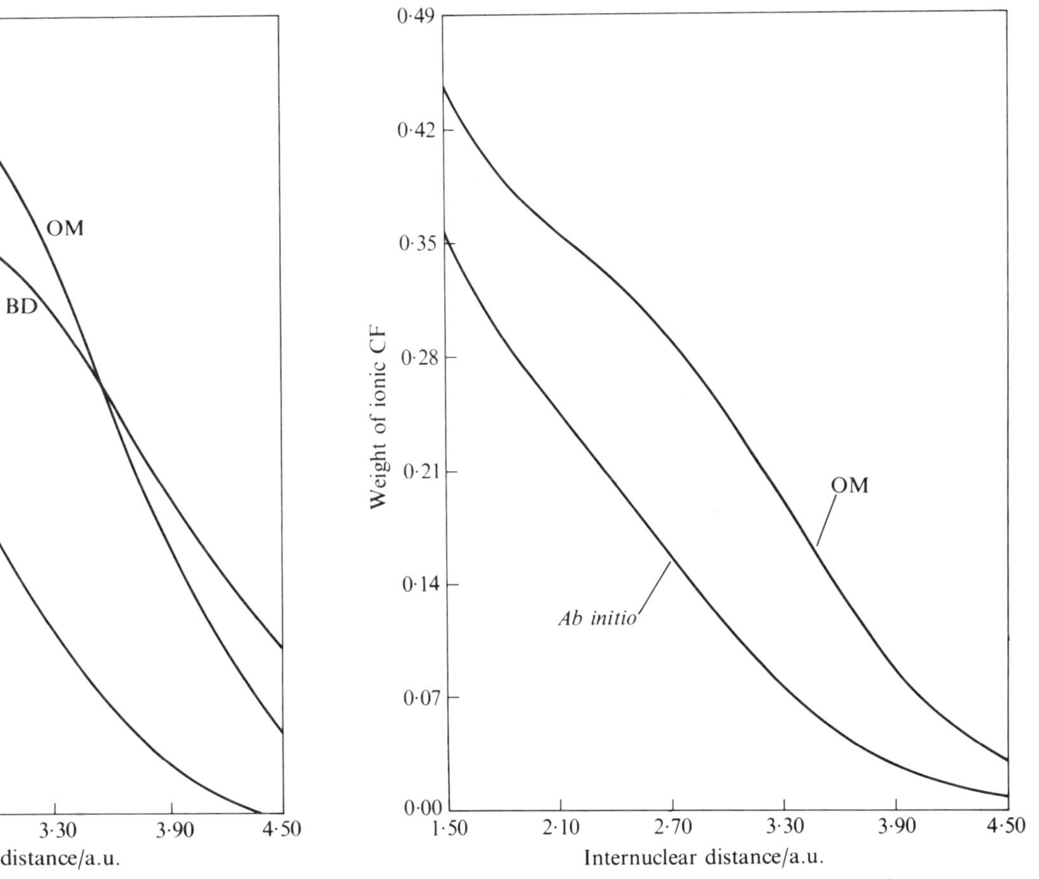

FIG. 6.5. Dipole moment for ground state of HF calculated using 100 CF basis; BD refers to Bender and Davidson (1968).

FIG. 6.6. Weight of ionic CF in HF ground state wavefunction calculated using 100 CF basis; see text for definition of weight.

dipole moment of HF, in contrast to that of LiF is always far smaller than that due to two point charges (see last column of Table 6.5).

6.5. Triatomic systems

The diatomic results reported in the previous section were obtained in pilot calculations for studies of triatomic potential energy surfaces. As an example of what is important in determining the nature of such surfaces, we consider the reaction

$$\mathrm{Li}+\mathrm{F}_2 \rightarrow \mathrm{LiF}+\mathrm{F}.$$

This reaction is the lightest member of a family of reactions, of which $\mathrm{K}+\mathrm{Br}_2 \rightarrow \mathrm{KBr}+\mathrm{Br}$ has been examined in most detail. It has long been assumed that such reactions proceed *via* an 'electron-jump' from Li to F_2 and that the resulting pair of ions attract each other very strongly (Magee 1940; Herschbach 1966). The position at which the electron-jump occurs depends on the electron affinity of the F_2 molecule (Balint–Kurti and Karplus 1971). Vertical electron affinity values for F_2 are not known,† except at infinite internuclear separations, where the electron affinity is the same as that of the F atom (0·127 a.u.) (Berry and Reinmann 1963). The *ab initio* calculation with the present atomic orbital set, predicts an electron affinity of $-0{\cdot}173$ a.u. This is a very large error and will clearly lead to a spurious potential-energy surface. The only way of improving the result appreciably within the *ab initio* framework is to extend the orbital and CF basis sets so as to include an intra-atomic configuration interaction for F^-. Use of such an extended basis for the triatomic system would have required a very large investment of computer time. A much simpler, and less expensive alternative is provided by the atoms-in-molecules methods, which include a correction for the electron affinity error; that is, both the AIM and OM methods yield the correct limiting ($R \rightarrow \infty$) vertical electron affinity for F_2. The poor potential-energy curve obtained from the AIM method for F_2 makes the electron affinities suspect at smaller internuclear distances. Since the OM method yields a reasonable F_2 potential curve, it is expected that the calculated electron affinity is more reliable. As to the ICC method, it is in error by 0·035 a.u., which is sufficient to significantly affect the electron-jump position.

Another criterion for assessing the suitability of the various methods for potential-energy surface calculations, is to compare the predicted reaction exothermicity. In Table 6.6 the calculated and experimental exothermicities

† The electron affinity of F_2 has recently been determined to be 3.08 eV (Chupka, Berkowitz, and Gutman 1971). It is not clear whether this value corresponds to an adiabatic electron affinity or not.

TABLE 6.6
Calculated and experimental exothermicities (kcal/mole).[a]

Reaction	*ab initio*	OM	AIM	ICC	Experiment
(A) $Li+F_2 \rightarrow LiF+F$	37·0	94·3	120·9	80·6	100·4
(B) $F+Li_2 \rightarrow LiF+Li$	54·5	126·7	—	—	114·2
(C) $F+LiH \rightarrow LiF+H$	22·5	100·8	—	76·8	80·4
(D) $Li+HF \rightarrow LiF+H$	−29·3	21·2	9·9	−16·7	−2·5

[a] All values are calculated ignoring zero-point energy contributions.

for the reactions

$$\text{(A)}\ Li+F_2 \rightarrow LiF+F$$

$$\text{(B)}\ F+Li_2 \rightarrow LiF+Li$$

$$\text{(C)}\ F+LiH \rightarrow LiF+H$$

$$\text{(D)}\ Li+HF \rightarrow LiF+H$$

are listed with the experimental values. The *ab initio* method values are very poor for the first three reactions, which all involve electron-jumps. The AIM method cannot be used for two of the reactions because it yields potential-energy curves with the wrong shape for one of the diatomic species involved. In the two cases where the AIM method does predict exothermicities, it is in good agreement with the experimental results. In both cases, however, the agreement with experiment stems from a cancellation of sizeable errors in the two diatomics involved. This comment also applies to the *ab initio* result for reaction (D). The OM and ICC methods generally yield reasonable values. The OM method on the whole, however, is more reliable for diatomics and is to be preferred on this count.

To determine the CF basis sets for the three-atom potential surface, calculations were performed on the component diatomics. When a sufficiently accurate description of these diatomics had been attained, the CF basis set for the triatomic system was constructed by multiplying all the diatomic CFs by all the atomic functions for the third atom and antisymmetrizing the product. Use of such a basis set ensured that when any of the atoms is removed to a large distance from the other two, the energies which result will be just those obtained in the diatomic calculation, plus the energy of the third atom. Additional CFs can be added to the set, if they do not contain diatomic CFs of the correct symmetry to contribute to the states of the isolated diatomics. In the triatomic calculations described below, the basis set was built up by including all possible CFs of specified ionicities (e.g. LiFF, Li^+F^-F, LiF^-F^+ for LiF_2). When any of the atoms was removed to a large distance, these basis sets gave the same results as those listed in Table 6.2 for the remaining diatomic.

6.6. The $Li+F_2 \rightarrow LiF+F$ reaction

This reaction is the lightest member of the family of reactions

$$M+X_2 \rightarrow MX+X$$

where M is an alkali metal atom and X is a halogen. These reactions have been extensively studied using the crossed molecular beam technique (Birley *et al*, 1967), and have played an important role in the dilute flame experiments of M. Polanyi (1932). They have also been the subject of several trajectory calculations (Blais 1968). The detailed interpretation of the experimental results ultimately depends on the potential-energy surface or surfaces which govern the reaction (Balint-Kurti and Karplus 1971). In the trajectory calculations, empirical potential-energy surfaces were used and the parameters in the surfaces were varied to obtain agreement with experiment. There is, of course, no assurance that a surface found in this manner will correspond to the true surface. Indeed, it is possible to find several different surfaces which lead to satisfactory agreement with the limited experimental data available (Karplus 1968). Any theoretical calculation which assists in choosing a potential-energy surface with the correct characteristics is therefore very important in achieving a more complete understanding of the reaction.

For the LiF_2 system a basis set of 204 CFs was used.† The lowest potential-energy surface calculated with the OM method for the Li–F–F linear configuration is shown in Fig. 6.7. The main features of the surface are:

(1) It is of the highly attractive, or 'early downhill' type (Polanyi and Rosner 1963).

(2) There is a well in the surface with respect to dissociation to $LiF+F$.

(3) The interaction between Li and F_2 is very long range and approximately coulombic.

It has been found from trajectory calculations that potential-energy surfaces of the 'attractive' type channel the exothermicity of the reaction into internal modes of the products, rather than into the relative kinetic energy of the final atom-molecule pair (Polanyi and Rosher 1963). The molecular beam studies for the $M+X_2$ family of reactions have shown that most of the exothermicity of the reaction goes into vibrational energy of the newly formed MX bond. This fact, taken together with the generalizations derived from trajectory calculations, leads to the conclusion that the potential-energy surface should be of the attractive type, in agreement with the OM results for $Li+F_2$. The trajectory studies of Godfrey and Karplus (1968) for $K+Br_2$ indicate that potentials having a substantial repulsion along the exit valley lead to calculated results which disagree with experiment.

† 204 CFs were used for non-linear nuclear configurations. For linear geometries the basis set broke down into one of Σ symmetry (112 CFs) and one of Π symmetry (92 CFs).

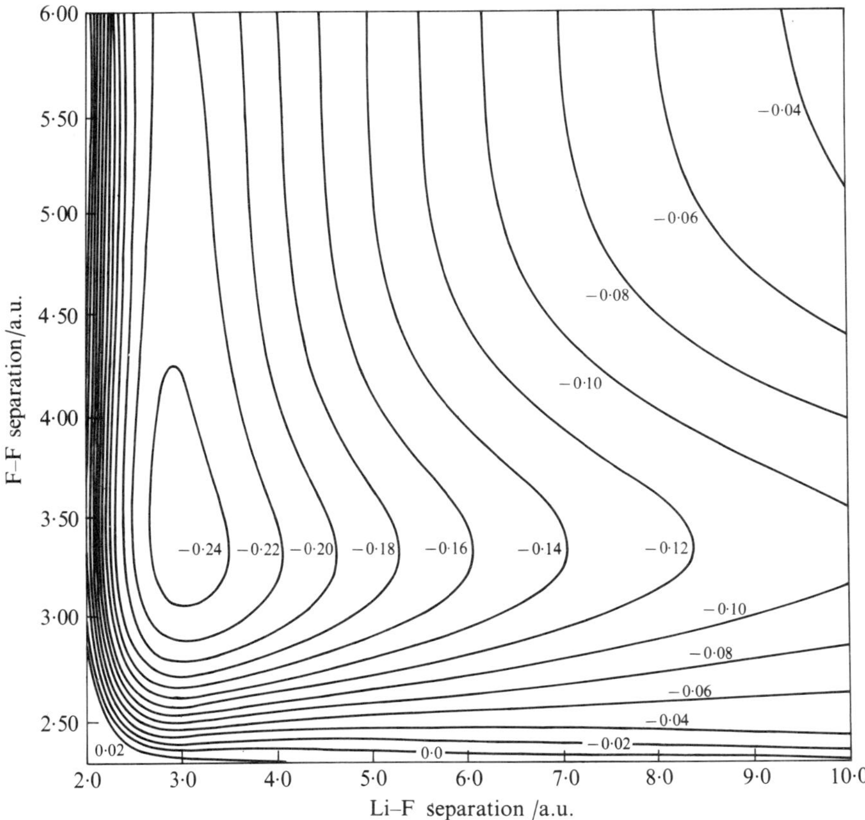

FIG. 6.7. Contour map of Li–F–F linear $^2\Sigma^+$ symmetry surface calculated using the OM method. The figure was drawn by fitting a polynomial to the calculated points.

This conclusion also is in agreement with the calculated $Li + F_2$ potential-energy surface, which has an attractive interaction along the exit valley.

The OM calculation yields a well of about 16 kcal/mole, with respect to dissociation to $LiF + F$, in the Li–F–F linear configuration. The depth of this well increases as the system is bent out of the linear geometry, and reaches a depth of 41 kcal/mole in the isosceles triangle configuration. The LiF_2 system may be regarded as an F_2^- ion perturbed by the presence of an Li^+ ion in the region of the well. The dissociation energy of F_2^- will therefore be reflected in the depth of the well (Balint-Kurti and Karplus 1969). While the actual depth of the well is subject to considerable uncertainty, the qualitative shape of the surface should be correct. Some nuclear configurations in which the Li atom lies between the F atoms also have lower energies than the products. Such nuclear configurations could, therefore, be accessible to the reacting systems, and the shape of the surface in this region should be carefully considered in future studies of the dynamics. It is not clear whether wells are artifacts of the calculation, a special feature of LiF_2, or whether they might occur more generally for $M+X_2$ systems.

As was to be expected from the discussion given in Section 6.4 the *ab initio* calculations yield very different results from those obtained by the OM method (Balint-Kurti 1969; Balint-Kurti and Karplus 1971). They predict a potential-energy surface which is of the 'repulsive' type, the exothermicity of the reaction being released along the exit valley.

The $M+X_2$ family of reactions are generally interpreted as proceeding by way of an electron-jump (Magee 1961; Herschbach 1966). The Li–F_2 separation at which the electron-jump occurs is a very important factor in determining the total reactive cross-section. The electron-jump distance has been examined as a function of both F–F separation and angle of approach. In Fig. 6.8 for the linear Li–F–F configuration, the electron-jump

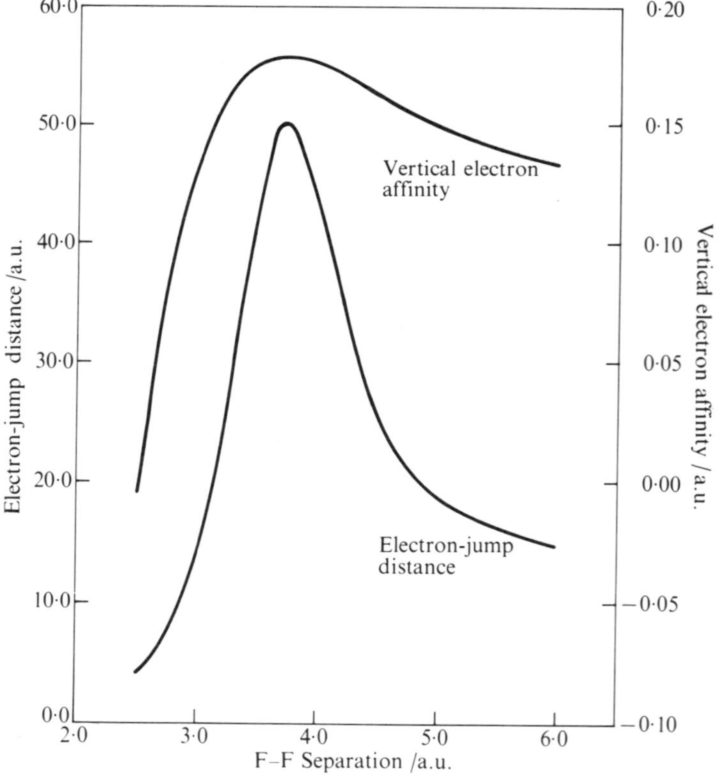

FIG. 6.8. Electron-jump distance for the linear Li, F_2 system and vertical electron affinity of F_2 as a function of F–F separation; both curves were calculated with the OM method.

distance, which is defined in this case as the separation of Li from the closest F atom, is shown as a function of the F–F separation. To illustrate the dominant role played by the vertical electron affinity of F_2 in determining the electron-jump distance, the calculated values for F_2 are also included in Fig. 6.8. It can be seen that the electron-jump distance reflects the vertical electron affinity curve.

The $M+X_2$ family of reactions provides an interesting example of the interplay between the potential-energy surface and the dynamics of the collisional process. By making alkali metal atoms and halogen molecules collide at sufficiently high energies, it is possible to observe the charge exchange process $M+X_2 \rightarrow M^+ + X_2^-$ (Baede, Moutinho, De Vries, Los 1969; Lacman and Herschbach 1970). Such a process can occur only via a non-adiabatic transition between two Born–Oppenheimer potential-energy surfaces (Berry 1957), so that a description of both surfaces is needed. At energies too low for the ions to be formed, the excited Born–Oppenheimer surface may still play a role in the dynamics of the collision. If the M atom is constrained to approach the X_2 molecule in an isosceles triangle configuration, the lowest covalent and ionic potential-energy surfaces are of different symmetries, and the electron-jump is symmetry forbidden. There is no interaction between the two zeroth-order states and the surfaces corresponding to them actually cross. In nuclear configurations slightly removed from the isosceles triangle symmetry, the ionic and covalent states interact and there is an avoided crossing. The splitting between the two adiabatic states might be expected to increase smoothly as the system is distorted out of the isosceles triangle configuration. In the vicinity of the isosceles triangle configuration, where the splitting is small, a non-adiabatic transition would have the largest probability (Zener 1932).

We have examined the splitting of the two lowest states in the region of the electron-jump for the Li–F–F linear geometry and for the geometry in which the Li atom approaches one of the F atoms at right angles to the F–F bond. The results of two such calculations are presented in Figs. 6.9 and 6.10. The F–F separation is held fixed at 2·8 a.u., which is close to the calculated F_2 equilibrium distance in both cases. The 'zeroth order' ionic curve is obtained from a variational calculation using only the six CFs of ionicity Li^+F^-F and the covalent curve was obtained from the remaining 198 CFs. As expected from the above discussion, the splitting of the two adiabatic curves is much greater in the linear than in the perpendicular configuration. The electron-jump is also found to be much more gradual in the linear configuration (Balint-Kurti 1969).

The lowest two surfaces also approach quite close to each other in F–Li–F linear configurations. If such configurations were to play a significant role in the reaction (see above), the possibility that nonadiabatic transitions could occur in this region would have to be examined.

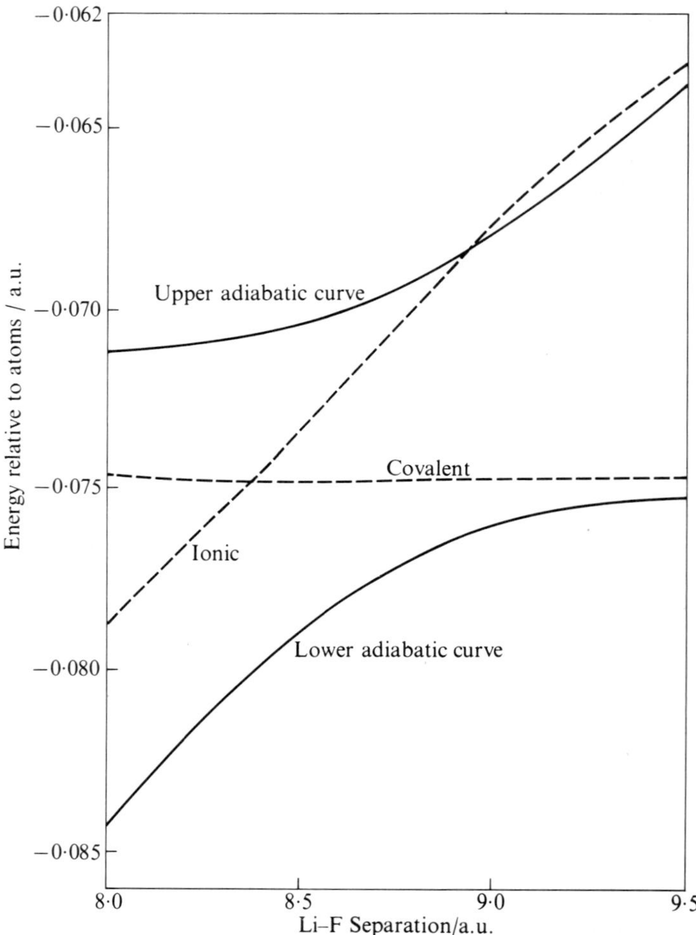

FIG. 6.9. Two lowest states in vicinity of electron-jump position in Li–F–F linear configuration with F–F separation fixed at 2·8 a.u. Also shown (dashed lines) are calculations using only ionic and only covalent CFs.

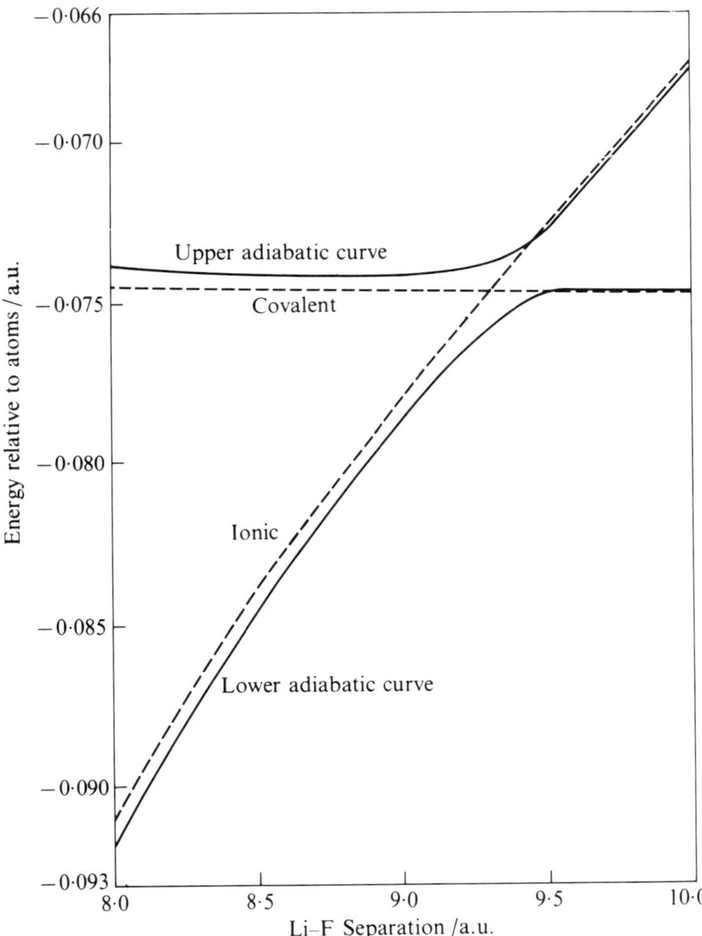

FIG. 6.10. Two lowest states in vicinity of electron-jump position in Li–F–F perpendicular configuration with F–F separation fixed at 2·8 a.u. Also shown (dashed lines) are calculations using only ionic and only covalent CFs.

6.7. The Li_2F system

The Li_2F potential-energy surface gives rise to two interesting reactions

$$F + Li_2 \rightarrow LiF + Li \tag{A}$$

$$Li' + LiF \rightarrow Li'F + Li \tag{B}$$

and to an important vibrational–electronic energy exchange process,

$$Li + LiF^{\dagger} \rightarrow Li^{*} + LiF \tag{C}$$

where (†) denotes vibrational and (*) electronic excitation. The Li_2F system is the lightest member of the M_2X family, which can give rise to processes analogous to A, B, and C. Processes A and C are important in the analysis of M. Polanyi's dilute alkali-metal–halogen flame experiments. The most spectacular aspect of these experiments was the strong luminescence they produced. The luminescence corresponded to the decay of an alkali-metal atom in its $^2P^0$ state to its ground ^2S state. The major problem in interpreting the experiments was to discover how the alkali-metal atoms came to be electronically excited. For the sodium and chlorine system, which was the one most thoroughly studied, the following reaction mechanism was proposed (Polanyi 1932). At the temperature of the experiment, the sodium existed mainly, but not entirely, in the form of atoms. A sodium atom was first assumed to react with a halogen molecule,

$$Na + Cl_2 \rightarrow NaCl + Cl. \tag{D}$$

The chlorine atom from this reaction then reacted with an Na_2 molecule producing a NaCl molecule in a highly excited vibrational state,

$$Cl + Na_2 \rightarrow NaCl^{\dagger} + Na. \tag{E}$$

A collision between a sodium atom and the vibrationally excited NaCl molecule yielded the electronically excited Na* atom

$$NaCl + Na(^2S) \rightarrow NaCl + Na^{*}(^2P^0). \tag{F}$$

A triatomic system always possesses a plane of symmetry which includes the three atoms. The potential-energy surface governing the reaction can be associated, therefore, with an electronic wavefunction which is either even or odd with respect to reflection in this plane. For the M_2X systems the lowest even symmetry surface dissociates to ground state $M(^2S) + MX$, while the lowest odd surface dissociates to give an electronically excited alkali-metal atom. As the hole in the 2p shell of a halogen atom can be oriented along any of the three coordinate axes with equal probability, two-thirds of $X + M_2$ collisions will occur on even symmetry surfaces and one-third on odd surfaces. As pointed out by Magee (1940), the reactions which proceed on surfaces of odd symmetry will give electronically, excited products directly. This provides an alternative mechanism for the production of electronically excited alkali-metal atoms.

Fig. 6.11 shows a contour map of the lowest linear F–Li–Li potential-energy surface calculated with the OM method and a basis of 204 CFs.†

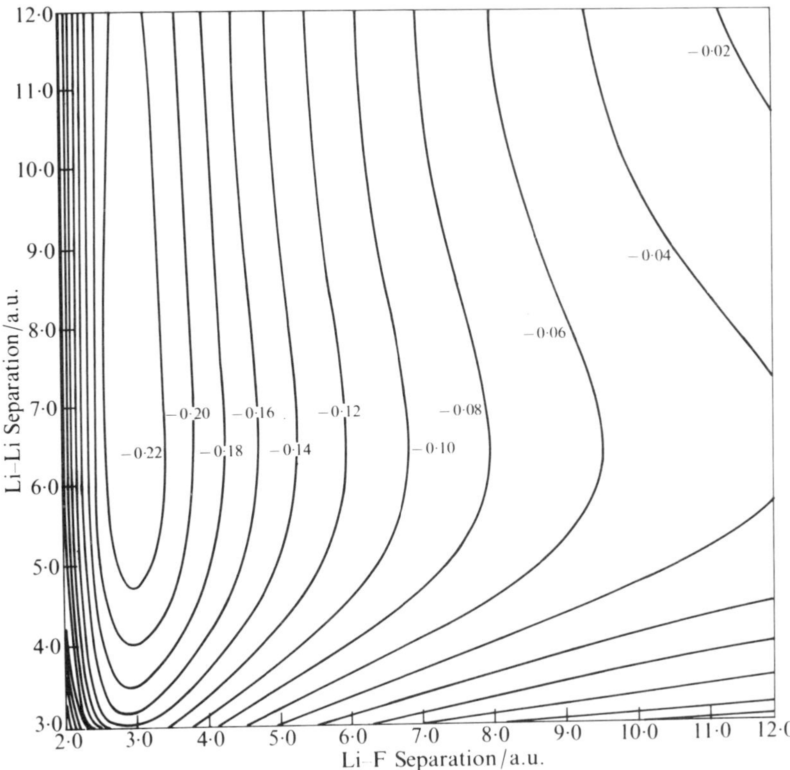

FIG. 6.11. Contour map of F–Li–Li linear $^2\Sigma^+$ symmetry surface calculated using OM method. The figure was drawn by fitting a polynomial to the calculated points.

The surface corresponds to a state of $^2\Sigma^+$ symmetry. It can be seen that the surface is of the very highly 'attractive' type (Polanyi and Rosner 1963), and would be expected to channel most of the exothermicity in the $F+Li_2$ reaction into vibrational energy of the product molecule. This agrees with the interpretation of the dilute alkali-metal–halogen flame experiments. Fig. 6.12 shows the ten lowest excited states of the system along the entrance valley for the $F+Li_2$ reaction. It is clear that the interaction in the lowest

† This is the size of the CF basis set for the non-linear nuclear geometries. In the linear case the basis set may be decomposed into two basis sets, one of Σ(112 CFs) and the other of Π(92 CFs) symmetry.

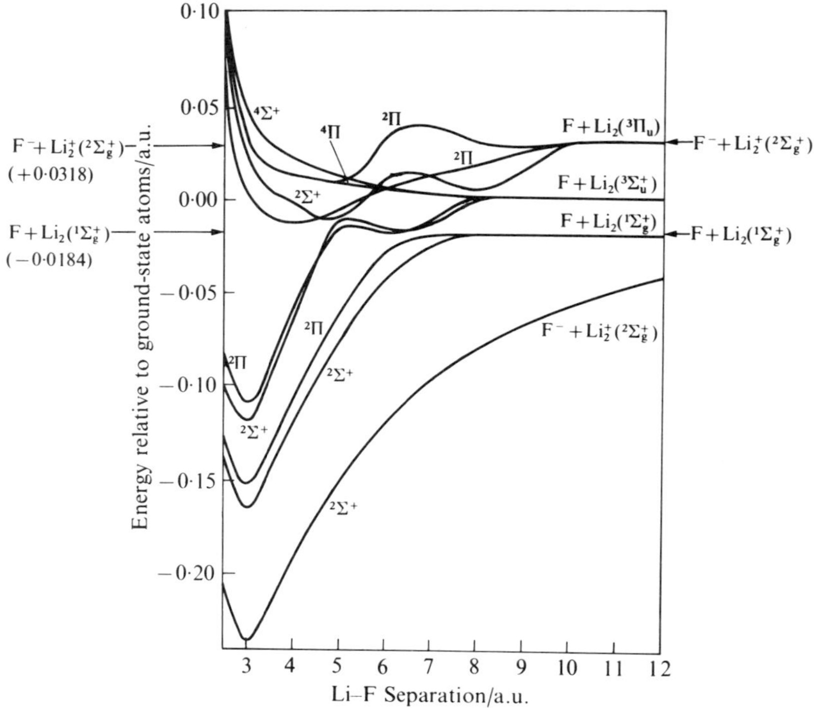

FIG. 6.12. Excited states calculated using OM method along a cut in the F–Li–Li linear surfaces with Li–Li separation fixed at 6·5 a.u. The humps in the curves are present in the calculations, but may be slightly exaggerated by the cubic spline-fitting procedure.

state has a much longer range than that in the first two excited states. As these excited states dissociate to yield electronically excited lithium atoms, we might expect the cross-section for the reaction which yields excited Li atoms directly to be much smaller than that for the reaction which yields ground-state products.

In Fig. 6.13 the lowest ten excited states of the system are shown along a cut in the exit valley of the linear F–Li–Li system. The first excited state is seen, even at an Li–Li separation of 12 a.u. to correspond to an Li^+ ion interacting with $LiF^-(^2\Sigma^+)$. As the Li–LiF distance increases still further this state will rise in energy and participate in an avoided crossing with the $^2\Sigma^+$ state immediately above it. The first excited $^2\Sigma^+$ surface is therefore expected to dissociate to $Li(^2P^0)+LiF(^1\Sigma^+)$.

Fig. 6.14 shows a contour map of the lowest Li–F–Li linear potential-energy surface as calculated by the OM method. Its principle feature is a deep well of 54 kcal/mole. The exchange reaction (B) takes place on this surface (though it is of course not restricted to the linear configuration). As the reaction is thermoneutral and the system does not therefore acquire a large amount of kinetic energy to help it escape from the well, we would expect the reaction to form a long-lived complex, if the collision energy is small. Miller, Safron, and Herschbach (1967) have examined some analogous exchange reactions of alkali atoms and alkali halides and have shown that they proceed via a collision complex mechanism.

The vibrational–electronic energy exchange process C (or F) involves the collision of an alkali-metal atom and an alkali halide molecule, both in the ground electronic state. An electronically excited alkali-metal atom can only be produced if there is a non-adiabatic transition which enables the system to 'jump' from one Born–Oppenheimer potential-energy surface to another. Such transitions occur with greatest probability in regions where two potential-energy surfaces of the same symmetry approach each other most closely. Magee (1940) has discussed the potential-energy surfaces governing the energy exchange process. He assumed that the two lowest potential-energy surfaces cross as a halogen atom approached an M_2 molecule in the isosceles triangle configuration. Our calculations do not agree with this assumption (Balint-Kurti 1969). On examining the excited

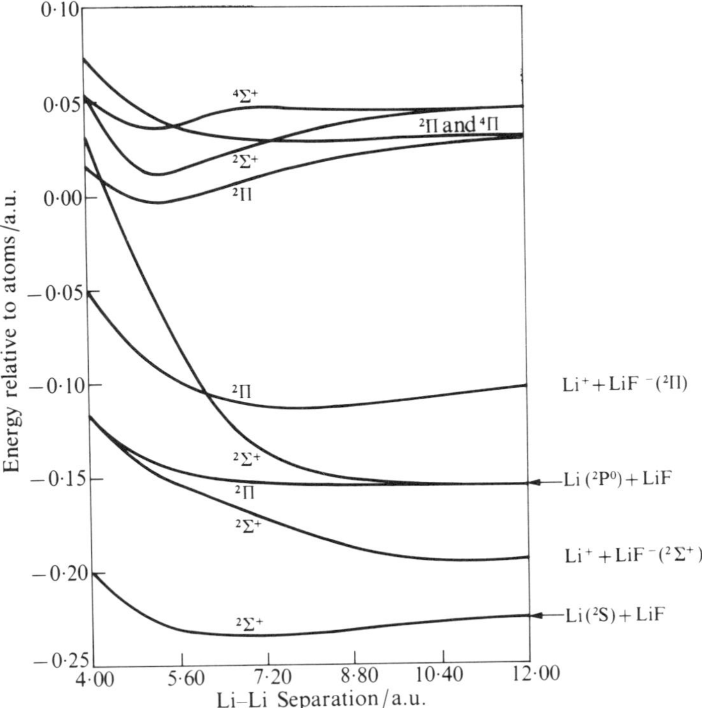

FIG. 6.13. Excited states calculated using OM method along a cut in the F–Li–Li linear surfaces with Li–F separation fixed at 3·0 a.u.

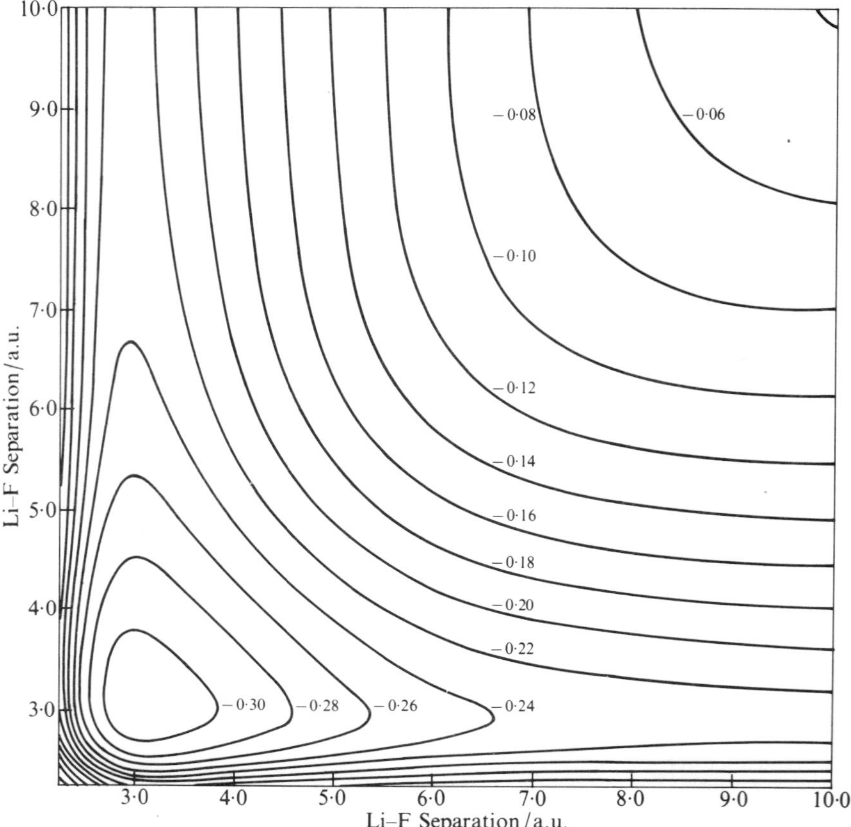

FIG. 6.14. Contour map of Li–F–Li linear $^2\Sigma^+$ symmetry surface calculated using OM method. The figure was drawn by fitting a polynomial to the calculated points.

states of the system in both linear and non-linear nuclear configurations, the ground state and the first excited state surfaces are found to approach each other most closely in highly stretched linear symmetric Li–F–Li configurations. For example, if both Li–F distances are equal to 7 a.u. in the linear symmetric configuration, the two lowest $^2\Sigma^+$ symmetry surfaces are separated by only about 0·4 kcal/mole. It is in the vicinity of such nuclear configurations, in which the two Li–F separations are both large and equal, that we consider the non-adiabatic transition most likely to occur. As an Li atom approaches an LiF molecule from the fluorine side, there must be a partial charge transfer at the point where the two Li–F separations become equal, if the system is to remain on the lowest potential-energy surface. The charge transfer process consists of an electron jumping from the approaching Li atom to the Li atom which was part of the LiF molecule. The non-adiabatic transition takes place if this charge transfer does not occur.

The proposed mechanism for the vibrational–electronic energy exchange process C can therefore be described as follows: An Li atom approaches the fluorine of an LiF molecule. The molecule is in a high vibrational state and consequently has a significant probability of being very stretched. As the system passes through the nuclear configuration in which the two Li–F distances are equal, the wavefunction does not change its character (as would be required if it were to stay on the lowest surface) and the system finds itself in its first excited state. If the energy exchange process proceeds via a direct mechanism, the system dissociates to give an electronically excited Li atom without again passing through a nuclear configuration in which the two Li–F separations are equal. It is possible that the process will proceed via a complex mechanism, in which case the system will pass through such nuclear configurations many times. An excited Li atom will be produced if the system undergoes an odd number of non-adiabatic transitions.

Moulton and Herschbach (1965) have studied both the reactive energy exchange

$$KBr^{\dagger}+Na \rightarrow K^{*}+NaBr$$

and the non-reactive process

$$NaBr^{\dagger}+K \rightarrow K^{*}+NaBr,$$

with an elegant triple beam technique. They find that, at the same total energy, the reactive process has a cross-section which is at least ten times greater than that of the non-reactive process. This result indicates that the energy exchange process proceeds via a direct mechanism, since a collision complex would be expected to yield cross sections which are of similar magnitudes for the two processes. Thus, the experimental results and the calculated potential energy surfaces point to the central role played by charge exchange in the vibrational–electronic energy transfer. The calculated potential is in qualitative agreement with experiment, if a direct mechanism and a large probability of occurrence for the non-adiabatic transition are assumed. Final confirmation of this point, and also of the relative importance of direct excitation of the alkali-metal atoms versus vibrational–electronic energy exchange must await a detailed study of the reaction dynamics.

In view of the low energy of $Li^{+}+LiF^{-}(^{2}\Sigma^{+})$ (Fig. 6.13) predicted by the present calculations, some calculations on the LiF^{-} diatomic ion have been performed. These calculations used two different basis sets. The smaller basis set corresponds to that of the triatomic calculations described above, while the larger basis set (which makes possible a two-fold configuration–interaction treatment of F^{-}) provides a much better value of the electron affinity of fluorine (0·105 a.u. versus −0·173 a.u., as compared with the experimental value of 0·1273 a.u.). The improved calculation also yields a significantly smaller binding energy for LiF^{-} (by about 0·08 a.u.) than the original basis set. The reason for this appears to be that the latter provides an inadequate description of the $F^-(^1S)$ state, so that the wavefunction gave a better description of the molecule than of the dissociation products. The results are sensitive to the choice of basis set in both the *ab initio* and OM formulations.† Calculations using improved basis sets are presently in progress for the Li_2F system. The depth of the well in the ground-state surface for this system (Fig. 6.14) may be reduced to some extent. The present treatment predicts the lowest point on the ground-state surface to lie in the Li–F–Li linear configuration. The force constant for bending out of this configuration is, however, predicted to be extremely small, so that the exact minimum-energy geometry may also be sensitive to the basis set.

6.8. The LiHF system

The LiHF system gives rise to three important reactions,

$$\mathrm{Li+HF \rightarrow LiF+H} \quad \text{(A)}$$

$$\mathrm{F+LiH} \begin{array}{l} \nearrow \mathrm{LiF+H} \quad \text{(B)} \\ \searrow \mathrm{HF+Li.} \quad \text{(C)} \end{array}$$

Reaction A is a prototype for the first chemical reaction which was successfully observed in a crossed molecular beam apparatus (Taylor and Datz 1955)

$$\mathrm{K+HBr \rightarrow KBr+H.} \quad \text{(D)}$$

The light weight of the H atom compared to the KBr makes accurate determination of the differential reactive cross-section in the centre-of-mass reference frame an extremely difficult task, and even the best available experimental results suffer from an unusually large uncertainty on this account (Gillen, Riley, and Bernstein *et al.* 1969; Martin and Kinsey 1967). In view of this difficulty, theoretical information concerning the potential-energy surfaces governing this family of reactions would seem to be especially useful.

The CF basis set used consisted of all possible CFs of ionicities LiHF, Li^+HF^-, LiH^+F^-, Li^-H^+F, and Li^+H^-F which could be formed from the same approximate atomic eigenfunctions as employed in Section 6.4. The CFs were combined to form valence-bond functions corresponding to doublet spin states. Calculations have been performed for linear nuclear configurations with basis sets consisting of 109 $^2\Sigma^+$ and 78 $^2\Pi$ symmetry

† LiF^- may well be unstable with respect to electron detachment. The present diatomic calculations did not include any extremely diffuse orbitals or any continuum functions which would be needed to examine this process. In the molecular environment the LiF^- ion is greatly stabilized by Coulomb attraction to the Li^+ ion, while at large LiF–Li separations Li+LiF lies at a lower energy than Li^++LiF+e. A similar situation occurs for HF^- (see below).

valence-bond functions. There are three different possible arrangements of the nuclei for the linear LiHF system, and preliminary results for all of them are described in this section.

In Fig. 6.15 the variation of the lowest $^2\Sigma^+$ potential-energy surface is shown for the LiFH configuration as the Li atom approaches the HF

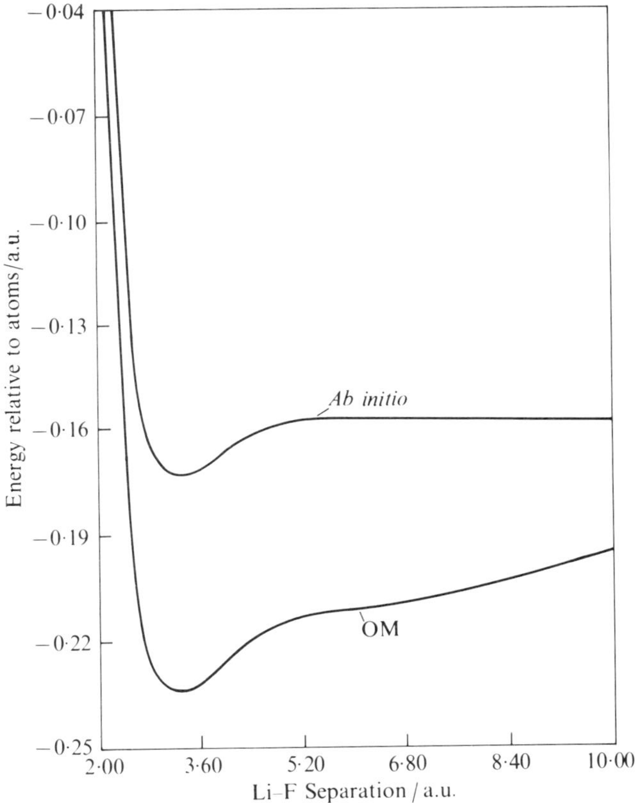

FIG. 6.15. Cut in OM and *ab initio* Li–F–H linear $^2\Sigma^+$ ground state potential energy surfaces with HF separation fixed at 1·9 a.u.

molecule, whose internuclear distance is held fixed at 1·9 a.u. Both the *ab initio* and OM methods are seen to predict deep wells along this cut in the potential-energy surface; the well depth is 25 kcal/mole in the OM method and 10 kcal/mole in the *ab initio* calculation. These wells are considerably deeper than those predicted by the recent SCF–LCAO calculations of Lester and Krauss (1970).

Fig. 6.16 shows the variation of the same $^2\Sigma$ symmetry Li–F–H linear surface as a hydrogen atom is removed, while the Li–F separation is held fixed near its equilibrium value. The figure approximates the lowest potential-energy surface along the exit 'valley' of reaction A. While the *ab initio* and

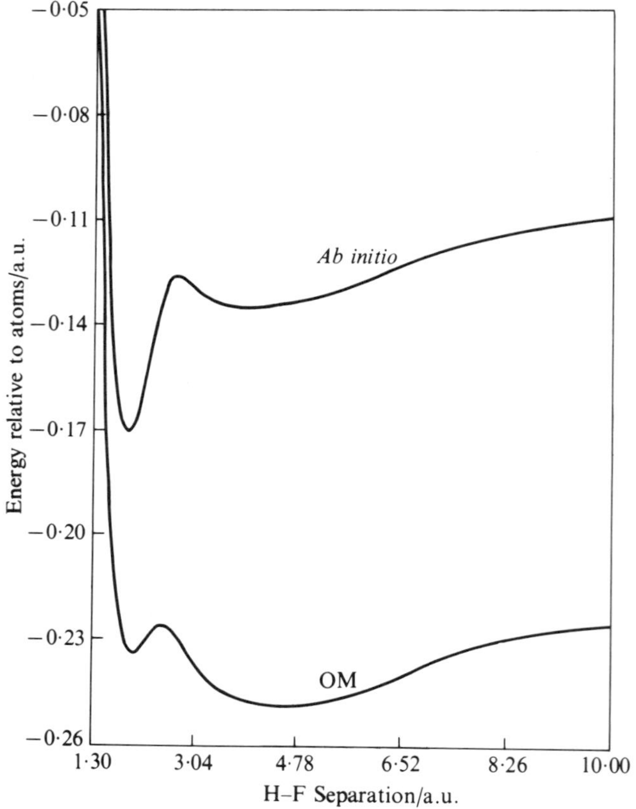

FIG. 6.16. Cut in OM and *ab initio* Li–F–H linear $^2\Sigma^+$ ground state potential energy surfaces with Li–F separation fixed at 3·0 a.u.

OM methods differ considerably, they both agree in predicting that the potential should have a double minimum along this cut in the surface, with the outer well depth about 16 kcal/mole below that of the LiF + H products. An examination of the valence-bond wavefunction in the region of the outer well shows that it corresponds mainly to an LiF diatomic interacting with an H atom; the $Li^+F^-H(1s)$ structure makes the dominant contribution. In the neighbourhood of the inner well, the wavefunction is much more 'molecular' in nature; structures of the type $Li(^2S)F^-H^+$ and $Li^+F^-H(1s)$ are both important.

The present calculations yield results that are qualitatively similar to the empirical potential used by Roach (1970) to rationalize the isotope effects in the K + HBr reaction (Gillen *et al.* 1969; Martin and Kinsey 1967), and which are also in agreement with the suggestion of Herschbach (1970) that the system has a substantial well. It was found necessary to assume that there is such a well in order to understand the experiments of Siska (1969) on the back reaction corresponding to reaction (D); that is,

$$H + KBr \rightarrow HBr + K.$$

The HBr product in this reaction was found to be scattered forwards in the centre-of-mass reference frame. This was an unexpected result, because reactions having small cross-sections (~ 1 Å^2 in the present case) are generally of the rebound type (Herschbach 1966; Karplus 1968). A potential-energy surface with the general characteristics predicted above for the Li–F–H linear $^2\Sigma^+$ symmetry surface may well explain the isotope effects in the K + HBr reactions and the forward scattering obtained for the H + KBr system. Before any definitive statement can be made, however, it is clear that a detailed study of the dynamics is required and perhaps also a more thorough analysis of the available experimental data in terms of non-separable centre-of-mass differential cross-sections in the product velocity space (Siska 1969).

In Fig. 6.17 the lowest $^2\Sigma^+$ and $^2\Pi$ potential-energy surfaces are shown for the F–Li–H geometry, with the F atom approaching an LiH molecule whose internuclear distance is fixed at the equilibrium value. The results for the F atom approaching the H end of LiH are shown in Fig. 6.18. In both cases, there is little interaction on the $^2\Sigma^+$ surface until an electron-jump occurs; after that, there is a strong attraction between $(LiH)^+$ and F^-. Examination of the wavefunctions in the ionic region shows that positive charge resides almost entirely on the Li atom. This charge distribution is dictated by the relatively high energy of the valence-bond structure corresponding to $F^-Li(^2S)H^+$.

As an F atom approaches either end of LiH on the $^2\Pi$ symmetry surface, there is a strong repulsion (Figs. 6.17 and 6.18). This repulsion will prevent reaction on the $^2\Pi$ surface at thermal energies. The situation contrasts markedly with that in $F + Li_2$ where there was no such repulsion (see Fig. 6.12) and the reaction of $F + Li_2$ to produce electronically excited $Li(^2P^0)$ molecules directly was possible. The difference is due to the relatively high $H(^2S) \rightarrow H(^2P^0)$ excitation energy. This energy is 0·375 a.u. as compared to 0·068 a.u. for lithium (the electron has to be excited from a 1s to a 2p orbital in hydrogen). As the fluorine atom approaches closer in the F–H–Li linear configuration, there is a hump followed by an inner attractive well (Fig. 6.18). In the more ionic region (inside the hump), the positive charge resides mainly on the H atom, the major ionic contribution being $F^-H^+Li(^2P^0)$. However, the neutral $F(^2P^0)H(1s)Li(^2S)$ structures also have large coefficients in this region.

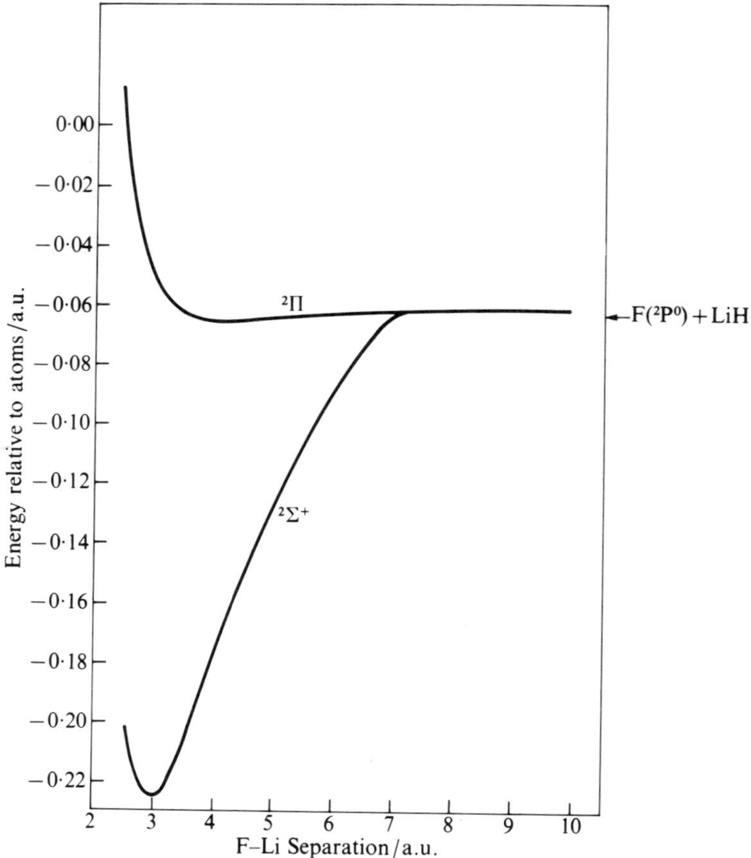

FIG. 6.17. Cut in lowest $^2\Sigma^+$ and $^2\Pi$ F–Li–H linear potential energy surfaces calculated using OM method. The Li–H separation is held fixed at 3·0 a.u.

The reason for the hump followed by an inner well on the F–H–Li linear $^2\Pi$ surface is that a curve crossing occurs only after a substantial repulsion of the electrons on the F and H atoms has set in. In the F–Li–H linear configuration the analogous curve crossing is prevented due to the repulsion of the core electrons in Li^+ for those on F^-. The presence of such a repulsion is apparent in the larger LiF equilibrium separation, as compared to HF.

Figs. 6.19 and 6.20 display the lowest $^2\Sigma^+$ and $^2\Pi$ potential surfaces along cuts in the exit valley of the reactions

$$\mathrm{F+LiH \rightarrow LiF+H} \quad \text{and} \quad \mathrm{F+LiH \rightarrow HF+Li}$$

respectively; in both cases only the linear geometries are considered. As either the H atom (Fig. 6.19) or the Li atom (Fig. 6.20) is removed to give

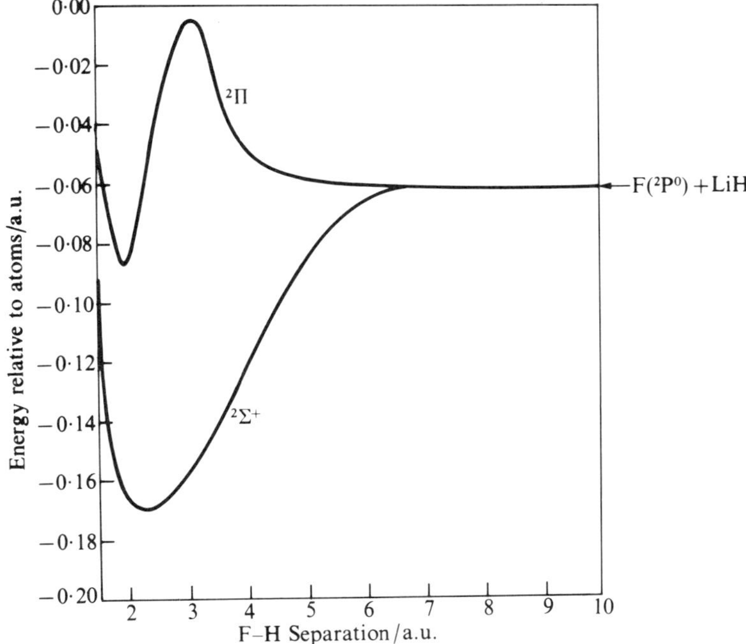

FIG. 6.18. Cut in lowest $^2\Sigma^+$ and $^2\Pi$ F–H–Li linear potential energy surfaces calculated using OM method. The Li–H separation is held fixed at 3·0 a.u.

the final products on the $^2\Sigma^+$ symmetry surface, there is a small well. This is similar to the situation found for $F + Li_2$ (see Fig. 6.13). The depth of the well as an Li atom approaches the H atom in an HF molecule is 10 kcal/mole by the OM method calculations and 4 kcal/mole by the *ab initio* method. Both of these values are much larger than the 0·2 kcal/mole well found in the SCF–LCAO calculation (Lester and Krauss 1970). The deep well in the F–Li–H linear $^2\Pi$ symmetry surface as an H atom approaches an LiF molecule is due mainly to the coulombic attraction in the ionic structure $F^-Li(^2P^0)H^+$. The system eventually dissociates to give $LiF(^3\Pi) + H(2S)$ which lies at a very high energy (Fig. 6.19).

From the calculations on the F + LiH potential-energy surfaces described above, it seems likely that the reactions

$$F + LiH \rightarrow LiF + H \quad \text{and} \quad F + LiH \rightarrow HF + Li$$

can occur only on surfaces of even symmetry ($^2\Sigma^+$ in the linear case) at thermal energies. This conclusion must be tested by calculating potential-energy surfaces for some non-linear nuclear configurations. The lowest, even symmetry surface is of the highly 'attractive' type and the newly formed molecule (LiF or HF) is expected to be in a highly excited vibrational state.

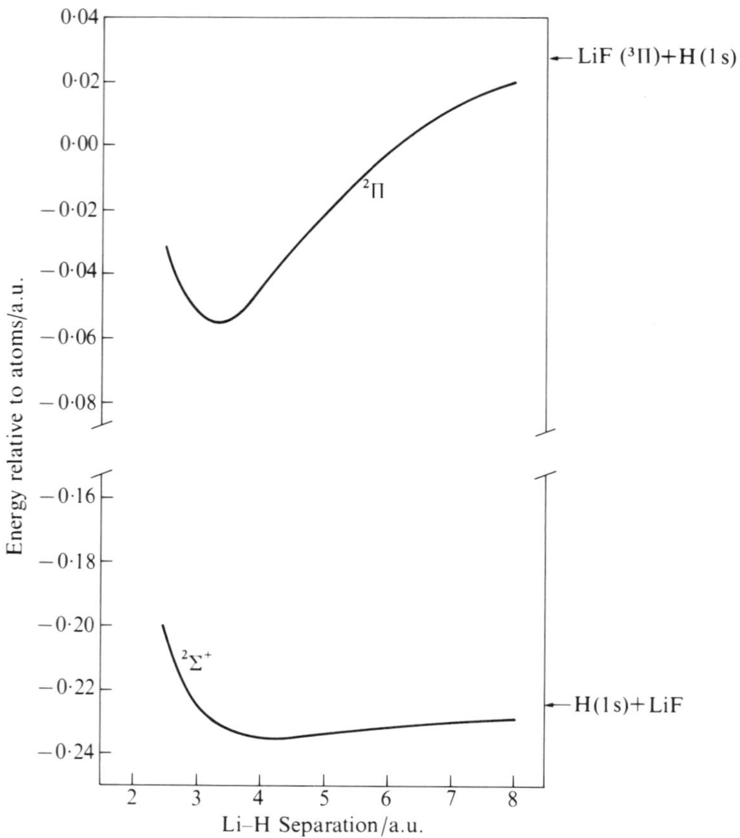

FIG. 6.19. Cut in lowest $^2\Sigma^+$ and $^2\Pi$ F–Li–H linear potential energy surfaces calculated using OM method. The Li–F separation is held fixed at 3·0 a.u.

The direct production of electronically excited Li($^2P^0$) atoms is expected to be much less probably than in the $F+Li_2$ case, since the only possibility of producing excited Li atoms in F+LiH collisions would be by means of non-adiabatic transitions between surfaces of even symmetry.

Just as in Li_2F, where $Li^+ + LiF^-$ structures were found to be very important in some nuclear configurations, so in the present case $Li^+ + HF^-$ structures play a greater role in the wavefunction than had been expected. Prompted by this observation, calculations on the HF^- diatomic ion have been performed. Two different basis sets were used in these calculations. The smaller basis set corresponds to that used in the LiHF triatomic calculations described above, while the larger basis set provides a greatly improved description of the electron affinity of fluorine. Both calculations predict the lowest $^2\Sigma^+$ state of HF^- to be bound; however, the binding energy

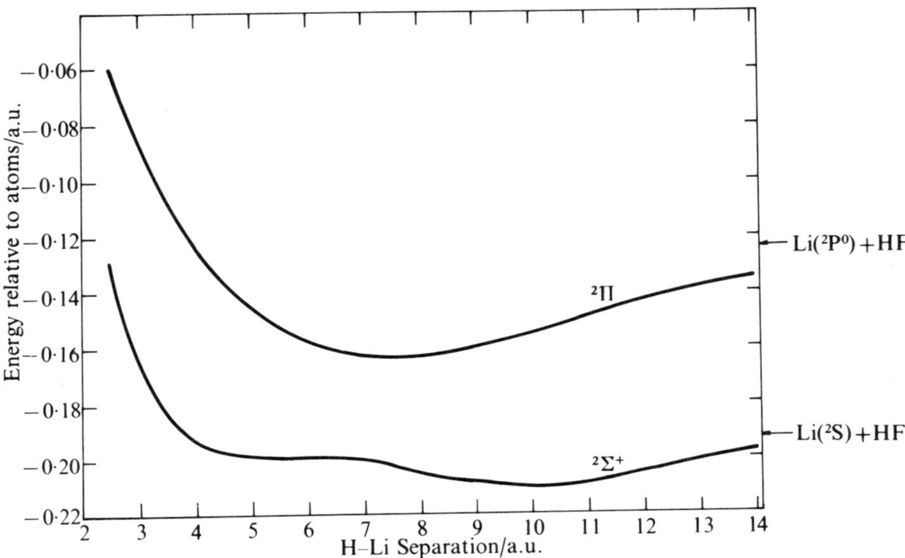

FIG. 6.20. Cut in lowest $^2\Sigma^+$ and $^2\Pi$ F–H–Li linear potential energy surfaces calculated using OM method. The H–F separation is held fixed at 1·9 a.u.

obtained in the small basis set calculations is much larger (by 0·15 a.u.) than that found by the calculations which employed the larger basis set. Both the *ab initio* and OM methods showed this effect.† It is therefore concluded that the cause of the problem was the poor description of F^- afforded by the smaller basis set and that this results in the basis set being better suited for describing the molecule than the dissociation products. The ion HF^- seems to be a singularly unfavourable case in this respect, because of the small bond length and the diffuse orbitals on the hydrogen. In the molecule these orbitals are used to improve the poor description of F^-.

The sensitivity of the energy of HF^- to the basis set will clearly affect the triatomic LiHF calculations. The main effects on the calculations described above are that the wells in the LiHF system are too deep and that the detailed shape of the potential along the exit valley of the reaction Li + HF → LiF + H (Fig. 6.16) is not correct. It may even prove to be the case that the double minimum predicted along this exit valley is an artifact of the basis set. The shape of the curves in Fig. 6.15 and the lower curve in Fig. 6.20 will also be

† The HF^- ion is unstable with respect to electron detachment at small internuclear separations (Michels, Harris, and Browne, 1968). The present diatomic calculations did not include any extremely diffuse orbitals or any continuum functions which would be needed to examine this process. In the molecular environment the HF^- ion is greatly stabilized by Coulomb attraction to the Li^+ ion, while at large HF–Li distances Li + HF has a lower energy than $Li^+ + HF + e$.

affected by the choice of basis set. Calculations with improved and much larger basis sets are in progress to check the LiHF results presented above.

6.9. Summary

In this chapter, we have shown how the introduction of atomic data can be used in atoms-in-molecules calculations to obtain useful results for diatomic molecules and triatomic potential surfaces. The importance of such an approach is that it can yield reasonable energies with limited basis set wavefunctions that would not be sufficiently accurate in an *ab initio* treatment. In this way, the speed of the calculation can be increased so as to permit exploration of the range of geometries required to characterize the potential surfaces for chemical reactions. As computing machines and the methods for molecular calculations are improved, extended basis set calculations will be performed in ever-increasing numbers. Nevertheless, an approach of the atoms-in-molecules type may well continue to be useful for organizing and refining *ab initio* results at any level. Of special importance in this regard is the fact that the atoms-in molecules method, as developed here, converges to the *ab initio* treatment as the basis set and multiconfiguration functions approach the exact result.

Relatively small basis sets were employed in the triatomic calculations reported here. They should consequently be regarded as being of an exploratory nature. While it is hoped that the present calculations correctly predict the qualitative form of the triatomic surfaces considered, calculations using larger basis sets will be required to obtain quantitatively accurate information concerning their detailed shapes.

Acknowledgement

We should like to thank Professor K. Morokuma, Professor L. Pedersen, and Dr R. M. Stevens for many discussions and much invaluable assistance, expecially with the computational aspects of this work.

Appendix: Atomic Orbital and CF basis sets for HF

Each atomic orbital is constructed as a linear combination of Gaussian-type orbitals (GTO). The orbitals used for fluorine were the result of exponent optimized SCF–LCAO calculations for the ground state $F(^2P^0)$ (Whitman and Hornback 1969); the reader is referred to Appendix c of Balint-Kurti and Karplus (1969) for details of the atomic orbitals of fluorine.

For the s orbitals of hydrogen, a linear combination of four GTOs was used. Three of the exponents were taken from Huzinaga's (1965) expansion of the 1s orbital of hydrogen in terms of three s-type GTOs. The remaining exponent had to be chosen so as to be suitable for the description of the 2s orbital in the H atom and also for the description of H^-. The H^- ion was described by a four-fold CI which included structures of orbital configuration $(1s)^2$, $(1s2s)$, $(2s)^2$ and $(2p)^2$. The fourth exponent was chosen so as

TABLE A6.1

Atomic orbitals for hydrogen; linear combinations of GTOs.

Exponent	1s	2s
4·50038	0·06974246	−0·02982539
0·681277	0·41083680	−0·19535583
0·151374	0·62247917	−0·54609970
0·032	0·03867312	1·2382051

Exponent	2p
0·04527	1·0

TABLE A6.2

Calculated and experimental atomic energies.[a]

Atomic state	F atom and ions: Calculated[b] energy	F atom and ions: Exponent[c] optimized SCF	Experimental[d]
$F(^2P^0)$	−99·0396	−99·0396	−99·8060
$F(^2S)$	−98·1845	−98·1859	−99·0386
$F^-(^1S)$	−98·8666	−99·0049	−99·9333[e]
$F^+(^3P)$	−98·4218	−98·5242	−99·1654
$F^+(^1D)$	−98·3284	−98·4174	−99·0711
$F^+(^1S)$	−98·1883	−98·2634	−98·9615
$F^+(^3P^0)$	−97·6424	(−97·719)[f]	−98·4146
$F^+(^1P^0)$	−97·2784	−97·3560	−98·0745

Atomic state	H atom and ion: Calculated[b] energy	H atom and ion: Experimental energy
H(1s)	−0·4973	−0·5
H(2s)	−0·1198	−0·125
H(2p)	−0·1132	−0·125
$H^-(^1S)$	−0·5080[g]	−0·5278[h]

[a] All energies are in a.u. (see Table 6.1).

[b] Energies calculated using orbitals of Balint–Kurti and Karplus (1969) and Table A6.1.

[c] Whitman *et al.* (1968).

[d] See Moore (1949). The experimental energy of $F(^2P^0)$, relative to which all the other fluorine atoms and ions are measured, is somewhat uncertain. As we are interested only in energies relative to the ground-state atom, this uncertainty does not effect the present calculations.

[e] The electron affinity of fluorine is taken to be 0·1273 a.u., see Berry and Reinmann (1963).

[f] This value is estimated.

[g] A four-fold configuration interaction was used to obtain this energy (see text). If no configuration interaction is used an energy of −0·3759 a.u. results.

[h] See Stewart (1964).

approximately to minimize the energy of H^-. The energies of the 1s and 2s orbitals in hydrogen, using this exponent, are not the lowest possible but are reasonable. The exponent of the single p-type GTO used is also taken from Huzinaga's (1965) work. The atomic orbitals for hydrogen are displayed in Table A6.1.

The energies of the hydrogen and fluorine atoms and ions are listed in Table A6.2. The energies under the heading 'Calculated' were obtained with the basis sets of Table A6.1 and those of Balint-Kurti and Karplus (1969). Those listed under the heading 'Exponent optimized SCF' are the result of SCF–LCAO calculations with exponent optimization performed by Whitman, Leyshon, and Hornback (1968) for each state of the atom or ion. The experimental energies (Moore 1949) quoted in the table are statistically weighted means of the multiplet values.

REFERENCES

AITKEN, A. C. (1959). *Determinants and matrices.* Oliver and Boyd, London.

ARAI, T. (1957*a*). *J. chem. Phys.* **26**, 435.

—— (1957*b*). *J. chem. Phys.* **26**, 451.

—— (1960). *Rev. mod. Phys.* **32**, 370.

—— and SAKAMOTO, M. (1958). *J. chem. Phys.* **28**, 32.

BAEDE, A. P. M., MOUTINHO, A. M. C., DE VRIES, A. E., and LOS, J. (1969). *Chem. Phys. Letts* **3**, 530; HELBING, R. K. B. and ROTHE, E. W. (1969). *J. Chem. Phys.* **51**, 1607.

BALINT-KURTI, G. G. (1969). *Ph.D. Thesis*, Columbia University.

—— (1971). *Molec. Phys.* **22**, 681.

—— and KARPLUS, M. (1969). *J. Chem. Phys.* **50**, 478.

—— —— (1971). *Chem. Phys. Letts* **11**, 203.

BATANA, A. and COHAN, N. V. (1962). *Proc. phys. Soc.* (*London*) **79**, 279.

BENDER, C. F. and DAVIDSON, E. R. (1968). *J. chem. Phys.* **49**, 4989.

BERRY, R. S. (1957). *J. Chem Phys.* **27**, 1288.

—— and REINMANN, C. W. (1963). *J. chem. Phys.* **38**, 1540.

BIRLEY, J. H., HERM, R. R., WILSON, K. R., and HERSCHBACH, D. R. (1967). *Discuss. Faraday Soc.* **44**, 176; GRICE, R. and EMPEDOCLES, P. B. (1968). *J. Chem. Phys.* **48**, 5352; PARRISH, D. D. and HERM, R. R. (1969). *ibid.* **51**, 5467; WARNOCK, T. T., BERNSTEIN, R. B., and GROSSER, A. E. (1967). *ibid.* **46**, 1685 and references therein.

BLAIS, N. (1968). *J. chem. Phys.* **49**, 9; GODFREY, M. and KARPLUS, M. (1968). *ibid.* **49**, 3602; KUNTZ, P. J., MOK, M. H., and POLANYI, J. C. (1969). *ibid.* **50**, 4623.

CHUPKA, W. A., BERKOWITZ, J., and GUTMAN, D. (1971). *J. chem. Phys.* **55**, 2724.

CLEMENTI, E. and DAVIS, D. R. (1965). The integral part of the computer program was based on 'IBMOL version 1A' written by these authors. The program is distributed by the Quantum Chemistry Program Exchange, Indiana University.

CLEMENTI, E. and MCLEAN, A. D. (1964). *Phys. Rev.* **133**, *A*419.

CONDON, E. U. and SHORTLEY, G. H. (1935). *The theory of atomic spectra* pp. 169 and 171. Cambridge University Press; *also* SLATER, J. C. (1960). *Quantum theory of atomic structure, vol.* 1, p. 291. McGraw-Hill, New York.

COPSEY, D. N., MURRELL, J. N., and STAMPER, J. G. (1971). *Molec. Phys.* **21**, 193.
COULSON, C. A. (1937). *Trans. Faraday Soc.* **33**, 1479.
—— (1938). *Proc. Cambridge phil. soc.* **34**, 204.
—— (1960). *Rev. Mod. Phys.* **32**, 170.
—— (1961). *Valence* p. 113. Clarendon Press, Oxford.
EISENHART, L. P. (1926). *Riemannian Geometry*, p. 101. Princeton University Press.
GILLEN, K. T., RILEY, C., and BERNSTEIN, R. B. (1969). *J. chem. Phys.* **50**, 4019.
GODFREY, M. and KARPLUS, M. (1968). *J. chem. Phys.* **49**, 3602.
HEITLER, W. and LONDON, F. (1927). *Z. Phys.* **44**, 455.
HERSCHBACH, D. R. (1966). *Advan. chem. Phys.* **10**, 319.
—— (1970). In *Proceedings of the conference on potential energy surfaces in chemistry* (ed. W. A. Lester). Publication RA 18, IBM Research Library, California.
HERZBERG, G. (1950). *Spectra of diatomic molecules*, p. 546. Van Nostrand, New York.
HIRSCHFELDER, J. O. (1967). *Chem. phys. letts* **1**, 325, 363, and references therein.
HURLEY, A. C. (1955). *Proc. phys. Soc. (London) A* **68**, 149.
—— (1956*a*). *Proc. phys. Soc. (London) A* **69**, 49.
—— (1956*b*). *Proc. phys. Soc. (London) A* **69**, 301.
—— (1956*c*). *Proc. phys. Soc. (London) A* **69**, 767.
—— (1958*a*). *J. chem. Phys.* **28**, 532.
—— (1958*b*). *Proc. R. Soc. A* **248**, 119.
—— (1959). *Proc. R. Soc. A* **249**, 402.
—— (1960). *Rev. Mod. Phys.* **32**, 400.
—— (1963). *Rev. mod. Phys.* **35**, 448.
HUZINAGA, S. (1965). *J. chem. Phys.* **42**, 1293.
JOHNS, J. W. C. and BARROW, R. F. (1959). *Proc. R. Soc. A* **251**, 504.
KARPLUS, M. (1958). In *Structural chemistry and molecular biology* (ed. A. Rich and N. Davidson). W. H. Freeman, San Fransisco.
KIM, K. C. and EDMINSTON, C. (1970). *J. chem. Phys.* **52**, 997.
KOLOS, W. and WOLNIEWICZ, L. (1964). *J. chem. Phys.* **41**, 3663.
KOTANI, M. (1961). *Handb. Phys.* **37**, 2.
KRAUSS, M. and WEHNER, J. F. (1958). *J. chem. Phys.* **29**, 1287. *See also* KRAUSS, M. and RANSIL, B. J. (1960). *ibid.* **33**, 840, where some poor ICC results are reported.
LACMAN, K. and HERSCHBACH, D. R. (1970). *Chem. phys. Letts* **6**, 106.
LESTER, W. A. and KRAUSS, M. (1970). *J. chem. Phys.* **52**, 4775.
LÖWDIN, P. O. (1955). *Phys. Rev.* **97**, 1474. *See also* KING, H. F., STANTON, R. E., KIM, H., WYATT, R. E., and PARR, R. G. (1967). *J. chem. Phys.* **47**, 1936; PROSSER, F. and HAGSTROM, S. (1968). *Int. J. quant. Chem.* **2**, 89; (1968). *J. chem. Phys.* **48**, 4807.
MAGEE, J. L. (1940). *J. Chem. Phys.* **8**, 687.
MARTIN, L. R. and KINSEY, J. L. (1967). *J. chem. Phys.* **46**, 4834.
MCLEAN, A. D. (1963). *J. chem. Phys.* **39**, 2653.
MICHELS, H. H., HARRIS, F. E., and BROWNE, J. C. (1968). *J. chem. Phys.* **48**, 2821.
MILLER, W. B., SAFRON, S. A., and HERSCHBACH, D. R. (1967). *Discuss. Faraday Soc.* **44**, 108.
MOFFITT, W. (1951*a*). *Proc. R. Soc. A* **210**, 245.
—— (1951*b*). *Proc. R. Soc. A* **210**, 224.
—— (1953). *Proc. R. Soc. A* **218**, 486.
—— (1954). *Rept. Progr. Phys.* **17**, 173.

MOORE, C. M. (1949). *Natl. Bur. Std. U.S.A. Circ. No.* 467.
MOROKUMA, K. and PEDERSEN, L.; this procedure was suggested by them and they have used it in CI calculations.
MOULTON, M. C. and HERSCHBACH, D. R. (1965). *J. chem. Phys.* **44**, 3010.
MURRELL, J. N. and SHAW, G. (1967) *J. chem. Phys.* **46**, 1768.
NESBET, R. K. (1963). *Rev. mod. Phys.* **35**, 552.
OHNO, K. (1957). *J. chem. Phys.* **26**, 1754.
PARR, R. G. (1964). *Quantum theory of molecular electronic structure*, p. 101. W. A. Benjamin, New York.
PAULING, L. (1960). *The nature of the chemical bond.* Cornell University Press, Ithaca, New York.
PAUNCZ, R. (1954). *Acta phys. hung.* **4**, 237; *see also* SCHERR, C. W. (1954). *J. chem. Phys.* **22**, 149; STEWART, E. T. (1960). *Proc. phys. Soc.* (*London*) **75**, 402.
POLANYI, M. (1932). *Atomic reactions.* Williams and Norgate, London.
—— and ROSNER, S. D. (1963). *J. Chem. Phys.* **38**, 1028; *see also* KUNTZ, P. J., NEMETH, E. M., POLANYI, J. C., ROSNER, S. D., and YOUNG, C. E. (1966). *ibid.* **44**, 1168.
POTTS, A. W. and PRICE, W. C. (1971). *Trans. Faraday Soc.* **67**, 1242.
RAHMAN, A. (1954). *Physica* **20**, 623.
ROACH, A. C. (1970). *Chem. phys. Letts* **6**, 389.
SHAVITT, I. (1963). *Methods in computational physics*, vol. 2, p. 1. Academic Press, New York; CLEMENTI, E. and DAVIS, D. R. (1966). *J. chem. Phys.* **45**, 2593.
SISKA, P. E. (1969). Ph.D. thesis, Harvard University and private communications.
STEWART, A. L. (1964). *Adv. Phys.* **12**, 229. The author calculates an energy of $-0{\cdot}5278$ a.u. for H^-. This was used as the experimental energy.
TAYLOR, E. H. and DATZ, S. (1955). *J. chem. Phys.* **23**, 1711.
VAN VLECK, J. H. (1936). *Phys. Rev.* **49**, 232: CHIRGWIN, B. H. and COULSON, C. A. (1950). *Proc. R. Soc. A* **201**, 196.
VELASCO, R. (1957). *Can. J. Phys.* **35**, 1024.
WAHL, A. C. (1964). *J. chem. Phys.* **41**, 2600.
WANG, S. C. (1928). *Phys. Rev.* **31**, 579; HIRSCHFELDER, J. O. and LINNETT, J. W. (1950). *J. chem. Phys.* **18**, 130.
WEISS, R. (1963). *Phys. Rev.* **131**, 659.
WHITMAN, D. R., LEYSHON, K., and HORNBACK, C. J. (1968) calculated most of the atomic energies and atomic orbitals; *see also* WHITMAN, D. R. and HORNBACH, C. J. (1969). *J. chem. Phys.* **51**, 398.
ZENER, C. (1932). *Proc. R. Soc.* (*London*) *A* **137**, 696.

7

INTERMOLECULAR ENERGIES IN THE REGION OF SMALL OVERLAP

J. N. MURRELL

7.1. General features of the theory

THE problem of the intermolecular potential is part of the more general problem of the nature of the chemical bond. The aspect that gives it a distinctive character is the weakness of intermolecular forces compared with normal valence forces. An intermolecular binding energy is typically 1–10 per cent of the strength of a typical chemical bond. There is, of course, a continual gradation between the two; the strength of the bond in some molecular complexes approaching that of weak valence bonds. We shall, however, just look at the theoretical approach to the weak intermolecular potential, because the answer for the strong potentials is to be found in normal valence theory.

The characteristic difference between *inter* and *intra*molecular interactions is that the overlap of orbitals of two atoms in different molecules is much smaller than that for atoms joined by a normal chemical bond. We shall rarely be dealing with overlap integrals that exceed 0·05 whereas 0·5 is more typical for a chemical bond. It is perhaps worth emphasizing that non-bonded interactions within molecules (steric forces for example) may be classified along with intermolecular forces and treated in a similar way.

If overlap integrals are small one might reasonably expect to construct wavefunctions for the system from the wavefunctions of the separate parts. If this is done one can expect to find that the total energy can be separated into a number of parts each of which has some physical significance. This approach can also be followed for the strong interactions associated with a chemical bond, but one finds that the 'mixing' of states becomes so large that it is very difficult to get a useful subdivision of the total bond energy.

If a subdivision of the total energy is made one can examine separately the attractive and repulsive parts. In general the repulsive energies are associated with the overlap of orbitals of the two molecules whereas the attractive energies are largely independent of overlap. In the region of the van der Waals' minimum the attractive terms are much the larger of the two, hence to calculate the binding energy it may be a reasonable approximation to calculate only the attractive terms and these may be obtained from a strictly zero-overlap theory. It must be emphasized however that one cannot deduce the equilibrium separation of two molecules by this method because this is determined by a balance of attractive and repulsive *forces* and the repulsive energies have therefore to be considered, if only to calculate their derivative.

A typical spherically averaged intermolecular potential as in Fig. 7.1, can be roughly divided up into weak, intermediate and strong-interaction

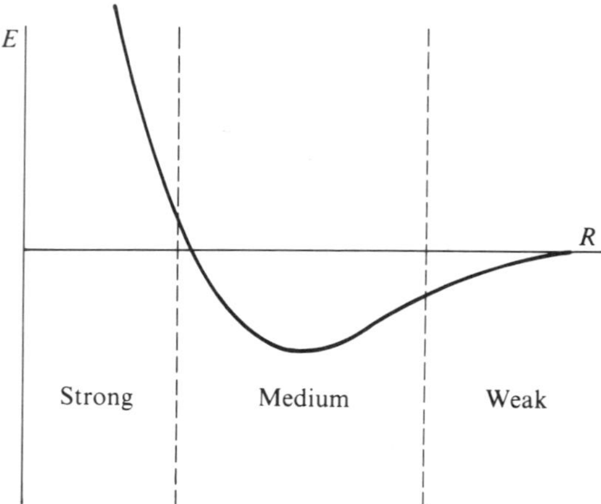

FIG. 7.1. The weak, medium, and strong interaction regions in the intermolecular potential.

regions. In the weak region, overlap can be neglected, in the strong region overlap is large as in a normal chemical bond problem. In the intermediate region there is small overlap and there are differing opinions as to how it should be tackled. My personal preference is for an extension of the weak interaction theory to allow for small overlap effects, but as we shall see later the development of such a theory is not without its problems.

In the zero-overlap theory the intermolecular energy can be divided into a first-order *electrostatic* or Coulomb term and second-order polarization terms which are of two types, *dispersion* and *induction.* When overlap is added one obtains an additional first-order term—the *exchange* energy—and second-order terms which can be described as *exchange-polarization.* An alternative formulation of the problem gives second-order overlap-dependent terms of the *charge-transfer* type. Before discussing these in detail it is perhaps worth drawing up a table (Table 7.1) of their most important properties; their range, whether they are attractive or repulsive, and whether for a many particle system they are pairwise additive. The latter property will not be discussed in detail in this chapter. A long-range energy will be defined as one that varies with the intermolecular separation roughly as R^{-n}, and a short-range energy as one that varies roughly as $\exp(-kR)$. It is important to note from Table 7.1 that the only energy term which is definitely not

TABLE 7.1
Characteristic properties of energy terms valid for weak and intermediate interactions.

Type	Range	Attractive or Repulsive	Pairwise Additive
Electrostatic or Coulomb	Long except for neutral spherical systems.	Either	Yes
Exchange	Short	Repulsive	Nearly
Dispersion	Long	Attractive	Yes
Induction	Long	Attractive	No
Exchange Polarization or Charge-Transfer	Short	Attractive	Nearly

pairwise additive is the induction energy and luckily this may not be very important in many cases.

For literature on intermolecular forces one has to go either to the original papers or to collected works which are at a rather advanced theoretical level. A list of these is given in the bibliography. There is a notable lack of a text which covers the theory at a level more suitable for the experimental physicist or chemist.

7.2. Perturbation theory in the zero-overlap approximation

We start with a zero-overlap theory as this is relatively straightforward. We will show later that the expressions we derive can be obtained from a more general perturbation theory which allows for overlap, when the zero-overlap limit is applied to the different orders of perturbation.

The basic equations of Rayleigh–Schrödinger perturbation theory start from a Hamiltonian

$$H = H^0 + \lambda U. \tag{7.1}$$

One then seeks a solution of

$$(H - E)\psi_q = 0, \tag{7.2}$$

by developing ψ and E as expansions in powers of λ as follows

$$\begin{aligned} \psi_q &= \psi_q^{(0)} + \lambda\psi_q^{(1)} + \ldots, \\ E_q &= E_q^{(0)} + \lambda E_q^{(1)} + \ldots. \end{aligned} \tag{7.3}$$

By equating powers of λ in eqn (7.2) one obtains equations like

$$(H^0 - E_q^{(0)})\psi_q^{(1)} + (U - E_q^{(1)})\psi_q^{(0)} = 0. \tag{7.4}$$

If one takes ψ_q to be normalized to unity to all orders of λ, then after multiplying eqn (7.4) from the left by $\psi_q^{(0)}$ and integrating over all coordinates of the system one obtains

$$E_q^{(1)} = \langle\psi_q^{(0)}|U|\psi_q^{(0)}\rangle \equiv U_{qq}, \tag{7.5}$$

and likewise from higher-order equations

$$E_q^{(2)} = \langle\psi_q^{(0)}|U|\psi_q^{(1)}\rangle, \quad \text{etc.} \tag{7.6}$$

If the functions $\psi_q^{(n)}$ are expanded in terms of the complete set of eigenstates of H^0 which are defined by

$$H^0\psi_j^{(0)} = E_j^{(0)}\psi_j^{(0)} \tag{7.7}$$

one obtains expressions such as

$$\psi_q^{(1)} = -\sum_j{}' \frac{\langle\psi_j^{(0)}|U|\psi_q^{(0)}\rangle}{E_j^{(0)}-E_q^{(0)}}\psi_j^{(0)} = -\sum_j{}' \frac{U_{jq}\psi_j^{(0)}}{E_j^{(0)}-E_q^{(0)}}, \tag{7.8}$$

$$E_q^{(2)} = -\sum_j{}' \frac{U_{jq}U_{qj}}{E_j^{(0)}-E_q^{(0)}}. \tag{7.9}$$

$\sum'$ represents a sum over all discrete states except $\psi_q^{(0)}$, plus an integral over the continuum states.

A more complete discussion of this standard perturbation method is given in an article by Hirschfelder, Byers-Brown, and Epstein (1964).

Let us now consider two molecules, A and B, separated by a large distance. Let the wavefunctions of their ground states be A_0 and B_0. If these are completely separate systems with Hamiltonians $H_a(i)$ and $H_b(j)$ respectively, then

$$(H_a(i)-E_a^{(0)})A_0(i) = 0 \tag{7.10}$$

$$(H_b(j)-E_b^{(0)})B_0(j) = 0. \tag{7.11}$$

Multiplying eqn (7.10) by $B_0(j)$ and eqn (7.8) by $A_0(i)$ and adding the two gives

$$\{H_a(i)+H_b(j)-E_a^{(0)}-E_b^{(0)}\}A_0(i)B_0(j) = 0. \tag{7.12}$$

If we write the complete Hamiltonian

$$H = H_a(i)+H_b(j)+U(a, b, i, j), \tag{7.13}$$

where U represents all the interaction terms between the nuclei and electrons of system A and the nuclei and electrons of system B

$$U = -\sum_a\sum_j Z_a/r_{aj} - \sum_b\sum_i Z_b/r_{bi} + \sum_i\sum_j 1/r_{ij} + \sum_a\sum_b Z_aZ_b/r_{ab}, \tag{7.14}$$

then

$$\psi_{00} = A_0(i)B_0(j) \tag{7.15}$$

can be considered as the zeroth-order wavefunction for the system under the perturbation U

The separation we have made of H is an arbitrary one based on a particular allocation of electrons to the two molecules. As electrons are indistinguishable we could have chosen any other allocation of electrons and taken an appropriate wavefunction $A_0(i')B_0(j')$ and perturbation term $U(a, b, i', j')$. One would normally take account of the indistinguishability of the electrons by using antisymmetrized wavefunctions,†

$$\Psi_{00} = \mathscr{A} A_0(i)B_0(j), \tag{7.16}$$

but for this function no single perturbation term can be formulated. We shall see that this leads to difficulties in formulating a theory for the small-overlap region. However, providing overlap contributions to the energy are insignificant we can work with simple product functions like ψ_{00}. This can be stated more precisely in the following way.

Let Ψ be the fully antisymmetrized lowest eigenfunction of H, and

$$E = \langle \Psi | H | \Psi \rangle \tag{7.17}$$

its energy. We now take a simple product wavefunction

$$\psi' = A'(i)B'(j) \tag{7.18}$$

and minimize its energy

$$E' = \langle \psi' | H | \psi' \rangle, \tag{7.19}$$

with respect to variation in A' and B'. If E' is insignificantly different from E then a zero-overlap theory is valid. It must be emphasized that if overlap is not negligible, then E' may be less than E because one places more restrictions on Ψ (its antisymmetry) than on ψ'. Another important point is that if A' and B' are expanded in terms of complete sets of functions

$$A' = \sum_{\mu} c_\mu \phi_\mu; \qquad B' = \sum_{\nu} c_\nu \eta_\nu \tag{7.20}$$

then although we may have zero-overlap between the functions A' and B' we may not have zero-overlap between individual terms in these expansions. This will certainly be the case if the ϕ and η are the complete sets of eigenfunctions of the two systems.

† Throughout this chapter antisymmetrized wavefunctions will be symbolized by capital letters and simple products by lower case letters.

7.3. The electrostatic energy

If we substitute the zeroth-order wavefunction ψ_{00} into eqn (7.5) we obtain the first-order energy

$$E^{(1)} = \langle A_0(i)B_0(j)|U(a, b, i, j)|A_0(i)B_0(j)\rangle. \tag{7.21}$$

This is the expectation value of the electrostatic energy between the particles of A and the particles of B. To calculate it one needs to know the positions of the nuclei and, from the wavefunctions A_0 and B_0, the electron density distributions on the two molecules which we write ρ_{00}^{A} and ρ_{00}^{B}. From these the energy is given by the same expression that one would use in classical electrostatics namely

$$E^{(1)} = -\sum_b Z_b \int \rho_{00}^{\mathrm{A}}(1) r_{b1}^{-1}\, \mathrm{d}v_1 - \sum_a Z_a \int \rho_{00}^{\mathrm{B}}(1) r_{a1}^{-1}\, \mathrm{d}v_1$$

$$+ \iint \rho_{00}^{\mathrm{A}}(1)\rho_{00}^{\mathrm{B}}(2) r_{12}^{-1}\, \mathrm{d}v_1\, \mathrm{d}v_2 + \sum_a \sum_b Z_{\mathrm{A}} Z_{\mathrm{B}} r_{ab}^{-1}. \tag{7.22}$$

Within a zero-overlap approximation the electron densities ρ_{00}^{A} and ρ_{00}^{B} will be non-overlapping, and one can hope to develop a convergent multipole expansion for $E^{(1)}$. This may be done by introducing the Neumann two-centre expansion for r_{12}^{-1}. The full mathematical details may be found in Hirschfelder, Curtiss, and Bird (1954).

7.4. Polarization energies

The second-order energy in the zero-overlap theory may be called a polarization energy because it involves a distortion or polarization of the electron clouds of the interacting molecules. It can be divided into induction and dispersion energies depending on whether one or both molecules are polarized. We first look at expressions for these in terms of the sum-over-eigenstates formula (7.9).

The complete set of eigenstates of H^0 are of the form

$$\psi_{rs} = A_r(i)B_s(j) \tag{7.23}$$

where either A_r or B_s may be the ground state. The summation in (6) may be divided into the following three parts:

$$r, s \neq 0; \qquad r \neq 0, s = 0; \qquad r = 0, s \neq 0.$$

We shall examine the matrix element $U_{rs,00}$ for each in turn.

7.4.1. *The dispersion energy* $(r, s \neq 0)$

We are concerned with the matrix element

$$\langle A_r(i)B_s(j)|U|A_0(i)B_0(j)\rangle. \tag{7.24}$$

If we make use of the orthogonality of the two sets of eigenfunctions

$$\langle A_r(i)|A_0(i)\rangle = \delta_{r0}; \qquad \langle B_s(j)|B_0(j)\rangle = \delta_{s0} \tag{7.25}$$

then we can see that the only part of U (eqn 7.14) that contributes to this matrix element is $\sum_{i,j} r_{ij}^{-1}$. The others can all be eliminated by integration over the coordinates of electrons i or j. If we take the functions $A_r(i)$ and $B_s(j)$ to be antisymmetric with respect to their set of electrons then we can argue that each of the terms in the summation r_{ij}^{-1} must make the same contribution to the matrix element. We can therefore take one such term, r_{12}^{-1} say, where 1 is in the set i and 2 in the set j, and multiply by the number of such terms N_iN_j

$$U_{rs,00} = N_iN_j\langle A_r(i)B_s(j)|r_{12}^{-1}|A_0(i)B_0(j)\rangle. \tag{7.26}$$

We can now integrate over the coordinates of all electrons except 1 and 2, and if we define the functions

$$\begin{aligned} \rho_{r0}^{\mathrm{A}}(1) &= N_i \int A_r^*(i)A_0(i)\,\mathrm{d}v_{i\neq 1}, \\ \rho_{s0}^{\mathrm{B}}(2) &= N_j \int B_s^*(j)B_0(j)\,\mathrm{d}v_{j\neq 2}, \end{aligned} \tag{7.27}$$

then

$$U_{rs,00} = \int \rho_{0r}^{\mathrm{A}}(1)\rho_{0s}^{\mathrm{B}}(2)r_{12}^{-1}\,\mathrm{d}v_1\,\mathrm{d}v_2. \tag{7.28}$$

Eqns (7.27) define important functions called transition densities which have physical significance through the fact that their dipole moments are the transition moments which determine the probabilities of the spectroscopic transitions $A_0 \to A_r$ and $B_0 \to B_s$, thus

$$\begin{aligned} \boldsymbol{\mu}_{r0}^{\mathrm{A}} &= \int A_r^*(\mathrm{i})\left(\sum_i \boldsymbol{\mu}_i\right)A_0(i)\,\mathrm{d}v_i, \\ &= N_i \int A_r^*(i)\boldsymbol{\mu}_1 A_0(i)\,\mathrm{d}v_i, \\ &= \int \rho_{r0}^{\mathrm{A}}(1)\boldsymbol{\mu}_1\,\mathrm{d}v_1. \end{aligned} \tag{7.29}$$

For the case $r = 0$, ρ_{r0}^{A} becomes the electron density in the ground state of A.

Eqn (7.28) represents the interaction between the two transition densities. This can be put in the form of a multipole expansion by defining the multipole moments of these densities. (We will examine the validity of this later.) The leading term in this case will be the dipole–dipole energy because as a result of the orthogonality of the two states involved the transition density has no net charge

$$\int \rho_{0r}(1)\,\mathrm{d}v_1 = 0. \tag{7.30}$$

If we now introduce eqn (7.28) into eqn (7.9) we get the dispersion energy in the form

$$E_{\mathrm{dis}} = \sum_r{}' \sum_s{}' \left\{\int\int \rho^{\mathrm{A}}_{0r}(1)\rho_{0s}(2)r_{12}^{-1}\,\mathrm{d}v_1\,\mathrm{d}v_2\right\}^2 (E^{(0)}_{00} - E^{(0)}_{rs})^{-1}, \tag{7.31}$$

the summations not including the ground state of either molecule. Taking the multipole expansion of this we write

$$-E_{\mathrm{dis}} = C_6R^{-6} + C_8R^{-8} + \ldots, \tag{7.32}$$

where the leading term has the form of a dipole–dipole interaction squared

$$C_6 = \sum_r{}' \sum_s{}' \{\boldsymbol{\mu}^{\mathrm{A}}_{r0} \cdot \boldsymbol{\mu}^{\mathrm{B}}_{s0} - 3(\boldsymbol{\mu}^{\mathrm{A}}_{r0} \cdot \mathbf{R})(\boldsymbol{\mu}^{\mathrm{B}}_{s0} \cdot \mathbf{R})R^{-2}\}^2(E^{(0)}_{rs} - E^{(0)}_{00})^{-1}. \tag{7.33}$$

The second term in the expansion (7.32) arises from a dipole–quadrupole interaction, and so on.

Note that the expression (7.33) is orientation-dependent for non-spherical systems so that dispersion forces have a strong orienting effect for very anisotropic molecules like CO_2 or benzene.

For spherically symmetric systems, or if we take a spherical average of the dispersion force in other cases, we get

$$C_6 = \sum_r{}' \sum_s{}' \tfrac{2}{3}(\mu^{\mathrm{A}}_{0r}\mu^{\mathrm{B}}_{0s})^2(E^{(0)}_{rs} - E^{(0)}_{00})^{-1}. \tag{7.34}$$

This is the mean of one end-to-end and two side-by-side dipole interactions and it is therefore important to note that the summations over r and s must include all components of a degenerate state.

The transition moments can be related to the electric dipole oscillator strengths of the separate molecules

$$f^{\mathrm{A}}_{0r} = \tfrac{2}{3}(E^{\mathrm{A}}_r - E^{\mathrm{A}}_0)|\mu^{\mathrm{A}}_{0r}|^2 \tag{7.35}$$

hence eqn (7.34) can be put in the form

$$C_6 = \sum_r{}' \sum_s{}' \tfrac{3}{2} f^{\mathrm{A}}_{r0} f^{\mathrm{B}}_{s0}(E^{\mathrm{A}}_r + E^{\mathrm{B}}_s - E^{\mathrm{A}}_0 - E^{\mathrm{B}}_0)^{-1}(E^{\mathrm{A}}_r - E^{\mathrm{A}}_0)^{-1}(E^{\mathrm{B}}_s - E^{\mathrm{B}}_0)^{-1}. \tag{7.36}$$

We therefore have a direct relationship between the R^{-6} component of the dispersion energy and experimental data. In most cases oscillator strengths are known for only a few of the low energy transitions, and although the high energy transitions may individually contribute very little to the total there are a large number of them. In particular, the contribution from the continuum states has to be included. However, as Dalgarno and his co-workers have shown (see Dalgarno 1967), accurate estimates of C_6 can be made if the distribution of f over excited states is restricted by consideration of the general summations

$$S(k) = \sum_s{}' f_{s0}(E_s - E_0)^{-k}. \quad (7.37)$$

For example, the well-known sum rule for oscillator strengths gives S(O) as the number of electrons in the molecule, and $S(1)$ and $S(2)$ are likewise related to measurable quantities. Thus although the individual oscillator strengths are not known very accurately their overall contribution to C_6 may be calculated fairly reliably. Dalgarno and co-workers have obtained coefficient for a wide range of interacting atoms and spherical molecules which are thought to be accurate to within 10 per cent.

For a direct quantum mechanical calculation of the dispersion energy eqn (7.31) is not very useful, as it involves a sum over an infinite number of states. A better approach is to use the Hylleraas variation principle and in particular to make use of the result that the finite sum

$$\tilde{E}^{(2)} = \sum_j{}' U_{qj}U_{jq}/(\tilde{E}_j^{(0)} - E_q^{(0)}), \quad (7.38)$$

is greater than or equal to the second-order energy provided that the set of states $\tilde{\psi}_j^{(0)}$ are orthogonal to each other and to the zeroth-order function $\psi_q^{(0)}$ and give a diagonal matrix of H_0

$$\langle\tilde{\psi}_j^{(0)}|H^0|\psi_k^{(0)}\rangle = \tilde{E}_k^{(0)}\delta_{jk}. \quad (7.39)$$

It must be emphasized that $\tilde{\psi}_j^{(0)}$ are not in general eigenstates of H^0, for if they are the convergence of the sum (7.38) to the exact second-order energy is slow.

It has been shown that very good results for the dispersion energy can be obtained from expression (7.38) using a small well-chosen set of functions (Murrell and Shaw 1968*b*). For example, if for two ground-state hydrogen atoms one considers only the states $\tilde{\psi}^{(0)}$ which are built up from 2*p* orbitals on both atoms and one optimizes (7.37) with respect to the orbital exponent ζ of the 2p orbital, then the calculated C_6 coefficient of the dispersion energy is (with $\zeta = 0{\cdot}86612$) 6·4457 a.u. This is within 1 per cent of the exact value, 6·4990. Note that a hydrogenic 2p orbital ($\zeta = 0{\cdot}5$) gives only $C_6 = 2{\cdot}46$.

The calculations we have so far described for dispersion energies have involved both molecules simultaneously. It is possible, however, to transform the calculation into two one-molecule problems, this having the advantage that once the A–A and B–B interactions are solved it is a trivial matter to do A–B.

The single-centre method makes use of the identity

$$\frac{1}{a+b} = \frac{2}{\pi}\int_0^\infty \frac{ab}{(a^2+u^2)(b^2+u^2)}\,\mathrm{d}u, \qquad (a, b > 0) \quad (7.40)$$

so that it is possible to write the energy denominator that appears in expression (7.34) as

$$\frac{1}{(E_r^A - E_0^A) + (E_s^B - E_0^B)} = \frac{2}{\pi} \int_0^\infty \frac{(E_r^A - E_0^A)(E_s^B - E_0^B)}{\{(E_r^A - E_0^A)^2 + u^2)((E_s^B - E_0^B)^2 + u^2\}} \, du. \quad (7.41)$$

This enables us to split up the double summation that occurs in eqn (7.34) into the product of two single summations as follows:

$$C_6 = \frac{4}{3\pi} \int_0^\infty \left\{ \sum_r{}' \frac{(\mu_{0r}^A)^2 (E_r^A - E_0^A)}{(E_r^A - E_0^A)^2 + u^2} \right\} \left\{ \sum_s{}' \frac{(\mu_{0s}^B)^2 (E_s^B - E_0^B)}{(E_s^B - E_0^B)^2 + u^2} \right\} du. \quad (7.42)$$

This can be written in the form

$$C_6 = \frac{3}{\pi} \int_0^\infty \alpha^A(iu) \alpha^B(iu) \, du, \quad (7.43)$$

where

$$\alpha^A(\omega) = 2 \sum_r{}' \frac{(E_r^A - E_0^A)(\mu_{0r}^A)^2}{(E_r^A - E_0^A)^2 - \omega^2} \quad (7.44)$$

is an expression for the frequency dependent dipolar polarizability. It is beyond the scope of this chapter to give a formal derivation of eqn (7.44) but the analogy with the static polarizability (7.53) can be seen if one puts $\omega = 0$.

The function $\alpha(iu)$ occurring in eqn (7.42) is a polarizability at imaginary frequencies (i.e. in an exponentially rising electric field). Although this is not a directly observable quantity it can be calculated by variation methods in just the same way as the polarizability at real frequencies. The important thing about eqn (7.43) is that it is one example of a very general type of formula which has been used to cover dispersion forces between molecules contained in dielectrics, between molecules and macroscopic bodies, and between molecules at very large distances where the so-called retardation effects of quantum electrodynamics give rise to an R^{-7} energy. These topics are, however, substantial in themselves and will not be covered here. They have been reviewed by Linder (1967).

Before leaving dispersion energies we return to the question of the multipole expansion (7.32). We shall see later that eqn (7.31) can be taken as a definition of dispersion energy even if there is significant overlap between the ground states of the two systems. To make a valid R^{-n} expansion on an individual term in eqn (7.31) would require that there is no overlap between the transition densities ρ_{0r}^A and ρ_{0s}^B. If there is strictly no overlap between the ground states then there is strictly no overlap between these transition densities which are zero at any point in space where the ground-state densities are zero. However, because of the diffuse nature of electron-wavefunctions and because excited states in general become more diffuse as their energies increase, we would have to say that if there is some overlap in the ground state then there will be more overlap between the transition densities, and this will in general increase as the excitation energies increase, and perhaps become most important for the continuum states. In short it is very difficult to decide if the multipole expansion is convergent, not only for an individual term in eqn (7.31) but for the overall dispersion energy.

It is impossible to do analytic calculations on dispersion energies between atoms or molecules and establish convergence conditions. However Brooks (1952) has established that the dispersion energy between two harmonic oscillators is a divergent series in R^{-n} for any finite value of R. Given that harmonic oscillator functions behave at large distances as $\exp(-kp^2)$ whereas electronic wavefunctions behave as $\exp(-kr)$, it is therefore likely that the multipole expansion of the dispersion energy between atoms and molecules is also a divergent series. This does not mean, however, that the multipole expansion is valueless. In practice it would appear that at distances a little larger than the van der Waals separation, the individual R^{-n} terms become very small after the first few members of the series and we get an accurate result by terminating the series at that point. We hopefully ignore the fact that at some large value of n the terms may start to diverge. We will return to the effect of overlap on dispersion energies later in this article.

7.4.2. *The induction energy*

We now consider the second-order terms associated with excited states like $A_r(i)B_0(j)$. These involve the matrix elements

$$U_{00,r0} = \langle A_0(i)B_0(j)|U|A_r(i)B_0(j)\rangle. \tag{7.45}$$

Following the procedure used to simplify eqn (7.26) we note that the terms in U which represent the repulsion of the nuclei, and the interaction between electrons j and the nuclei of A will make no contribution. The reduction of the many-electron integrals then follows the same procedure as in eqn (7.26)–(7.28) to give

$$U_{00,r0} = \int \rho^{\mathrm{A}}_{r0}(1)\rho^{\mathrm{B}}_{00}(2)r_{12}^{-1}\,\mathrm{d}v_1\,\mathrm{d}v_2 + \int \rho^{\mathrm{A}}_{r0}(1)\left(-\sum_b Z_b r_{1b}^{-1}\right)\mathrm{d}v_1. \tag{7.46}$$

The first integral represents the interaction of the transition density ρ^{A}_{r0} with the electrons of B, and the second represents the interaction with the nuclei of B, overall therefore, we have the interaction between the transition density of A and the electrostatic field of B.

We can again make a multipole expansion of the matrix element (7.46). The first term will be the interaction between the transition dipole moment and the field of B,

$$\mathbf{F}^{\mathrm{B}} = -\mathrm{grad}\, V^{\mathrm{B}}, \tag{7.47}$$

the second will be the quadrupole moment with the field gradient, and so on. If we retain only the dipolar term, eqn (7.46) can be written

$$U_{00,r0} = -\boldsymbol{\mu}^{\mathrm{A}}_{r0} \cdot \mathbf{F}^{\mathrm{B}}. \tag{7.48}$$

Suppose now we consider the wavefunction of molecule A when it is perturbed by the field $\mathbf{F}^{\mathrm{B}}$. To first order we have from eqns (7.8) and (7.47)

$$A_0^{(1)} = A_0 + \sum_r{}' \left(\frac{\boldsymbol{\mu}^{\mathrm{A}}_{r0} \cdot \mathbf{F}^{\mathrm{B}}}{E^{\mathrm{A}}_r - E^{\mathrm{A}}_0}\right) A_r. \tag{7.49}$$

The dipole moment induced by this field is

$$\boldsymbol{\mu}^{\mathrm{A}}_{\mathrm{ind}} = \langle A_0^{(1)} | \boldsymbol{\mu}^{\mathrm{A}} | A_0^{(1)} \rangle - \langle A_0 | \boldsymbol{\mu}^{\mathrm{A}} | A_0 \rangle \tag{7.50}$$

which to first order is

$$\boldsymbol{\mu}^{\mathrm{A}}_{\mathrm{ind}} = 2 \sum_r{}' \frac{(\boldsymbol{\mu}^{\mathrm{A}}_{r0} \cdot \mathbf{F}^{\mathrm{B}}) \mu_{r0}}{E^{\mathrm{A}}_r - E^{\mathrm{A}}_0}. \tag{7.51}$$

The polarizability of a molecule, α, is defined by

$$\boldsymbol{\mu}^{\mathrm{A}}_{\mathrm{ind}} = \boldsymbol{\alpha}^{\mathrm{A}} \cdot \mathbf{F}^{\mathrm{B}} \tag{7.52}$$

where $\boldsymbol{\alpha}^{\mathrm{A}}$ is a second rank tensor. It can be seen from eqn (7.51) that the components of $\boldsymbol{\alpha}$ are given by

$$\alpha^{\mathrm{A}}_{\alpha\beta} = 2 \sum_r \frac{\mu^{\mathrm{A}}_{r0,\alpha} \mu^{\mathrm{A}}_{r0,\beta}}{E^{\mathrm{A}}_r - E^{\mathrm{A}}_0}. \tag{7.53}$$

In the dipole approximation therefore, the induction energy is related to the polarizabilities of the two molecules by

$$E_{\mathrm{ind}} = -\tfrac{1}{2}[\alpha^{\mathrm{A}}_{\alpha\beta} F^{\mathrm{B}}_{\alpha} F^{\mathrm{B}}_{\beta} + \alpha^{\mathrm{B}}_{\alpha\beta} F^{\mathrm{A}}_{\alpha} F^{\mathrm{A}}_{\beta}), \tag{7.54}$$

where we use the tensor notation of implied summation over the three components of the vector field $\mathbf{F}$.

The field outside a dipolar molecule varies as R^{-3} hence from expression (7.48) the leading term in the induction energy varies as R^{-6}, which is the same form as the leading term of the dispersion energy. The average squared field outside a dipole is $2\mu^2 R^{-6}$ hence the average induction energy is in this approximation (and for both molecules)

$$E_{\mathrm{ind}} = -\{\alpha^{\mathrm{A}}(\mu^{\mathrm{B}})^2 + \alpha^{\mathrm{B}}(\mu^{\mathrm{A}})^2\} R^{-6}. \tag{7.55}$$

The same problems of convergence occur for the multipole expansion of the induction energy as for the dispersion energy. Brooks (1952) calculated the induction energy of a hydrogen atom in the field of a proton, and the multipole expansion was found to be divergent at all finite values of R: but as for the dispersion energy the multipole expansion will no doubt be extensively used because if the series is terminated at the smallest term good approximations to the total energy will usually be found.

It is perhaps useful at this point to make a comparison between the magnitudes of the induction and dispersion energies that we may typically find. In order to do this we first get a rough relationship between the dispersion energy and the static polarizability of a molecule.

If we replace the energy denominator in (7.34) by a sum of average excitation energies for the two molecules

$$E_{rs}^{(0)} - E_{00}^{(0)} = E_r^{\mathrm{A}} - E_0^{\mathrm{A}} + E_s^{\mathrm{B}} - E_0^{\mathrm{B}} = \Delta E^{\mathrm{A}} + \Delta E^{\mathrm{B}} \tag{7.56}$$

then the leading term of the dispersion energy (for the spherical average case) is

$$E_{\mathrm{dis}} \sim -\tfrac{2}{3} R^{-6} (\Delta E^{\mathrm{A}} + \Delta E^{\mathrm{B}})^{-1} \Big\{ \sum_r{}' (\mu_{r0}^{\mathrm{A}})^2 \Big\} \Big\{ \sum_s{}' (\mu_{s0}^{\mathrm{B}})^2 \Big\}. \tag{7.57}$$

Making a similar approximation in eqn (7.53) gives the polarizability in the form

$$\alpha_{\alpha\beta}^{\mathrm{A}} \sim 2(\Delta E^{\mathrm{A}})^{-1} \sum_r{}' \mu_{r0,\alpha}^{\mathrm{A}} \mu_{r0,\beta}^{\mathrm{A}}, \tag{7.58}$$

which in the spherical average is

$$\alpha^{\mathrm{A}} \sim \tfrac{2}{3} (\Delta E^{\mathrm{A}})^{-1} \sum_r{}' (\mu_{r0}^{\mathrm{A}})^2. \tag{7.59}$$

Substituting eqn (7.58) into (7.57) gives the well-known London formula for the dispersion energy

$$E_{\mathrm{dis}} \sim -\frac{3}{2} \left(\frac{\Delta E^{\mathrm{A}} \Delta E^{\mathrm{B}}}{\Delta E^{\mathrm{A}} + \Delta E^{\mathrm{B}}} \right) \alpha^{\mathrm{A}} \alpha^{\mathrm{B}}. \tag{7.60}$$

ΔE is usually taken to be somewhere between the first excitation energy and the first ionization potential of the molecule. If the first excitation energy is chosen, then it can be shown that eqn (7.60) gives a lower bound to the C_6 coefficient. For two hydrogen atoms the exact value of the dispersion energy is obtained by taking ΔE as five-sixths of the ionization potential.

It is also worth emphasizing in passing that after making the average energy approximation for the polarizability we can reduce the summation in eqn (7.58) to the form

$$\alpha_{\alpha\beta}^{\mathrm{A}} = 2(\Delta E^{\mathrm{A}})^{-1} \langle A_0 | \mu_\alpha \mu_\beta | A_0 \rangle. \tag{7.61}$$

Thus the components of the polarizability are related to the mean square radii ($\boldsymbol{\mu} = e \sum_i \mathbf{r}_i$) of the electron density in the ground state of the molecule.

If we combine eqns (7.55) and (7.60) we get

$$\frac{E_{\text{dis}}}{E_{\text{ind}}} \sim \frac{3}{2}\left(\frac{\Delta E^{\text{A}}\Delta E^{\text{B}}}{\Delta E^{\text{A}}+\Delta E^{\text{B}}}\right)\frac{\alpha^{\text{A}}\alpha^{\text{B}}}{(\mu^{\text{B}})^2\alpha^{\text{A}}+(\mu^{\text{A}})^2\alpha^{\text{B}}}, \tag{7.62}$$

which for two like molecules is

$$\frac{E_{\text{dis}}}{E_{\text{ind}}} \simeq \left(\frac{3\Delta E}{8}\right)\left(\frac{\alpha}{\mu^2}\right). \tag{7.63}$$

Suppose we take a molecule with a small dipole moment and polarizability such as NH_3($\mu = 1{\cdot}4$ D,† $\alpha = 2{\cdot}26\times10^{-30}$ cm^3) then an average excitation energy of 10 eV gives $E_{\text{dis}}/E_{\text{ind}} \simeq 7$. If the dipole moment goes up to say 4 D without much increase in polarizability, then the induction energy starts to be relatively as important as the dispersion energy. For non-polar molecules the induction energy is very much smaller than the dispersion energy.

7.5. The theory of intermolecular energies in the small overlap approximation

If the electron clouds of two systems overlap, then the wavefunctions of the pair must be completely antisymmetric to exchange of all electrons. The basis set of antisymmetrized products of eigenfunctions of the two systems would seem to be a natural starting point for a perturbation theory:

$$\Psi_{rs} = \mathscr{A}A_r(i)B_s(j). \tag{7.64}$$

If $A_r(i)$ and $B_s(j)$ are separately normalized functions and we normalize Ψ_{rs} at infinite separation of the two systems, then we can write

$$\mathscr{A} = \left(\frac{N_i!N_j!}{(N_i+N_j)!}\right)^{\frac{1}{2}}(1+P) = N^{\frac{1}{2}}(1+P) \tag{7.65}$$

where N_i and N_j are the number of electrons associated with the two systems, and

$$P = -\sum_i\sum_j P_{ij}+\sum_i\sum_{i'>i}\sum_j\sum_{j'>j} P_{ij}P_{i'j'}\ldots \tag{7.66}$$

is the operator that exchanges electrons between the two sets with appropriate signs to make the whole antisymmetric.

Suppose now that we evaluate the overlap integral between two of these functions:

$$S_{rs.tu} = \langle\Psi_{rs}|\Psi_{tu}\rangle. \tag{7.67}$$

If we make use of the equivalence of each of the N^{-1} terms in Ψ then we can

† 1 debye (D) = $3{\cdot}336\times10^{-30}$ coulomb metre (Cm).

follow the standard technique for reducing many-electron integrals and get

$$\begin{aligned} S_{rs,tu} &= \langle (1+P)A_r(i)B_s(j)|A_t(i)B_u(j)\rangle \\ &= \delta_{rt}\delta_{su} + \langle PA_r(i)B_s(j)|A_t(i)B_u(j)\rangle. \end{aligned} \tag{7.68}$$

If there is overlap between the electron clouds of the two systems then the second term in this expression is not zero. If we take two one-electron systems, for example, then

$$\begin{aligned} S_{rs,tu} &= \delta_{rt}\delta_{su} - \langle A_r(2)B_s(1)|A_t(1)B_u(2)\rangle \\ &= \delta_{rt}\delta_{su} - \langle B_s|A_t\rangle \langle A_r|B_u\rangle. \end{aligned} \tag{7.69}$$

Thus the overlap matrix is a unit matrix plus a matrix whose elements depend on the square of the overlap integrals between orbitals of the two systems.

The set of antisymmetrized functions is therefore not orthogonal, and as such they will not be eigenfunctions of the same Hamiltonian. Thus no unique zeroth-order Hamiltonian can be found for the perturbation problem if antisymmetrized functions have to be used for the basis states.

At this point it is worth making a few comments about the definition of a complete set of states in quantum mechanics. This is a set ϕ_r such that any function θ which satisfies the same boundary conditions and has the same dimensions can be written exactly in the form

$$\theta = \sum_r c_r\phi_r \tag{7.70}$$

with the c_r being uniquely determined. If a set is incomplete then a function can be found which is orthogonal to all the ϕ_r. If the set is overcomplete then the coefficients are not uniquely determined, or, put another way, members of the set can be found which are themselves exactly expandable in the other members of the set:

$$\phi_s = \sum_r{}' c_r\phi_r. \tag{7.71}$$

The eigenfunctions of a quantum mechanical Hamiltonian are a complete orthogonal set for the expansion of any eigenfunction of another Hamiltonian of the same dimensions; this is the basis of the perturbation expressions (7.8) and (7.9). It follows that the set of simple product functions which we used for the zero-overlap perturbation theory are complete for the expansion of the wavefunction at all separations. Why then should we bother to work with antisymmetrized products? One reason is that the perturbation operator U cannot develop the full wavefunction even in infinite order, starting from the zeroth-order function (7.15).

To show this we consider the following Slater determinant of spin orbitals:

$$|a_\alpha b_\beta| \equiv \sqrt{\tfrac{1}{2}}[a_\alpha(1)b_\beta(2) - b_\beta(1)a_\alpha(2)]. \tag{7.72}$$

If orbital b is expanded in the set a_r and orbital a in the set b_s, we can write the function (7.72) in terms of simple product functions as follows:

$$a_\alpha(1)b_\beta(2) - \sum_{r,s} k_{rs} a_{r\beta}(1) b_{s\alpha}(2). \tag{7.73}$$

However, as U is a spin free operator the matrix of U is factorizable into $\alpha(1)\beta(2)$ and $\beta(1)\alpha(2)$ parts. Thus if the simple product $a_\alpha(1)b_\beta(2)$ is the correct zeroth-order function there is no way that U can build up the complete antisymmetrized wavefunction even if the product functions are a complete set.†

Until very recently it has been the custom to develop the wavefunction as an expansion using the antisymmetrized products as a basis. The fact that these are not orthogonal (7.69) is inconvenient, but not a serious barrier to the successful solution of the variational problem. However, from the point of view of perturbation theory there is a serious difficulty arising from the fact that the antisymmetrized functions are an overcomplete set.

To show this we first expand one member of the set in terms of the simple-product functions (for brevity we use a single suffix s to represent the state of both systems A and B)

$$\Psi_r = \sum_s c_s \psi_s = \sum \langle \Psi_r | \psi_s \rangle \psi_s. \tag{7.74}$$

We now operate on both sides with the antisymmetrizer

$$\mathscr{A}\Psi_r = N^{\frac{1}{2}}(1+P)\Psi_r = \sum_s \langle \Psi_r | \psi_s \rangle \Psi_s. \tag{7.75}$$

However, if $(1+P)$ operates on a function that is already antisymmetric it just multiplies it by the total number of permutations of electrons, that is by N^{-1}, hence

$$N^{-\frac{1}{2}}\Psi_r = \sum_s \langle \Psi_r | \psi_s \rangle \Psi_s$$

or

$$\Psi_r = \sum{}' (\langle \Psi_r | \psi_s \rangle \Psi_s)(N^{\frac{1}{2}} - \langle \Psi_r | \psi_r \rangle)^{-1}. \tag{7.76}$$

Thus one member of the set is expandable in the other members of the set, and the set is overcomplete.

An alternative perturbation procedure is one based on a zeroth-order antisymmetrized wavefunction but with an expansion of the perturbation terms in the basis of simple product functions. We write

$$\Psi = \Psi_{00} + \psi', \tag{7.77}$$

† An alternative view is that unless the antisymmetry condition is imposed on the wavefunction the resulting solutions of the Schrödinger equation may violate the variation theorem. An article by Claverie (1971) makes this point within the framework of group theory.

where $$\psi' = \sum_{r,s}' c_{rs}\psi_{rs} \tag{7.78}$$

and solve $$(H-E)\Psi_{00}+(H-E)\psi' = 0. \tag{7.79}$$

The function ψ_{00} may be omitted from the sum in (7.78) because ψ_{00} is already part of Ψ_{00}.

As H is symmetric in the electrons ψ' must be an antisymmetric function. If, however, ψ' is divided into parts then each part need not necessarily be antisymmetric and this is the main difficulty with an expansion of this type.

We can define the first-order energy by the expectation value of the zeroth-order function with the full Hamiltonian, relative to the energy of the separate molecules.

$$\begin{aligned} E^{(1)} &= \frac{\langle\Psi_{00}|H-E^{(0)}_{00}|\Psi_{00}\rangle}{\langle\Psi_{00}|\Psi_{00}\rangle} = \frac{\langle\Psi_{00}|H-E^{(0)}_{00}|\psi_{00}\rangle}{\langle\Psi_{00}|\psi_{00}\rangle} \\ &= \frac{\langle(1+P)\psi_{00}|H^0+U-E^{(0)}_{00}|\psi_{00}\rangle}{\langle(1+P)\psi_{00}|\psi_{00}\rangle} \\ &= \frac{\langle\psi_{00}|U|\psi_{00}\rangle+\langle P\psi_{00}|U|\psi_{00}\rangle}{1+\langle P\psi_{00}|\psi_{00}\rangle}. \end{aligned} \tag{7.80}$$

For small overlap between the two molecules $\langle P\psi_{00}|\psi_{00}\rangle$ will be small, hence on expanding $(1+\langle P\psi_{00}|\psi_{00}\rangle)^{-1}$ we get (writing ψ_{00} in the form of (7.15))

$$\begin{aligned} E^{(1)} \simeq \langle A_0(i)B_0(j)|U|A_0(i)B_0(j)\rangle &+ \{\langle PA_0(i)B_0(j)|U|A_0(j)B_0(j)\rangle \\ &- \langle A_0(i)B_0(j)|U|A_0(i)B_0(j)\rangle \, . \, \langle PA_0(i)B_0(j)|A_0(i)B_0(j)\rangle\}. \end{aligned} \tag{7.81}$$

The first part of eqn (7.81) is the electrostatic energy (the first-order energy in the zero-overlap model given by eqn (7.21) and the second may be called the exchange energy. This itself consists firstly of an exchange integral as usually defined in a valence-bond terminology, and a correction to the electrostatic energy arising from re-normalization. There may be some advantages in treating these two together, as they are likely to have a similar dependence on R.

To develop the higher order terms we first multiply eqn (7.79) from the left by ψ^*_{00} and integrate, and use the hermitian property of H.

$$\langle\Psi_{00}|H-E|\psi_{00}\rangle+\langle\psi'|H-E|\psi_{00}\rangle = 0. \tag{7.82}$$

From (7.78) it follows that ψ' is orthogonal to ψ_{00}, hence from eqn (7.82) we can obtain using (7.80)

$$E' = \frac{\langle\psi'|U|\psi_{00}\rangle}{\langle\Psi_{00}|\psi_{00}\rangle}, \tag{7.83}$$

where

$$E' = E - E^{(0)}_{00} - E^{(1)}. \tag{7.84}$$

We now need to find ψ'. If we introduce (7.78) into (7.79), multiply from the left by ψ^*_{tu} and integrate, we get

$$\langle\psi_{tu}|H-E|\Psi_{00}\rangle + \sum_{r,s}{}' c_{rs}\langle\psi_{tu}|H-E|\psi_{rs}\rangle = 0. \tag{7.85}$$

This can be written

$$-c_{tu} = \frac{\langle\psi_{tu}|H-E|\Psi_{00}\rangle}{\langle\psi_{tu}|H-E|\psi_{tu}\rangle} + \sum_{r,s}{}'' c_{rs}\frac{\langle\psi_{tu}|H-E|\psi_{rs}\rangle}{\langle\psi_{tu}|H-E|\psi_{tu}\rangle}, \tag{7.86}$$

where the summation now excludes ψ_{00} and ψ_{tu}. This formula can be the basis of an iterative expansion. If we want the nth approximation to the coefficients, then

$$-c^{(n)}_{tu} = \frac{\langle\psi_{tu}|H-E|\Psi_{00}\rangle}{\langle\psi_{tu}|H-E|\psi_{tu}\rangle} + \sum_{r,s}{}'' c^{(n-1)}_{rs}\frac{\langle\psi_{tu}|H-E|\psi_{rs}\rangle}{\langle\psi_{tu}|H-E|\psi_{tu}\rangle}. \tag{7.87}$$

If the coefficients are small the convergence should be rapid. For the first approximation we can take

$$-c^{(1)}_{tu} = \frac{\langle\psi_{tu}|H-E|\Psi_{00}\rangle}{\langle\psi_{tu}|H-E|\psi_{tu}\rangle}. \tag{7.88}$$

Thus as a first approximation to E' we have, from (7.84)

$$-E' \simeq \sum_{r,s}{}' \frac{\langle\psi_{rs}|H-E|\Psi_{00}\rangle\langle\psi_{rs}|U|\psi_{00}\rangle}{\langle\psi_{rs}|H-E|\psi_{rs}\rangle\langle\Psi_{00}|\psi_{00}\rangle}. \tag{7.89}$$

After expansion this becomes

$$-E' = \sum{}' \frac{\langle\psi_{rs}|U|\psi_{00}\rangle}{E^{(0)}_{rs}-E^{(0)}_{00}}\{\langle\psi_{rs}|U|\psi_{00}\rangle + \langle P\psi_{rs}|U|\psi_{00}\rangle - (E'+E^{(1)})\langle P\psi_{rs}|\psi_{00}\rangle\}$$
$$\times\left\{1+\frac{\langle\psi_{rs}|U|\psi_{rs}\rangle}{E^{(0)}_{rs}-E^{(0)}_{00}} - \frac{(E^{(1)}+E')}{E^{(0)}_{rs}-E^{(0)}_{00}}\right\}^{-1}\{1+\langle P\psi_{00}|\psi_{00}\rangle\}^{-1}. \tag{7.90}$$

If we retain only terms which are quadratic in U and no higher than linear in P then we get

$$-E' = \sum_{r,s}{}' \frac{\langle\psi_{rs}|U|\psi_{00}\rangle}{E^{(0)}_{rs}-E^{(0)}_{00}}\{\langle\psi_{rs}|U|\psi_{00}\rangle + \langle P\psi_{rs}|U|\psi_{00}\rangle$$
$$-\langle\psi_{rs}|U|\psi_{00}\rangle\langle P\psi_{00}|\psi_{00}\rangle - \langle\psi_{00}|U|\psi_{00}\rangle\langle P\psi_{rs}|\psi_{00}\rangle\}. \tag{7.91}$$

The term which is independent of P is the polarization energy of the zero-overlap theory; the others may be called second-order-exchange or exchange-coulomb energies. Relatively little has been said about these, but we will discuss them briefly later in this chapter.

Eqn (7.91) has been obtained by an iterative perturbation scheme. Although we have picked out terms in U^2 and U^2P it does not strictly represent one of the terms in a power series in U for it is a relatively simple matter to show there is no strict order of separability for the potential U. For example, we can expand the Hamiltonian matrix elements H_{rs} and H_{sr} as follows (we again revert to a single suffix to identify a state):

$$\begin{aligned}\langle\Psi_r|H|\Psi_s\rangle &= \langle(1+P)\psi_r|H|\psi_s\rangle\\ &= \langle(1+P)\psi_r|U|\psi_s\rangle+E_s^{(0)}\langle P\psi_r|\psi_s\rangle\\ \langle\Psi_s|H|\Psi_r\rangle &= \langle(1+P)\psi_s|U|\psi_r\rangle+E_r^{(0)}\langle P\psi_s|\psi_r\rangle,\end{aligned} \tag{7.92}$$

and equating the two gives

$$\begin{aligned}(E_r^{(0)}-E_s^{(0)})\langle P\psi_r|\psi_s\rangle &= \langle P\psi_r|U|\psi_s\rangle-\langle P\psi_s|U|\psi_r\rangle\\ &= \langle\psi_r|PU-UP|\psi_s\rangle.\end{aligned} \tag{7.93}$$

As the left-hand side is not zero, this shows that the difference of two U matrix elements can be independent of U. In other words the right-hand side of the identity (7.93) cannot be considered as a linear function of a parameter U. In addition one notes that the operators P and U do not commute.

One can likewise show that P cannot strictly be used as an expansion parameter from the following manipulations. Firstly, from the quantum mechanical sum rule we have

$$\sum_r\langle P\psi_s|\psi_r\rangle\langle\psi_r|P\psi_t\rangle = \langle P\psi_s|P\psi_t\rangle. \tag{7.94}$$

The right-hand side may then be written

$$\langle(1+P)\psi_s|(1+P)\psi_t\rangle-2\langle P\psi_s|\psi_t\rangle-\delta_{st} = (N^{-1}-2)\langle P\psi_s|\psi_t\rangle+(N^{-1}-1)\delta_{st}. \tag{7.95}$$

Thus a sum of P^2 terms can be reduced to an expression in which P only occurs in a linear form. I think it is clear from these examples where the difficulty arises in formulating a unique rigorous perturbation expansion.

The iterative perturbation formula (7.91) for the pseudo second-order energy has also been obtained by other methods (an operator perturbation technique) and other formulae have been proposed which are formally equivalent but which may be less useful because of the rate of convergence of the summation. Such formulae can readily be produced by making use of identities such as (7.93) and (7.94)–(7.95). As an example we can start from eqn (7.79) and take the scalar product with Ψ_{00}. The expression for E' analogous to (7.83) is then

$$E' = \frac{\langle\psi'|H-E|\Psi_{00}\rangle}{\langle\Psi_{00}|\Psi_{00}\rangle}$$

$$= N^{\frac{1}{2}}\frac{\langle\Psi_{00}|H-E|\psi'\rangle}{\langle\Psi_{00}|\Psi_{00}\rangle}, \tag{7.96}$$

and using (7.88) we get an expression derived by several authors

$$E' = -\sum_{r,s}{}' N^{\frac{1}{2}}\frac{\langle\Psi_{00}|H-E|\psi_{rs}\rangle^2}{\langle\Psi_{00}|\psi_{00}\rangle\langle\psi_{rs}|H-E|\psi_{rs}\rangle},$$

$$= -\sum_{r,s}{}' N\frac{\langle(1+P)\psi_{00}|H-E|\psi_{rs}\rangle^2}{\langle(1+P)\psi_{00}|\psi_{00}\rangle\langle\psi_{rs}|H-E|\psi_{rs}\rangle}. \tag{7.97}$$

If one compares this with eqn (7.89) it can be seen that the terms independent of P are different by a factor of N where ($N^{-1} \geqslant 2$) and therefore do not correspond with the polarization terms of the zero-overlap theory. Thus if eqn (7.97) is used to calculate the long-range energies a large part must arise from nominally exchange energies. Musher and Amos (1967) have shown that this is so for two hydrogen atoms, with contributions from the continuum states being particularly important. For this reason eqn (7.97) does not seem to be as useful as eqn (7.89) either for a qualitative or quantiative discussion of intermolecular forces.

7.6. The exchange energy

There are two problems to the calculation of the exchange energy: firstly it requires some knowledge of the exact ground-state wavefunctions of the two molecules and secondly the integrals can be difficult to evaluate. We will look at the second problem first by considering the interaction of two hydrogen atoms. For two one-electron systems with the same spins the exchange energy is, from eqns (7.14) and (7.81), given by the following expression

$$E_{\text{ex}} = -\langle a(2)b(1)|-Z_b r_{b1}^{-1} - Z_a r_{a2}^{-1} + r_{12}^{-1} + Z_a Z_b r_{ab}^{-1}|a(1)b(2)\rangle$$
$$+S_{ab}^2\langle a(1)b(2)|-Z_b r_{b1}^{-1} - Z_a r_{a2}^{-1} + r_{12}^{-1} + Z_a Z_b r_{ab}^{-1}|a(1)b(2)\rangle. \tag{7.98}$$

The nuclear repulsion has been included in both terms although it cancels overall. If orbitals a and b are orthogonal then the exchange energy reduces to the one term

$$-\langle a(2)b(1)|r_{12}^{-1}|a(1)b(2)\rangle, \tag{7.99}$$

and this will be a negative quantity: that is there is an exchange stabilization.

One can conceive of situations involving excited states where this might be important, but it is not the typical exchange energy.

If a and b are not orthogonal then the largest terms in eqn (7.98) are usually the nuclear attraction terms such as

$$S_{ab}\langle b(1)|Z_b r_{b1}^{-1}|a(1)\rangle, \tag{7.100}$$

which are positive. It is important to realize, however, that all the terms that contribute to eqn (7.98) can have about the same value and the total exchange energy is therefore likely to be much smaller than any one of these, after cancelation.

As an example we take two ground-state hydrogen atoms and show the individual contributions to the exchange energy, and the total, in Fig. 7.2.

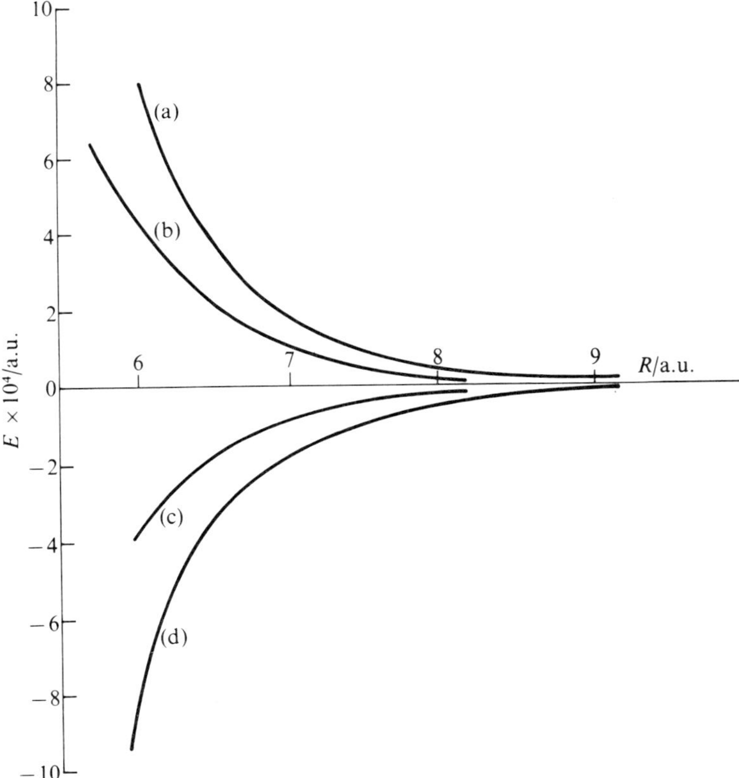

FIG. 7.2. Contributions to the exchange energy between two hydrogen atoms: (a) the electron–nuclear interaction integral (7.100); (b) the total; (c) the nuclear–nuclear repulsion S^2/R; (d) the electron–electron repulsion. The scaled coulomb energy, which is the second term in eqn (7.98) is negligible in the range of R considered.

We have covered the range $6 < R < 11$ a.u. as the van der Waals' minimum for H_2 in its lowest triplet state is approximately 8 a.u. The valence-bond exchange integral, which is the first integral in eqn (7.98), is much larger than the scaled coulomb energy for this example because the coulomb energy is small for neutral spherical systems unless there is large overlap. For polar systems this may not be the case, but in such circumstances the exchange energy as a whole may be negligible in comparison with the normal coulomb energy.

Exchange energies are troublesome to evaluate because of the two-electron integrals. For this reason any approximation which proved reasonably accurate would be of considerable value. The one suggested in the past for the two-electron case is

$$E_{\text{ex}} \sim A S_{ab}^2 R^{-1}. \tag{7.101}$$

This relationship can be rationalized on the basis of the well-known Mulliken approximation for two-centre molecular integrals in which the overlap density ab is replaced by a suitable normalized sum of one-centre densities namely $0{\cdot}5S_{ab}(a^2+b^2)$. Certainly the exchange energy and S_{ab}^2 both show a similar exponential dependence on R if overlap is small. The relationship (7.101) is quite accurate for the 1s–1s exchange energy we have just considered, with a value of $A = 1{\cdot}23$. Thus if the exchange energy is evaluated at one internuclear distance, to determine the constant A, then it is only necessary to calculate the overlap integral at other distances.

Unfortunately it would appear that this simple relationship is not sufficiently flexible in other cases. Fig. 7.3 shows the behaviour of the exchange energy between hydrogen 2s and 2p atomic orbitals. It is seen, for example, that the $2p_\sigma$–$2p_\sigma$ exchange energy actually changes sign in the vicinity of 20 a.u. Murrell and Texeira-Dias (1970) examined other approximations for the exchange energy and found that if one wished to encompass both orthogonal and non-orthogonal orbitals then the expression

$$E_{\text{ex}} \simeq (C_{ab}^0 - C_{ab})(A+BR) \tag{7.102}$$

was more successful than any overlap-dependent expression. In this C_{ab} is the two-electron coulomb integral

$$C_{ab} = \langle a(1)a(1)|r_{12}^{-1}|b(2)b(2)\rangle \tag{7.103}$$

and C_{ab}^0 is that part of this which does not have an exponential dependence on R. In other words $C^0 - C$ is the overlap dependent part of the two-electron coulomb integral, which for the case of two hydrogen 1s orbitals is given by the following expression

$$C^0 - C = (R^{-1} + \tfrac{11}{8} + \tfrac{3}{4}R + \tfrac{1}{6}R^2)\exp(-2R). \tag{7.104}$$

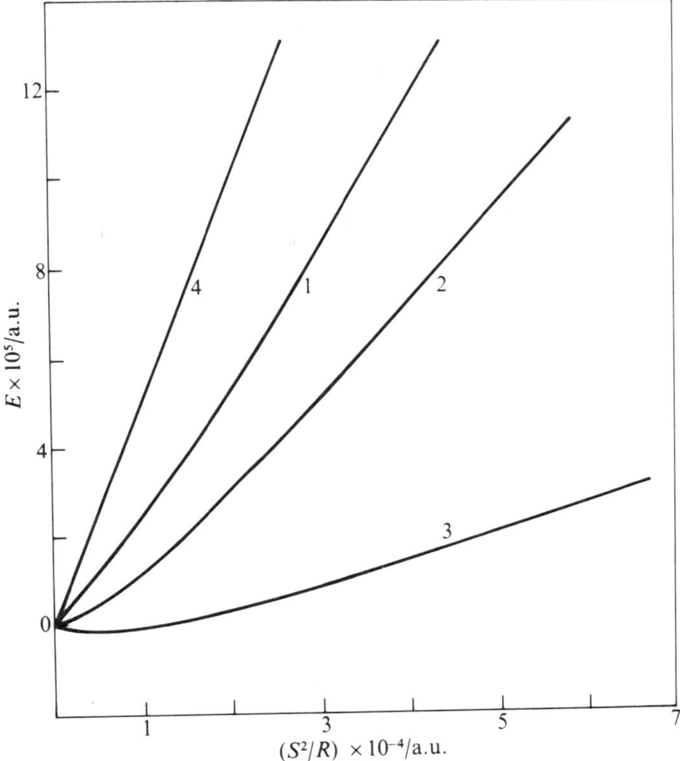

FIG. 7.3. The relationship between exchange energy and S^2/R for hydrogenic orbitals: 1. 2s–2s; 2. 2s–$2p_\sigma$; 3. $2p_\sigma$–$2p_\sigma$; 4. $2p_\pi$–$2p_\pi$.

Fig. 7.4 shows the relationship between exchange energy and $C^0 - C$ for the cases already referred to and for 2s–$2p_\pi$ and $2p_\sigma$–$2p_\pi$ exchange.

The integrals making up the exchange energy between two one-electron systems eqn (7.98) (with the same electron spin) may be collected together in the form

$$-E_{\text{ex}} = \int\int \rho_{ab}^{\text{A}}(1) r_{12}^{-1} \rho_{ab}^{\text{B}}(2)\,\mathrm{d}v_1\,\mathrm{d}v_2 + S_{ab}\int \rho_{ab}^{\text{A}}(1) V^{\text{B}}(1)\,\mathrm{d}v_1$$

$$+ S_{ab}\int \rho_{ab}^{\text{B}}(2) V^{\text{A}}(2)\,\mathrm{d}v_2. \tag{7.105}$$

We have here introduced the one-electron densities

$$\rho_{ab}^{\text{A}} = ab - S_{ab}a^2$$

$$\rho_{ab}^{\text{B}} = ab - S_{ab}b^2 \tag{7.106}$$

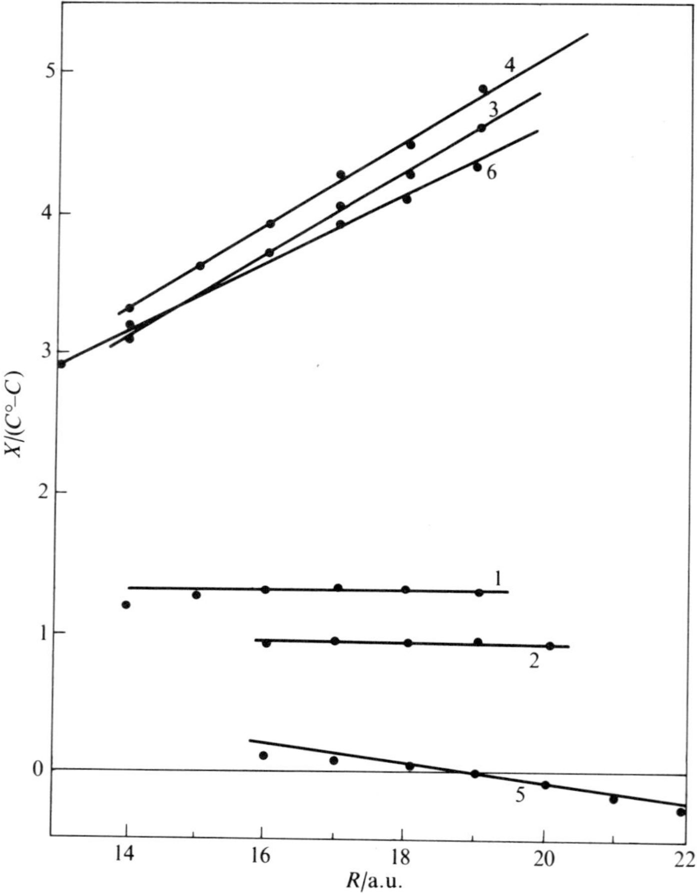

FIG. 7.4. An examination of the relationship between exchange energy and the penetration part of the two-electron coulomb integral (eqn 7.102): 1. 2s–2s; 2. 2s–$2p_\sigma$; 3. $2p_\sigma$–$2p_\pi$; 4. 2s–$2p_\pi$; 5. $2p_\sigma$–$2p_\sigma$; 6. $2p_\pi$–$2p_\pi$.

which may be called overlap transition densities in so far as they satisfy eqn (7·30) in the sense

$$\int \rho^{A}_{ab}(1)\,dv_1 = \int \rho^{B}_{ab}(2)\,dv_2 = 0. \tag{7.107}$$

V^A and V^B are the potentials of the separate A and B systems given by

$$V^{B}(1) = -Z_b r_{b1}^{-1} + \int b^2(2) r_{12}^{-1}\,dv_2$$

$$V^{A}(2) = -Z_a r_{a2}^{-1} + \int a^2(1) r_{12}^{-1}\,dv_1. \tag{7.108}$$

The reason for introducing the exchange energy in this form is that it can now be generalized fairly easily to many-electron systems. If we take the ground-state wavefunctions of the two molecules as normalized Slater determinants of spin-orbitals

$$A_0 = |a_1 \dots a_m|$$
$$B_0 = |b_1 \dots b_n| \tag{7.109}$$

then the exchange energy is

$$-E_{\mathrm{ex}} = \sum_{i=1}^{m}\sum_{j=1}^{n}\left\{\iint \rho_{ij}^{\mathrm{A}}(1)r_{12}^{-1}\rho_{ij}^{\mathrm{B}}(2)\,\mathrm{d}\tau_1\,\mathrm{d}\tau_2 + S_{ij}\int \rho_{ij}^{\mathrm{A}}(1)V^{\mathrm{B}}(1)\,\mathrm{d}\tau_1 + S_{ij}\int \rho_{ij}^{\mathrm{B}}(2)V^{\mathrm{A}}(2)\,\mathrm{d}\tau_2\right\} \tag{7.110}$$

where the integration is over the space and spin coordinates of the electrons, and where the densities are defined by

$$\rho_{ij}^{\mathrm{A}} = a_i b_j - \sum_{k=1}^{m} S_{kj} a_i a_k$$
$$\rho_{ij}^{\mathrm{B}} = a_i b_j - \sum_{l=1}^{n} S_{il} b_j b_l. \tag{7.111}$$

These definitions are consistent with (7.106) for the one-electron systems.

Thus the total exchange energy has one term for each possible exchange between an orbital of A and one of B. Now a determinantal function is invariant to a unitary transformation of the orbitals, and hence the exchange energy must be also. We can use this property in such a way that in most cases just one term of the double summation in eqn (7.110) dominates. If we make use of the fact that each term in the exchange energy is roughly proportional to the square of the relevant overlap integral than it may be sufficient to choose two orbitals

$$a'_r = \sum_k c_{kr} a_k$$
$$b'_s = \sum_l c_{ls} b_s \tag{7.112}$$

such that $(S'_{rs})^2$ is a maximum,† and to calculate just the contribution to the exchange energy from this pair.

† By making use of the relationship

$$\sum_i \sum_j S_{ij}^2 = tr\mathbf{S}\mathbf{S}^{\dagger}$$

we can easily see that a unitary transformation of the two sets of orbitals ($\mathbf{a} = \mathbf{U}\mathbf{a}'$, $\mathbf{b} = \mathbf{U}\mathbf{b}'$) leaves this double summation unchanged.

The other problem we have referred to is the fact that to calculate the exchange energy one should strictly know the exact wavefunctions of the interacting systems. Eqn (7.80) has been derived on the basis that $A_0(i)B_0(j)$ is an eigenfunction for the separated systems. If we cannot make this assumption then we have to add to $E^{(1)}$ the quantity

$$\Delta = \frac{\langle P\psi_{00}|H^0|\psi_{00}\rangle - E^{(0)}_{00}\langle P\psi_{00}|\psi_{00}\rangle}{\langle(1+P)\psi_{00}|\psi_{00}\rangle} \tag{7.113}$$

which one might refer to as the zeroth-order perturbation.

Table 7.2 shows values of Δ and $E^{(1)}$ (given by eqn (7.80)) calculated for the He_2 system for different SCF wavefunctions of He. These all have orbitals constructed as linear combinations of Slater-type functions ($r^n \exp(-kr)$) and are respectively the best one-term, two-term, and five-term functions previously reported in the literature.

TABLE 7.2
He_2 *energies calculated from the best n-term (Slater) expansions of the SCF orbitals of* He *at a separation of* 5·5 *a.u.* (*Murrell and Conway*, 1972). *The functions with n* = 2 *and* 5 *are those given by Clementi* (1965).

n	1	2	5
$E^{(1)}$	0·634	3·538	3·927
Δ	1·361	0·332	0·003
Total	1·995	3·870	$3·930 \times 10^{-5}$ a.u.

It is seen that for the one-term function, which is a very poor approximation to the Hartree–Fock orbital, the quantity Δ is larger than $E^{(1)}$. For the two-term function however Δ is down to about 10 per cent of $E^{(1)}$ at the van der Waals distance but increases in relative importance as R increases. On going to the four-term function there is a further decrease in the importance of Δ. One notes further that the difference in $E^{(1)}$ between the two- and five-term functions is comparable to Δ for the two-term function. We conclude that Δ will be small for a wavefunction based on Hartree–Fock orbitals, but that minimum-basis SCF orbitals would generally be quite unreliable for calculating exchange energies.

The next question to resolve is whether exchange energy calculated from Hartree–Fock wavefunctions of the two systems is approximately equal to the exchange energy between exact wavefunctions. To study this we have examined the exchange energy for a multi-configuration wavefunction of

He. This function was constructed from the natural orbitals which were themselves obtained from the Weiss 35-term He function and which gave 85 per cent of the correlation energy. Table 3 shows $E^{(1)}$ and Δ for a single s^2 configuration and the effect of adding p^2 and $(s')^2$ configurations. Note that the single s^2 configuration is much poorer than the near Hartree–Fock function investigated above, as measured by the appropriate value of Δ.

TABLE 7.3

Effect of intra-atomic correlation on the first-order energy of He_2 (R = 5·5 *a.u.*). *The results correspond to the configuration interaction expansion in terms of the first three natural orbitals of* He(s, s′, p). (*Murrell and Conway*, 1972.)

	s^2	$s^2+s'^2$	s^2+p^2	$s^2+s'^2+p^2$
$E^{(1)}$	3·594	3·527	3·717	3·647
Δ	0·305	0·924	−0·058	0·553
Total	3·899	4·451	3·659	4·200 × 10^{-5} a.u.

It is seen that the $(p)^2$ configuration (left–right correlation) changes the energy by about 6 per cent and the $(s')^2$ configuration (in–out correlation) by 14 per cent. However, these changes are in opposite directions so that there is some cancellation overall. We conclude that the first-order energy will be reasonably accurate when calculated from a near Hartree–Fock function.

7.7. The effect of overlap on polarization energies

We have seen that part of the 'second-order' energy E' defined by eqn (7.91), consists of the polarization energies obtained in the zero-overlap model. We can therefore take eqns (7.31) and (7.47) to define dispersion and induction energies respectively even when exchange contributions to the energy are important. The factor that we need to worry about is the accuracy of the multipole expansions to these energies under such circumstances.

Although the R^{-n} expansions of the long-range energy are valuable because they show the relationship between the energy and observable properties of the interacting systems such as polarizability it is clear that these expansions diverge even from the leading term at small R. If for example the dispersion energy is calculated from C_6R^{-6} (eqn 7.32) then one may get an overestimate even at van der Waals' separations. One can show this by a relatively simple calculation for two hydrogen atoms.

If one calculates the contribution to the dispersion energy from all states in which both hydrogen atoms are in p-states, then from the fact that the transition densities ρ_{sp} transform like the spherical harmonics Y_{lm}, it follows that these states will contribute only to the C_6 term in the multipole expansion. We have already shown in Section 7.4 that one can calculate this term variationally with considerable accuracy, and such a calculation can be done in both the zero-overlap and overlap regions. One can compare this contribution to the dispersion energy with its R^{-6} approximation, and this is done in Table 7.4. We compare both the exact and approximate energies and also an effective C_6 defined such that $\tilde{C}_6 R^{-6}$ equals the exact energy. It is seen that at the van der Waals separation ($\sim$8 a.u.) $C_6 R^{-6}$ overestimates the exact contribution of these terms to the dispersion energy by approximately 4 per cent.

TABLE 7.4

Comparison of C_6R^{-6} *energy with a variational calculation of the* p–p *dispersion energy for two hydrogen atoms.* $\tilde{C}_6$ *is the effective* C_6 *at different values of* R. *The exact value of* C_6 *is* 6·499. (*Murrell and Shaw* (1968).)

R(a.u.)	C_6R^{-6}	Variational energy	$\tilde{C}_6$
4	0·001586	0·000571	2·340
5	0·000416	0·000262	4·089
6	0·000139	0·000114	5·349
7	0·000055	0·000051	6·015
8	0·000025	0·000024	6·296
9	0·000012	0·000012	6·398
∞	0	0	6·446 (6.499)

Kreek and Meath (1969) have made a similar analysis of the overlap dependence on induction energies for the H–H$^+$ system. They find for example that at $R = 5$ a.u. an expansion of the induction energy to order R^{-10} differs from the exact value by 6 per cent.

7.8. Exchange-polarization energy and charge-transfer energy

At the risk of being over-optimistic I would say that higher-order exchange energies are generally less important than the other energies we have so far discussed. However, there are indications that some phenomena are predominantly due to these terms, so they cannot be ignored. In particular the difference in energy between the hexagonal and cubic close packed structures of the inert gases (see Jansen and Lombardi 1965) and the stabilities of donor–acceptor complexes (see Mulliken and Person 1968) have been attributed to these energies.

Considering two one-electron systems with the antisymmetrized ground state

$$\Psi_0 = \mathscr{A}a(1)b(2),$$

then an excited state

$$\psi' = a'(1)b'(2),$$

gives a contribution to the exchange-dependent parts of E' (eqn 7.91) of the following form

$$\frac{-\langle a'(1)b'(2)|U|a(1)b(2)\rangle}{E'-E_0}\Big\{\langle b'(1)a'(2)|U|a(1)b(2)\rangle$$

$$-S_{ab}^2\langle a'(1)b'(2)|U|a(1)b(2)\rangle - S_{ab'}S_{a'b}\langle a(1)b(2)|U|a(1)b(2)\rangle\Big\}. \quad (7.114)$$

This may be called exchange-polarization or second-order exchange energy. For H–H interactions at least, terms of this type have been found to be not very large near the van der Waals' minimum (Murrell and Shaw 1968*a*; Murrell and Texeira-Dias 1971). Another way of looking at this is that from accurate calculations on H_2 in the region 4–8 a.u. it can be shown that the first-order exchange energy accounts for 90 per cent of the difference between the ${}^1\Sigma_g^+$ and ${}^3\Sigma_u^+$ ground states (Certain, Hirschfelder, Kolos, and Wolniewicz 1968). Thus the higher-order terms associated with antisymmetrization only give 10 per cent of this difference.

One would expect second-order exchange energies to be large in a situation where normal chemical bond behaviour is appearing—this would tie in with the view that the chemical bond is associated with a stabilizing exchange energy. It may well be that the type of perturbation theory we have described is not very suitable for describing the energy in these circumstances because the expansion is very slowly convergent. It is for this reason that the usual approach is to try a variational wavefunction of a different form.

If there is no overlap between the orbitals of two molecules then an electron can be said to lie on one molecule or the other. If there is overlap, then the electron may be shared between the two, so that as two neutral molecules come together there may be a partial transfer of charge between the two. In very general terms we can represent the situation by a wavefunction of the type

$$\Psi(\mathrm{A,B}) + \lambda\Psi(\mathrm{A}^-, \mathrm{B}^+), \quad (7.115)$$

where the transfer of charge depends on the magnitude of λ.

We can easily show that the introduction of the ionic term in the wavefunction leads to a stabilization of the system which is roughly second-order in overlap between orbitals of the two molecules.

To illustrate this we consider two one-electron systems and write the wavefunctions

$$\Psi(\mathrm{A, B}) = \mathscr{A}a(1)b(2) \tag{7.116}$$

and

$$\Psi(\mathrm{A}^-, \mathrm{B}^+) = \mathscr{A}a(1)a'(2) \tag{7.117}$$

where we will assume orbitals a and a' to be orthogonal.

To arrive at the stabilization energy it is probably easiest to examine the interaction between $\Psi(\mathrm{A, B})$ and the orthogonalized function

$$\Psi'(\mathrm{A}^-, \mathrm{B}^+) = \Psi(\mathrm{A}^-, \mathrm{B}^+) - \langle\Psi(\mathrm{A, B})|\Psi(\mathrm{A}^+, \mathrm{B}^-)\rangle\Psi(\mathrm{A, B}). \tag{7.118}$$

If we split up the Hamiltonian as in eqn (7.13) we have

$$\begin{aligned}\langle\Psi'(\mathrm{A}^-, \mathrm{B}^+)|H|\Psi(\mathrm{A, B})\rangle \\ &= \langle(1+P)a(1)a'(2)|U|a(1)b(2)\rangle - S_{a'b}\langle(1+P)a(1)b(2)|U|a(1)b(2)\rangle \\ &= \int(a'b - S_{a'b}b^2)V^{\mathrm{A}}\,\mathrm{d}v - [\langle a'(1)a(2)|U|a(1)b(2)\rangle \\ &\quad - S_{a'b}\langle b(1)a(2)|U|a(1)b(2)\rangle]\end{aligned} \tag{7.119}$$

where V^{A} is the potential of the acceptor. The first term arises generally for matrix elements of this type if there is transfer of an electron from an orbital b of the donor to a' of the acceptor. The terms in square brackets arise from the P operator and are probably smaller—in general their exact form depends on the detailed wavefunctions of the two molecules.

If we confine our attention to the first term in (7.119) we can see that it is first-order in the overlap of orbitals of the two molecules, using order here in an inexact sense. The stabilization of the ground state arising from such a variation wavefunction will be proportional to the square of the Hamiltonian matrix element, that is of second-order in overlap. We therefore have a formally similar dependence on overlap to the exchange-polarization energy although it must be emphasized again that strict orders of perturbation cannot be based on overlap.

The other important point about this overlap is that it involves an orbital a' of the acceptor which may well be more diffuse than the ground-state orbitals a, so that the overlap integral $S_{a'b}$ may be much larger than S_{ab} which, as we have seen, roughly determines the magnitude of the exchange repulsion. On the other hand for the matrix element (7.119) to be large it is necessary not only for there to be a large overlap of a' and b but that the overlap density $a'b$ shall penetrate to a region in which the potential field of A is large. For neutral spherical systems the latter condition seems to be met only for rather close distances. In most cases it appears that the first-order exchange energy is more important than the charge-transfer energy in the region of the van der Waals' minimum.

The energy of a charge-transfer state may be related to observable properties of the molecules if we use Koopmans' approximation that the energy of an orbital can be equated to minus the ionization potential for removing an electron from that orbital. At large separations the charge-transfer state has an energy relative to the ground state of $I_b - A_{a'}$. I_b is the ionization potential for removing an electron from b, $A_{a'}$ is the electron affinity of orbital a' (or the ionization potential of A^-). Providing there is no large overlap between the orbitals of the two systems, we have only to allow for the difference in coulomb energy between the ground and charge-transfer states. If A and B and neutral spherical systems, and A^- and B^+ are charged spherical systems, the coulomb energy stabilises the charge-transfer state by approximately R^{-1}. If the orbitals a' and b do overlap appreciably then exchange energies of the charge-transfer state may have to be included.

If system B has a low ionization potential and A a high electron affinity, the charge-transfer state will have a low energy and the variational wavefunction (7.115) will give a low energy. For this reason the charge-transfer model introduced by Mulliken is thought to be a good approach to the bonding in the so-called donor–acceptor complexes; systems such as quinone–hydroquinone and iodine–benzene. However, the model has also been used successfully when the two systems are equivalent. In this case there is no net charge-transfer in the ground state of the complex, and a wavefunction of the type

$$\Psi(\mathrm{A}, \mathrm{B}) + \lambda\{\Psi(\mathrm{A}^-, \mathrm{B}^+) + \Psi(\mathrm{A}^+, \mathrm{B}^-)\} \tag{7.120}$$

must be used to represent the situation. The term charge-resonance stabilization is more appropriate in this case.

If the charge-transfer model is a good basis for evaluating second-order exchange energies then we need to know whether one can just add the charge-transfer energy to the first-order and polarization energies that are obtained from perturbation theory. The problem arises when we improve our wavefunction by adding those terms which we have already seen are responsible for induction and dispersion energy.

Mulliken and Person (1969) have defined a no-bond state by

$$\Psi(\mathrm{A}, \mathrm{B}) = \mathscr{A} A_{\mathrm{mod}}(i) B_{\mathrm{mod}}(j) \tag{7.121}$$

where the suffix mod means a wavefunction for the ground state of the system in the nuclear configuration it has in the complex and subject to exchange repulsion, dispersion, and classical electrostatic attraction forces. However, the function (7.121) is not as flexible as it might appear at first sight.

If we introduce the polarization of the two systems by

$$\begin{aligned} A_{\mathrm{mod}} &= A_0 + \lambda A' \\ B_{\mathrm{mod}} &= B_0 + \mu B' \end{aligned} \tag{7.122}$$

then on expansion the function (7.121) has the form

$$\Psi(\mathrm{A}, \mathrm{B}) = \mathscr{A} A_0 B_0 + \lambda A' B_0 + \mu \mathscr{A} A_0 B' + \lambda\mu \mathscr{A} A' B'. \qquad (7.123)$$

This function has within it all terms which are of the type discussed earlier to represent first- and second-order energies, both long-range and exchange; however it lacks the flexibility of separate optimization of the induction ($A'B_0$ and A_0B') and dispersion ($A'B'$) terms. Thus one cannot have simultaneously low-induction and high-dispersion energies, which is what is usually required for non-polar systems. The widely-held view that a wave-function such as (7.121) gives all energy terms except a charge-transfer contribution is incorrect.

For the H–H system in its lowest triplet state a variational calculation has been carried out which includes in the basis states that give almost all the induction and dispersion energies (Murrell and Texeira-Dias 1971). On adding a charge-transfer state to the basis a further lowering of the energy was obtained, but not as much as when the charge-transfer state and the unperturbed ground state alone were used in the calculation. As a measure of the non-additivity of the second-order energies a quantity P was defined by

$$E(D+I) + PE(C) = E(D+I+C) \qquad (7.124)$$

where $E(D+I)$ is the stabilization of the system through states which give induction and dispersion energies, $E(C)$ the stabilization through charge-transfer state alone and $E(D+I+C)$ when all three were used. For $P = 1$ we would have perfect additivity and for $P = 0$ it would mean that the charge-transfer states were completely expandable in terms of the dispersion and induction states, and hence contributed nothing to the energy of the system; this would be the case if a complete set of dispersion and induction states were used.

The calculation showed that, although the basis gave almost all of the dispersion and induction energies, the second-order exchange terms from this basis was relatively small, so that the charge-transfer terms gave an additional contribution for which $P \sim 0{\cdot}65$, the result being almost independent of the internuclear separation. One would therefore conclude that charge-transfer energies cannot simply be added to other energies, but that they are an efficient way of allowing for second-order exchange effects in a variational calculation.

7.9. Conclusion

I have attempted in this chapter to show that most of the terms in frequent usage in the language of intermolecular forces can be rigorously defined in the medium-to-small-overlap region and that perturbation or variation calculations can be made on the separate contributions to the energy.

However, I have also emphasized that care must be taken in adding separate contributions to the total energy, particularly if charge-transfer energy is included in the calculation.

BIBLIOGRAPHY

General references on intermolecular forces.

Advances in chemical Physics, vol. 12, 1967, Intermolecular Forces, Ed. Hirschfelder, J. O., Interscience.

Discussions of the Faraday Society No. 40, 1965.

Hirschfelder, J. O., Curtiss, C. F., and Bird, R. B., 1964, *Molecular theory of gases and Liquids*, Wiley.

Margenau, H. and Kestner, N. R., 1969, *Theory of intermolecular forces*, Pergamon Press.

REFERENCES

BROOKS, F. C. (1952). *Phys. Rev.* **86**, 92.

CERTAIN, P. R., HIRSCHFELDER, J. O., KOLOS, W., and WOLNIEWICZ, L. (1968). *J. chem. Phys.* **49**, 24.

CLAVERIE, P. (1971). *Int. J. Quantum Chem.* **5**, 273.

CLEMENTI, E. (1965). *IBM Jl. Res. Dev.* **9**, 2.

CONWAY, A. and MURRELL, J. N. (1972). *Mol. Phys.* **23**, 1143.

DALGARNO, A. (1967). *Adv. chem. Phys.* **12**, 143.

HIRSCHFELDER, J. O., BYERS-BROWN, W., and EPSTEIN, S. T. (1964). *Adv. Quantum Chem.* **1**, 255.

JANSEN, L. and LOMBARDI, E. (1965). *Discuss. Farad. Soc.* **40**, 78.

KREEK, H. and MEATH, W. J. (1969). *J. chem. Phys.* **50**, 2289.

LINDER, B. (1967). *Adv. Chem. Phys.* **12**, 225.

MULLIKEN, R. S. and PERSON, W. B. (1969). *Molecular complexes*. Wiley.

MURRELL, J. N. and SHAW, G. (1968*a*). *Theoret. Chim. Acta* **11**, 434.

—— —— (1968*b*). *J. chem. Phys.* **49**, 4731.

—— and TEIXEIRA-DIAS, J. J. C. (1970). *Mol. Phys.* **19**, 521.

—— —— (1971). *Mol. Phys.* **22**, 535.

MUSHER, J. I. and AMOS, A. T. (1967). *Phys. Rev.* **164**, 31.

8

ELECTRONIC ENERGY LEVELS OF MOLECULAR SOLIDS

D. P. CRAIG

IN 1948 Professor Coulson proposed an assignment of the two lower electronic band systems of the linear polycyclic aromatic hydrocarbons on the basis of Hückel molecular orbital theory (Coulson 1948). There had been some experimental indications, but they were inconclusive, and a direct observation (Craig and Hobbins 1955*a*) was first made by measuring the spectrum of anthracene crystals in which the molecules were held in orientations known from X-ray crystallography. It was shown that the first and second band systems were polarized in-plane, respectively along the shorter and longer molecular axes, exactly as he had deduced. This connection with an early interest of Charles Coulson's is the stimulus for this chapter, which is about electronic transitions between energy states of crystals, and the relationship of the transitions to those of free molecules.

The forces acting to maintain the ordered arrangements in molecular crystals do not cause large changes in the molecular electronic structure in the ground state. In this sense intermolecular forces are weak (cf. the discussion by Professor Murrell in the preceding chapter). In excited states, the coupling between molecules can be large; it is a resonance interaction between identical molecules, one of which is excited and the other unexcited. Knowledge of this field has grown up along two distinct lines of development, and it is perhaps an allowable exaggeration to say that the adherents to one barely understand what the others are about, even though they study the same objects. In the one, we think of free molecules having their spectra modified by incorporation into a lattice structure of identical molecules, and consider how these spectral changes promote understanding of the interactions between the molecules. Emphasis on this molecular aspect is typical of the chemist's viewpoint. The other is to see in molecular crystals a new way to study the physics of the solid state, and to concentrate upon problems of band structure, on the dynamical properties of excitons, and their coupling to other pseudo-particles such as phonons and photons of the electromagnetic field, with no particular interest in the molecular aspect. Most of the original papers fall clearly into one of these two classes, and so too does much of the text book literature (for example, Knox 1963; Craig and Walmsley 1968; Davydov 1962, 1971).

It is not unexpected to find important points of connection between the microscopic and macroscopic viewpoints. Where molecular crystals have been viewed simply as a means for gaining new insights into the structure and properties of free molecules, it has been helpful to go beyond the simplest picture of the interaction of radiation with molecular electrons by using the methods of quantum electrodynamics. It then becomes natural to investigate the consequences of the retardation of the interaction between different molecules and to look at interactions at very long range. At another extreme peripheral area of chemical application, one recalls that the constituent molecules in a crystal may be photochemically reactive and that light absorption may be the first of a series of steps leading to chemical change (Schmidt 1965; Craig and Sarti-Fantoni 1966). There are examples in the photodimerization of crystalline 9-cyanoanthracene and the photo-oxidation of anthracene itself by reaction with oxygen included in the lattice. Energy transport by the exciton mechanism from the absorption site to the reaction site is an important step, and the molecular motions from the ordered lattice toward the final configuration of the reaction products is another. Thus if its results are to be of value in a wide sense for the chemistry and physics of the molecular solid state, the study of the energy states must itself be given a wide span.

The plan of the chapter is to begin (Section 8.1) with a description of the resonance interaction of two identical atoms, as an introduction to the resonance problem in a crystal lattice. In Section 8.2 the energy levels of crystals are discussed in the simplest way, by extending the two-atom treatment to a three-dimensional lattice, with electrostatic interactions between molecules. Section 8.3 outlines the possibilities of second quantization in exciton problems. In Section 8.4 the electrostatic theory is connected with that of the excitation of dielectrics, and the difficulties of the model elaborated. In Section 8.5 the retarded interaction is introduced with the help of the Hertz vector, from the point of view of classical electrodynamics, while in Section 8.6 the exciton-photon system is described using quantum electrodynamics, and the connections brought out between this more exact description of the crystal states and the simpler models.

8.1. Resonance coupling of two atoms

The ground-state coupling of two atoms, leading to chemical bonding, is often introduced with the example of two hydrogen atoms a and b interacting at short distances. In Heitler–London form the molecular wavefunction is

$$(2+2\sqrt{S})^{-1}\{a(1)b(2)+a(2)b(1)\}\{\alpha(1)\beta(2)-\alpha(2)\beta(1)\} \tag{8.1}$$

expressed in terms of 1s atom wavefunctions $a(v)$ and $b(v)$, their overlap S, and spin functions α and β. The wavefunction is a spin singlet and has full symmetry in its space part, belonging to species ${}^1\Sigma_g^+$. We begin the discussion of excitation resonance similarly, by taking two hydrogen atoms of which one is in the ground state and the other is excited to the $2p_\sigma$ state, i.e., with zero angular momentum about the interatomic axis. The wavefunctions for the $2p_\sigma$ states are $a'(v)$ and $b'(v)$. At separation distances allowing overlap of the atomic orbitals there are four degenerate configurations, corresponding to $a'(1)b(2)$, $a(1)b'(2)$, $a'(2)b(1)$, and $a(2)b'(1)$. We describe the members of the first pair, and the members of the second pair, as degenerate to excitation interchange, while the first and third, and the second and fourth, are degenerate to electron interchange. There are four states in each one of which the molecule would dissociate into a 1s and a 2p hydrogen atom. The wavefunctions are given in expression (8.2), without normalization,

$$\left.\begin{array}{l}[a(1)b'(2)+a(2)b'(1)\pm\{a'(1)b(2)+a'(2)b(1)\}]\{\alpha(1)\beta(2)-\alpha(2)\beta(1)\}\\ [a(1)b'(2)+a(2)b'(1)\pm\{a'(1)b(2)+a'(2)b(1)\}]\{\alpha(1)\beta(2)+\alpha(2)\beta(1)\}\end{array}\right\} \quad (8.2)$$

The four, in order, belong to species ${}^1\Sigma_g^+$, ${}^1\Sigma_u^+$, ${}^3\Sigma_u^+$, ${}^3\Sigma_g^+$. Anticipating the situation in molecular crystals, we assume that the separation between the atoms is large enough to allow neglect of orbital overlap and electron exchange. In that limit we assign electrons definitely to one centre or the other, and extract from expressions (8.2) the appropriate terms. The space factors of the remaining wavefunctions are given in (8.3),

$$\frac{1}{\sqrt{2}}\{a(1)b'(2)\pm a'(1)b(2)\} \quad (8.3)$$

the two signs giving functions symmetric and antisymmetric to excitation interchange. There is no spin correlation in the absence of electron exchange, and the spin-wavefunction (8.4) describes as equally probable that the electrons have spins α and β,

$$\tfrac{1}{2}\{\alpha(1)+e^{i\phi_1}\beta(1)\}\{\alpha(2)+e^{i\phi_2}\beta(2)\} \quad (8.4)$$

where the ϕs are arbitrary phase angles. This is the spin state of a 'double doublet'. Because of the absence of spin correlation it is unnecessary to specify the spin-wavefunctions in what follows. We now find the energy of the states (8.3) from the Hamiltonian (8.5) for fixed nuclei,

$$H = H_0+H' \quad (8.5)$$

and

$$H_0 = -\frac{1}{2}\nabla_1^2-\frac{1}{2}\nabla_2^2-\frac{1}{r_{a1}}-\frac{1}{r_{b2}}; \qquad H' = -\frac{1}{r_{b1}}-\frac{1}{r_{a2}}+\frac{1}{R}+\frac{1}{r_{12}},$$

H_0 is the sum of the separated atom Hamiltonians, H' the interatomic part of the energy and R the internuclear separation. The energies of the states (8.3) consist of free atom energies and perturbation corrections in H', namely

$$W^{\pm} = 2W_0+\langle a(1)b'(2)|H'|a(1)b'(2)\rangle\pm\langle a(1)b'(2)|H'|a'(1)b(2)\rangle. \quad (8.6)$$

The first correcting term is always small, because at distances large enough for the neglect of exchange the attractions between one nucleus and the electron on the other nucleus are almost exactly balanced by the repulsion terms. The second correcting term is the coupling of transition densities at the two centres. The transition density for 1s and $2p_\sigma$ wavefunctions is dipolar in character (Fig. 8.1) with the dipole axis along **R**. Thus the final

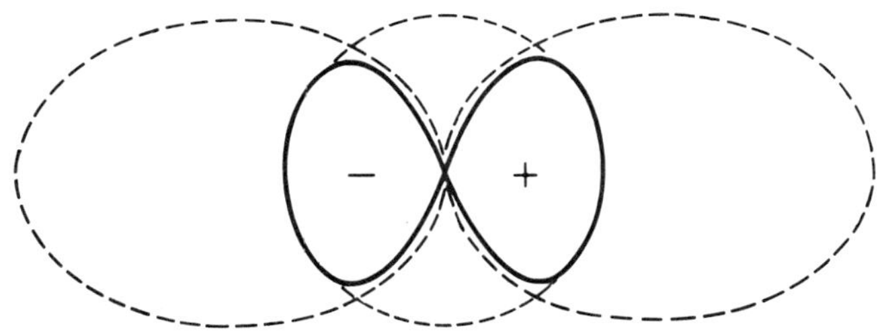

FIG. 8.1. Transition density for $2p_\sigma \leftarrow 1s$ excitation in the hydrogen atom. Dotted lines show 1s and $2p_\sigma$ orbitals. The full lines show the transition density with strongly dipolar character.

term of (8.6) is equal to the interaction energy of two dipolar charge distributions. Since the primed and unprimed functions at one centre are orthogonal, the only surviving term of H' is the electron repulsion, giving a splitting of the states (8.7),

$$2\langle a(1)b'(2)|r_{12}^{-1}|a'(1)b(2)\rangle. \qquad (8.7)$$

This matrix element can be evaluated exactly, but exact evaluation is not a possibility in practical examples in crystals, and we examine instead a common approximation. The Hamiltonian H' in (8.5) is expanded in a series of multipole–multipole interactions. The first term is the charge–charge interaction. It vanishes for neutral systems. The second is the dipole–dipole interaction which dominates in many of the most interesting examples, allowing further terms to be dropped. It is given in three equivalent forms in (8.8),

$$H' = R^{-3}\{\mathbf{r}^{(1)}\,.\,\mathbf{r}^{(2)} - 3(\mathbf{r}^{(1)}\,.\,\hat{\mathbf{R}})(\mathbf{r}^{(2)}\,.\,\hat{\mathbf{R}})\} \qquad (8.8a)$$

$$= R^{-3} r_i^{(1)} r_j^{(2)} (\delta_{ij} - 3\hat{R}_i\hat{R}_j) \qquad (8{\cdot}8b)$$

$$= R^{-3}(x_1x_2 + y_1y_2 - 2z_1z_2). \qquad (8{\cdot}8c)$$

The operators $\mathbf{r}^{(1)}$ and $\mathbf{r}^{(2)}$ are displacements of the electrons from origins at the nuclei of a and b, $\hat{\mathbf{R}}$ is a unit vector along the a–b axis. In 8.8(b) the components r_i refer to any rectangular system (i, j, k), and the convention of summation over indices repeated inside and outside the brackets applies.

In the form (8.8c) the cartesian displacements refer to an axis system with z chosen along **R**. The electronic charge is set equal to unity. Substitution of form 8.8(b), for example, in the last term of (8.6), or equally in eqn (8.7), leads to the splitting

$$\Delta W = 2R^{-3}\mu_i^{(a)}\mu_j^{(b)}(\delta_{ij} - 3\hat{R}_i\hat{R}_j) \tag{8.9}$$

where the μ_i are now components of the transition moment of the $2p_\sigma \leftarrow 1s$ transition of atomic hydrogen. In this particular case the moment lies along **R**. We note that the form of (8.9) is exactly that of the interaction of *permanent* electric dipoles with moments equal to the *transition* moments of the real problem. This feature is characteristic of excitation resonance based on the Hamiltonians of type (8.5) built on the electrostatic potentials of the charged particles.

We see now that two hydrogen atoms, one excited to an allowed state, couple at distances beyond overlap range by the dipole mechanism to give two excited states; the pair spectrum has two transitions both related to the single $2p_\sigma \leftarrow 1s$ transition of a free hydrogen. In order to get a more acceptable account of the spectrum, and not merely of the spectral splitting ΔW, we need to return to eqn (8.6) which includes a term common to the split levels, as well as one that distinguishes them. We saw that this term was small, and notice that in the dipole approximation it vanishes altogether. However in the same approximation we could not account for the dispersion interaction of two ground-state hydrogen atoms which, while small, is generally counted as chemically significant. It appears as a second-order perturbation correction to the ground state. We expect a corresponding second-order term to appear as a displacement of the averaged energy of the split levels; it is the dispersion interaction of 1s and $2p_\sigma$ hydrogens. For consistency we must also investigate the second-order term in the splitting energy.

For two ground-state atoms the calculation is familiar (Eyring, Walter, and Kimball 1944). The second-order perturbation to the ground-state energy is, expression (8.10),

$$\sum_{k\neq 0}\sum_{l\neq 0} \frac{\langle a_0(1)b_0(2)|H'|a_k(1)b_l(2)\rangle\langle a_k(1)b_l(2)|H'|a_0(1)b_0(2)\rangle}{2E_0 - E_k - E_l} \tag{8.10}$$

where the subscript is an index of the atomic quantum level. To avoid the variable denominator in the sum, it is usual to note that the highest possible value of the denominator is $-\frac{3}{4}$ in atomic units (1 a.u. = 27·205 eV) and the lowest -1, the first for both atoms in the $n = 2$ state and the second for both ionized. To get a lower limit to (8.10) we use the second value, and find the energy correction (8.11),

$$-\sum_p{}' H'_{0p}H'_{p0} \tag{8.11}$$

where p is a joint index running over k and l in (8.10). The primed sum excludes $k, l = 0$. Then going to the dipole approximation (8.8), and using a closure relation we reduce (8.11) to the expression (8.12),

$$-R^{-6}\langle a_0(1)b_0(2)|x_1^2x_2^2+y_1^2y_2^2+4z_1^2z_2^2|a_0(1)b_0(2)\rangle. \qquad (8.12)$$

Since the 1s state is spherically symmetrical we can replace $\langle|x_1^2|\rangle$ by $(\frac{1}{3})\langle|r_1^2|\rangle$, complete the integrations, and find for the ground-state dispersion correction the expression

$$\Delta W_G \simeq -6/R^6.$$

For the excited states (8.3) the second-order correction is:

$$\frac{1}{4}\sum_{k,l}\frac{\langle a_0b_1 \pm a_1b_0|H'|a_kb_l \pm a_lb_k\rangle\langle a_kb_l \pm a_lb_k|H'|a_0b_1 \pm a_1b_0\rangle}{E_0+E_1-E_k-E_l} \qquad (8.13)$$

in which all the upper signs are taken together. There are limitations on the values of k and l to keep the virtual states distinct from the states (8.3); thus if $l(k) = 0$, $k(l) \neq 1$, and if $l(k) = 1$, $k(l) \neq 0$. Consideration of the selection rules for optical transitions at once convinces us that, in dipole approximation, there are no contributions to the splitting in this order. Splitting terms could only come from terms in (8.13) which change sign between $(+)$ and $(-)$ states, typically,

$$\frac{\langle a_0b_1|H'|a_lb_k\rangle\langle a_lb_k|H'|a_1b_0\rangle}{E_0+E_1-E_l-E_k}. \qquad (8.14)$$

Since $a_1 \leftarrow a_0$ is allowed, the states are of opposite parity so that $a_l \leftarrow a_0$ and $a_1 \leftarrow a_l$ cannot *both* be dipole allowed, and the energy (8.14) is zero. Terms such as (8.15) which do not change sign between the plus and minus states (8.3) contribute to the dispersion correction, affecting the states equally,

$$\frac{\langle a_0b_1|H'|a_kb_l\rangle\langle a_kb_l|H'|a_0b_1\rangle}{E_0+E_1-E_k-E_l} \qquad (8.15)$$

They may be evaluated and summed over the allowed values of k and l to give a dispersion energy again varying as R^{-6}. The difference between the ground-state and excited-state dispersion energies (8.12) and (8.13) is equal to the difference between the transition energy of free hydrogen atoms, and the average of the transition to energies in the states of the atom pair (Fig. 8.2).

Thus to the second order of perturbation theory the splitting appears only in the first order, and the displacement of the average transition energy only in the second order, and

$$\Delta W^{\pm} = D \pm R^{-3}\mu_i^{(1)}\mu_j^{(2)}(\delta_{ij}-3\hat{R}_i\hat{R}_j); \qquad (8.16)$$

where D is the dispersion energy difference. Such a two-atom system can undergo spectroscopic transitions. Where the atom separation R is small compared to the absorbed wavelength, the intensity depends on the

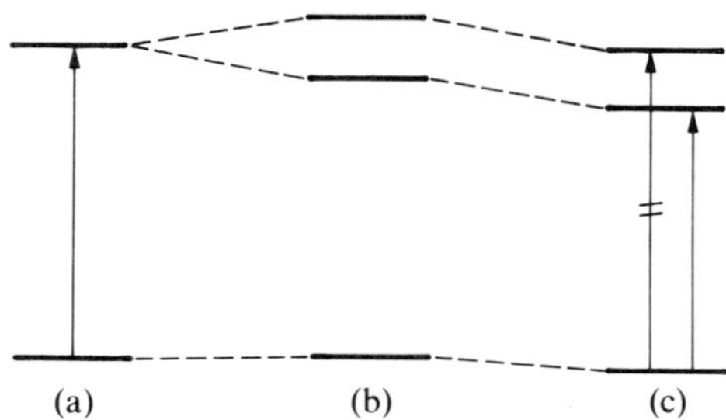

FIG. 8.2. Schematic diagram of transition energies of a pair of hydrogen atoms for a $2p_\sigma \leftarrow 1s$ excitation, at distances beyond the overlap region: (a) free atom; (b) atom-pair with energies corrected to the first order of perturbation theory, showing the resonance splitting; (c) energies corrected to the second order, showing dispersive corrections to ground and excited states. In (c) only the lower transition is spectroscopically allowed.

transition moments $\boldsymbol{\mu}^+$ and $\boldsymbol{\mu}^-$ joining the ground state to the states (8.3), calculated by eqn (8.17),

$$\boldsymbol{\mu}^{\pm} = \frac{1}{\sqrt{2}} \left\langle a(1)b(2) \middle| \sum_i \mathbf{r}_i \middle| a(1)b'(2) \pm a'(1)b(2) \right\rangle = (\boldsymbol{\mu}^{(a)} \pm \boldsymbol{\mu}^{(b)})\frac{1}{\sqrt{2}}, \quad (8.17)$$

where $\boldsymbol{\mu}^{(a)}$ and $\boldsymbol{\mu}^{(b)}$ are the moments for the $2p_\sigma \leftarrow 1s$ transition in the free atoms. These quantities have already appeared in the splitting expression (8.9). Since the atomic moments lie along **R** the minus combination in (8.17) vanishes, and transitions to the minus state (8.3) are forbidden. The plus state is allowed in spectral transitions with moment equal to $\sqrt{2}$ times the free atom moment, and with oscillator strength (8.18)

$$f = (\tfrac{1}{3})\mu^2 \nu \quad (8.18)$$

equal to twice that of the free molecule. It is thus a 'superradiant' state. In (8.18) ν is the frequency converted to energy in Rydbergs. The oscillator strength is that for the atom pair, and the value per atom is unchanged. However the oscillator strength is concentrated in one of the two transitions, the other being forbidden. The allowed transition is polarized along the interatomic direction, a result which follows from the assumption that the allowed atomic transition was itself polarized in that direction. It is easily shown that if the coupled systems are anisotropic, and if the moments $\boldsymbol{\mu}^{(a)}$ and $\boldsymbol{\mu}^{(b)}$ point in general directions, transitions to both levels in Fig. 8.2 are allowed.

8.2. Crystals

The model to be used for the excitation of a molecular crystal is an extension of the two-atom system to take account of large numbers of interacting systems in a regular spatial arrangement, and of their anisotropic properties (Knox 1963; Craig and Walmsley 1968; Davydov 1971).

A molecule in the crystal will be distinguished by two indices: φ_{ij} is the ground-state wavefunction of a molecule at the ith site of the jth unit cell. The ground-state wavefunction of the crystal is written as the product (8.19),

$$\varphi_{11}\varphi_{12} \cdots \varphi_{21}\varphi_{22} \cdots \varphi_{31} \cdots \varphi_{hN}. \tag{8.19}$$

For N unit cells and h sites the total number of molecules in the crystal is hN. In writing the wavefunction in this way we are already imposing certain approximations, namely that the free-molecule ground-state wavefunctions are unmodified by the crystal environment and that electron exchange between molecules in unimportant. The role of exchange in molecular crystals has been variously estimated. Its influence on the position of the absorption observed in a crystal of one of the smaller aromatics is probably a few tens of wave numbers, and may be neglected only if spectral displacements from other causes are greatly in excess. In certain examples, such as those of spin-forbidden singlet–triplet transitions, exchange must be included. In allowed singlet–singlet transitions a correction of this order is usually not important.

In finding the form of the crystal excited wavefunctions belonging to an excited free molecule state φ'_{ij} we first form the product functions (8.20) for localized excitation,

$$\phi_{ij} = \varphi_{11}\varphi_{12} \cdots \varphi_{21}\varphi_{22} \cdots \varphi'_{ij} \cdots \varphi_{hN}. \tag{8.20}$$

The hN-fold degeneracy is resolved in two stages. Let us consider the set of molecules occupying a given unit cell site i. The members of this translationally equivalent set combine to transform like the representations $\boldsymbol{\kappa}$ of the group of translations, according to expression (8.21),

$$\Phi_i(\boldsymbol{\kappa}) = N^{-\frac{1}{2}} \sum_j \exp(\mathrm{i}\boldsymbol{\kappa} \cdot \mathbf{r}_{ij})\phi_{ij} \tag{8.21}$$

$\mathbf{r}_{ij}$ is the coordinate of the (ij) molecule and $\boldsymbol{\kappa}$, the wave vector, is a vector of the reciprocal lattice limited to values in the first Brillouin zone, namely, for its component κ_a,

$$a\kappa_a = 0, \pm\frac{2\pi}{N_a}, \pm\frac{4\pi}{N_a}, \ldots, \pi, \tag{8.22}$$

where a is a lattice spacing and N_a the number of unit cells along the a direction. Next, by combining the various sites, we get the crystal wavefunctions (8.23),

$$\Psi(\boldsymbol{\kappa}) = \sum_i c_i(\boldsymbol{\kappa})\phi_i(\boldsymbol{\kappa}), \tag{8.23}$$

in terms of coefficients as yet undetermined. For special $\boldsymbol{\kappa}$ values, of which $\boldsymbol{\kappa} = 0$ is of particular importance, the c_i are fixed by symmetry. Otherwise, in general, the coefficients are found by solving a secular equation of degree h for each $\boldsymbol{\kappa}$ value. The levels are separated into a band characterized by the range of allowed values of the wave vector. The significance of the $\boldsymbol{\kappa} = 0$ states (h in number) is that, according to the naive view that the wavelength of the absorbed light is much greater than any dimension of the crystal, they are the only states that can be spectroscopically active, and by calculating their energies we can get an approximation to the crystal absorption spectrum. In most cases of interest $c_i(0) = \pm h^{-\frac{1}{2}}$ in the wavefunctions (eqn 8.23) and no secular equation has to be solved. Thus where there are two crystallographically-equivalent molecules in the unit cell (e.g., naphthalene and anthracene, space group $P2_1/a$) the spectroscopic states are

$$\Psi^{\pm}(0) = \frac{1}{\sqrt{2}}\{\Phi_1(0) \pm \Phi_2(0)\}, \tag{8.24}$$

1 and 2 denoting the two unit cell sites.

Let us now calculate the energies in the framework used in Section 8.1 for the hydrogen atom-pair. For the present the Hamiltonian will be given the form

$$H = \sum h_{\text{mol}} + V$$
$$V = \tfrac{1}{2} \sum_{p \neq q} V_{pq} \tag{8.25}$$

where h_{mol} stands for a free molecule Hamiltonian, and the perturbation V is a sum of pairwise intermolecular interactions like the term H' in (8.5), but now including the Coulomb energy of every pair of particles made up of one from molecule p and one from q. Evidently the unperturbed energies of the ground and first excited states are NW and $(N-1)W + W'$ where W and W' are the molecular ground and excited states, and the unperturbed part of the transition energy is $\Delta\omega = W' - W$. The first-order correction to $\phi_i(\boldsymbol{\kappa})$, namely $\langle \Phi_i(\boldsymbol{\kappa}) | V | \Phi_i(\boldsymbol{\kappa}) \rangle$ contains the representative term (8.26),

$$\langle \phi_{ip} | V_{pq} | \phi_{iq} \rangle = \langle \varphi'_{ip} \varphi_{iq} | V_{pq} | \varphi_{ip} \varphi'_{iq} \rangle. \tag{8.26}$$

With the help of eqn (8.14) to give the coefficient of the term (8.25), we readily find

$$\langle \Phi_i(\boldsymbol{\kappa}) | V | \Phi_i(\boldsymbol{\kappa}) \rangle = \sum_{q \neq p} \exp\{-\mathrm{i}\boldsymbol{\kappa} \cdot (\mathbf{r}_{ip} - \mathbf{r}_{iq})\} \langle \phi'_{ip} \phi_{iq} | V_{pq} | \phi_{ip} \phi'_{iq} \rangle = I_{ii}(\boldsymbol{\kappa}) \tag{8.27a}$$

and, for the cross term between different sites,

$$\langle \Phi_i(\boldsymbol{\kappa}) | V | \Phi_j(\boldsymbol{\kappa}) \rangle = \sum_{q} \exp\{-\mathrm{i}\boldsymbol{\kappa} \cdot (\mathbf{r}_{ip} - \mathbf{r}_{jq})\} \langle \phi'_{ip} \phi_{jq} | V_{pq} | \phi_{ip} \phi'_{jq} \rangle = I_{ij}(\boldsymbol{\kappa}) \tag{8.27b}$$

the multiplier $(1/N^{\frac{1}{2}})^2$ from (8.21) being cancelled by summing over the N equivalent sites in finding (8.27a) and (8.27b). These are modulated lattice sums over every molecule pair in the crystal. For $\boldsymbol{\kappa} = 0$, as in the spectroscopically active states in (8.24) the modulation factor is unity. It is easy to see that V_{pq} can be expanded in a multipole series as before, and that its dipole–dipole part, used in (8.27*a*) and (8.27*b*), will give the modulated sum of dipole–dipole interactions between transition dipoles located at each of the molecules in the lattice. We are now able to write down the secular eqn (8.28) for the transition energies ΔW belonging to wave vector $\boldsymbol{\kappa}$ in a crystal of any number of crystallographically equivalent molecules in the unit cell,

$$\begin{vmatrix} \Delta\omega + D + I_{11}(\boldsymbol{\kappa}) - \Delta W & I_{12}(\boldsymbol{\kappa}) & I_{13}(\boldsymbol{\kappa}) & \cdots \\ I_{21}(\boldsymbol{\kappa}) & \Delta\omega + D + I_{22}(\boldsymbol{\kappa}) - \Delta W & I_{23}(\boldsymbol{\kappa}) & \cdots \\ \vdots & \vdots & \vdots & \end{vmatrix} = 0. \tag{8.28}$$

Values of the sums $I_{ij}(\boldsymbol{\kappa})$ for $\boldsymbol{\kappa} = 0$ are obtained without difficulty and are quoted in texts (e.g. Craig and Walmsley 1968). The sums are conditionally convergent, and thus depend on the shape of the zone of crystal within which the sum is found (Section 8.4). For a spherical zone in an anthracene crystal the lattice dipole sum ($\boldsymbol{\kappa} = 0$) for the coupling of a long-axis oriented transition dipole at one lattice site to its environment of molecules at the second site, also with long-axis dipoles, is $1{\cdot}382 \times 10^5\ \mathrm{cm}^{-1}$ per nm^2. The transition in free anthracene at 250 nm was at first estimated to have a dipole length of 0·23 nm, and the calculation of splitting of the two crystal components then gives $2 \times (2{\cdot}3)^2 \times 1382 \doteqdot 14\,600\ \mathrm{cm}^{-1}$. The experimental splitting is about $16\,000\ \mathrm{cm}^{-1}$. This rough agreement was used to assign the 250 nm system to the long-axis polarized species and has been part of the evidence underpinning the theoretical model. In fact the transition is probably weaker (dipole length $\sim 0{\cdot}207$ nm) and agreement with experiment is better if the zone is slab-shaped, for which the splitting is $16\,300\ \mathrm{cm}^{-1}$. The assignment to long-axis polarization remains unambiguous. Later theories taking account of the interaction of light incident normally on a plane crystal face point to the slab as the correct model in general (Philpott 1969), as well as in very thin crystals (Craig and Dissado 1971). The model has been successfully used to assign molecular transitions, and must be broadly correct, but there are a number of difficult points, as will be seen in Section 8.4.

8.3. Interband mixing by the second quantization method

The band structure calculated in Section 8.2 by solving eqn (8.28) for the allowed energy values for each wave vector $\boldsymbol{\kappa}$ derives from a single excited state of the free molecule. Bands may similarly be calculated for other free molecule states. The parent free molecul wavefunctions are orthogonal, but the bands related to them in the crystal are coupled to one another by the intermolecular parts of the crystal Hamiltonian. In the dipole limit the transition moment to one state in molecule p is coupled to the moment of a second state in molecule q. The bands are decoupled in the next approximation by diagonalizing the Hamiltonian in the enlarged basis built on two or more free molecule excited states. The decoupled bands have mixed parentage; the absorption of polarized light in transitions to the levels may not correspond precisely to that expected for any single molecular state but to a mixture of two or more. A typical example in the spectrum of anthracene is that of the system in the near u.v. (Craig and Hobbins 1955*b*). The 'polarization ratio' of the crystal transition in the 380 nm region is about 4:1, in the sense that the absorption intensity for light polarized along the b monoclinic axis is four times that along the a axis. If this transition had as its pure parent the free molecule 380 nm transition polarized along the shorter in-plane molecular axis, the ratio would be near 7:1. The polarization ratio measured indicates a mixed parentage, the participating free molecule states being the 380 nm state and the long-axis polarized state appearing in the free molecule transition at 250 nm.

In calculating crystal induced mixing of this type, we note that levels, to be capable of mixing, must belong to the same representation of the translation group, i.e., have the same value of the wave vector $\boldsymbol{\kappa}$. The problem of diagonalizing the energy, including the mixing of states, reduces to that of solving a set of secular equations, each belonging to one $\boldsymbol{\kappa}$ value. The form of such a secular equation is shown in eqn (8.29) applying to an example in which there are two molecules in the unit cell and in which treatment of inter-band mixing is limited to two bands, each having its parentage in a free molecule excited state.

$$\begin{vmatrix} \Delta\omega^f + D^f + I_{11}^{ff}(\kappa) - \Delta W & I_{12}^{ff}(\kappa) & I_{11}^{fg}(\kappa) & I_{12}^{fg}(\kappa) \\ I_{21}^{ff}(\kappa) & \Delta\omega^f + D^f + I_{22}^{ff}(\kappa) - \Delta W & I_{21}^{fg}(\kappa) & I_{22}^{fg}(\kappa) \\ I_{11}^{fg}(\kappa) & I_{22}^{fg}(\kappa) & \Delta\omega^g + D^g + I_{11}^{gg}(\kappa) - \Delta W & I_{12}^{gg}(\kappa) \\ I_{21}^{fg}(\kappa) & I_{22}^{fg}(\kappa) & I_{21}^{gg}(\kappa) & \Delta\omega^g + D^g + I_{22}^{gg}(\kappa) - \Delta W \end{vmatrix} = 0. \quad (8.29)$$

f and g refer to two molecular states, 1 and 2 to the two molecules in a unit cell, and the off-diagonal elements of the secular equation are defined as follows:

$$I_{ij}^{fg}(\kappa) = \sum_q \exp\{-\mathrm{i}\boldsymbol{\kappa}\,.\,(\mathbf{r}_{ip} - \mathbf{r}_{jq})\}\,\langle \phi_{ip}^f \phi_{jq} | V_{pq} | \phi_{ip} \phi_{jq}^g \rangle.$$

The 2×2 block of elements in the top left places of eqn (8.29), and the similar block in the bottom right places refer separately to the two electronic states, and have the form of eqn (8.28). The remaining off-diagonal elements are dipole–dipole matrix components, coupling the two different transitions.

The solutions of eqn (8.29) are excitation energies in a model in which only one molecule at a time is excited. The method may be adapted to deal with multiple excitation, and higher order corrections, but it rapidly becomes clumsy. An alternative (Agranovitch 1961; Hoffmann 1963; Anderson 1968; Tyablikov 1968; Davydov 1971) is to use second-quantization, in which multiple excited levels are included quite naturally, and which is especially appropriate to the treatment of exciton–photon coupling to be given in Section 8.6. In this section we introduce the second quantization method by applying it to interband mixing of excitons and comparing the results with the perturbation method. To proceed by second quantization we introduce creation and annihilation operators $B_n^\dagger$ and B_n for excitation of a molecule at site n to the rth excited state. Each molecule can have only one such excitation and the excitations (excitons) should be treated as Fermi particles. The appropriate commutator is given in (8.30), the quantity $N_n(=0, 1)$

$$[B_n, B_n^\dagger] = 1 - 2N_n \tag{8.30}$$

being the number of excitations at the nth site. So long as there are many sites and the total excitations few, the average of N_n is negligibly different from zero, and the averaged commutator over all sites may be put equal to unity, as for Bose particles. Thus the excitons are treated as bosons in the low illumination limit.

The illustrative problem is to be limited, as before, to the example of two molecules per cell (1 and 2), each molecule having two excited states (f and g).

The Hamiltonian for non-interacting molecules is, expression (8.31),

$$h = \Delta\omega^f \sum_n (B_{nf1}^\dagger B_{nf1} + B_{nf2}^\dagger B_{nf2}) + \Delta\omega^g \sum_n (B_{ng1}^\dagger B_{ng1} + B_{ng2}^\dagger B_{ng2}). \tag{8.31}$$

In this approximation the total excitation energy of the crystal is equal to the summed excitation energy of the free molecules, each molecule making contributions according to the expectation values of the number operators $B_{nf1}^\dagger B_{nf1}$ in (8.31). The single-site operators are now transformed to apply to delocalized excitations in a manner similar to that used in eqn (8.21) to go from localized to delocalized crystal wavefunctions,

$$\left.\begin{aligned} B_{nf1}^\dagger &= N^{-\frac{1}{2}} \sum_\kappa \exp(-\mathrm{i}\boldsymbol{\kappa}\,.\,\mathbf{n}) B_{\kappa f1}^\dagger \\ B_{nf1} &= N^{-\frac{1}{2}} \sum_\kappa \exp(\mathrm{i}\boldsymbol{\kappa}\,.\,\mathbf{n}) B_{\kappa f1} \end{aligned}\right\} \tag{8.32}$$

with similar expressions for the other operators. The complete Hamiltonian includes the transform of (8.31), according to (8.32), together with the cross terms,

$$\begin{aligned}H = \sum_{\kappa} \{&\Delta\omega_1^f(\kappa)B^\dagger_{\kappa f1}B_{\kappa f1} + \Delta\omega_2^f(\kappa)B^\dagger_{\kappa f2}B_{\kappa f2}\\ &+\Delta\omega_1^g(\kappa)B^\dagger_{\kappa g1}B_{\kappa g1} + \Delta\omega_2^g(\kappa)B^\dagger_{\kappa g2}B_{\kappa g2}\}\\ &+\sum_{\kappa} I_{12}^{ff}(\kappa)(B^\dagger_{\kappa f1} + B_{-\kappa f1})(B^\dagger_{-\kappa f2} + B_{\kappa f2})\\ &+\sum_{\kappa} I_{11}^{fg}(\kappa)(B^\dagger_{\kappa f1} + B_{-\kappa f1})(B^\dagger_{-\kappa g1} + B_{\kappa g1})\\ &+\sum_{\kappa} I_{12}^{fg}(\kappa)(B^\dagger_{\kappa f1} + B_{-\kappa f1})(B^\dagger_{-\kappa g2} + B_{\kappa g2})\\ &+\sum_{\kappa} I_{11}^{fg}(\kappa)(B^\dagger_{\kappa f1} + B_{-\kappa f1})(B^\dagger_{-\kappa g1} + B_{\kappa g1})\\ &+\sum_{\kappa} I_{22}^{fg}(\kappa)(B^\dagger_{\kappa f2} + B_{-\kappa f2})B^\dagger_{-\kappa g2} + B_{\kappa g2})\\ &+\sum_{\kappa} I_{12}^{gg}(\kappa)(B^\dagger_{\kappa g1} + B_{-\kappa g1})(B^\dagger_{-\kappa g2} + B_{\kappa g2})\end{aligned} \tag{8.33}$$

in which, for example,

$$\Delta\omega_1^f(\kappa) = \Delta\omega^f + D^f + I_{11}^f(\kappa). \tag{8.34}$$

Each term of the sum in eqn (8.33) must transform according to the $\boldsymbol{\kappa} = 0$ representation of the translation group, and the particular form of the interaction operator is fixed by this requirement, in that $B^\dagger_\kappa$ and B_κ transform like the $+\boldsymbol{\kappa}$ and $-\boldsymbol{\kappa}$ representations respectively.

We proceed to transform (8.33) to diagonal form, following the Bogolyubov–Tyablikov method (Tyablikov 1968), by introducing new creation and annihilation operators $\xi^\dagger$ and ξ with the property that the Hamiltonian becomes (apart from a term giving the dispersion energy of the ground-state) (Mahan 1965; Power and Thirunamachandran 1971)

$$H = \sum_{\kappa\lambda} E_{\kappa\lambda}\xi^\dagger_{\kappa\lambda}\xi_{\kappa\lambda};$$

the energies $E_{\kappa\lambda}$ are characteristic of new excitations formed by combining the primitive excitations $B^\dagger_{\kappa f1}$, etc. appearing in (8.32). The index κ gives the wave vector, and the second index distinguishes the several levels of each κ. The second index is omitted from the next few equations, to simplify the presentation, and re-introduced later where necessary. The new particle operators $\xi^\dagger_\kappa$, ξ_κ are introduced in eqn (8.35) in terms of amplitudes u, v, w, and x,

$$\left.\begin{aligned}B_{\kappa f1} &= \xi_\kappa(u_{\kappa f1} + v_{\kappa f2}) + \xi^\dagger_{-\kappa}(w_{-\kappa g1} + x_{-\kappa g2})\\ B_{\kappa a2} &= \xi_\kappa(u_{\kappa f2} + v_{\kappa f1}) + \xi^\dagger_{-\kappa}(w_{-\kappa g2} + x_{-\kappa g1})\\ B_{\kappa b1} &= \xi_\kappa(u_{\kappa g1} + v_{\kappa g2}) + \xi^\dagger_{-\kappa}(w_{-\kappa f1} + x_{-\kappa f2})\\ B_{\kappa b2} &= \xi_\kappa(u_{\kappa g2} + u_{\kappa g1}) + \xi^\dagger_{-\kappa}(w_{-\kappa f2} + x_{-\kappa f1})\end{aligned}\right\} \tag{8.35}$$

together with the creation operators found by taking complex conjugates. The amplitudes are to be determined to effect the desired reduction to diagonal form, and to ensure the commutation property $[\xi_\kappa\, \xi_\kappa^\dagger] = 1$. It is obvious from the form of the relations (8.35) that the right-hand sides must transform under translations according to the wave vector $\boldsymbol{\kappa}$ appearing in the operators $B^\dagger$, B. Thus the new excitations $\xi^\dagger$, ξ are characterized by the same wave vectors. There are four such operator pairs for each $\boldsymbol{\kappa}$, later distinguished by an extra index λ. A convenient method of reduction is to evaluate the commutators (8.36),

$$\begin{aligned}
[B_{\kappa f1}, H] &= \Delta\omega_1^f(\kappa)B_{\kappa f1} + I_{12}^{ff}(\kappa)(B_{\kappa f2}^\dagger + B_{\kappa f2}) \\
&\quad + I_{11}^{fg}(\kappa)(B_{\kappa g1}^\dagger + B_{\kappa g1}) + I_{12}^{fg}(\kappa)(B_{\kappa g2}^\dagger + B_{\kappa g2}) \\
[B_{\kappa f2}, H] &= \Delta\omega_2^f(\kappa)B_{\kappa f2} + I_{12}^{fg}(\kappa)(B_{\kappa f1}^\dagger + B_{\kappa f1}) \\
&\quad + I_{21}^{fg}(\kappa)(B_{\kappa g1}^\dagger + B_{\kappa g1}) + I_{22}^{fg}(B_{\kappa g2}^\dagger + B_{\kappa g2}) \\
[B_{\kappa g1}, H] &= \Delta\omega_1^g(\kappa)B_{\kappa g1} + I_{11}^{fg}(\kappa)(B_{\kappa f1}^\dagger + B_{\kappa f1}) \\
&\quad + I_{21}^{fg}(\kappa)(B_{\kappa f2}^\dagger + B_{\kappa f2}) + I_{12}^{gg}(\kappa)(B_{\kappa g2}^\dagger + B_{\kappa g2}) \\
[B_{\kappa g2}, H] &= \Delta\omega_2^g(\kappa)B_{\kappa g2} + I_{12}^{fg}(\kappa)(B_{\kappa f1}^\dagger + B_{\kappa f1}) \\
&\quad + I_{22}^{fg}(\kappa)(B_{\kappa f2}^\dagger + B_{\kappa f2}) + I_{12}^{gg}(\kappa)(B_{\kappa g1}^\dagger + B_{\kappa g1})
\end{aligned} \tag{8.36}$$

and then to substitute the expressions (8.35) into the commutator relations such as

$$[B_{\kappa f1}, H] = \mathrm{i}\dot{B}_{\kappa f1}. \tag{8.37}$$

Finally, the condition for diagonal form is introduced by requiring the equations of motions (8.38) to be obeyed,

$$[\xi_\kappa, H] = \mathrm{i}\dot{\xi}_\kappa = E_\kappa\xi_\kappa; \qquad [\xi_\kappa^\dagger, H] = -E_\kappa\xi_\kappa. \tag{8.38}$$

In the case of $B_{\kappa f1}$ we find

$$\begin{aligned}
&\Delta\omega_1^f(\kappa)\{\xi_\kappa(u_{\kappa f1} + v_{\kappa f2}) + \xi_{-\kappa}^\dagger(w_{-\kappa g1} + x_{-\kappa g2})\} \\
&\quad + I_{12}^{ff}(\kappa)\{\xi_{-\kappa}(w_{-\kappa g2} + x_{-\kappa g1}) + \xi_\kappa^\dagger(u_{\kappa f2} + v_{\kappa f1}) \\
&\quad + \xi_\kappa(u_{\kappa f1} + v_{\kappa f1}) + \xi_{-\kappa}^\dagger(w_{-\kappa g2} + x_{-\kappa g1})\} \\
&\quad + I_{11}^{fg}(\kappa)\{\xi_{-\kappa}(w_{-\kappa f1} + x_{-\kappa f2}) + \xi_\kappa^\dagger(u_{\kappa g1} + v_{\kappa g2}) \\
&\quad + \xi_\kappa(u_{\kappa g1} + v_{\kappa g2}) + \xi_{-\kappa}^\dagger(w_{-\kappa f1} + x_{-\kappa f2})\} \\
&\quad + I_{12}^{fg}(\kappa)\{\xi_{-\kappa}(w_{-\kappa f2} + x_{-\kappa f1}) + \xi_\kappa^\dagger(u_{\kappa g2} + v_{\kappa g1}) + \xi_\kappa(u_{\kappa g2} + v_{\kappa g1}) \\
&+ \xi_{-\kappa}^\dagger(w_{-\kappa f2} + x_{-\kappa f1})\} = \dot{\xi}_\kappa(u_{\kappa f1} + v_{\kappa f2}) + \dot{\xi}_{-\kappa}^\dagger(w_{-\kappa g1} + x_{-\kappa g2}) \\
&\qquad = E_\kappa\xi_\kappa(u_{\kappa f1} + v_{\kappa f2}) - E_\kappa\xi_{-\kappa}^\dagger(w_{-\kappa g1} + x_{-\kappa g2}).
\end{aligned} \tag{8.39}$$

By equating the coefficients of ξ_κ, and then of $\xi^\dagger_{-\kappa}$ we get two equations (8.40a) and (8.40b),

$$\{\Delta\omega_1^f(\kappa) - E_\kappa\}\alpha + I_{12}^{ff}(\kappa)\alpha' + I_{11}^{fg}(\kappa)\beta + I_{12}^{fg}(\kappa)\beta' = 0 \tag{8.40a}$$

$$\{\Delta\omega_1^f(\kappa) - E_\kappa\}\gamma + I_{12}^{ff}(\kappa)\gamma' + I_{11}^{fg}(\kappa)\delta + I_{12}^{fg}(\kappa)\delta' = 0 \tag{8.40b}$$

where the amplitudes have been combined according to

$$\begin{aligned} \alpha &= u_{\kappa f1} + v_{\kappa f2}, & \alpha' &= u_{\kappa f2} + v_{\kappa f1} \\ \beta &= u_{\kappa g1} + v_{\kappa g2}, & \beta' &= u_{\kappa g2} + v_{\kappa g1} \\ \gamma &= w_{-\kappa g1} + x_{-\kappa g2}, & \gamma' &= w_{-\kappa g2} + x_{-\kappa g1} \\ \delta &= w_{-\kappa f1} + x_{-\kappa f2}, & \delta' &= w_{-\kappa f2} + x_{-\kappa f1}. \end{aligned}$$

Eqns (8.40a) and (8.40b) are readily combined to give (8.41),

$$[\{\Delta\omega_1^f(\kappa)\}^2 - E_\kappa^2](\alpha+\gamma) + 2\Delta\omega_1^f(\kappa)I_{12}^{ff}(\kappa)(\alpha'+\gamma') + 2\Delta\omega_1^f(\kappa)I_{11}^{fg}(\kappa)(\beta+\delta) + 2\Delta\omega_1^f(\kappa)I_{12}^{fg}(\kappa)(\beta'+\delta') = 0. \tag{8.41}$$

Likewise, with the commutator $[B_{\kappa f2}, H]$ as starting point, we find eqn (8.42),

$$[\{\Delta\omega_2^f(\kappa)\}^2 - E_\kappa^2](\alpha'+\gamma') + 2\Delta\omega_2^f(\kappa)I_{12}^{ff}(\kappa)(\alpha+\gamma) + 2\Delta\omega_2^f(\kappa)I_{22}^{fg}(\kappa)(\beta'+\delta') + 2\Delta\omega_2^f(\kappa)I_{12}^{fg}(\kappa)(\beta+\delta) = 0. \tag{8.42}$$

There are similar equations from $[B_{\kappa g1}, H]$ and $[B_{\kappa g2}, H]$, giving in all four simultaneous equations in the amplitudes $\alpha+\gamma$, $\alpha'+\gamma'$, $\beta+\delta$, and $\beta'+\delta'$. The condition for non-trivial solutions leads to the determinantal eqn (8.43),

$$\begin{vmatrix} \{\Delta\omega_1^f(\kappa)\}^2 - E_\kappa^2 & 2\Delta\omega_1^f(\kappa)I_{12}^{ff}(\kappa) & 2\Delta\omega_1^f(\kappa)I_{11}^{fg}(\kappa) & 2\Delta\omega_1^f(\kappa)I_{12}^{fg}(\kappa) \\ 2\Delta\omega_2^f(\kappa)I_{12}^{ff}(\kappa) & \{\Delta\omega_2^f(\kappa)\}^2 - E_\kappa^2 & 2\Delta\omega_2^f(\kappa)I_{12}^{fg}(\kappa) & 2\Delta\omega_2^f(\kappa)I_{22}^{fg}(\kappa) \\ 2\Delta\omega_1^g(\kappa)I_{11}^{fg}(\kappa) & 2\Delta\omega_1^g(\kappa)I_{12}^{fg}(\kappa) & \{\Delta\omega_1^g(\kappa)\}^2 - E_\kappa^2 & 2\Delta\omega_1^g(\kappa)I_{12}^{gg}(\kappa) \\ 2\Delta\omega_2^g(\kappa)I_{12}^{fg}(\kappa) & 2\Delta\omega_2^g(\kappa)I_{22}^{fg}(\kappa) & 2\Delta\omega_2^g(\kappa)I_{12}^{gg}(\kappa) & (\Delta\omega_2^g(\kappa))^2 - E_\kappa^2 \end{vmatrix} = 0. \tag{8.43}$$

There are eight solutions $E_{\kappa\lambda}$, $\lambda = 1, \ldots, 4$, for each κ value, the second index now being restored to the specification of levels. Of the eight, four are negative and of no interest. The remaining four belong to the two branches of each of two bands, and for each one may find values of the amplitudes $\alpha+\beta, \ldots$, giving the mixture of 'pure' parentage excitations in each case. This equation is the counterpart in the second quantization method of eqn (8.29) in coordinate representation. The relationship of the two methods is closer than appears from a superficial inspection of eqns (8.29) and (8.43). Thus if the diagonal energies $\Delta\omega_1^f(\kappa) \ldots \Delta\omega_2^g(\kappa)$ differ little from their mean, and if the four solutions $E_{\kappa\lambda}$ are also close to this mean, we may make the approximations

$$E_{\kappa\lambda}+\Delta\omega_1^f(\kappa) \approx 2\Delta\omega_1^f(\kappa); \qquad E_{\kappa\lambda}+\Delta\omega_2^f(\kappa) \approx 2\Delta\omega_2^f(\kappa);$$

$$E_{\kappa\lambda}+\Delta\omega_1^g(\kappa) \approx 2\Delta\omega_1^g(\kappa); \qquad E_{\kappa\lambda}+\Delta\omega_2^g(\kappa) \approx 2\Delta_2^g(\kappa),$$

then, by dividing successive rows of eqn (8.43), by these quantities we reduce eqn (8.43) to a form identical to that of (8.29), removing at the same time the unwanted negative solutions of eqn (8.43). The conditions required to validate the approximate equalities are unlikely to occur in problems of interest, and in practice the eigenstates from (8.43) differ appreciably from those of (8.29).

8.4. Problems of the simple model

There are two related points of difficulty in the work of Section 8.2. The first is the use of the static interaction Hamiltonian (8.5) in crystals to describe the coupling of molecules at all distances, even up to many wavelengths of light. Static interactions apply to cases where the propagation time of the field of one particle to the positions of other particles, whatever their distance, is unimportant. This is true of fixed permanent charge dispositions, as in the coupling of permanent electric dipoles, but the essence of the treatment given in Section 8.2 is that the coupling is between transition dipoles oscillating at the transition frequency. Thus we must expect that a static potential will be wrong for the resonance interaction of molecules sufficiently far separated in a crystal. This leads to the second difficulty: the assumption of instantaneous interactions is tantamount in the present context to making the velocity and wavelength of light infinite. Consistency then demands that only the $\boldsymbol{\kappa} = 0$ crystal mode can be excited in an optical transition since, whatever the size of a crystal, it is *a fortiori* negligible compared with the wavelength of light. The transition energy in such a mode depends on the shape of the crystal specimen, as was first pointed out by Fox and Yatsiv (1956). The reason is that the dipole lattice sum that determines the energies for $\boldsymbol{\kappa} = 0$ is simply the sum of the interactions of a point dipole at the origin with a set of dipoles (parallel within the set but not necessarily to the origin dipole) at the lattice sites. The sum may be divided into two parts. One is the sum over molecules in an inner zone (local order zone) near enough for the detailed crystal structure to be important. The second sum is over the outer zone, where the molecules are far enough from the origin to be treated as a continuum; this part can be converted to an integral. The critical separation is usually 30–50 Å, and so one draws a sphere, centre at the origin and radius about 50 Å, and sums the inner molecular couplings term-by-term, and integrates outside the sphere. The problems of size and shape thus arise only in the outer zone, which we now discuss.

In the outer zone we replace the molecule transition moment $\boldsymbol{\mu}$ by the dielectric polarization $\boldsymbol{\mu}/V_0$, V_0 being the volume of the crystalline unit cell. If there is one molecule per unit cell, this is the sole contribution; where there is more than one we consider separately the contribution by each translationally equivalent set, and sum the results. The problem is the interaction of the origin dipole with the surrounding polarized continuous medium. Since the behaviour of transition dipoles is no different from permanent dipoles at this stage we have lost all trace of the characteristically spectroscopic feature that the dipoles are oscillating. We suppose at first that the surrounding medium is in the form of a spherical shell, with inner and outer radii ρ_1 and ρ_2. Polar coordinates are chosen with polar (z) axis along $\boldsymbol{\mu}$ and the electric field calculated at the origin. With the use of the expressions (8.44) for the radial and tangential components of the field at distance r produced by moment $\mathbf{m}$

$$E_r = \frac{2m}{r^3}\cos\theta; \qquad E_\theta = \frac{m}{r^3}\sin\theta \tag{8.44}$$

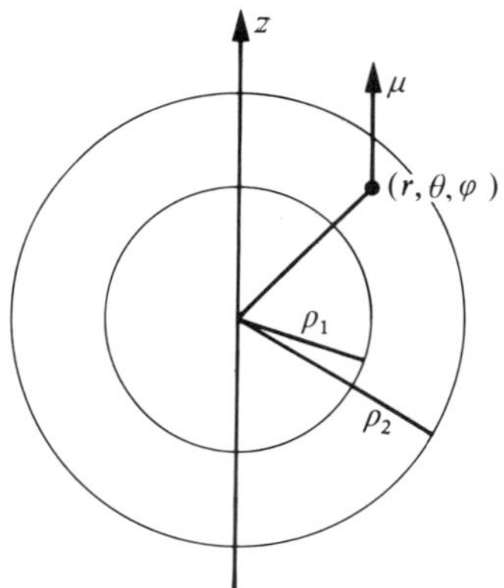

we find for the z component produced by the polarization element at (r, θ, ϕ),

$$\frac{\mu}{V_0 r^3}(2\cos^2\theta - \sin^2\theta) r^2\, \mathrm{d}r \sin\theta\, \mathrm{d}\theta\, \mathrm{d}\phi. \tag{8.45}$$

This vanishes on integration over the angles. The x and y components also vanish. Thus the spherical shell makes no contribution to the field at the origin and, if the excitation zones within crystals were spherical, the region of local order near the origin would alone give a contribution to the lattice sum.

For other shapes, such as the more realistic disc and rectangular slab, it is easier to proceed via the dielectric polarization charges. It is well known (see, for example, Coulson 1961) that for a uniformly polarized dielectric, the electric field is the same as that produced by a surface charge-density equal in magnitude to the normal component of the polarization. We first take the sphere from this point of view; the normal component of polarization at the outer surface is $\mu/V_0 \cos\theta$, and the z component of the electric field at the origin is, expression (8.46)

$$-\int_0^{2\pi}\int_0^{\pi} \rho_2^{-2}\frac{\mu}{V_0}\cos^2\theta\rho_2^2 \sin\theta \, d\theta \, d\phi = -\frac{4\pi\mu}{3V_0}. \tag{8.46}$$

The x and y components vanish, so that the field (the 'depolarizing field') is $-4\pi\boldsymbol{\mu}/3V_0$. The field of the inner spherical surface (the 'cavity field') is $4\pi\boldsymbol{\mu}/3V_0$ and the resultant zero, as before.

Where the outer surface is flat, as in a disc, an appropriate depolarizing field must be combined with the field of a spherical cavity. For a disc with axis along $\boldsymbol{\mu}$ the depolarizing field is $-4\pi\boldsymbol{\mu}/V_0$ and the resultant of cavity plus depolarizing field is $8\pi\boldsymbol{\mu}/3V_0$; with the disc axis at right angles to $\boldsymbol{\mu}$ the values are 0 and $+4\pi\boldsymbol{\mu}/3V_0$. The total field is the sum of these terms and the field of the molecules in the zone of local order, and the result shows a very strong dependence on the shape of the excitation zone. If this theory were correct, the spectrum would show effects of sample shape that would be unmistakable in experimental measurements. However, large shape dependences are not seen though small effects on shape and thickness may be real.

Within the descriptive framework of static interactions and infinite speed of light there was no reason in principle for choosing one zone shape rather than another. The state of polarization produced by a light wave travelling infinitely fast necessarily belonged to zero wave vector and carried no 'memory' of the properties of the exciting field, such as the propagation direction. The choice of zone shape was thus seen as a matter to be settled empirically, by comparison of calculated results with experimental spectroscopic findings. As explained in Section 8.2, spherical zones seemed to fit well, but better values of quantities such as free molecule oscillator strengths might well have indicated the transverse slab, as required by later and improved models.

The improvements involve detailed study of the coupling of the light wave to the crystal. The wave vector $\boldsymbol{\kappa}$ is non-zero, and has the direction of the incident ray, usually taken to be normal to a crystal face. In the simplest terms (see Section 8.6 for a refinement), the wavefunction for one translationally equivalent set is now given by eqn (8.47),

$$\Phi(\boldsymbol{\kappa}) = N^{-\frac{1}{2}}\sum_p \exp(\mathrm{i}\boldsymbol{\kappa}\cdot\mathbf{p})\phi_p \tag{8.47}$$

ϕ_p being localized excitation function. The energy of this state depends upon the modulated lattice sum of dipole interactions (8.48),

$$\sum_q \exp\{i\boldsymbol{\kappa}\,.\,(\mathbf{p}-\mathbf{q})\}R_{pq}^{-3}\beta_{ij}\mu_i\mu_j \tag{8.48}$$

where $\beta_{ij} = \delta_{ij}-3\hat{R}_i\hat{R}_j$, namely the tensor for dipolar coupling. Here, as before, (8.48) is to be summed over repeated indices. In the outer (continuum) zone the sum is converted to an integral similar to that for the field in (8.45) but now with modulation of the polarization $\boldsymbol{\mu}/V_0$. The equivalent classical electrostatic problem is the coupling to a modulated polarization of a dipole at the centre of a spherical cavity in a dielectric. In a slab specimen

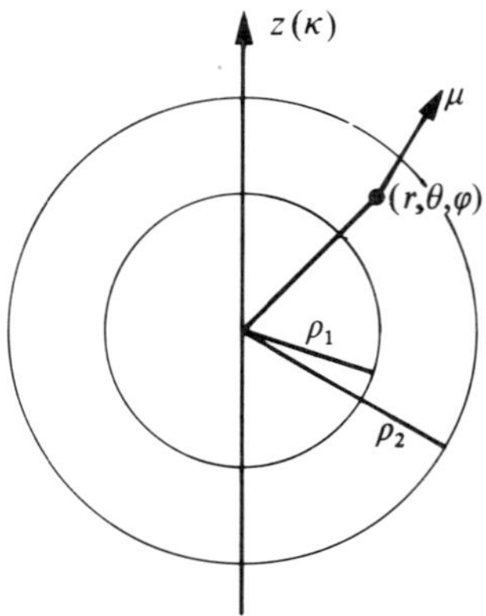

the direction of $\boldsymbol{\kappa}$ is normal to the slab surface, for excitation by light incident normally. We examine the contribution by a spherical shell, of radii ρ_1 and ρ_2, as before. The z polar axis is chosen to be along $\boldsymbol{\kappa}$. The required integral is, expression (8.49),

$$V_0^{-1}\int_0^{2\pi}\int_0^{\pi}\int_{\rho_1}^{\rho_2} R^{-3}\mu_i\mu_j\beta_{ij}\exp(i\boldsymbol{\kappa}\,.\,\mathbf{R})R^2\,\mathrm{d}R\sin\theta\,\mathrm{d}\theta\,\mathrm{d}\phi. \tag{8.49}$$

After some manipulation the integration over ϕ gives

$$\frac{2\pi}{V_0}\int_0^{\pi}\int_{\rho_1}^{\rho_2}\mu_i\mu_j\beta_{ij}\frac{\exp(i\kappa R\cos\theta)}{R^3}(1-3\cos^2\theta)\sin\theta\,\mathrm{d}\theta R^2\,\mathrm{d}R$$

where the tensor $B_{ij} = \delta_{ij}-3\hat{\kappa}_i\hat{\kappa}_j$ differs from β_{ij} by referring all components to the direction of the polarization wave $\boldsymbol{\kappa}$. Evidently $\mu_i\mu_jB_{ij} = \mu_1^2+\mu_2^2-2\mu_3^2$. Finally we get for the energy expression (8.50),

$$\frac{4\pi}{V_0}\mu_i\mu_jB_{ij}\left|\frac{\sin\kappa R}{(\kappa R)^3}-\frac{\cos\kappa R}{(\kappa R)^2}\right|_{\rho_1}^{\rho_2}. \tag{8.50}$$

For $\boldsymbol{\kappa} \neq 0$, the contribution at the outer limit can be made to vanish by choosing a large enough sphere, so that the interaction is no longer shape dependent. The contribution at the inner limit is best analysed by supposing (unphysically) that the continuum approximation applies down to the shortest distances. We consider the limit $\kappa\rho_1 \to 0$, $\kappa \neq 0$. The R dependent part becomes $\frac{1}{3}$, and expression (8.33) reduces to $(-4\pi/3V_0)\mu_i\mu_j B_{ij}$. The first term of B_{ij} gives

$$-\frac{4\pi}{3V_0}\delta_{ij}\mu_i\mu_j = -\frac{4\pi}{3V_0}\boldsymbol{\mu}^{(1)}\cdot\boldsymbol{\mu}^{(2)}, \tag{8.51}$$

which is the contribution to the energy by the Lorentz cavity field $(4\pi/3V_0)\boldsymbol{\mu}^{(2)}$. The second term gives

$$4\pi\hat{\kappa}_i\hat{\kappa}_j\mu_i\mu_j = (\boldsymbol{\mu}^{(1)}\cdot\hat{\boldsymbol{\kappa}})(\boldsymbol{\mu}^{(2)}\cdot\hat{\boldsymbol{\kappa}}). \tag{8.52}$$

The form of the right-hand side shows that this term is a coupling of the origin dipole to the longitudinal component of the field of the polarized medium. It is exactly the contribution to the energy by the depolarizing field of a disc cut transversely to the $\boldsymbol{\kappa}$ axis. Thus the modulated cavity field is the Lorentz field modified by the depolarization of the transverse slab and to adapt the static model the excitation zone is taken to be a transverse slab with $\boldsymbol{\kappa} = 0$, this being a limiting value approached in the direction $(\boldsymbol{\kappa}/\kappa)$.

8.5. The retarded interaction.

Features of the simple treatment so far given are the use of the electrostatic Hamiltonian (8.5) and the restriction for consistency to the $\boldsymbol{\kappa} = 0$ states. It has long been known that while the simple model gives acceptable agreement with experiment for splittings and polarization properties in a number of examples, it is unsatisfactory from a stricter point of view. The reason why the simple theory gives good results in calculations of these quantities appears to be that they are dominated by contributions made by molecular interactions within the zone of local order, say up to distances of 50 Å and by the macroscopic field. This zone is small compared with the wavelength of the absorbed light and both the electrostatic approximation and the $\boldsymbol{\kappa} = 0$ approximation apply. The outer zone, where these approximations fail, does not play a large part. In other crystal properties, such as the refractive index, it is course essential to make proper allowance for the velocity of light.

The formulae for the interaction of transition dipole moments used in Sections 8.1 and 8.2 are the same as those for permanent electric moments, a result implicit in assuming the instantaneous propagation of electric influences. There is obviously a limit of distance beyond which this cannot be expected to apply, but if the coupled system is an isolated pair of atoms or molecules, the critical distance is too great for corrections to be of significance; at such distances the interaction energies are already negligible before the effect of retardation is felt. Where, as in a crystal, the coupling is that of one molecule to a three-dimensional environment of other molecules, the summed contributions from long range cannot be ignored, unless the local order zone is dominant. For an interaction falling off as R^{-3} as in eqn (8.9) the decrease with distance is offset by the increase as R^3 of the number of molecules in the spherical shell round a given site. The overall distance law would thus be R^0, which shows that in general one cannot drop long-range terms.

In this section we discuss the form of the retarded interaction between transition dipoles, using an analogy with classical electrodynamics. The argument applies to isolated pairs of dipoles, but does not by itself provide an adequate basis for interactions in solids, a point to which we return later.

In quantum mechanics, we should begin a discussion of the retarded interaction by adding to the Hamiltonian (8.5) terms for the coupling of the atoms to the electromagnetic field, as in (8.53), for the full Hamiltonian,

$$
\begin{aligned}
H &= H_0 + H', \\
H_0 &= -\frac{1}{2}\nabla_1^2 - \frac{1}{2}\nabla_2^2 - \frac{1}{r_{a1}} - \frac{1}{r_{b2}} + H_{rad} \\
H' &= -\frac{1}{r_{a2}} - \frac{1}{r_{b1}} + \frac{1}{r_{12}} + \frac{1}{R} - \sum_i \mathbf{p}_i \cdot \mathbf{A}_i + \frac{1}{2}\sum_i \mathbf{A}_i^2
\end{aligned}
\tag{8.53}
$$

where H_{rad} is the Hamiltonian for the free electromagnetic field, $\mathbf{p}_i$ is the momentum operator of the ith electron, and $\mathbf{A}_i$ is the vector potential of the field at the position of the ith electron. This Hamiltonian includes both static and retarded terms (through $\mathbf{A}$). A form equivalent to (8.53) in the dipole approximation is the Power–Zienau operator (1959) (8.54) which is fully retarded, involving operators for the coupling of the transition dipole moment of the molecules to the transverse electric field of the radiation,

$$
H' = -\boldsymbol{\mu}_1 \cdot \mathbf{E}_1^{\perp} - \boldsymbol{\mu}_2 \cdot \mathbf{E}_2^{\perp}. \tag{8.54}
$$

In this form the interaction between atoms appears only through the mediation of the electromagnetic field, there being no direct term. The calculation thus involves diagonalizing a Hamiltonian that includes free atoms, quantized radiation field, and coupling between the two. It can be found in the literature (McLone and Power 1964). We now use the classical electrodynamical analogy to obtain the same result.

In quantum mechanics the atoms (or molecules) have allowed stationary states, and absorption or emission occur by transitions between them. Let us suppose that the ground and an excited state differ by energy $\Delta W = \hbar\nu$

$= \hbar ck$, c being the velocity of light and k the wavenumber. The classical analogue for the excited state is an oscillating electric dipole with frequency ν and amplitude equal to the transition dipole length of the real system. We calculate the interaction energy of the excited system 1 with system 2 in its ground state by calculating the electric field at 2 produced by the oscillating dipole at 1, and coupling this field with a dipole at 2. In the first place retardation is neglected and no phase difference allowed between the source dipole and the field at a distant point. Then according to standard methods (Stratton 1941) we define the Hertz vector $\mathbf{\Pi}$ in terms of the polarization $\boldsymbol{\mu}^{(1)}$ at system 1 and find the electric field by taking the curl curl with respect to the coordinates of system 2;

$$\mathbf{P}^{(1)} = \mathrm{e}^{-\mathrm{i}\omega t}\boldsymbol{\mu}; \qquad \mathbf{\Pi} = \mathrm{e}^{-\mathrm{i}\omega t}\boldsymbol{\mu}/R$$
$$\mathbf{E}^{(2)} = \text{curl curl } \mathbf{\Pi}. \tag{8.55}$$

In an axis system (i, j, k) we find for the i component of the curl of $\boldsymbol{\mu}/R$

$$(\text{curl } \boldsymbol{\mu}/R)_i = -R^{-3}(x_j\mu_k - x_k\mu_j)$$

and from the ith component of the curl curl we have for the electric field (written without time dependence)

$$E_i^{(2)} = \frac{\partial}{\partial x_j}\{R^{-3}(-x_i\mu_j + x_j\mu_i)\} - \frac{\partial}{\partial x_k}\{R^{-3}(-x_k\mu_i + x_i\mu_k)\}$$
$$= R^{-3}(-\mu_i + 3\hat{R}_i^2\mu_i + 3\hat{R}_i\hat{R}_j\mu_j + 3\hat{R}_i\hat{R}_k\mu_k)$$

or, with summation over repeated indices,

$$= -R^{-3}\mu_j^{(1)}(\delta_{ij} - 3\hat{R}_i\hat{R}_j) \tag{8.56}$$

where x_i are the cartesian displacements of system 2 relative to 1. The interaction of this field with a moment at system 2 gives the familiar energy expression (8.57), (cf. 8.9)

$$W = R^{-3}\mu_i^{(1)}\mu_j^{(2)}(\delta_{ij} - 3\hat{R}_i\hat{R}_j). \tag{8.57}$$

In this framework retardation may be included by using the retarded in place of static Hertz vector. A field propagated from system 1 at time t is retarded by R/c at system 2, where c is the velocity of light, so that if the time dependence of the field at its source is $\mathrm{e}^{-\mathrm{i}\omega t}$ that at system 2 is $\mathrm{e}^{-\mathrm{i}\omega(t-R/c)} = \mathrm{e}^{-\mathrm{i}\omega t}\,\mathrm{e}^{ikR}$, k being the wave number for frequency ω. The retarded Hertz vector for the point dipole source is given in (8.58)

$$\mathbf{\Pi} = \mathrm{e}^{-\mathrm{i}\omega t}\boldsymbol{\mu}\,\mathrm{e}^{ikR}/R. \tag{8.58}$$

Proceeding as before to find the electric field at system i by use of (8.55) we have

$$(\text{curl}\,\mathbf{\Pi})_i = \left(\frac{ik}{R^2}-\frac{1}{R^3}\right) e^{ikR}(\mu_k x_j - \mu_j x_k)$$

and

$$(\text{curl curl}\,\mathbf{\Pi})_i = e^{ikR}\Bigg[\left(\frac{ik}{R^2}-\frac{1}{R^3}\right)\{-2\mu_i+3\mu_i(1-\hat{R}_i^2)$$
$$-3\mu_k\hat{R}_i\hat{R}_k-3\mu_j\hat{R}_i\hat{R}_j\}+\frac{k^2}{R}\{\mu_i(1-\hat{R}_i^2)-\mu_k\hat{R}_i\hat{R}_k-\mu_j\hat{R}_i\hat{R}_j\}\Bigg].$$

This result may be expressed as in (8.59) for the component of the electric field at system 2,

$$E_i^{(2)} = \mu_j^{(1)}\,e^{-i\omega t}\,e^{ikR}\left\{\left(\frac{ik}{R^2}-\frac{1}{R^3}\right)(\delta_{ij}-3\hat{R}_i\hat{R}_j)+\frac{k^2}{R}(\delta_{ij}-\hat{R}_i\hat{R}_j)\right\} \quad (8.59)$$

with summation over the repeated index j. The corresponding expression for the retarded interaction energy is given by

$$W = -\mu_i^{(1)}\mu_j^{(2)}\left\{\frac{k^2\cos kR}{R}\alpha_{ij}-\left(\frac{\cos kR}{R^3}+\frac{k\sin kR}{R^2}\right)\right\}\beta_{ij} \quad (8.60)$$

$$\alpha_{ij} = \delta_{ij}-\hat{R}_i\hat{R}_j; \qquad \beta_{ij} = \delta_{ij}-3\hat{R}_i\hat{R}_j$$

summation now being over both indices i and j. The result (8.60) of this classical derivation is identical with the quantum electrodynamical result of McLone and Power (1964). At short distances, $R \ll \lambda/2\pi$, the interaction is that of static dipoles, as discussed earlier. At very long range, $R \gg \lambda/2\pi$, the terms in β_{ij} cease to have importance, and only the wave-zone part remains. This falls off very slowly with distance ($\sim R^{-1}$) and depends on angle through the tensor α_{ij}, which couples only the transverse components of the dipoles at the two centres. One rationalizes by saying that this part of the coupling is the result of transverse photon emission and reabsorption. The tensor β_{ij} on the other hand couples the complete dipole fields, longitudinal as well as transverse. The interaction energy in (8.60) thus has a very long-range component, and is modulated with distance (through the appearance of $\sin kR$ and $\cos kR$) with a period depending on the characteristic transition frequency $\omega = ck$ of the identical coupled systems.

8.6. Mixed exciton–photon states

We have already seen that two transition dipoles are coupled by the retarded interaction, the energy of coupling being given in (8.60). It depends upon the characteristic frequency $\omega = ck$ of the free molecule transition, and so is altered when the transition energy is changed by the interaction itself. Thus an accurate calculation must be carried out self-consistently, the new frequencies of transitions of the system being used in (8.60) instead of those of the free molecules. In the instance of a pair of molecules, at long range, the coupling energy is too small to make this of significance, but in a crystal the energy displacements (e.g., the Davydov splitting) can be 20 per cent or more of the free molecule transition frequency, and self-consistent solutions are essential. If one keeps cyclic boundary conditions, so that the allowed values of the wave vector $\boldsymbol{\kappa}$ are known, the requirement is to establish the eigenfrequency $\nu(\boldsymbol{\kappa})$ for each level in a self-consistent way. Viewed in this light, the problem is closely related to that of finding the polarization modes of a dielectric slab as solutions of the classical equation of motion (8.61) (Born and Wolf 1965),

$$\left(\frac{\partial^2}{\partial t^2}+\nu_r^2\right)\mathbf{P}_r(n) = \frac{e^2}{m}f_r\hat{\mathbf{p}}_r\hat{\mathbf{p}}_r \cdot \mathbf{E}_0(n), \tag{8.61}$$

where $\mathbf{P}(n)$ is the polarization at site n produced in the slab by an external electromagnetic field which is then switched off. The polarisation belongs to a mode $\boldsymbol{\kappa}$ and frequency ν

$$\mathbf{P}_r^{(n)} = \mathbf{p}_r\,\mathrm{e}^{\mathrm{i}\boldsymbol{\kappa}.\mathbf{n}}\,\mathrm{e}^{-\mathrm{i}\nu_r t} \tag{8.62}$$

$\mathbf{p}_r$ being a vector amplitude associated with the rth state of the system. ν_r is the characteristic frequency of the system in the absence of coupling, m the mass of the electron, f_r the oscillator strength, and $\mathbf{E}_0(n)$ the internal electric field at the site n produced by the polarization. This field is calculated as the summed retarded field of the polarizations at all sites, in the manner of Section 8.5. The desired solution is the frequency ν for the chosen mode $\boldsymbol{\kappa}$. We shall subsequently see that the quantum mechanical problem is closely related to that of eqn (8.61).

The energy of the $\boldsymbol{\kappa}$ exciton mode thus depends on the photon field (through the appearance of photon frequencies in the interaction) as well as on the properties of 'pure' excitons (as determined, e.g. by boundary conditions). It is therefore quite natural in a physical sense to regard the modes of crystal polarization as having a mixed exciton–photon character. This viewpoint appeared first (for Frenkel-type excitons) in the work of Hopfield (1958) and has been developed by a number of other authors (Davydov and Onyshuk 1968; Philpott 1969; Ball and MacLachlan 1964), and especially Agranovitch (1960). Their work has been based on Hamiltonians of type (8.53) in which the radiation field appears as an addition to the electrostatic interaction terms. The presentation however is somewhat easier by the use of the fully retarded operator of type (8.54), as in a recent paper (Craig and Dissado 1971) and will be given here in that form.

The Hamiltonian to be used is expression (8.63) consisting of the sum of Hamiltonians of free molecules, of the free radiation field, and an interaction term between the molecular electrons and the field,

$$H = \sum H_{mol} + H_{rad} - \sum_{i} \boldsymbol{\mu}_i \cdot \mathbf{E}^{\perp}(i) \tag{8.63}$$

$\boldsymbol{\mu}_i$ is the dipole operator for the ith molecule and $\mathbf{E}^{\perp}(i)$ the transverse electric field at its position. Only dipole interactions are included. The method to be used depends upon second quantization in which the excitons and photons appear first independently and are combined into new pseudo-particles modified by exciton–photon interaction as required to diagonalize the Hamiltonian (8.63). The method has been outlined in Section 8.3.

The primitive excitation is the electronic excitation of one molecule to its rth state. Since only one such excitation can 'occupy' a molecule, it should be treated according to Fermi statistics. However in a large crystal in a low density of illumination the probability of two excitons on one molecule is so low that, from that point of view, the Fermi or Bose character is unimportant. From a formal standpoint it becomes useful to treat the excitations as bosons because this enables the key step in the diagonalization to be done in a convenient way.

By using the results given in (8.31) and (8.32) of Section 8.3, we find the Hamiltonian for the free molecule excitations in the form (8.64),

$$H_1 = W_0 + \Delta\omega \sum_{\boldsymbol{\kappa}} B_{\kappa}^{\dagger} B_{\kappa} \tag{8.64}$$

where $\Delta\omega$ is the free molecule excitation energy and W_0 the summed ground-state energies. No molecular interactions have been included, and the excitations $\boldsymbol{\kappa}$ are all of the same energy. By assuming Bose character we neglect the possibility that two excitations can have the same $\boldsymbol{\kappa}$ value.

The free-field Hamiltonian is given in eqn (8.65). The creation and annihilation operators now refer to photons of wave vector $\mathbf{q}$ and polarization j,

$$H_2 = \sum_{\mathbf{q}j} \hbar q c (a_{qj}^{\dagger} a_{qj} + a_{-qj}^{\dagger} a_{-qj}). \tag{8.65}$$

The summation runs over positive $\mathbf{q}$ only, whereas in (8,64) the sum is over all $\boldsymbol{\kappa}$. The third term of the Hamiltonian (8.63) giving the interaction between the molecules and the radiation field must now be put into second-quantized form. The transverse field $\mathbf{E}^{\perp}(n)$ is given in (8.66)

$$\begin{aligned} \mathbf{E}^{\perp}(n) = \mathrm{i} \sum_{\mathbf{q}j} (2\pi\hbar c q/V)^{\frac{1}{2}} \{ & \mathbf{e}_j (a_{\mathbf{q}j}\, \mathrm{e}^{\mathrm{i}\mathbf{q}.\mathbf{n}} + a_{-\mathbf{q}j}\, \mathrm{e}^{-\mathrm{i}\mathbf{q}.\mathbf{n}}) \\ & - \mathbf{e}_j (a_{\mathbf{q}j}^{\dagger}\, \mathrm{e}^{-\mathrm{i}\mathbf{q}.\mathbf{n}} + a_{-\mathbf{q}j}^{\dagger}\, \mathrm{e}^{\mathrm{i}\mathbf{q}.\mathbf{n}}) \} \end{aligned} \tag{8.66}$$

in which V is the volume in which the radiation is supposed to be confined for purposes of quantization (see for example Power 1964) and $\mathbf{e}_j$ is a polarization vector. The interaction energy (8.67) is then found to be

$$-\sum_n \boldsymbol{\mu}_n \cdot \mathbf{E}_n^{\perp}(n) = -\sum_{\mathbf{q}j}\sum_n (2\pi\hbar cq/V)^{\frac{1}{2}}\{(\mathbf{e}_j \cdot \boldsymbol{\mu}_n^r)(a_{qj}\,\mathrm{e}^{\mathrm{i}\mathbf{q}.\mathbf{n}} + a_{-qj}\,\mathrm{e}^{-\mathrm{i}\mathbf{q}.\mathbf{n}})$$
$$-(\mathbf{e}_j \cdot \boldsymbol{\mu}_n^r)(a_{qj}^{\dagger}\,\mathrm{e}^{-\mathrm{i}\mathbf{q}.\mathbf{n}} + a_{-qj}^{\dagger}\,\mathrm{e}^{\mathrm{i}\mathbf{q}.\mathbf{n}})\}\;; \qquad (8.67)$$

$\boldsymbol{\mu}_n^r$ is the dipole moment operator for the molecule at site n. Again,

$$\boldsymbol{\mu}_n^r = (B_n^{\dagger} + B_n)\boldsymbol{\mu}^r \qquad (8.68)$$

$\boldsymbol{\mu}^r$ being the molecular transition moment, and

$$\sum_n (\mathbf{e}_j \cdot \boldsymbol{\mu}_n^r) = N^{-\frac{1}{2}}\sum_n \sum_{\boldsymbol{\kappa}} (\mathrm{e}^{-\mathrm{i}\boldsymbol{\kappa}.\mathbf{n}}B_{\boldsymbol{\kappa}}^{\dagger} + \mathrm{e}^{\mathrm{i}\boldsymbol{\kappa}.\mathbf{n}}B_{\boldsymbol{\kappa}})(\mathbf{e}_j \cdot \boldsymbol{\mu}^r)$$
$$= N^{-\frac{1}{2}}\sum_n \sum_{\boldsymbol{\kappa}} \mathrm{e}^{\mathrm{i}\boldsymbol{\kappa}.\mathbf{n}}(B_{\kappa}^{\dagger} + B_{-\kappa})(\mathbf{e}_j \cdot \boldsymbol{\mu}^r) \qquad (8.69)$$

with the use of eqns (8.32) and a rearrangement of the sum over $\boldsymbol{\kappa}$ values, for which a sum over $-\boldsymbol{\kappa}$ includes the same terms as a sum over $+\boldsymbol{\kappa}$. We now collect terms and write the interaction part of the Hamiltonian as follows:

$$-\sum_n \boldsymbol{\mu}_n \cdot \mathbf{E}^{\perp}(n) = -\sum_n \sum_{\boldsymbol{\kappa}} \sum_{\mathbf{q}} \{A(n, \kappa, \mathbf{q}, j)(a_{qj} + a_{-qj}^{\dagger})(B_{\kappa}^{\dagger} + B_{-\kappa})$$
$$+ A(n, -\kappa, -\mathbf{q}, j)(a_{-qj} - a_{qj}^{\dagger})(B_{-\kappa}^{\dagger} + B_{\kappa}) \qquad (8.70)$$

where

$$A(n, \mathbf{q}, j) = \mathrm{i}(2\pi\hbar qc/NV)^{\frac{1}{2}}\,\mathrm{e}^{-\mathrm{i}\boldsymbol{\kappa}.\mathbf{n}}\,\mathrm{e}^{\mathrm{i}\mathbf{q}.\mathbf{n}}(\mathbf{e}_j \cdot \boldsymbol{\mu}^r). \qquad (8.71)$$

Examination of terms (8.70) and (8.71) shows that the distance modulation is governed by $\exp\{\mathrm{i}(\boldsymbol{\kappa}-\mathbf{q})\cdot\mathbf{n}\}$, and if it were the case that the ranges of the vectors $\boldsymbol{\kappa}$ and $\mathbf{q}$ were the same, the sum over sites n could be done according to $\sum_n \exp\{-\mathrm{i}(\boldsymbol{\kappa}-\mathbf{q})\cdot\mathbf{n}\} = \delta(\boldsymbol{\kappa}-\mathbf{q})$ giving on the right-hand side of eqn (8.70), a new expression (8.72) for the interaction operator

$$-\mathrm{i}\sum_{\boldsymbol{\kappa}} (2\pi\hbar\kappa c/NV)^{\frac{1}{2}}(\mathbf{e}_j \cdot \boldsymbol{\mu}^r)(a_{\kappa j}B_{\kappa}^{\dagger} - a_{\kappa j}^{\dagger}B_{\kappa}) \qquad (8.72)$$

showing that the only coupling of an exciton state $\boldsymbol{\kappa}$ is to the photon of the same wave vector. It has sometimes been supposed that this is the situation in a crystal (Silbey 1967). The argument depends on choosing the quantization box for the radiation field to be the crystal itself. In that case, apart from some differences that are unimportant in principle, arising from the detail of the crystal lattice, the range of wave vectors $\mathbf{\kappa}$, is included in the range of the $\mathbf{q}$. If the ranges were identical the interaction would be confined to the term (8.72). There are reasons for believing that this is incorrect (Craig and Dissado 1968). Since the retarded interaction energy (8.60) follows in quantum electrodynamics from the use of the interaction (8.67) by integration over the photon variables $\mathbf{q}$ and j, conceived as independent of $\mathbf{\kappa}$ (*vide infra*), it is easy to see that the limited interaction (8.72) does not imply the known intermolecular dipole force system. Moreover, conceptually, we do not expect the quantization box to be more than a convenient artifice; it should be independent of any physical dimensions of the particular problem, and should be able to be made as large as necessary to take its quantized levels into a continuum within some appropriate definition. Also, for a description of absorption and emission, the radiation field needs to exist outside the crystal so that photons can be interchanged between the external field and the internal modes. Such a broader physical model brings some extra complication, particularly the need to consider boundary conditions at the crystal surfaces, but seems essential for a proper treatment of crystal properties.

There is one point in the foregoing argument that has not always been properly discussed. Even if the radiation and excitons are quantized in the same box (i.e. the crystal specimen) it is not true that each exciton mode couples to *one* radiation mode. In the delocalization of excitons, each $\mathbf{\kappa}$ value applies to one state of the Brillouin zone. There are N in all, N being the total number of molecules in the crystal. For the photons there is no similar limitation in the number of modes. If for convenience one classifies the photon wave vectors by reference to the Brillouin zone of the crystal, one sees that while the levels of this zone have the same wave vectors as the excitons, the photons of the other zones (sometimes called the second, third, Brillouin zones) have different energies, and are distinct. Thus every exciton $\mathbf{\kappa}$ wave couples to a *family* of photons, the wave vectors for which have components differing from those of $\mathbf{\kappa}$ by integral numbers of π/a, π/b, π/c, where a, b, and c are the primitive lattice translations of the direct lattice. The arguments however leading to the rejection of this type of box quantization remain.

Returning to the main discussion, we note that the Hamiltonian is the sum of the three terms (8.64), (8.65), and (8.70) given by the sum of exciton energies, photon energies, and cross terms connecting the two. The next stage of the calculation, as in Section 8.3, is to find pseudo-particles, described by operators $\zeta_\rho^\dagger$ and ζ_ρ, with energies E_ρ, in which the Hamiltonian is diagonal, in the form (8.73). The calculation (Tyablikov 1968)

$$H = \sum_\rho E_\rho \zeta_\rho^\dagger \zeta_\rho \tag{8.73}$$

is again made by the Bogolyubov–Tyablikov method, in which the starting point is the set of transformations (8.74) in terms of amplitude functions u and v,

$$\left.\begin{aligned} B_\kappa &= \sum_\rho (\zeta_\rho u_{\kappa\rho} + \zeta_\rho^\dagger v_{\kappa\rho}^*) \\ B_\kappa^\dagger &= \sum_\rho (\zeta_\rho^\dagger u_{\kappa\rho}^* + \zeta_\rho v_{\kappa\rho}) \\ a_{qj} &= \sum_\rho (\zeta_\rho u_{qj,\rho} + \zeta_\rho^\dagger v_{qj,\rho}^*) \\ a_{qj}^\dagger &= \sum_\rho (\zeta_\rho^\dagger u_{qj,\rho}^* + \zeta_\rho v_{qj,\rho}) \end{aligned}\right\}. \tag{8.74}$$

The range of the index ρ includes the ranges of κ and $\mathbf{q}j$. Where there is no photon-exciton coupling, the operators ζ_ρ are identical either with one of the operators B_κ, or one of the a_{qj}, according to the index ρ. Bose statistics are secured by the relations

$$\sum_\kappa (u_{\kappa\rho} u_{\kappa\rho'}^* - v_{\kappa\rho} v_{\kappa\rho'}^*) = \delta(\rho - \rho')$$

$$\sum_\kappa (u_{\kappa\rho} v_{\kappa\rho'} - u_{\kappa\rho'} v_{\kappa\rho}) = 0$$

and correspondingly for sums over ρ. There are inverse transformations for the two pairs of expressions (8.74), of which the first is

$$\left.\begin{aligned} \zeta_\rho &= \sum_\kappa (u_{\kappa\rho}^* B_\kappa - v_{\kappa\rho}^* B_\kappa^\dagger) \\ \zeta_\rho^\dagger &= \sum_\kappa (u_{\kappa\rho} B_\kappa^\dagger - v_{\kappa\rho} B_\kappa) \end{aligned}\right\} \tag{8.75}$$

and the second

$$\left.\begin{aligned} \zeta_\rho &= \sum_{\mathbf{q}j} (u_{qj,\rho}^* a_{qj} - v_{qj,\rho}^* a_{qj}^\dagger) \\ \zeta_\rho^\dagger &= \sum_{\mathbf{q}j} (u_{qj,\rho} a_{qj}^\dagger - v_{qj,\rho} a_{qj}) \end{aligned}\right\}. \tag{8.76}$$

We now see from the terms (8.75) the pseudo-particles ζ are exciton-like, and in (8.76), belonging to a different set of ρ values, they are photon-like. We are here interested only in the first, although both need to be brought into the analysis. We now find equations of motion with the help of (8.77), and with the eigenvalue equations (8.78),

$$[B_\kappa, H] = \mathrm{i}\frac{\mathrm{d}B_\kappa}{\mathrm{d}t}; \qquad [a_{qj}, H] = \mathrm{i}\frac{\mathrm{d}a_{qj}}{\mathrm{d}t} \tag{8.77}$$

$$[\zeta_\rho, H] = \mathrm{i}\frac{\mathrm{d}\zeta_\rho}{\mathrm{d}t} = E_\rho \zeta_\rho; \qquad [\zeta_\rho^\dagger, H] = -E_\rho \zeta_\rho^\dagger. \tag{8.78}$$

It is straightforward to obtain the following equations in the energies E_ρ,

$$\left.\begin{aligned}
&(\Delta\omega - E_\rho)u_{\kappa\rho} - \sum_{\mathbf{q}jn} A(n\kappa q)(u_{qj,\rho} - v_{-qj,\rho}) \\
&\qquad + A(n\kappa - q)(u_{-qj,\rho} - v_{qj,\rho})\} = 0 \\
&(\Delta\omega + E_\rho)v_{-\kappa\rho} - \sum_{\mathbf{q}jn} \{A(n\kappa q)(u_{qj,\rho} - v_{-qj,\rho}) \\
&\qquad + A(n\kappa - q)(u_{-qj,\rho} - v_{qj,\rho})\} = 0 \\
&(\hbar qc - E_\rho)u_{qj,\rho} + \sum_{n\kappa} A(n - \kappa - q)(u_{\kappa\rho} + v_{-\kappa\rho})\} = 0 \\
&(\hbar qc + E_\rho)v_{-qj,\rho} - \sum_{n\kappa} A(n - \kappa - q)(u_{\kappa\rho} + v_{-\kappa\rho})\} = 0.
\end{aligned}\right\} \tag{8.79}$$

The second pair of equations enable the elimination of the amplitudes $u_{qj,\rho}$ and $v_{-qj,\rho}$ from the first pair, which may be combined into a single eqn (8.80), with the help of the new amplitude $w_{\kappa\rho} = u_{\kappa\rho} + v_{-\kappa\rho}$

$$\begin{aligned}
&\{(\Delta\omega)^2 - E_\rho^2\}w_{\kappa\rho} + 2\Delta\omega \sum_{\substack{qjj'n\\ \kappa m}} \{A(n\kappa q_j)A(m - \kappa' - qj' + A(n\kappa - qj)A(m - \kappa' qj)\} \\
&\qquad \times \{\hbar cq - E_\rho)^{-1} + (\hbar cq + E_\rho)^{-1}\}w_{\kappa'\rho} = 0.
\end{aligned} \tag{8.80}$$

Substitution for the expressions A, from (71), gives

$$\begin{aligned}
&\{(\Delta\omega)^2 - E_\rho^2\}w_{\kappa\rho} + (2\Delta\omega/N) \sum_{nm\kappa'} \mathrm{e}^{-\mathrm{i}(\boldsymbol{\kappa}.\mathbf{n} - \boldsymbol{\kappa}'.\mathbf{m})} \left[\sum_{\mathbf{q}jj'} (\mathbf{e}_j \cdot \boldsymbol{\mu}^r)(\mathbf{e}_{j'} \cdot \boldsymbol{\mu}^r)(2\pi\hbar cq/V)\right. \\
&\qquad \left. \times 2\cos q(n - m)\{(E_\rho - \hbar cq)^{-1} - (E_\rho + \hbar cq)^{-1}\}\right] w_{\kappa'\rho} = 0.
\end{aligned} \tag{8.81}$$

By eliminating the amplitudes $u_{qj,\rho}$ and $v_{qj,\rho}$ we have arrived at equations for the exciton-like waves. Those for photon-like waves may be treated in the same way, but are of less interest.

The expression inside square brackets is familiar. It is the quantum electrodynamical expression for the retarded interaction between two molecules at sites n and m, in the form first found by McLone and Power (1964). It is evaluated by conversion of the $\mathbf{q}$ sum to an integral, and gives the real part of the interaction $V(k, R)$ in (8.82) for $R = |\mathbf{n} - \mathbf{m}|$,

$$V(k, R) = \mu_i^{(1)}\mu_j^{(2)}\,\mathrm{e}^{\mathrm{i}kR}\{-\alpha_{ij}k^2/R + \beta_{ij}(-\mathrm{i}k/R^2 + 1/R^3)\}. \tag{8.82}$$

Thus the eqn (8.81) becomes

$$\{(\Delta\omega)^2 - E_\rho^2\}w_{\kappa\rho} + (2\Delta\omega/N) \sum_{nm\kappa'} \mathrm{e}^{-\mathrm{i}(\boldsymbol{\kappa}.\mathbf{n} - \boldsymbol{\kappa}'.\mathbf{m})}\,\mathrm{Re}\,V(k, R)\omega_{\kappa'\rho} = 0. \tag{8.83}$$

The result of the quantum electrodynamical work, after integrating over the photon states, is that the intermolecular coupling (8.82) has the form already found in eqn (8.60), by classical electrodynamics and the Hertz vector.

Since this is the expected equivalence, we can have confidence that the result is correct. An important difference of detail is that the classical result (8.60) contains the wave number k for the free molecule transition, while the interaction $V(k, R)$ depends on a wave number displaced from the free molecule value of $\Delta\omega/\hbar c$ by the interaction itself.

The content of eqn (8.83) can be brought out by looking at a sequence of approximations. Exact solution would require inclusion of the coupling between excitons of different wave vectors, so that each of the energies E_ρ should apply to an exciton involving not only modification by the radiation field but also formed by a mixture of $\boldsymbol{\kappa}$ waves. Let us first approximate by dropping the cross-terms, supposing that each of the exciton-like eigenstates ρ belongs to a single $\boldsymbol{\kappa}$ value, which can then be used as a label without confusion. We are then left with eqn (8.84), from which the amplitudes have disappeared, after setting $\boldsymbol{\kappa}' = \mathbf{k}$

$$(\Delta\omega)^2 - E_\kappa^2 + 2\Delta\omega \sum_{\mathbf{m}} \mathrm{e}^{-\mathrm{i}\boldsymbol{\kappa}.(\mathbf{n}-\mathbf{m})} \operatorname{Re} V(k, R) = 0. \tag{8.84}$$

Again, accepting that the allowed values of $\boldsymbol{\kappa}$ are known, being determined by the boundary conditions, the lattice sum over m can be evaluated as a function of $k = E_\kappa/hc$ and the equation solved self-consistently for E_κ, for each $\boldsymbol{\kappa}$ level of the exciton band.

In a further simplifying approximation, we can suppose that the energy eigenvalues E_κ for the crystal are displaced only slightly from the excitation energy $\Delta\omega$ of the free molecules. This applies where the exciton bandwidth is narrow compared with $\Delta\omega$. The wave number k appearing in the retarded interaction can now be replaced by $\Delta\omega/hc$ and the equation can be simplified using

$$(\Delta\omega)^2 - E_\kappa^2 = (\Delta\omega - E_\kappa)(\Delta\omega + E_\kappa) \approx 2\Delta\omega(\Delta\omega - E_\kappa).$$

Eqn (8.84) now reduces to (8.85)

$$E_\kappa = \Delta\omega + \sum_{\mathbf{m}} \mathrm{e}^{-\mathrm{i}\boldsymbol{\kappa}.(\mathbf{n}-\mathbf{m})} V(\Delta\omega/hc, R) \tag{8.85}$$

which differs from the perturbation result given in the diagonal places of the secular eqn (8.28) only by the substitution of the retarded interaction and the disappearance of the dispersion correction D. Finally, by replacing $V(\Delta\omega/hc, R)$ by the static interaction we return to the simplest form of energy expression for the levels of the band, namely,

$$E_\kappa = \Delta\omega + \sum_{\mathbf{m}} \mu_i^{(n)} \mu_j^{(m)} \mathrm{e}^{\mathrm{i}\boldsymbol{\kappa}.(\mathbf{n}-\mathbf{m})} \beta_{ij} |\mathbf{n}-\mathbf{m}|^{-3}. \tag{8.86}$$

The disappearance of the dispersion correction term D from the expressions given in (8.28) is a result of the restriction to one-photon interactions. In this framework the dispersion corrections arise from two-photon terms.

There is a further interesting property of eqn (8.84), namely, its relationship to the classical equation (8.61) connecting the polarization of a crystal slab maintained by its own electric field. Suppose the polarization at site n, $\mathbf{P}_r(n)$, has the plane-wave form (8.87), with frequency ν,

$$\mathbf{P}_r(n) = \mathbf{p}_r\, \mathrm{e}^{\mathrm{i}\boldsymbol{\kappa}.\mathbf{n}}\, \mathrm{e}^{-\mathrm{i}\nu t} \tag{8.87}$$

where $\mathbf{p}_r$ is a dipole moment. After substitution in (8.61), and taking the scalar product with $\hat{\mathbf{p}}_r$ on both sides,

$$|\mathbf{p}_r|(\nu_r^2 - \nu^2) = \frac{e^2}{m} f_r \hat{\mathbf{p}}_r \cdot \mathbf{E}_0(n). \tag{8.88}$$

The oscillator strength is now expressed in terms of $|\mathbf{p}_r|$, the latter being equated to the molecular transition moment,

$$f_r = \frac{2m\nu_r|\mathbf{p}_r|^2}{e^2 h} \tag{8.89}$$

and we have from (8.88)

$$\nu_r^2 - \nu^2 = \frac{2\nu}{h}\, \mathrm{e}^{-\mathrm{i}\boldsymbol{\kappa}.\mathbf{n}} \mathbf{p}_r \cdot \mathbf{E}_0(n). \tag{8.90}$$

The relation of this equation to (8.84) depends on the field $\mathbf{E}_0(n)$, which must now be determined to be the field which, acting on the molecule at site n, produces an interaction energy, expression (8.91),

$$\sum_m \mathrm{e}^{-\mathrm{i}\boldsymbol{\kappa}.(\mathbf{n}-\mathbf{m})} \operatorname{Re} V(k, R) \tag{8.91}$$

as in eqn (8.84). The ith component of the required field is given in (8.92)

$$E_i^{(1)}(n) = -\mathrm{e}^{\mathrm{i}\boldsymbol{\kappa}.\mathbf{n}} \mu_j^{(2)} \sum_m \mathrm{e}^{\mathrm{i}\boldsymbol{\kappa}.(\mathbf{m}-\mathbf{n})} \left\{ \frac{k^2 \cos kR}{R} \alpha_{ij} - \left(\frac{\cos kR}{R^3} + \frac{k \sin kR}{R^2} \right) \beta_{ij} \right\} \tag{8.92}$$

with summation over the repeated index j. The separation $R = |\mathbf{m} - \mathbf{n}|$. We see that apart from the modulation in $\boldsymbol{\kappa}$ this is identical with the real part of the classical retarded field in eqn (8.59).

We have thus shown that the eqn (8.84) derived with the help of a second-quantized Hamiltonian and quantum electrodynamics is identical with the classical result, if the site dipole moment is identified as the molecular transition moment, and if the driving field $\mathbf{E}_0$ is identified as the modulated summed field of the molecules on lattice sites, each acting as a source of strength $\boldsymbol{\mu}_r$. It is interesting and typical of many situations in this field of enquiry that the result is exactly that expected from classical electromagnetism.

Acknowledgements

I am grateful to my colleagues, Professor E. A. Power, Dr T. Thirunamachandran, and Dr K. Roby for their comments on the manuscript.

REFERENCES

AGRANOVITCH, V. M. (1961). *Sov. Phys.: Sol. State* **3**, 593.
—— (1960). *Sov. Phys.: J.E.T.P.* **10**, 307.
ANDERSON, P. W. (1968). *Concepts in solids*, p. 132. Benjamin, New York.
BALL, M. A. and MACLACHLAN, A. D. (1964). *Proc. R. Soc. A* **282**, 433.
BORN, M. and WOLF, E. (1965). *Principles of optics*. Pergamon, Oxford.
COULSON, C. A. (1948). *Proc. phys. Soc.* **60**, 257.
—— (1961). *Electricity*, p. 52 *et seq*. Oliver and Boyd, London.
CRAIG, D. P. and DISSADO, L. A. (1968). *J. Chem. Phys.* **48**, 576.
—— —— (1971). *Proc. R. Soc. A* **325**, 1.
—— and HOBBINS, P. C. (1955*a*). *J. chem. Soc.* 539.
—— —— (1955*b*). *J. chem. Soc.* 2309.
—— and Walmsley, S. H. (1968). *Excitons in molecular crystals*. Benjamin, New York.
—— and SARTI-FANTONI, P. (1966). *Chem. Comm.* 742.
DAVYDOV, A. S. (1962). *Theory of molecular excitons*. McGraw Hill, New York.
—— (1971). *Theory of molecular excitons*, p. 127. Plenum Press, New York.
—— and ONYSHUK, V. A. (1968). *Sov. Phys.: Doklady* **13**, 139.
EYRING, H., WALTER, J., and KIMBALL, G. E. (1944). *Quantum chemistry*, p. 351. John Wiley, London.
FOX, D. and YATSIV, S. (1956). *J. Chem. Phys.* **24**, 1103.
HOFFMANN, R. (1963). *Radiation Res.* **20**, 140.
HOPFIELD, J. J. (1958). *Phys. Rev.* **112**, 1555.
KNOX, R. S. (1963). *Theory of excitons*. Academic Press, New York.
MAHAN, G. D. (1965). *J. chem. Phys.* **43**, 1569.
MCLONE, R. and POWER, E. A. (1964). *Mathematika* **11**, 91.
PHILPOTT, M. R. (1969). *J. chem. Phys.* **50**, 5117 and references therein.
POWER, E. A. (1964). *Elementary quantum electrodynamics*, p. 67. Longmans, London.
—— and THIRUNAMACHANDRAN, T. (1971). *Phys. Rev. B* **3**, 3546.
—— and ZIENAU, S. (1959). *Phil. Trans. R. Soc. A* **251**, 427.
SCHMIDT, G. M. J. (1965). 13th Solvay Conference, Brussels.
SILBEY, R. (1967). *J. Chem. Phys.* **46**, 4029.
STRATTON, J. A. (1941). *Electromagnetic theory*, p. 28. McGraw Hill.
TYABLIKOV, S. V. (1968). *Methods in the theory of magnetism*, p. 105. Plenum Press, New York.

AUTHOR INDEX

SUBJECT INDEX